Lecture Notes in Computer Science 12511

More information about this subseries at http://www.springer.com/series/7409

Chaoyi Pang · Yunjun Gao · Guanliang Chen ·
Elvira Popescu · Lu Chen · Tianyong Hao ·
Bailing Zhang · Silvia Margarita Baldiris Navarro ·
Qing Li (Eds.)

Learning Technologies and Systems

19th International Conference on Web-Based Learning, ICWL 2020
and 5th International Symposium on Emerging Technologies for Education, SETE 2020
Ningbo, China, October 22–24, 2020
Proceedings

Springer

Editors
Chaoyi Pang
Ningbo Institute of Industrial Technology
Ningbo, China

Guanliang Chen
Monash University
Clayton, VIC, Australia

Lu Chen
Department of Physics
Zhejiang University
Hangzhou, China

Bailing Zhang
Zhejiang University
Ningbo, China

Qing Li
Department of Computing
Hong Kong Polytechnic University
Kowloon, Hong Kong

Yunjun Gao
College of Computer Science
Zhejiang University
Hangzhou, China

Elvira Popescu
Faculty of Automation, Computers
and Electronics
University of Craiova
Craiova, Romania

Tianyong Hao
School of Computer Science
South China Normal University
Guangzhou, China

Silvia Margarita Baldiris Navarro
Universidad Internacional De La Rioja
Logroño, Spain

ISSN 0302-9743 ISSN 1611-3349 (electronic)
Lecture Notes in Computer Science
ISBN 978-3-030-66905-8 ISBN 978-3-030-66906-5 (eBook)
https://doi.org/10.1007/978-3-030-66906-5

LNCS Sublibrary: SL3 – Information Systems and Applications, incl. Internet/Web, and HCI

This Springer imprint is published by the registered company Springer Nature Switzerland AG
The registered company address is: Gewerbestrasse 11, 6330 Cham, Switzerland

Preface

This volume presents the contributions of the 19th edition of the annual International Conference on Web-based Learning (ICWL). The first edition of ICWL was held in Hong Kong in 2002. Since then, it has been held 16 more times, on three continents: Australia (2003); Beijing, China (2004); Jinhua, China (2008); Shanghai, China (2010); Hong Kong, China (2005, 2011, 2015); Malaysia (2006); UK (2007); Germany (2009); Romania (2012); Taiwan, China (2013); Estonia (2014); Italy (2016); South Africa (2017); Thailand (2018); and Germany (2019); This year, ICWL 2020 was held in Ningbo, China, organized by Ningbo University of Technology, China.

Furthermore, the conference continued the traditional initiative, started by ICWL 2016, of holding the 5th International Symposium on Emerging Technologies for Education (SETE) at the same location. SETE collected the traditional workshop activities managed by ICWL in the past years and additionally was organized in tracks. Workshops and tracks added new and hot topics on technology-enhanced learning, providing a better overall conference experience to the ICWL and SETE attendees.

The topics proposed in the ICWL&SETE Call for Papers included several relevant issues, ranging from Semantic Web for E-Learning, through Learning Analytics, Computer-Supported Collaborative Learning, Assessment, Pedagogical Issues, E-learning Platforms, and Tools, to Mobile Learning. Due to the impact of the COVID-19 pandemic, we decided to combine papers from ICWL 2020 and SETE 2020 into one proceedings volume this year.

We received 16 submitted contributions for ICWL 2020, and received 56 submitted contributions for SETE 2020. All of the submitted papers were assigned to three members of the Program Committee (PC) for peer review. All reviews were checked and discussed by the PC chairs, and additional reviews or meta-reviews were elicited if necessary. Finally, we accepted seven full papers for ICWL 2020 with an acceptance rate of 43.75%, and accepted 23 full papers for SETE 2020 with an acceptance rate of 41%.

In addition to regular papers, SETE 2020 also featured a special workshop on Educational Technology for Language Learning (ETLL-20). ETLL-20 is a forum for presenting and discussing novel ideas and solutions related to educational technology for language and translation learning from the perspectives of both theory and application. We received 21 submitted contributions, and finally accepted 15, where each submitted paper was assigned to two reviewers. As the General Chair of ETLL-20, Prof. Shili Ge would like to thank all the authors for their enthusiastic high-quality submissions, the reviewers (PC members) for their careful and timely reviews, and the organizing committee for their excellent publicity.

We would like to sincerely thank our keynote and invited speakers:

- Prof. Chin-Chung Tsai, a Chair Professor and Dean of the School of Learning Informatics and the Director of the Institute for Research Excellence in Learning Sciences at National Taiwan Normal University, Taiwan, China;

- Prof. Dragan Gašević, Director of the Centre for Learning Analytics at Monash University, a founder and former President (2015–2017) of the Society for Learning Analytics Research, Austria;
- Prof. Kaizhu Huang, an Associate Dean of research in the School of Advanced Technology at Xi'an Jiaotong-Liverpool University, the founding director of Suzhou Municipal Key Laboratory of Cognitive Computation and Applied Technology, China.

Many contributions made the conference possible and successful. First of all, we would like to thank all the authors who considered ICWL&SETE for their submission. We thank the PC members, and the additional reviewers, for their evaluations, which made the selection of the accepted papers possible. Additional thanks go to the publicity co-chairs An Liu from Soochow University, Michael A. Herzog from Magdeburg-Stendal University of Applied Sciences, Jianmin Zhao from Zhejiang Normal University, and Mart Laanpere from Tallinn University.

For the organization, the hosting Smart IoT Engineering and Application Lab under the direction of Prof. Genlang Chen served in multiple functions: local organization chair Shiting Wen, also as web chair. Their strong and continuous commitment made it a remarkable conference event with excellent design, smooth orientation, personal service, and warm hospitality. We also thank the sponsor Springer, for the enlightened and much appreciated support that helped to dedicate the Best Paper Awards.

We expect that the ideas that have emerged in ICWL&SETE 2020 will result in the development of further innovations for the benefit of scientific, industrial, and social communities. We hope that the reader of this volume will be pleased by the relevance of the topics and contents of the papers, and thus be enticed to contribute to the next editions of ICWL&SETE.

November 2020

Chaoyi Pang
Yunjun Gao
Guanliang Chen
Elvira Popescu
Lu Chen
Tianyong Hao
Bailing Zhang
Silvia Margarita Baldiris Navarro
Qing Li
Ralf Klamma

Organization

ICWL Honorary Conference Chairs

Xianmin Zhang	President of NIT, Zhejiang University, China
Ramamohanarao (Rao) Kotagiri	The University of Melbourne, Australia

ICWL General Co-chairs

Chaoyi Pang	NIT, Zhejiang University, China
Elvira Popescu	University of Craiova, Romania

ICWL Program Co-chairs

Yunjun Gao	Zhejiang University, China
Lu Chen	Aalborg University, Denmark

ICWL Finance Chair

Shiting Wen	NIT, Zhejiang University, China

ICWL Organization Co-chairs

Genlang Chen	NIT, Zhejiang University, China
Xuan Liu	NIT, Zhejiang University, China

ICWL Publicity Co-chairs

An Liu	Soochow University, China
Michael A. Herzog	Magdeburg-Stendal University of Applied Sciences, Germany
Jianmin Zhao	Zhejiang Normal University, China
Mart Laanpere	Tallinn University, Estonia

ICWL Steering Committee

Qing Li	PolyU, Hong Kong, China
Ralf Klamma	RWTH Aachen University, Germany

ICWL Program Committee

Lorna Uden	Staffordshire University, UK
Charoenchai Wongwatkit	Mae Fah Luang University, Thailand
Hans Põldoja	Tallinn University, Estonia
Matteo Lombardi	Griffith University, Australia
Carla Limongelli	Universita' Roma Tre, Italy
Adriana Botha	University of Pretoria, South Africa
Sergej Zerr	L3S Research Center, Germany
Maiga Chang	Athabasca University, Canada
Yunhui Zhuang	City University of Hong Kong, China
Fu Lee Wang	The Open University of Hong Kong, China
Ivana Marenzi	L3S Research Center, Germany
Michael Kickmeier-Rust	Graz University of Technology, Austria
Maria-Iuliana Dascalu	University POLITEHNICA of Bucharest, Romania
Harald Mayer	JOANNEUM RESEARCH, Austria
Kyparissia Papanikolaou	School of Pedagogical and Technological Education, Greece
Flippo Sciarron	Roma Tre University, Italy
Keyu Yang	Zhejiang University, China
Zuzana Kubincová	Comenius University, Bratislava
Pierpaolo Vittorini	University of L'Aquila, Italy
Tomaž Klobučar	Jozef Stefan Institute, Slovenia
Wei Chen	Agricultural Information Institute, Chinese Academy of Agricultural Sciences, China
Carlos Vaz de Carvalho	GILT - ISEP, Portugal
Amruth Kumar	Ramapo College of New Jersey, US
Giovanni Fulantelli	National Research Council of Italy, Italy
Tianyong Hao	South China Normal University, China
Denis Gillet	Swiss Federal Institute of Technology in Lausanne (EPFL), Switzerland
Wolfgang Nejdl	L3S and University of Hannover, Germany
Ralf Klamma	RWTH Aachen University, Germany
Umberto Nanni	Univ. "La Sapienza" of Rome, Italy
Tak-Lam Wong	Douglas College, Canada
Wolfgang Mueller	University of Education Weingarten, Germany
Riina Vuorikari	Institute for Prospective Technological Studies, Spain
Tanja Mitrovic	University of Canterbury, New Zealand
Damiano Distante	University of Rome Unitelma Sapienza, Italy
Zhenguo Yang	City University of Hong Kong, China
Marcus Specht	Delft University of Technology and Open University of the Netherlands, the Netherlands
Andreas Harrer	University of Applied Sciences and Arts Dortmund, Germany
Frederick Li	University of Durham, UK
Elise Lavoué	iaelyon, Université Jean Moulin Lyon 3, LIRIS, France

Luigi Laura	International Telematic University Uninettuno, Italy
Dickson K. W. Chiu	The University of Hong Kong, Hong Kong, China
Milos Kravcik	DFKI GmbH, Germany
Alke Martens	University of Rostock, Germany
Sabine Graf	Athabasca University, Canada
Olga C. Santos	aDeNu Research Group (UNED), Spain
Michael A. Herzog	Magdeburg-Stendal University, Germany
Elvira Popescu	University of Craiova, Romania
Demetrios Sampson	Curtin University, Australia
Mirjana Ivanovic	University of Novi Sad, Serbia
Francesca Pozzi	Istituto Tecnologie Didattiche - CNR, Italy
Ioannis Kazanidis	Eastern Macedonia and Thrace Institute of Technology, Greece
Marco Temperini	Sapienza University of Rome, Italy
Qing Liu	Hong Kong Baptist University, Hong Kong, China
Jianqiu Xu	Nanjing University of Aeronautics and Astronautics, China
Xiaoye Miao	Zhejiang University, China
Detian Zhang	Soochow University, China
Yang Yang	University of Science and Technology of China, China
An Liu	Soochow University, China

SETE General Co-chairs

Bailing Zhang	Ningbo Institute of Technology, Zhejiang University, China
Guanliang Chen	Monash University, Australia

SETE Technical Programme Committee Co-chairs

Silvia Margarita Baldiris Navarro	Universidad Internacional De La Rioja, Spain
Tianyong Hao	South China Normal University, China

SETE Workshop Co-chairs

Bailing Zhang	Ningbo Institute of Technology, Zhejiang University, China
Quanlong Guan	Jinan University, China

SETE Publicity Chairs

Xiaolu Wang	NIT, Zhejiang University, China
Chiu-Pin Lin	National Tsing Hua University, Taiwan

SETE Track Co-chairs

Fachao, Li	Hebei University of Science and Technology, China
Jusheng Mi	Hebei Normal University, China
Xiaomin Wang	Northeastern University at Qinhuangdao, China
Degang Chen	North China Electric Power University, China
Jianying Liang	Hebei University of Economics and Business, China
Zhibin Hu	South China University of Technology, China
Jian Chen	South China University of Technology, China
Jing Huang	South China Normal University, China
Yan Yan	University of Iowa, US
Yizhou Fan	The University of Edinburgh, UK
Jingmin Miao	The Open University of China, China
Zehui Zhan	South China Normal University, China
Bailing Zhang	NIT, Zhejiang University, China
Han Liu	Cardiff University, UK
Fangyu Wu	Xi'an Jiaotong-Liverpool University, China
Zhao Xin	University of Sheffield, UK
Zhou Qijia	Huaibei Normal University, China
Tianchong Wang	The Education University of Hong Kong, Hong Kong, China
Xiang Zhao	National University of Defense Technology, China
Jia Xu	Guangxi University, China
Cai Liang	Ningbo Institute of Technology, Zhejiang University, China
Mohammad Khalil	University of Bergen, Norway
Jacqueline Wong	Erasmus University Rotterdam, the Netherlands
Jalal Nouri	Stockholm University, Sweden

Contents

Online and Web-Based Learning

Intelligent Mentoring Bots in Learning Management Systems

Concepts, Realizations and Evaluations

Alexander Tobias Neumann[1](✉), Peter de Lange[1], Ralf Klamma[1], Norbert Pengel[2], and Tamar Arndt[2]

[1] RWTH Aachen University, Aachen, Germany
{neumann,lange,klamma}@dbis.rwth-aachen.de
[2] Leipzig University, Leipzig, Germany
{norbert.pengel,tamar.arndt}@uni-leipzig.de

Abstract. Mentoring students in online courses is an essential task to keep students on track. Often, they ask simple questions or have repetitive issues. However, due to a high resource demand, this task is difficult to achieve on a massive scale. Digitizing mentoring processes has a high potential to fill this resource gap. In this paper, we present the concept, implementation and evaluation of intelligent mentoring bots. These bots are implemented as chatbots that can be integrated into learning management systems. With these bots, we enable 24/7 mentoring support for students. Students can ask Frequently Asked Questions (FAQs) regarding their general studies or about one of their current classes. Further, we integrate external learning applications with the bots to provide personalized feedback to students. The realization of the chatbots is based on Natural Language Understanding (NLU) to identify the respective intent of the student. We apply those chatbots as an additional offer in university courses and evaluate them in this context. We see the potential of bots as a scalable solution for mentoring in online courses.

Keywords: Chatbots · Mentoring · Learning management systems

1 Introduction

In their basic definition, *Social Bots* are software programs that automate tasks on behalf of users. Recently, with the advance of artificial intelligence, they are able to be trained for advanced tasks like customer support at help desks and human resource agents in job interviews. The rapid growth of their popularity has also resulted in controversies, in particular when used in social media [8,15]. The deployment of social bots in intelligent tutoring systems, especially in the context of Massive Open Online Courses (MOOCs) [3], is a recent development. In educational contexts, bots are already used for the medical sector [1], language learning [9] and meta-cognitive support [11]. However, in informal learning situations like mentoring, the use of social bots is seldom, if ever. At universities

C. Pang et al. (Eds.): SETE 2020/ICWL 2020, LNCS 12511, pp. 3–14, 2021.
https://doi.org/10.1007/978-3-030-66906-5_1

where lectures are attended by several hundred students, a one to one mentoring by a teacher with personal feedback is hardly feasible [4]. Therefore, social bots might prove to be a useful alternative, because it is possible to scale social bot interaction within large learner communities. With that perspective, a large national research project[1] in Germany has started at the end of 2018 to research the many aspects of scaling mentoring processes in informal and formal learning at universities. Researchers from different disciplines like pedagogy, psychology, computer science, and mathematics are working together on the necessary technical infrastructure, the conceptual models and the organizational implementation. Based on formal mentoring and domain models, the use of artificial intelligence as well as machine learning and deep learning facilitates the development of new organizational and inter-organizational support structures for mentors and mentees. One way to support mentees is the deployment of social bots in Learning Management Systems (LMSs). This way students can benefit from mentoring support through well-known applications, such as the daily used messenger. By providing a chat partner, the conversation becomes more personal, but can be kept anonymous and confidential. For example, students may disclose sensitive information [7] that requires individual mentoring. The students are able to learn in their own environment while social bots support them, e.g., by being available all the time for content related questions or by sending reminders for homework and arrival of new learning materials. In the context of this paper, we make the following contributions; our mentoring bots have been deployed as chatbots in the LMSs as well as in commonly used instant messaging and chat tools to provide three functions:

- They can answer frequently asked questions based on an extendable catalog of questions and answers.
- They provide writing assignments to the students.
- They provide feedback to submissions by using an analysis tool from a partner in the consortium called T-MITOCAR [14].

In this way, students receive personalized feedback without additional workload for teachers. The bots can respond to student requests both rule-based and via NLU. We present the concept and the implementation of our open-source learning environment that allows to create such bots in a scalable manner. We also present two empirical studies, whereby the first one tested the usability of the framework. The second evaluation refers to the developed bot based on existing requirements. The chat interface was integrated into an LMS and was provided to educators to interact with it. The main goal was to understand the acceptance criteria of teachers. The bot is used for courses at a university with several hundred students. The acceptance and use of the created bots is currently being carried out in MOOCs. The bot was well accepted by the participants and considered as a feasible solution, despite some minor technical problems in such an early phase of the project.

[1] https://tech4comp.de.

The further course of the paper is structured as follows: In the following the related work is described (Sect. 2). Subsequently, we introduce the use case (Sect. 3), followed by the concept (Sect. 4), while the next section explains the implementation (Sect. 5). In Sect. 6 we present the evaluation, before the conclusion consisting of a summary and an outlook (Sect. 7).

2 Related Work

2.1 (Social) Bots

A bot is an automated computer program which is able to perform repetitive tasks that are usually performed by humans [10]. The main benefit of bots is that they do not require any direct interaction with a human to realize specific tasks. In this paper, we refer to social bots in learning environments as mentoring bots, because we see a learning environment also as a social platform for knowledge discovery, in which bots should behave like social bots. The term "Social Bot" encompasses a broad number of different bot concepts. Social Bots are computer programs which are able to act on platforms where humans can interact with each other [8]. This can be achieved by either "mimicking the actions of a real Online Social Network (OSN) user or by simulating such a user using artificial intelligence" [2]. For example, there are bots that directly communicate with humans, simulating a human conversational partner [17]. Daniel et al. [5] classify common actions that can be performed by bots into four categories: The first is *chatting*, which is directly talking to a user. This can also be used for redirecting users, e.g., by suggesting another service. The second category is *posting*, which encompasses writing posts oneself as well as commenting on posts, for example by replying to them, or forwarding posts, which would correspond to actions such as retweeting in the case of Twitter. The third category is *endorsing*, which means either liking specific posts or following users. The fourth category is *participation*, e.g., by creating new users on a social platform. In this paper we mainly focus on the first category. In a classical sense, chatbots are programs designed to simulate a human conversational partner in a chat room.

2.2 Bots as Mentors for MOOCs

MOOCs are an evolution of traditional courses, being completely offered online and serving a far greater audience at once. Having this huge audience comes at a cost: Personal interactions between students and teachers have to be sacrificed in order to make these courses viable. Since there is no local attendance in this type of course, students can easily become isolated [3]. Utilizing bots in such a context consists of three main goals:

1. Provide 24/7 service to students, which would be expensive or in some cases outright impossible to offer without bots [12].
2. Reduce the workload of human teachers by letting bots answer frequently asked questions.

3. Help students to get in touch with other fellow students or teachers and to encourage conversations between them [3].

MOOC-Bot is a chatbot for a MOOC website with the purpose to answer frequently asked questions about the course, provide knowledge about the course and to provide a chatbot interface [12]. With our contribution we prioritize the solution of the aforementioned objectives by using a modeling environment [13] instead of scripting languages (e.g. AIML) to facilitate the creation for non-technical users. *Teacherbot* is a Twitter bot for the MOOC "E-Learning and Digital Cultures" on the Coursera platform [3]. The role of the bot is to start discussions between students and answer questions regarding the organization of the course itself. The bot tweets quotes related to the course's content in regular intervals and responds to tweets containing a certain hashtag. It pulls other students or teachers into the discussion to help overcome shyness, since it follows the idea that it is easier for shy people to start a conversation with a bot than with another human. The quantitative evaluation shows that a bot can attract "lurkers" who usually do not interact with others in a course. Their evaluation conveyed that those who interacted with the bot had an increased likelihood of being drawn into further conversation and interaction. In this case, the bot works on a public channel (Twitter) and can only react to keywords with simple rules. However, a public channel limits the types of requests and therefore we focus on chat interfaces to allow private and public messages. The *Link Student Assistant (LiSA)*, is a virtual assistant type chatbot for helping students with common tasks on a university campus [6]. The bot was used to survey the students via chat about their requirements for a bot and how the bot's "personality" affects the user experience. Therefore, the bot conducted a survey with 100 students, including both university and high school students. After they determined inadequate behavior, LiSA's personality was changed to a more sensitive one, which led to students to apologize for their behavior once LiSA showed feelings of being hurt. This type of interaction has potential for leading discussions back onto the right track by using "filters and emphatic answers". Here the chatbot was implemented via Google's Dialogflow framework and was accessed via Facebook Messenger. Eicher et al. show with *Jill Watson* that sensitive information can also be exchanged, they highlight the advantage of using a private and confidential channel [7]. Therefore, we attach great importance to data protection in our work and only use open-source software which can be hosted by ourselves.

3 Use Case

The use case results from an accumulation of requirements[2] that emerged during the course of the national research project mentioned in the introduction. One of the universities offers a seminar course in Moodle for more than 400 students of educational science. For this course, the lecturers offer the students a chatbot for support. It gives the students the possibility to ask comprehension questions via

[2] https://requirements-bazaar.org/projects/393/categories/808.

a chat. The chat is integrated within the LMS. The lecturer can add questions and respective answers to the chatbot's repertoire. Within the seminar, students have to read literature throughout the semester and acquire key concepts. For selected topics of the seminar, students complete optional writing assignments on a concept of their choice. The chatbot is also used to analyze the submission of the students. Therefore, the chatbot forwards the submission to the analysis tool T-MITOCAR and replies with the generated personal feedback document. The students have the freedom to choose the topic they want to receive feedback for at any time and can send texts as often as they want, thereby they can iteratively improve their knowledge on a certain topic. The feedback is a comparison with an expert text and reveals contained and missing key concepts. It is presented to the student as a feedback document.

4 Concept

Within this section we refer to the communication between the person and the bots. For the implementation of the bots we build on an existing open-source bot modeling tool [13]. The framework is adapted with the function to model incoming messages and to react to them. Figure 1 shows the communication flow between a user and a bot. The communication flow is based on the Responder-Classifier-Graphmaster architecture [16]. The *Messenger* and *Bot Action* are responsible for the communication transfer, the *NLU* unit is the classifier and the derived *Intent* and the *Bot Action* corresponds to the graphmaster which handles the responses. The requests of a user are sent to the bot as *Incoming Messages*. The bot has access to the respective *Messenger* and can collect the *Incoming Messages*. The *Messenger* object contains the type of chat service that is going to be used, the required information to authenticate with the chat service. Every *Incoming Message* contains an *Intent* keyword. During the creation of the bot, several sample *Incoming Messages* can be predefined and assigned to an *Intent.* The Bot uses the *NLU* component, analyzes the *Incoming Messages* and filters out the *Intent* with its *Entities.* This information can be used to perform *Bot Actions*. With this component, the bot can access external services while reusing and processing the *Entities* of the *Intent* for the *Chat Response* to be generated. A *Chat Response* can be defined as a static text message. To ensure response diversity in static responses, multiple *Chat Responses* can be linked to a single *Incoming Message.* Subsequently, a random static response is chosen at runtime when the intent is recognized. The *Chat Response* is determined using the *Bot Action,* in which the *Intent* is assigned to a reasonable response. The recognized *Intents* can also be specially linked so that a certain state is present and a conversation with follow-up messages can be held. For more complex tasks, *Bot Actions* can also be used to execute HTTP requests, for example to access a REST API of an external service or to receive content dynamically. This is where relevant information can be passed to external services using an *Entity* of an *Intent.* For example, if a message is received and classified as the *Intent* `needed_support`, the *Entity* can specify the exact system on which the user needs help. In order to enable bot developers to handle unrecognized or badly recognized intents, the intent keyword "default" is reserved. The *NLU* component classifies the *Intent* with a certain confidence. If this confidence is below a certain threshold, the "default" *Intent* is used as a fallback for otherwise non handled messages. This is useful for telling users that their input was not understood correctly.

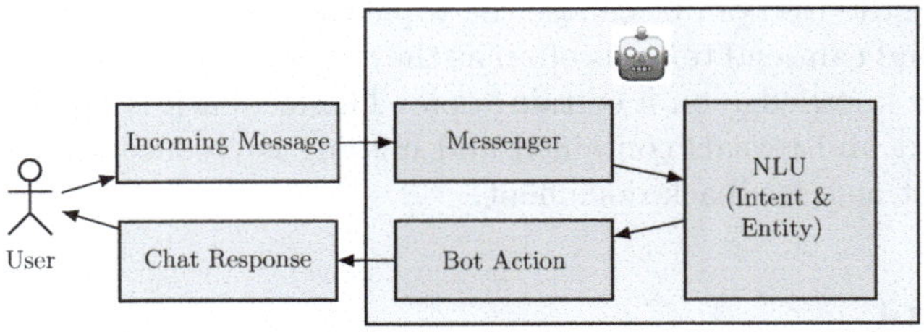

Fig. 1. Pipeline of a human-bot conversation flow.

5 Realization

The new chat connection features were implemented within an existing bot creation framework [13]. We developed a *chat* component for dealing with chat services. The component is kept as basic as possible so that it can be reused for multiple chat services. The most fundamental part of this component is the *ChatMessage*, which is used for storing information about a message that is received. It contains sender's user ID, the channel it originates from and the text of the message. Building on top of this is an abstract class called *ChatMediator*. Concrete implementations of this class are supposed to extend the connection to a chat service. The *ChatMediator* contains methods to send messages to a channel, to retrieve messages received by the *ChatMediator* and a method for retrieving the direct message channel ID of a user (matching the email address). The connection to the chat service itself is designed to be established within the constructor of the actual *ChatMediator* implementation. Currently, we provide implementations for Slack[3] and Rocket.Chat[4]. Incoming messages can be received using a continuous WebSocket connection to Slack or Rocket.Chat. After the bot has successfully signed in to the messenger, it listens for incoming messages.

5.1 Processing Chat Messages in the Bot Manager

The intent classification and chat connection are combined as one element in the modeling view. For this, the meta-model had to be extended with the new modeling elements—*Messenger*, *IncomingMessage* and *IntentEntity*.

[3] https://slack.com/.
[4] https://rocket.chat/.

Once a bot model is parsed and at least one bot is running, a thread checks for any new messages received every second. If a message starts with an exclamation mark, the NLU component is bypassed and the first word after the exclamation mark is chosen as the *Intent* and the rest of the message is used as an *Entity*. This was designed to provide functionality similar to so-called slash commands which are used to invoke applications with parameters. Otherwise, the message is sent to a Rasa[5] NLU server to retrieve the *Intent* and *Entity* from it. The definition of intents and the corresponding exemplary statements are given in markdown format.

```
## intent:showtasks
- Show me the writing assignment
- Show me the tasks
- Which tasks should I process?

## intent:submission
- How can I enter my answer?
- How can I submit my answer
- I want to turn in my writing assignment
```

Listing 1.1. Specification of intents *showtasks* and *submission* with sample requests.

To give an example of a model, Fig. 2 shows the realized bot model for the use case described in Sect. 3. The example shows the *IncomingMessage* objects (with *Intent*) that are received by the Messenger. These messages are paired with matching replies. Sent files are also considered as incoming messages and thus student submissions can be analyzed. As an analysis tool we use T-MITOCAR which is a heuristic for specific, language-oriented questions of knowledge and learning diagnosis [14]. It can represent individual association networks from texts in graphical form. Figure 3 shows the resulting chatbot with a sample discussion, integrated in a Moodle instance.

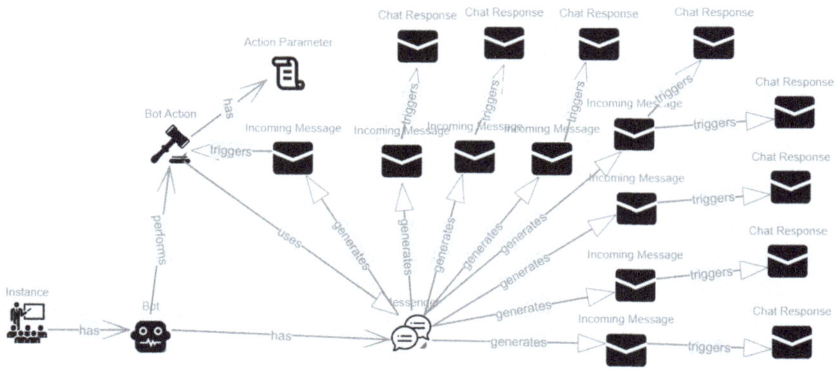

Fig. 2. Realized bot model from the use case.

[5] https://rasa.com/.

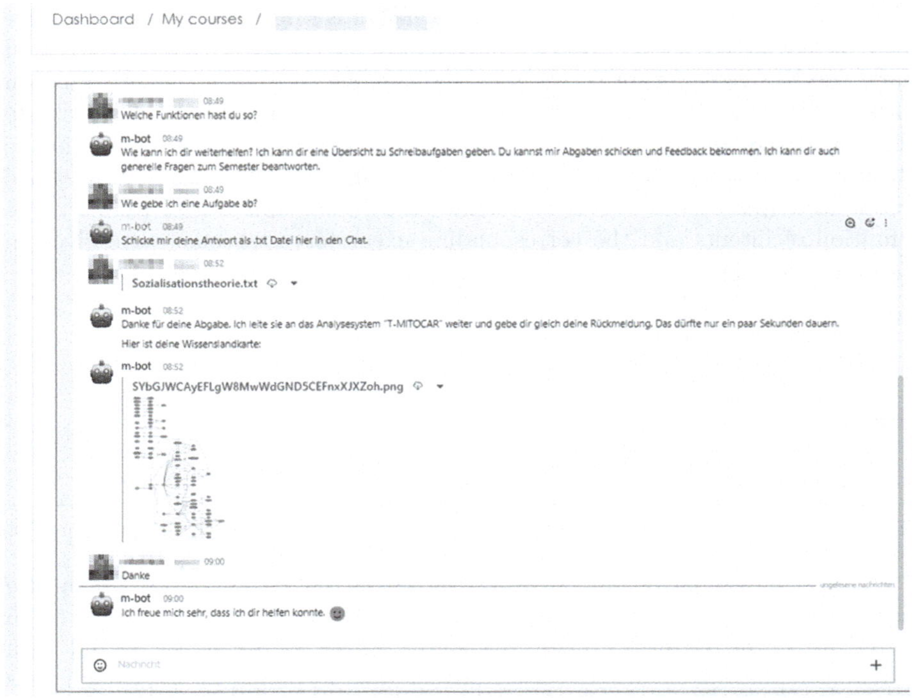

Fig. 3. Chat integrated in Moodle with a sample conversation.

6 Evaluation

We performed two evaluations. In the developer evaluation, the creation of bots with the new features was evaluated and in the pedagogical evaluation the usage of these bots was evaluated. The software setup encompassed the same constellation as described in Sect. 5.

6.1 Developer Evaluation

The main goals of the first evaluation were to verify whether the creation of bots functioned as expected and to examine the usability and the usefulness of the framework's features. It took place within one week. We conducted eight sessions, each session having one participant who created and interacted with a chatbot. Six of the participants were researchers or students of a computer science department, the other two were former computer science students. A short introductory presentation was held before each session to explain the functionality of the framework. The first task was to train a Rasa NLU model using the example training data that was provided. Then, the participants were asked to model a simple bot that responded to each intent present in the training data with at least one message. Afterwards, the participants had access to a sample slack workspace where they had to test messages with handled and unhandled intents, and observe how the bot reacted. The next task was to obtain logs and identify handled and unhandled intentions. Then, new intents should be added to fix the undetected requests. Finally, these changes were tested by the participant. In the end, participants were asked to fill out a questionnaire with six usability questions. All of these questions were based on a Likert rating scale of one to five, ranging from "strongly disagree" to "strongly agree". Additionally, a "further comments and suggestions" field was part of the questionnaire. A boxplot of the results for the statements rated on an ordinal scale can be found in Fig. 4.

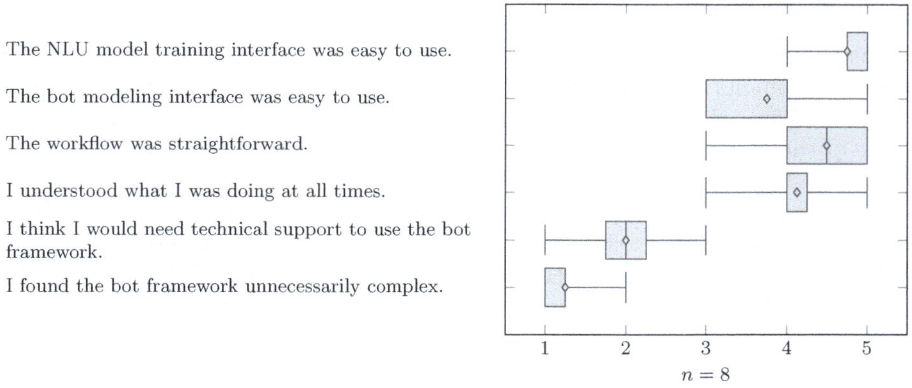

Fig. 4. Results of the usability questionnaire presented as a boxplot. Score is an ordinal scale where 1 ≙ "strongly disagree" and 5 ≙ "strongly agree".

The results show that participants generally liked the Rasa NLU model training interface and found it easy to use, with the question having a median score of 5 and an average score of 4.75. This is potentially due to the combination of both the very simple structure of the markdown training data with the ability to edit it directly in the browser with syntax highlighting. The question about ease of use of this interface was rated with a median score of 4, with an average of 3.75. In general, participants found the workflow straightforward, even though the Rasa NLU model training front end was separate from the bot modeling tool. Overall, participants did not think that the chat conversation extensions of the framework were overcomplicated. However, most answered that they would at least need a little bit of technical support to use it.

6.2 Pedagogical Evaluation

For the second evaluation, we used the bot implemented in the use case described in Sect. 3. 11 educators from the project described in the introduction and use case interacted with the bot during a continuous use phase of three days.

They had the possibility to access the bot directly via the messenger or via the integrated interface in the LMS. At the beginning the participants were asked to contact the bot and send a greeting. Afterwards, the participants were asked if they would like to provide their demographic data (age, gender, etc.). The provision of the data is voluntary and could be revoked afterwards. One task was to ask the bot questions about the course that the students would commonly ask the lecturers directly. Additionally, participants were provided with a sample text submission, which should be sent to the bot to be analyzed by the text analysis tool we described in Sect. 5.1. After a few seconds the participants received a feedback in form of a visual knowledge map representation about their submission document (see also Fig. 3). After the participants completed all tasks, they were instructed to ask the bot how to evaluate it. This request should give the participant a link with a questionnaire with six questions about the usefulness of the bot and its usage. At the end, the participants could add further comments and suggestions for improvement in form of a free text field. Figure 5 shows the results of the pedagogical evaluation. The lecturers were aware of the importance of being able to obtain an answer at any time and that the data will be treated confidentially. As side comments, it became clear that the idea of a digital accompaniment as support was very positively received. The possibility of informal exchange and communication with an anonymous chat partner was also well received. However, it has to be mentioned that the bot did not always provide the right answers to the requests and some participants had to rephrase their request to receive the desired response. Misunderstood requests were a potential cause of frustration. Hence, it is important that the recognition of the intent of a request is accompanied by a high confidence level. Nevertheless, the last two statements show that bots have been positively received and can certainly be a feasible mentoring assistance.

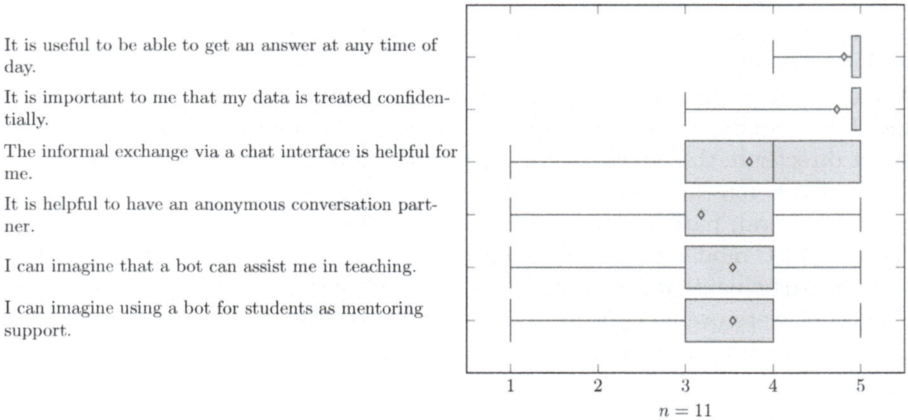

Fig. 5. Results of the acceptance and usage questionnaire, presented as a boxplot. Score is an ordinal scale where 1 ≙ "strongly disagree" and 5 ≙ "strongly agree".

7 Conclusion and Future Work

In this paper, we presented a framework that allows the creation of intelligent mentoring chatbots in learning environments. Our bots provide answers to simple FAQ at any time of the day and are able to give direct feedback on text submissions. The intention of each incoming message is determined by NLU and a corresponding answer is taken from a pool of possible answers. As the initial user evaluation shows, the core chat functionality is well implemented from a usability perspective. The indication of intents is intuitive and is also well recognized in the English model. The second evaluation aimed at the pedagogical usage of chatbots as mentors. Here, the requirements of everyday accessibility and the sensitive treatment of data are important. The German intent recognition model was used to implement the use case of a nationwide research project. However, the recognition rate of the intents in the implemented model needs to be improved with more data and more training. Overall, the use of such mentoring bots is an approach that has been well received. To conclude, we have introduced a system which makes it possible to create mentoring bots and integrate them into LMSs. Due to its scalability, the system has great potential for MOOCs. The evaluation also showed the usability as a personal mentor that can be reached at any time. Nevertheless, there are some drawbacks in this early phase of the project such as problems with the recognition of intents. Although some of the intents were recognized incorrectly or were only solved by similar requests, the evaluation concluded that the presented system is a viable approach to aid teachers/instructors. We currently apply our intelligent mentoring bots in large courses with several hundred students to cope with the very recent issue of missing face-to-face teaching. During this extended utilization phase, we relentlessly evaluate and improve the intent recognition and functionality the bot provides aid in overcoming shortfalls in education due to the current situation. With this, we are convinced mentoring bots will find their everyday habitat in LMSs.

Acknowledgments. The authors would like to thank the German Federal Ministry of Education and Research (BMBF) for their kind support within the project "Personalisierte Kompetenzentwicklung durch skalierbare Mentoringprozesse" (tech4comp) under the project id 16DHB2110.

References

1. Bickmore, T.W., Schulman, D., Sidner, C.: Automated interventions for multiple health behaviors using conversational agents. Patient Educ. Couns. **92**(2), 142–148 (2013)
2. Boshmaf, Y., Muslukhov, I., Beznosov, K., Ripeanu, M.: Design and analysis of a social botnet. Comput. Netw. **57**(2), 556–578 (2013)
3. Bozkurt, A., Kilgore, W., Crosslin, M.: Bot-teachers in hybrid massive open online courses (MOOCs): a post-humanist experience. Australas. J. Educ. Technol. **34**(3), 39–59 (2018)

4. Brinton, C.G., Rill, R., Ha, S., Chiang, M., Smith, R., Ju, W.: Individualization for education at scale: MIIC design and preliminary evaluation. IEEE Trans. Learn. Technol. **8**(1), 136–148 (2015)
5. Daniel, F., Cappiello, C., Benatallah, B.: Bots acting like humans: understanding and preventing harm. IEEE Internet Comput. **23**(2), 40–49 (2019)
6. Dibitonto, M., Leszczynska, K., Tazzi, F., Medaglia, C.M.: Chatbot in a campus environment: design of LiSA, a virtual assistant to help students in their university life. In: Kurosu, M. (ed.) HCI 2018. LNCS, vol. 10903, pp. 103–116. Springer, Cham (2018). https://doi.org/10.1007/978-3-319-91250-9_9
7. Eicher, B., Polepeddi, L., Goel, A.: Jill Watson doesn't care if you're pregnant: grounding AI ethics in empirical studies. In: Proceedings of the 2018 AAAI/ACM Conference on AI, Ethics, and Society, pp. 88–94. AIES 2018. Association for Computing Machinery (ACM), New York (2018)
8. Ferrara, E., Varol, O., Davis, C., Menczer, F., Flammini, A.: The rise of social bots. Commun. ACM **59**(7), 96–104 (2016)
9. Fryer, L., Carpenter, R.: Bots as language learning tools. Lang. Learn. Technol. **10**, 8–14 (2006)
10. Geiger, R.S.: Bot-based collective blocklists in Twitter: the counterpublic moderation of harassment in a networked public space. Inf. Commun. Soc. **19**(6), 787–803 (2016)
11. Kerly, A., Ellis, R., Bull, S.: CALMsystem: a conversational agent for learner modelling. In: Ellis, R., Allen, T., Petridis, M. (eds.) Applications and Innovations in Intelligent Systems XV. LNCS, pp. 89–102. Springer, London (2008). https://doi.org/10.1007/978-1-84800-086-5_7
12. Lim, S.L., Goh, O.S.: Intelligent Conversational Bot for Massive Online Open Courses (MOOCs) (2016)
13. Neumann, A.T., de Lange, P., Klamma, R.: Collaborative creation and training of social bots in learning communities. In: 2019 IEEE 5th International Conference on Collaboration and Internet Computing (CIC), pp. 11–19. IEEE (12 Dec 2019–14 Dec 2019)
14. Pirnay-Dummer, P., Ifenthaler, D.: Automated knowledge visualization and assessment. In: Ifenthaler, D., Pirnay-Dummer, P., Seel, N.M. (eds.) Computer-Based Diagnostics and Systematic Analysis of Knowledge, vol. 1, pp. 77–115. Springer, US, Boston, MA (2010). https://doi.org/10.1007/978-1-4419-5662-0_6
15. Shao, C., Ciampaglia, G.L., Varol, O., Yang, K.C., Flammini, A., Menczer, F.: The spread of low-credibility content by social bots. Nat. Commun. (Nat. Commun.) **9**(1), 4787 (2018)
16. Stoner, D.J., Ford, L., Ricci, M.: Simulating Military Radio Communications Using Speech Recognition and Chat-Bot Technology. The Titan Corporation (2003)
17. Weizenbaum, J.: ELIZA–a computer program for the study of natural language communication between man and machine. Commun. ACM **9**(1), 36–45 (1966)

A Review on Visualization of Educational Data in Online Learning

M. Ali Akber Dewan(✉), Walter Moreno Pachon, and Fuhua Lin

School of Computing and Information Systems, Athabasca University, Edmonton, Canada
{adewan, oscarl}@athabascau.ca, waltmorpa@hotmail.com

Abstract. Higher educational institutions capture huge amounts of educational data, especially in online learning. Data mining techniques have shown promises to interpret these data using different patterns. However, understanding the mining patterns and extracting meaningful information from the data require reasonable skills and knowledge for the users. Information visualization, due to its potential to display large amount of data, may fill this gap. In this paper, we present a short review of such visualization systems that focus on extracting meaningful information from the educational data. Visualizations have been used in different applications dealing with educational data, especially for *monitoring student performance, understanding learning style, analyzing course and program status,* and *dropout prediction.* In this paper, we reviewed the existing visualization systems, their design considerations, and their strengths and weaknesses to analyze educational data in the context of online learning. Research findings indicate that although some progress has been achieved in educational data mining and visualizations, designing and developing effective and easy to understand visualizations and having the functionalities of interactivity and time-series analysis are still challenging. This review provides insight into how to build a learner and instructor focused effective visualization system for an online learning environment.

Keywords: Information visualization · Educational data mining · Pattern recognition · Learning management system · Visual analytics

1 Introduction

The number of students in the online courses has grown considerably in the last decade. There is a great demand for the online courses delivered in computer-supported learning management systems (LMS), such as Moodle, WebCT and Blackboard, or massive open online courses (MOOC) delivery systems, such as Coursera, Udacity, and EdX [1]. These systems capture huge amounts of data through the interactions among learners, instructors and administrators of the educational institutions. Understanding these data can provide insights about *student performance, learning style, course and program status,* and *tendency for students to dropout from the course or program.* This can help in fine tuning the curricula, increasing in teaching quality, better learning outcomes, and overall improving of student success [2]. Its timely

C. Pang et al. (Eds.): SETE 2020/ICWL 2020, LNCS 12511, pp. 15–24, 2021.
https://doi.org/10.1007/978-3-030-66906-5_2

analysis is very crucial for improvement of the teaching and learning process especially in online course delivery.

Data mining provides a great support for extracting meaningful patterns from the educational data. The patterns are useful to study the data of what went on in the past and to make informed predictions for the future. Although the extracted patterns are much more meaningful than the raw data, the number of patterns can still be intimidating and not conducing to a comprehensive analysis. To understand the patterns also often requires a reasonable knowledge of the underlying data mining algorithms and statistical models, which many users may not possess [2].

To identify the commonalities and differences in the data, it is important to provide an overall view of all patterns as a coherent whole. It is also necessary to interpret individual patterns and establish relations between them. The system should allow interactive functionalities and allow users to take advantage of their creativity, flexibility and domain-specific knowledge. A possible solution to this problem is to use *information visualization*. This can help display and understand large amounts of data to discover trends, patterns and outliers [3]. A well-designed visualization can effectively represent large amounts of data and alleviate the cognitive load associated with interpreting it [4]. A visualization of the results from data mining has the potential to provide the insights it needs.

In the past several years, a wide range of review papers have been published on educational data mining. Slater et al. [5] presented a review on commonly used and accessible tools available for educational data mining research. Shahiria et al. [6] presented a review on predicting student performance by using data mining to improve student achievement. Dutt et al. [7] provided a systematic review of literature (from 1983 till 2016) on clustering algorithms and their applications in the context of educational data mining. Aldowah et al. [8] presented a comprehensive review of educational data mining and learning analytics (from 2000 till 2017) in higher education. Although a good number of review papers have been published in the field of educational data mining, not many papers could be found in the field of educational data visualization, especially in the context of online learning.

In this paper, we present a review on visualization of educational data from online learning perspective. This paper provides insight into how to build an effective visualization system for an online learning environment. The visualization systems are found to be promising for extracting and presenting meaningful information for both the learners and instructors. The rest of the paper is organized as follows: Section II provides the literature search procedure and criteria. Section III details about the existing systems and related trends dealing with educational data visualization. Section IV concludes the paper with some critical discussion and recommendations.

2 Literature Search Procedure and Criteria

The main purpose of this review is to examine existing visualization systems, their design considerations and applications, and their effect on online learning and teaching. The research questions that we have addressed in this review are the following: *what existing systems are being used for educational data visualization in online learning?*

what design considerations went through when designing the visualization systems? what applications the visualization systems are being used in online courses in terms of teaching and learning?

The search methodology we used in this review identifies the papers that are focused on visualizations of the educational data. We have considered the research papers published between 2000 to 2019 in this review. To search for related visualization papers, we identify the search keywords, such as "distance learning", "online learning", "distance education", "online courses", "course performance", "data mining", "learning analytics", "learning outcome", "eLearning", "educational data" and "visualization". We also searched for alternative synonyms and keywords. We used Boolean operators like AND, OR, NOT in our search strings. We search in the *IEEE Xplore*, *Vispubdata*, *EuroGraphics*, *Springer*, *Elsevier* and *ACM* digital libraries. We also search in Google Scholar's with the "cited by" feature to find literature citing for any given research paper. The related work section of each paper is also examined for sources of visualization papers in the context of online learning.

The papers that focus on examining educational data with the aid of visualization techniques are considered within the scope of this review. The studies that focus on visualization of online course discussion forums are considered out of scope of this research, as we have addressed this topic in a separate review paper [9]. The studies that focus on visual representations for a topic to teach students showing the material in an easy to understand way also have not been considered in this review.

3 Educational Data Visualization

Although there are some overlaps, the applications of the educational data visualization systems reviewed in this paper can be divided into four main categories: *monitoring student performance, understanding learning style, analyzing course and program status,* and *dropout prediction.* The visualization systems with their categorization are shown in Fig. 1.

3.1 Monitoring Performance

Visualization systems that focus on presenting educational data to give views of students' performance are included in this category. These systems often use course grades, scores in quizzes and exams, and points or badges earned in gamification systems for analyzing and visualizing the students' performance.

Interactive visualization allows data exploration and visualization of many variables simultaneously according to the user's need. A parallel coordinate-based interactive representation is used by Xiaoya et al. [10], where the results of students of an English course is visualized from a multi-dimensional dataset. This representation include *classification*, *averaging*, *boxplot*, *axis permutation*, *correlation*, *association*, and *roll-up and drill.* The *classification* allows user to divide the data into sets corresponding to different lectures, where the *averaging* depicts average values for each set by interacting with the main chart. The *boxplots* assess data dispersion and the *axis permutation* allows changing their order to find new insights. The *correlation* presents the correlation of two data sets along the different axis and the *association* estimates value of an attribute based on another attribute. The *roll-up and drill-down* represent the data hierarchically.

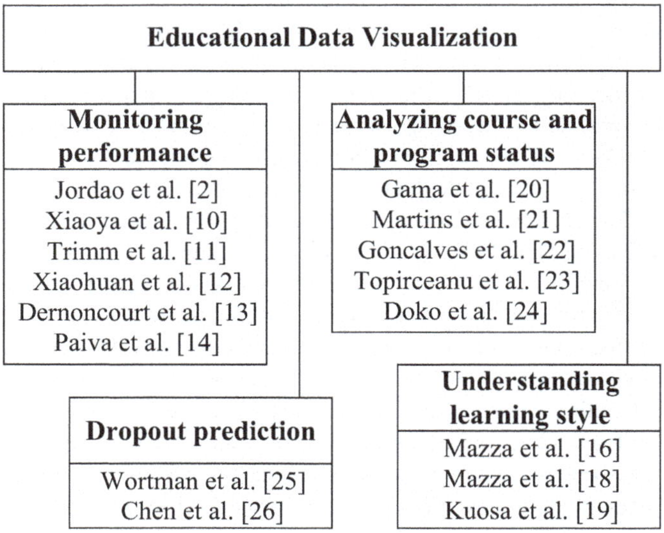

Fig. 1. Categorization of educational data visualization systems

Some visualization systems focus on general trends for a group instead of focusing on the performance data of an individual student. Trimm et al. [11] presented such a system where the students are grouped according to their grades and the evolution of each group is visualized over time. Historical data from student courses are represented using two dimensional trajectories, where these are colored by semester grades. Although the visualization representations are very appealing, it is hard to observe individual student's progress with the cluttered trajectory over time. However, using a color blending technique and color weaving, the trajectories are merged which provides a better perception of the overall trend and the dispersion of data. In another research, Xiaohuan et al. [12] presented a system called AVOJ aims at highlighting the different capabilities of students by visualizing performance data (e.g., exam grades, age, attendance, and participation). By grouping students according to grades and similar factors, the system allows direct comparison of general trends, based on bar charts displaying real-time statistics, and makes it possible to understand some specific things, such as the ways students manage study sessions. The system uses a color scheme with maximum brightness and saturation values which is found to be overwhelming and makes it difficult to find patterns.

Dernoncourt et al. [13] presented MOOCViz that provides visual representations to help teachers understand the performance of their students using choropleth and stacked bar chart. Choropleth summarizes the certificates delivered in all countries where the students are, and it is represented by a white (fewer certificates) to red (more certificates) scale. The stacked bar chart allows the user to see a more detailed information, regarding the time dedicated to the degree, which include access to books, exams, problems, lectures and tutorials. This allows the teachers to get a view of what kind of content is visited by each country. However, due to the lack of interactive functionalities, it is hard to explore all the available information.

Jordao et al. [2] presented a visualization system, called EduVis, which uses data mining techniques to identify patterns and make performance predictions of the students. A multi-matrix visualization mechanism is used that allows to observe the academic journey of students as well as to infer the various relationships and precedence of existing subjects. In this visualization, the matrices represent the subjects of the course, and each subject corresponds to a square, which is divided into two triangles: the upper triangle represents approvals and the lower one the failures. The system allows some interaction functionalities including filtering patterns and comparing selected patterns. The system found to be useful to explore the educational data in temporal dimension, which corresponds to the choice of various disciplines, held by students at a given instant of time, giving rise to immediate visualization of the various routes followed by them, throughout the course.

To help instructors understand the learning analytics and data mining results on the student data, Paiva et al. [14] presented a visualization system, called DataViz. The system comprises with three separate graphical representations. The first visualization uses segmented bar chart to represent the amount of interactions of a group of students with each educational resource. The second visualization shows the most impactful interactions on students' performance using ordered weights. The third visualization shows the most impactful combination of interactions on students' performance using combined interactions. The combined interactions apply association rules based on frequent and relevant patterns in the data [15]. The visualizations were found to be effective for teachers in a simple and objective way to interpret what is going on with their groups and assist teachers for daily evidence-based decision-making.

3.2 Understanding Learning Style

In this category, visualization systems focus on identifying students' behaviour and learning style analyzing student log-data in the learning management systems. The attributes like frequency and duration of access to the course, quiz and assignment submissions, participation in the group activities are visualized and analyzed. The discovered learning styles are useful to measure the awareness of an individual student or a group within the online courses.

To analyze student learning styles, Mazza et al. [16] developed CourseVis that visualize log data generated by the course management systems. To support instructors in online learning courses, the visualization considers three aspects: *social*, *cognitive* and *behavioural*. The *social* aspect uses two representations. The first representation maps the discussion board variables (e.g., originator, date, time) onto a three-dimensional scatterplot. The second representation maps the number of threads started and the number of postings by student onto a two-dimensional diagram. The *cognitive* aspect is represented using a cognitive matrix based on students' performance in quizzes. The students are mapped onto x-axis, the concepts are mapped onto the y-axis, and the performance values are mapped onto square using colours. The *behavioural* aspect is visualized using some 2D graphical representations which include accesses to content pages by topics, global accesses to the course, progress with the course schedule, messages, quiz, and assignment submission. OpenDX [17] is used for the development of CourseVis. The system found to be useful for the instructors to quickly grasp information about social, cognitive, and behavioural aspects of students for managing successful distance courses. However, establishing a link among the cognitive, social and behavioral aspects would be useful to investigate reasons for students struggling with the various course contents. This system also has some limitations. Some representations are found to be excessive and confusing during the system evaluation. To further improve the system, authors proposed the Graphical Interactive System for Monitoring Students (GISMO) [18], which uses two-dimensional plots and allows the exploration of students' behaviour over time. The system is found to be useful in understanding the students' behavior as well as the redesign of the course according to the students' needs.

Kuosa et al. [19] presented a system that include two interactive visualizations to analyze learning style. The *first* visualization is used to analyze student activity from automatically recorded user log data, which provides valuable insights into the learning process and participation of students in the course. The *second* visualization extends navigation and search functionalities in the discussion forum of an LMS with a topic-driven paradigm. The four main perspectives used in the *first* visualizations include showing student activity, finding relevant contents, showing effectiveness of the activity according to scores, and motivating students to participate and be active. The interactive visualizations represent the activity and participation of each member of the student group. Basically, three types of interactive visualizations are connected by means of linking and brushing as follows: set of interactive live histograms that users click and drag in order; a pie chart representing types of user content; and a bubble chart in which each bubble represents a different student. The main visualization views are created using D3.js and Raphael.

3.3 Analyzing Course and Program Status

Visualization systems that focus on identifying the course or program status analyzing educational data are considered in this category. The attributes like interrelations among courses, number of students completed a particular set of courses, percentage of passing or failing students in a course or program and difficulty levels of course topics are often used in these analyses.

Visualization systems are used to analyze interrelation among courses in a program. To analyze interdependences among courses in a university program, Gama and Goncalves [20] applied educational data mining techniques followed by a multilevel visualization. With the multilevel visualization, each level depicts a semester with corresponding courses. Visual connectors are used to display a high number of interrelations among the courses. The connector combines visual merging techniques with Bezier curves to represent course interrelation. In another study, Martins et al. [21] presented an interactive multiple visualization system that allows analyzing crossed data to understand the academic path and the evolution of degrees and courses over the years. The visualization consists of an interactive matrix where each column represents the year and semester and each cell fills with a range of colors from red (low approval rating) to green (high approval rating). A scatter plot is used to represent the correlations among the attributes. To make the system interactive, a dropdown list is used to filter each degree and course. Another view was added to show the data related to the teachers that teach the courses. To represent the filtered course or degree, a pill is used with its name that can be removed when the user decides to. The system allows teachers to identify certain problems of the different degrees and courses and improve teaching and learning in the online courses. To develop this system, MySQL, Apache, and Python have been used. The customizable visual representation is developed using D3.js framework.

Goncalves et al. [22] developed a visual educational data mining framework (FMDEV) with the intend to visualize the data mining patterns using graphs in a few clicks for the Moodle courses. The system has three modules—previsualization, visualization, and framework administration. The framework administration module allows integrating a growing number of data mining tools that provide various types of analysis and graphing of user interaction. The system is found to be useful to support course instructors/administrators for decision making how to improve students' learning and organize instructional resources more efficiently. FMDEV is developed using HTML5, CSS, and JavaScript and integrated into Moodle. However, the system still needs to be tested for some real courses and monitor the use of the tutors/teachers, giving support and adding new modules of data mining.

To analyze impact of student demographic specific to online courses, Topirceanu et al. [23] presented a system using decision trees. The system uses various information from students' profile, such as age, gender, group, university, faculty, specialization, study year, participation history and desire to participate in future in online courses. Based on the extracted statistics, the system presented six decision trees for classifying the finalization and participation rates of online courses based on the students' individual traits. Authors state that the study is useful in understanding the needs of modern students, as well as in optimizing the way the online learning is further developed. In another study, Doko et al. [24] developed a system for understanding and learning the most difficult topics through using sequential pattern mining. The system collects learners' video watching logs consisting of data from stop, replay, and backward activities. The system then generates sequence from the collected learning histories and extracts important patterns to find learners' most difficult/important topic. To visualize the patterns, 2D plots have been used. The system is useful for understanding and learning the most difficult topics in an online course for the learners.

3.4 Dropout Prediction

The visualization systems that apply data mining and visualization techniques to predict dropout tendency in students and identify the students at risk by analyzing the educational data are considered in this category. Attributes such as difficulty level of courses and programs, loosing interests, learners' activity logs, clickstream, assignment records, behavior of learners and the resulting outcomes are typically used in this category.

Wortman et al. [25] presented a such system to visualize success and failure patterns to understand students' dropout tendency in a computational science degree. A node-like representation is used, where the nodes represent events and their sizes represent the number of students that participated in it. Edges are used to represent students' trajectories along those events, where edge thickness represents the number of students and the color represents the students' performance. This allows to identify and visualize the student groups with similar characteristics according to the above measures. The visualization system allows users to select students of a similar profile, such as "students that never failed in a certain course". As a result, instructors can gain insights about the causes for dropout predictions. However, the clutter representations of the overlapping links make it hard to do any deeper visual analysis using the system.

Chen et al. [26] presented a visual analytic system DropoutSeer to predict dropout tendency in students by analyzing three different learner activity logs — clickstream, assignment records, and forum posts. The system integrates four linked visual designs to enable analysts to identify learning patterns related to dropout behavior at multiple scales. Along with this, a novel temporal data visualization design is included to uncover the learners' detailed learning activities. This system not only helps instructors and education experts to understand the reasons for dropout, but also allows researchers to identify crucial features which can further improve the performance of the students. To improve further the DropoutSeer, a more flexible analytical framework could be adopted which could address the structure of courses and the metric of evaluation that evolves over time. The provision for analyzing certain types of learners could also be accommodated. Along with dropout prediction, building a classification model for recognizing specific types of learners would be a good addition to this system.

4 Conclusions

This paper presents a review on educational data visualization to provide insights and inform instructional decisions to the instructors and the students in online learning. The applications of educational data visualization were found in main four categories that include *monitoring student performance, understanding learning style, analyzing course and program status,* and *dropout prediction.* This review provides useful insights about the existing visualization systems, their design considerations, and applications. We believe that the review can contribute to building effective visualization systems to enhance learning experience for the online learners.

Research indicates that the current visualization systems lack several important aspects to reach a mass level of adoption by the learning management systems (LMS). Different visualization systems are developed targeting different LMS, which limits the ability to adopt a particular visualization system by all LMS in general. Visualizations are created with the intent to represent the higher dimensional educational data in an easy to understand way, however in reality, some are difficult to interpret and have a steep learning curve. Future research should focus on making complex visualizations simple yet meaningful.

The design of interactive visualizations brings the ability to slice and dice information for the learners and the instructors. Visualizations provide the ability to group, classify, and associate the data at different levels of aggregation, providing better understanding of the information. Though some improvement has occurred, designing and developing interactive visualizations focusing on the needs of learners and instructors for the time-series analysis is still challenging. Further research is still needed to improve interactive visualizations in online learning.

Acknowledgement. This research was supported by NSERC Discovery Grant, and Academic and Professional Development Grant, Athabasca University, Canada.

References

1. Rabbany, R., Elatia, S., Takaffoli, M., Zaïane, O.R.: Collaborative learning of students in online discussion forums: a social network analysis perspective. In: Peña-Ayala, A. (ed.) Educational Data Mining. SCI, vol. 524, pp. 441–466. Springer, Cham (2014). https://doi.org/10.1007/978-3-319-02738-8_16
2. Jordao, V.R., Gama, S., Goncalves, D.: Visualizing sequential educational datamining patterns. Int. J. Creative Interfaces Comput. Graph. **7**, 1–8 (2016)
3. Heer, J., Bostock, M., Ogievetsky, V.: A tour through the visualization zoo: a survey of powerful visualization techniques, from the obvious to the obscure. ACM Queue **8**(5), 1–22 (2010)
4. Ware, C.: Information Visualization, Perception for Design, 2nd edn. Morgan Kaufmann, Burlington (2004)
5. Slater, S., Joksimovic, S., Kovanovic, V., Baker, R.S., Gasevic, D.: Tools for educational data mining: a review. J. Educ. Behav. Stat. **42**(1), 1–21 (2016)
6. Shahiria, A.M., Husaina, W., Rashida, N.A.: A review on predicting student's performance using data mining techniques. Proc. Comput. Sci. **72**, 414–422 (2015)
7. Dutt, A., Ismail, M.A., Herawan, T.: Asystematic review of educational data mining. IEEE Access **5**, 15991–16005 (2017)
8. Aldowah, H., Al-Samarraie, H., Fauzy, W.M.: Educational data mining and learning analytics for 21st century higher education: a review and synthesis. Telemat. Inform. **37**, 13–49 (2019)
9. Wu, M., Dewan, M.A.A., Lin, F., Murshed, M.: Visualization of course discussion forums: a short review from online learning perspective. In: Canadian Conference on Electrical and Computer Engineering, Edmonton, Canada (2019)
10. Xiaoya, G., Kan, L., Ping, L.: Visual analysis of college students' scores in English test. In: International Conference on Computer Science and Education, Nanning, China (2009)
11. Trimm, D., Rheingans, P., desJardins, M.: Visualizing student histories using clustering and composition. IEEE Trans. Visual. Comput. Graph. **18**(12), 2809–2818 (2012)
12. Xiaohuan, W., Guodong, Y., Huan, W., Wei, H.: Visual exploration for time series data using multivariate analysis method. In: International Conference on Computer Science and Education, Colombo, Sri Lanka (2013)
13. Dernoncourt, F., Do, C., Halawa, S.: MoocViz: a large scale, open access, collaborative, data analytics platform for MOOCs. In: NIPS Education Workshop, Nevada, USA (2013)
14. Paiva, R., Bittencourt, I.I., Dermeval, D.: Visualizing learning analytics and educational data mining outputs. In: International Conference on Artificial Intelligence in Education, London, UK (2018)
15. Cohen, W.W.: Fast effective rule induction. In: Internationation Conference on Machine Learning (1995)

16. Mazza, R., Dimitrova, V.: Generation of graphical representations of student tracking data in course management system. In: IEEE International Conference on Information Visualization, London, UK (2005)
17. OpenDX, Open Source Visualization Data explorer. https://en.wikipedia.org/wiki/IBM_OpenDX
18. Mazza, R.: Monitoring an online course with the GISMO tool: a case study. J. Interact. Learn. Res. **18**(2), 251–265 (2007)
19. Kuosa, K., et al.: Interactive visualization tools to improve learning and teaching in online learning environments. Int. J. Distance Educ. Technol. **14**(1), 1–21 (2015)
20. Gama, S., Goncalves, D.: Visualizing large quantities of educational datamining information. In: International Conference on Information Visualization, Paris, France (2014)
21. Martins, T., Goncalves, D., Gama, S.: Visualizing historical patterns in large educational datasets. Int. J. Creative Interfaces Comput. Graph. **9**(1), 32–48 (2018)
22. Goncalves, A.F.D., Maciel, A.M.A., Rodrigues, R.L.: Development of a data mining education framework for data visualization in distance learning environments. In: International Conference on Software Engineering and Knowledge Engineering (2017)
23. Topirceanu, A., Grosseck, G.: Decision tree learning used for the classification of student archetypes in online courses. Proc. Comput. Sci. **112**, 51–61 (2017)
24. Doko, E., Bexheti, L.A., Hamiti, M., Prevalla, B.: Sequential pattern mining model to identify the most important or difficult learning topics via mobile technologies. Int. J. Interact. Mob. Technol. **12**(4), 109–122 (2018)
25. Wortman, D., Rheingans, P.: Visualizing trends in student performance across computer science courses. In: SIGCSE Technical Symposium on Computer Science Education, Kentucky, USA (2007)
26. Chen, Y., Chen, Q., Zhao, M., Boyer, S., Veeramachaneni, K., Qu, H.: DropoutSeer: visualizing learning patterns in massive ppen online courses for dropout reasoning and prediction. In: IEEE Conference on Visual Analytics Science and Technology, Baltimore, USA (2016)

Technology Affordances in the Web-Based Learning Environment for SLA: A Case Study of a Chinese English Learner in the Time of COVID-19

Jin Wang[1], Zili Chen[2], and Nana Jin[1](✉)

[1] School of Foreign Languages, Shenzhen University, Shenzhen, China
szuwangjin@szu.edu.cn, nanajin7@163.com

[2] School of Professional Education and Executive Development, CPCE, The Hong Kong Polytechnic University, Hong Kong, China
spczili@speed-polyu.edu.hk

Abstract. This paper reports on a case study exploring how technology affordances were perceived and acted on by a Chinese English learner in the time of COVID-19 from an ecological perspective. Qualitative data for the study came from the participant's learning journals, a questionnaire and a semi-structured interview. Findings of the study indicated that learning emancipation, anxiety reduction and sociocultural identity construction were three most salient dimensions in the focal student's perceptions and that the emergence of technology affordances in the interaction between the learner and the learning environment was a complex one, not only mediated by the learner but also sensitive to the specific knowledge area where learning was taking place. The ecological perspective and the notion of affordance adopted in the study have proved to be informative and efficient in exploring students' perceptions and interaction with their learning environments.

Keywords: Learning environment · Affordance · SLA · Ecological perspective · Educational technologies

1 Introduction

In the past 50 years, learning environment research has attracted increased attention from researchers and this research tradition has developed from the very strong emphasis on the use of a variety of questionnaires and inventories that measure students' perceptions of different types of learning environments [1, 2]. One recent development of the research field is to explore a variety of Web-based learning environments as responses to the development of Web technologies and increased use of computers in educational settings [3–5].

Web-based learning or online learning refers to the situation when students are using the Internet to interact with content, other students and their tutors [6]. Allen and Seaman [7] once categorized three types of online courses offered by educational institutions: Web course (course that uses Web technology to facilitate what is

C. Pang et al. (Eds.): SETE 2020/ICWL 2020, LNCS 12511, pp. 25–37, 2021.
https://doi.org/10.1007/978-3-030-66906-5_3

essentially a face-to-face course, using Blackboard or WebCT to post the syllabus and assignments), Blended/Hybrid course (course that is a blended of the asynchronous online and face-to-face course with substantial proportion of the content delivered online, typically supplemented by some face-to-face meetings) and Fully Online course (course where content, assignments and activities are all delivered online, typically asynchronously). However, this typology should be expanded to include live streaming courses, which were widely used to replace face-to-face teaching during the 2020 COVID-19 lockdown period. During this period, most school teaching and learning activities were made possible because of the Web technologies. Furthermore, the massive lockdown made it a special period of time when we particularly need to combine together two learning environments in our discussion as Dorman [1] once called for: classrooms and families, where learning activities again usually involved Web technologies especially for adult learners. A learner consequently faced a very rich and complex Web-based learning environment during the pandemic-caused lockdown. The present inquiry has aimed to examine this Web-based learning environment and in doing so, we focused on SLA (Second Language Acquisition) as the area where learning was taking place and investigated how the Web-based learning environment influenced a Chinese college student's English learning. The theoretical orientation of the study has been guided by the ecological perspective on SLA.

2 Literature Review

2.1 Researching Web-Based Learning Environment

Although learning environment seems to be an ambiguous concept, it is generally understood as the learning climate, the psychological or emotional climate and the social setting in which students are expected to learn [8]. While learning environment research can be traced back to the work of early social psychologists [9, 10], the modern era of learning environment research began in the late 1960s [1, 2]. Moor's [11] seminal and enduring tripartite framework of human environments, namely Relationship, Personal Development and System Maintenance and Change, has been underlying a wide body of learning environment research literature until today. Walberg's work [12] showed that student perceptions about their classrooms should be used in learning environment research. These studies generated two important questionnaires, i.e. Learning Environment Inventory (LEI) and Classroom Environment Scale (CES). Inspired by these pioneer work, learning environment researchers developed several widely used questionnaires, among which were Individualized Classroom Environment Questionnaire (ICEQ) [13], Science Laboratory Environment Inventory (SLEI) [14], Constructivist Learning Environment Survey (CLES) [15], and the What is Happening in this Class (WIHIC) [16]. Later researchers continued to adapt and use these questionnaires in different countries with different languages to investigate a variety of learning environments. Research interests have been expanded from students' perception to teachers' perception of the learning environments, comparisons of actual learning environment and preferred learning environment, correlation between students' perception and learning outcomes.

More relevant to the present study are three technology-related environment questionnaires. Chang and Fisher's [4] Web-based Learning Environment Inventory (WEBLEI) includes four scales of Emancipatory Activities, Co-participatory Activities, Qualia and Information Structure and Design Activities. Taking four features of online activities as the basis, i.e. student-interface, student-tutor, student-student and student-content [17], Clayton's [5] Online Learning Environment Survey (OLLES) includes seven scales (Computer Competence, Material Environment, Student Collaboration, Tutor Support, Active Learning, Information Design and Appeal, and Reflective Thinking). Aldridge and Fraser [3] designed the Technology-Rich Outcomes-Focused Learning Environment Inventory (TROFLEI), which can be seen as an extended version of WIHIC (seven scales as Student Cohesiveness, Teacher Support, Involvement, Task Orientation, Investigation, Cooperation and Equity), with three more scales (Differentiation, Computer Usage and Young Adult Ethos).

With the quantitative method as the mainstream in the field of learning environment research, studies usually involve a large-scale sampling. While these studies have produced very informative and inspiring results, an increased number of recent studies [18, 19] resorted to mix-methods approaches, where quantitative components containing the traditional inventories or questionnaires were mostly the main part, supplemented by qualitative components, which usually include observations, short answer questionnaires, interviews, field notes, etc.

Although these studies have provided insightful understanding of learners and their interaction with the learning environments, there are unfortunately still important issues to be adequately addressed. Despite the fact that many aspects and dimensions of the learning environments have been attended to and conceptualized, the normative approach of questionnaire surveys says nothing about which aspects or dimensions might be more important than others in what way, under what circumstances, and for what reasons. Furthermore, little has been known about how exactly perceptions of the learning environments prompt individual learners' further actions in the learning process. One more gap identified here is the lack of studies in the literature that focus on SLA [20, 21]. Since different disciplines have different ways of organizing the knowledge, it is argued that the way learners interact with the environment is discipline-specific and thus more attention should be paid to less explored disciplines. As an effort to fill the gaps, we decided to conduct an in-depth qualitative study and focus on one Chinese learner of English, with an aim to provide a "thick description" [22] of the interaction between the learner and the Web-based learning environment.

2.2 An Ecological Perspective on SLA

Also interested in the interaction between learners and their learning environments, SLA researchers had for many years been following the positivist tradition and correlating different factors in the environments to learning outcomes. Resonating with the recent call for a situated and holistic approach to SLA [23, 24], an ecological perspective on language learning foregrounds the context-embeded, constructive and emergent nature of the learning progress [25]. Adopting an ecological perspective means investigating the learning environment as it were an ecosystem. Language development, from an ecological perspective, occurs as a result of meaningful participation in human events and the participation involves perception, action and joint construction of meaning [26].

One crucial concept here is that of affordance, which has its roots in Gibson's work, meaning "what [the environment] offers the animal, what it provides or furnishes, either for good or ill" [27]. Instead of as a property, affordance is a relationship that holds between the object and the organism that is acting on the object [28], a relationship between an organism (a learner, in our case) and the environment, that signals an opportunity for or inhibition of action [26]. The concept is central to the ecological perspective on language learning in that it is at the roots of the relationship between the person and the physical, social and symbolic world [26]. Affordance emerges as we interact with the environment and the preconditions for the emergence are action, perception and interpretation, in a continuous cycle of mutual reinforcement [26].

From an ecological perspective, technology affordances for SLA, as will be discussed in this paper, is not properties of Web technologies or Web tools, but what a learner "detected, picked up, and acted upon as part of a person's resonating with, or being in tune with, her or his environment" [26]. As affordance can be "good or ill", Web-based technologies can possibly attract and distract learners from learning process and thus promote or hinder learning. This makes the construct "affordance" particularly pertinent to discussing the role technologies play in education.

Although van Lier [26] maintained that both qualitative and quantitative methods can be adopted in an ecological approach to language learning, we gave up a framework-guided and deductive method as usually used in technology affordances studies [29, 30] for a inductive bottom-up approach, in the hope of highlighting the holistic, emergent and contextulized nature of an ecological exploration.

3 Methodology

The present study is part of a broader inquiry into autonomy in English learning by Chinese students at the tertiary level from an ecological perspective. During the massive lockdown period, which basically covered the spring school semester, Web technologies became the most obvious new factor in the students' learning environments, which explains why the present topic was proposed. The investigation was conducted from March to August, 2020. To narrow down the focus of the study, we focused on one participant Xiaomei (pseudonym) and the research questions are as follows:

(1) What are the learner's perceptions of the Web-based learning environment for English learning during the pandemic-caused lockdown period?
(2) How does this Web-based learning environment influence her English learning?

The focal student Xiaomei of the present study was a female first-year college student who majored in English. She was recruited as a volunteer participant in the above-mentioned broader study, for which she wrote learning journals from March to August, 2000 as required by the researchers. Compared with other focal students, Xiaomei's journals contain more descriptions and evaluations concerning Web-based technologies in her English learning, making her an ideal participant for the present study. For the Spring semester 2020, Xiaomei had five compulsory courses that focused on different aspects of English proficiency, all of which were offered as live streaming online courses: “Comprehensive English”, “English reading”, “English listening comprehension”, “English writing” and “Oral English”. Web-based tools involved included three live streaming platforms (Tencent Classroom, Zoom and Tencent Meeting), the education platform Blackboard where course syllabuses, e-copies of textbooks, supplementary learning materials and assignments were handled, two exam-oriented online tools (Wenjuanxing and Itest) and WeChat as a convenient social net-working tool for after-class communication. Besides these university-based courses, towards the end of the semester, Xiaomei subscribed three other paid courses offered by educational services establishments: “English grammar” from a Bilibili channel, “Learn how to learn” form Coursera, and “English listening and speaking” from a WeChat subscription official account. Meanwhile, other websites and applications such as Google search, Oulu Dictionary, ChinaDaily, Bilibili, Instagram, Youtube, Library Genesis, Podcast Addict, etc. were also used by Xiaomei frequently to learn English. It is this rich landscape of Web-based learning environment that we set out to explore in the present inquiry.

Data of the study came from Xiaomei’s learning journals, a questionnaire and a semi-structured interview (see Table 1). Language for the three instruments is Chinese for the sake of expression clarity and all excerpts in the following section were translations by the authors of this paper. The questionnaire was basically two tables collecting basic information about Web-based courses Xiaomei was taking and Web-based platforms and applications she was using plus short comments on them. The interview was composed of two parts. In the first part, we checked with participants questions arising from reading the journals. The second part included self-developed open-ended questions designed based on previous online learning environment survey scales [3–5], also taking the specific learning environment of our investigation into consideration. These open-ended questions represented a spectrum of learning environment dimensions discussed in previous literature and can be categorized into seven types: ability to use Web-based technologies, technical and operational issues, emancipatory activities, communication and collaboration, pedagogic helpfulness, system maintenance, and affect orientation. Since these questions were theoretically rather than empirically grounded, the interview was conducted after the collection of all the journal entries in order to ensure that journal contents were not elicited from theses questions. For the same reason, the interview transcript and the questionnaire data were not included in the following data coding and theme inducing procedure, but for the purpose of achieving research triangulation [31] only.

Table 1. The data

Instruments	Indicated as	Dates
Learning journals	J1, J2,…J15	March–July 2020 (15 entries)
Questionnaire	QTN	July 31, 2020
Interview	ITV	August 7, 2020

We started the data analysis by reading through all the journal entries. The whole set of data were read thoroughly and independently by the three authors of this paper. Everything in the journal entries related to the research questions were noted down in the form of meaning unit. For the second reading, these meaning units were then inductively coded for repeated themes or patterns to emerge. After that, we compared notes and sorted out discrepancies. Any disagreements with regards to the coding or themes were discussed and checked with the journal entries until a consensus was reached. We finally classified and conceptualized three central themes, but we also resorted to the other two data sources, i.e. the questionnaire answers and interview transcript for interpretation and explanation of the three central themes.

4 Findings

In this section, we will describe how technology affordances in the Web-based learning environment was perceived, evaluated and acted on by the participant Xiaomei and how this learning environment influenced her English learning. Presentation of the findings centers on the following three central themes, which are meant not only to identify the technology affordances but also to interpret and explain how these affordances were made possible in the participant's learning.

4.1 An Emancipatory Space

In her first journal, which was written two weeks after the online teaching began, Xiaomei voiced her willing embrace of the Web-based learning of the courses, saying that "the outbreak of the pandemic made the online teaching and learning an unusual experience". She summarized benefits of this new teaching mode, several of which were concerning what Tobin [19] called the emancipatory activities.

One obvious benefit was that the online teaching mode was time-saving. Xiaomei could now "use the time saved in traveling for on-campus class attendance to learn more things" (J1). She used this saved time to preview her lessons and she could immediately review the lessons, which was usually impossible because she always chatted with classmates when the face-to-face classes were over.

Another thing that she liked was that during this period, her teachers seemed to use electronic materials more frequently, most of which were uploaded timely onto Blackboard. Xiaomei felt that this also gave her freedom to arrange her English learning. She added in the interview that she really hated to be told what to study and when and how to do the learning tasks all the time, like what it had been in her senior high school life. She determined to make use of the freedom to work harder, harder than her classmates and harder than before. And so she did and she was satisfied with the learning outcome: very positive comments from the English writing teacher and praise from the Comprehensive English course teacher. These learning results, as she said, then reinforced her determination to better organize her study plans.

The massive lockdown also gave Xiaomei plenty of time to explore English learning materials from the Internet. She first read quite a few English learning experience sharing posts and videos and then took online courses offered by educational services establishments and utilized internet resources to learn English:

Excerpt 1

I was lucky that I found a listening and speaking program taught by Ma Sirui [Chris]. It was 249 yuan and would take a month and I registered. Then from 21st June, I got started and I attended the lessons every day, listening to Laoma [Chris] speaking as a native English speaker and making notes about native ways of pronunciation and about where linking is occurring. I have to say that after ten classes I am now better able to recognize linking, the flap t and the glottal stops. I really learned a lot!... Several days ago I decided to learn English grammar with the uploader Xuan Yuanyou. I think I can finish the 400 pages grammar book with his teaching. I quite look forward to attending his class every day, which let me feel that I have a private tutor helping me... (J14).

The English department where Xiaomei studied provided its students an English grammar course, but Xiaomei said she was told by other fellow students that the teacher responsible for this grammar course was not helpful enough and she was happy that she found a useful course from the Internet. Actually Xiaomei had so many online tools and resources to help her learn English that she simply could not try them all, and the emancipatory space made possible by the Web technologies also required more self-control. Xiaomei admitted that she had problems of classroom distraction and that it was not easy to keep regular learning with online resources although they were quite helpful. Have technologies provided the solutions? Xiaomei did not recall any efforts that her teachers made to utilize technologies to make their classrooms more attracting, nor did she offer any suggestions for improving the interface of the learning platforms and applications she used. As a matter of fact, Xiaomei did not believe it to be a very important factor. "If you don't like an application, you simply stop suing it without thinking about how to improve it."(ITV) Instead, she believed that technology is not the solution. After considering the pros and cons of buying herself an Ipad, she wrote, "people always expect a new electronic device to help bring better learning outcomes. But more often than not, the truth is that the people who see better outcomes are those who have efficient ways of learning at the first place."(J14).

With the emancipatory space provided by the Web technologies and with so many learning resources at her proposal, we discussed with Xiaomei the possibility of replacement of campus face-to-face learning by online courses. She believed that online teaching cannot replace face-to-face teaching completely, "I think the most important thing about campus face-to-face learning is not about what the teachers teach you, but that you have classmates around you and you have teachers to guide you. They give you kind of emotional support."(ITV).

4.2 Reducing Second Language Anxiety

While Xiaomei expressed inconvenience of interaction with her teachers and classmates in the live streaming classroom, she, however, also saw the benefit behind it:

Excerpt 2

... I think interaction can be said to have increased in some way. For people like me who are quite unpretentious and don't like putting yourself forward before the classmates, I feel more at ease to answer the teacher's questions by posting in the comment area because people don't see me when I'm typing online, which avoids much embarrassment. It's also easier for me to use microphone and speak before my classmate without being watched (J1).

With this computer mediated environment of the online courses to help reduce second language anxiety, Xiaomei was trying to build up her confidence to speak, which she believed to be vital in language learning. Several times in her journals, she encouraged herself to take initiative in learning activities:

Excerpt 3

Today I was thrilled! In the English Reading class, we had an online quiz. The teacher could see who submit the answer at what time on her operating interface and she found that I was the first one who did all the questions right. .... If it is not for the online environment, I would not have had the courage to be the first one to hand in an exam paper (J3).

Previous studies [32, 33] found that asynchronous computer-mediated communication environment helps reduce foreign language anxiety. While van Lier and McNeil write for asynchronous computer-mediated learning environment, the ideas lying behind their findings have an equivalence in the present inquiry. In a live streaming classroom environment with 20-40 students as was the case with Xiaomei's class, usually not all students could use webcam to see each other because of the network bandwidth pressure. The restricted mode of the synchronous online learning was fortunately perceived and acted upon positively by Xiaomei to enhance her engagement in learning activities.

4.3 "It's More Than Learning the Language": The Construction of a Sociocultural Identity

To Xiaomei, however, the Web-based learning environment is "more than learning the language". A case in point is her use of the photo sharing application Instagram, as illustrated in the following excerpt:

Excerpt 4

It can be said that viewing posts on Instagram is purely entertaining, and you don't expect to really learn English there. If you are preparing for a coming exam, Instagram can't help you. But, it has magic, and it really makes me want to learn English well! (QTN).

While Xiaomei admitted in the interview that she did learn some colloquial and cyber language by visiting Instagram, the magic was that Instagram enhanced her English learning motivation, and the magic, according to Xiaomei, actually lied in the whole genuine English environment the social- networking application has provided for her. Similarly, she liked Youtube because of the feeling of getting totally immersed in the English-only environment, which Bilibili the Chinese counterpart of Youtube cannot otherwise provide, although a large amount of English learning videos originally from Youtube could also be found now on Bilibili. Posting comments and private correspondence with celebrities and stars she followed on Instagam gave her a taste of being a true user of English, or a dialogical communicator of English [34] rather than simply being a learner of English. This perceived sociocultural identity in turn changed into affordances, which was essentially made possible by the Web technologies, and promoted her determination in English learning.

The case of Xiaomei is in line with our previous study [35] that Chinese college students described themselves as user of English much less frequently than as learners of English in their learning journals. However, relating language learning process to the broader sociocultural context was what Xiaomei was becoming increasingly aware of. She reflected on her lack of self-control in implementing her English learning plans, relating it to what she learned in "Learn How to Learn", an online course she was taking on Coursera.

Excerpt 5

I still remember what a linguistic genius said to the professor in "Learn How to Learn" when the professor asked him what made it possible for him to speak so many languages. The answer given by the man was, I can't remember the exact wordings, but it's like this: never learn a language in order to pass an exam or to chat up somebody; learn the language out of the love of the language and the culture behind it (J15).

She figured that the reason why she could not fulfill her learning plan was her pragmatism; in other words, she placed too much stress on learning the language per se without enjoying and embracing the language and the culture. It then dawned on her that she did not have to feel guilty when she spent time viewing English videos simply for fun and not for a specific learning purpose, and that loving the culture behind was part of the language learning process. In so doing, Xiaomei perceived the English contents provided by Web-based applications, interpreted them in a positive way for construing her new sociocultural identity and enhancing her English learning, and finally took action to make them real affordances for her English learning, forming the perception-interpretation-action continuous cycle proposed by van Lier [26]. This process, from perception to interpretation and to action, was nevertheless not without difficulties and problems. Self-control again became an issue because "you have to be cautious. Applications like Youtube always push you eye-catching videos and it's easy to get lost and you just spend lot's time watching them." (ITV).

5 Discussion

The present inquiry has explored how a Chinese learner of English perceived and acted on the Web-based learning environment in the time of the COVID-19 pandemic. Analysis of the learning journals as the primary data showed that learning emancipation, anxiety reduction and sociocultural identity construction were three most salient dimensions in the focal student's perception. While the specific live streaming online course platforms had helped reduce the focal student's second language anxiety, actuated more speaking practices and hence facilitated her English learning, the technology affordances provided in terms of the emancipatory space and sociocultural identity construction were a complicated and mixed picture. Web technologies made possible more learning opportunities at the student's proposal but also imposed challenges for her self-control, leading to positive or negative affordances for her English learning.

Findings of the study indicate that students' perception of the learning environment is complicated and aspects of it are interwoven into each other, among which some are more significant than others. For the focal student in the present study, emancipation seemed to be the most salient aspect of Web-based learning environment she perceived and the design of interface of online tools was never at the center of her perception. Furthermore, it seems that the relation between education technology and learners' anxiety, which has been the interest for SLA researchers for many years and might be significant for adult learners, be it learners of language or not, as we speculate it, has not yet been given proper attention by learning environment researchers and education technology providers. It all boils down to mean that turning the perceived technology affordances into real language learning offordances is a complicated process mediated by the learner and sensitive to the specific knowledge area where learning is taking place. Reducing the students' perception of the learning environment to normative checklist-like categories is not enough to bring the holistic picture and to arrive at a profound understanding of how exactly those potential affordances are mediated into real ones.

The present study has also demonstrated that while new education technologies usually appeal to students and motivate them towards learning, how to maintain students' learning interest remains an issue to be addressed. With researchers having long ago proposed system maintenance as one of the three dimensions of learning environments to be considered [11], we argue that this dimension is foregrounded in Web-based learning environments. Xiaomei in this case study seldom complaint about the Web-based learning tools for her university courses and we interpret this as having a bearing on the learning culture in China, where students seldom challenge and propose suggestions to improve their learning environments, paying more attention to making the best out of what has been provided. However, taking a ecological perspective to look at the holistic picture, they are still active agents in searching for affordances in the environments: when the applications are not good enough, they quit and turn to other ones; when the courses provide by the university are not desirable, they resort to other online resources. Learners do not treat classroom affordance and out-of-classroom affordance separately, and they do not treat university offordance and other Web-based affordance separately. They treat their leaning environments as an integrated whole with an abundance of Web-based resources and opportunities.

It is confirmed by the present inquiry that teachers' role in language learning is undergoing a substantial change in a Web technology rich learning environment. With Web technologies making available online most contents students are supposed to learn or are keen to learn, learners expect teachers to offer "emotional support" (ITV). In this sense, blended learning might be a more appropriate teaching and learning mode, which provides integrated affordances from teachers and technologies.

The data also show that technologies are expected to provide more interaction possibilities for learners in this Web-based learning environment. This is particularly important for language learners because language knowledge emerges in the process of meaning negotiation in situated contexts. To take one step further, this meaning negotiation takes place not only when learners interact with other people, but also with other possible resources including social and cultural artefact. In other words, education technologies in the field of SLA should be able to help learners construe new identities to match their language learning process.

Limitations of the present study should be discussed. Because of the massive lockdown policy and technical problems, we were unable to observe the participants learning behaviour both in and out of classroom. The focus of the study on only one focal student might also be seen as another limitation. We would defend ourselves by saying that the significance of the present study lies in its thick description of the studied topic, and that as one of the first attempts to provide a holistic picture of how Web technologies can empower SLA in the special context of COVID-19 lockdown, the present study should be deemed as a starting point. The pandemic has forever changed the learning landscape of SLA, which is now much heavier-loaded with technologies than ever before. With an ever-increasing number of different forms of Web-based courses available and blended learning possibly becoming the mainstream, it would be worthwhile for future research to investigate technology affordances for SLA in the post COVID-19 massive lockdown peirod among large groups of students and over a longer time span.

6 Conclusion

The present study demonstrated a holistic and detail-rich picture of a Chinese English learner's perceptions of the Web-based learning environment and how she acted on the Web-based technology to afford her English learning in the time of the COVID-19 pandemic. Highlighted in this inquiry were three salient dimensions in the focal student's perceptions: learning emancipation, anxiety reduction and sociocultural identity construction. It was elucidated in the paper that the emergence of technology affordances in the interaction between the learner and the learning environment was a complex one, not only mediated by the learner but also sensitive to the specific knowledge area where learning was taking place. The ecological perspective and the notion of affordance adopted in the present study have proved to be informative and efficient in exploring students' perceptions and interaction with their learning environments.

References

1. Dorman, J.: Classroom environment research: progress and possibilities. Queensland J. Educ. Res. **18**(2), 112–140 (2002)
2. Fraser, B.J.: Learning environments research: yesterday, today, and tomorrow. In: Goh, S.C., Khine, M.S. (eds.) Studies in Educational Learning Environments: An International Perspective, pp. 1–26. World Scientific, Singapore (2002)

3. Aldridge, J.M., Fraser, B.J.: Outcomes-focused Learning Environments: Determinants and Effects. Sense Publishers, Rotterdam (2008)
4. Chang, V., Fisher, D.: Students' perceptions of the efficacy of web-based learning environment: the emergence of a new learning instrument. In: Paper Presented at Annual International Conference, Melbourne, 12–15 July 1999
5. Clayton, J.: Validation of the online learning environment survey. In: ICT: Providing Choices for Learners and Learning. Proceedings Ascilite, Singapore (2007)
6. Siragusa, L.: Identification of effective instructional design principles and learning strategies for students studying in web-based learning environments in higher education. Unpublished Doctor of Philosophy Thesis, Curtin University of Technology, Perth (2005)
7. Allen, I.E., Seaman, J.: Going the Distance: Online Education in the United States. Babson Survey Research Group and Quahog Research Group (2011)
8. Hiemstra, R.: Creating Environments for Effective Adult Learning. Jossey Bass, San Francisco (1991)
9. Lewin, K.: Principles of Topological Psychology. McGraw, New York (1936)
10. Murray, H.A.: Explorations in Personality. Oxford University Press, Oxford (1938)
11. Moos, R.H.: The Social Climate Scales: An Overview. Consulting Psychologists Press, Palo Alto (1974)
12. Walberg, H.J., Anderson, G.J.: Classroom climate and individual learning. J. Educ. Psychol. **59**, 414–419 (1968)
13. Fraser, B.J.: Individualized Classroom Environment Questionnaire. Australian Council for educational Research, Melbourne (1990)
14. Fraser, B.J., Giddings, G.J., McRobbie, C.J.: Evolution and validation of a personal form of an instrument for assessing science laboratory classroom environments. J. Res. Sci. Teach. **32**, 399–422 (1995)
15. Taylor, P.C., Fraser, B.J., Fisher, D.L.: Monitoring constructivist classroom learning environments. Int. J. Educ. Res. **27**, 293–302 (1997)
16. Aldridge, J.M., Fraser, B.J.: A cross-cultural study of classroom learning environments in Australia and Taiwan. Learn. Environ. Res. **3**, 101–134 (2000)
17. Haynes, D.: The social dimensions of online learning: perceptions, theories and practical responses. In: Paper Presented at the Distance Education Association of New Zealand, Wellington, April 2002
18. Lee, S.S.U., Fraser, B.J.: High school science classroom learning environments in Korea. In: Paper Presented at the Annual Meeting of the American Educational Research Association, New Orleans, April 2002
19. Tobins, K.: Qualitative perceptions of learning environments on the world wide web. In: Fraser, B.J., Tobins, K.G. (eds.) International Handbook of Science Education, pp. 581–594. Kluwer Academic Publishers, Boston (1998)
20. Luo, S.: Evaluating the learning environment of MOOC-based college English flipped classrooms. Technol. Enhanced Foreign Lang. Educ. **182**, 16–22 (2018)
21. Sun, Y.M.: A study of college comprehensive English classroom environment. Foreign Lang. Teach. Res. **42**(6), 438–444 (2010)
22. Holliday, A.: Issues in validity in progressive paradigms of qualitative research. TESOL Q. **38**(4), 731–734 (2004)
23. Lantolf, J.: SLA theory building: "Letting all the flowers bloom!". Lang. Learn. **46**(4), 713–749 (1996)
24. Larsen-Freeman, D.: Chaos/complexity science and second language acquisition. Appl. Linguist. **18**(2), 141–165 (1997)

25. Kramsch, C., Steffensen, S.V.: Ecological perspectives on second language acquisition and socialization. In: Duff, P.A., Hornberger, N.H. (eds.) Encyclopedia of Language and Education, Language Socialization, 2nd Edn., vol. 8, pp. 17–28. Springer, New York (2008)
26. van Lier, L.: The Ecology and Semiotics of Language Learning: A Sociocultural Perspective. Kluwer Academic Publishers, Boston (2004)
27. Gibson, J.: The Ecological Approach to Visual Perception. Lawrence Erlbaum Associates, Hillsdale (1979)
28. Norman, D.A.: The Invisible Computer. MIT Press, Cambridge (1998)
29. Li, X.M., Song, S.Q.: Mobile technology affordance and its social implications: a case of "Rain Classroom". Brit. J. Educ. Technol. **49**(2), 276–291 (2018)
30. Mesgari, M., Faraj, S.: Technology affordances: the case of wikipedia. In: The proceedings of the Americas' AMCIS Proceedings, vol. 13 (2012). http://aisel.aisnet.org/amcis2012/proceedings/VirtualCommunities/13
31. Denzin, N.K.: The Research Act: A Theoretical Introduction to Sociological Methods. Mc Graw-Hill, New York (1978)
32. McNeil, L.: Ecological affordance and anxiety in an oral asynchronous computer-mediated environment. Lang. Learn. Technol. **18**, 142–159 (2014)
33. van Lier, L.: From input to affordance: social-interactive learning from an ecological perspective. In: Lantolf, J. (ed.) Sociocultural Theory and Second Language Learning, pp. 245–259. Oxford University Press, New York (2000)
34. Gao, Y.H.: Faithful imitator, legitimate speaker, playful creator and dialogical communicator: shift in English learners' identity Prototypes. Lang. Intercultural Commun. **14**(1), 59–75 (2014)
35. Wang, J.: Beliefs and affordances: a discourse analysis of learning stories by Chinese English learners. In: Paper presented at the 16th National Conference on Discourse Analysis, East China Normal University, Shanghai, 17–18 December (2018)

HTML 5-Based Learning for 3D Computer Graphics and GIS

Changbin Yu[1,2,3,4,5(✉)], Mingming Fu[3], and Chaoyi Pang[3]

[1] Key Laboratory of Urban Land Resources Monitoring and Simulation, Ministry of Natural Resources, No. 69 Xinwen Road, Lianhua Street, Futian District, Shenzhen City, Guangdong Province, China
240212388@qq.com

[2] Shenzhen Research Center of Digital City Engineering, No. 69 Xinwen Road, Lianhua Street, Futian District, Shenzhen City, Guangdong Province, China

[3] School of Computing and Data Engineering, Ningbo Tech University, No. 1 Qianhu South Road, Ningbo City, Zhejiang Province, China

[4] Key Laboratory of Digital Mapping and Land Information Application Engineering, National Administration of Surveying, Mapping and Geoinformation, Beijing, China

[5] School of Resource and Environmental Sciences, Wuhan University, Wuhan, China

Abstract. The online learning becomes more and more popular where visualization of 3D objects can be employed to lots of fields, e.g., 3D building information modeling for architecture, 3D city model for urban planning, 3D animation for film, 3D skeleton visualization of medical treatment, etc.. In tradition, the most popular tool for 3D modeling and visualization is OpenGL which is based on Client/Server. In another word, C++ code for OpenGL must be installed and configured firstly which is not convenient for new beginners. By contrast, this paper focus on visualization of 3D objects in Computer Graphics and GIS (Geographic Information System) purely depending on JavaScript technology where no additional software or plugins, applet, ActiveX control is needed. Typical JavaScript libraries covers ThreeJS for Computer Graphics, OpenLayers for 2D GIS and CesiumJS for 3D GIS. Based on the above libraries, e-learners could view and manipulate 3D objects in web browser (e.g., Google Chrome). Such intuitive and convenient 3D visualization and manipulation results in much more interesting for learning.

Keywords: WebGL · 3D computer graphics · 3D GIS · E-learning

1 Introduction

Online learning/teaching is an excellent approach to supplement traditional student learning/teacher teaching or business tutoring. MOOC (abbreviation for Massive Open Online Courses), as a typical application of e-learning, is especially popular in developed and developing countries, because not so many people are allowed to receive enough higher education mostly due to financial limitation.

C. Pang et al. (Eds.): SETE 2020/ICWL 2020, LNCS 12511, pp. 38–46, 2021.
https://doi.org/10.1007/978-3-030-66906-5_4

In some occasions, online learning/teaching (also called e-learning) could even be employed as a primary method for learning/teaching. In this year, COVID-19 (Corona Virus Disease 2019) are spreading all over the world, students of kindergartens, primary schools, junior schools, senior schools, colleges or universities are not suggested to gather offline, and in some regions gathering offline is even forbidden. In stead, learning/teaching online using multiple approaches including textual messaging, voice talking, face-to-face video communication, etc. is greatly encouraged. In situations similar to the above, e-learning shows its great advantages.

In this paper, HTML5-based online learning/teaching using pure JavaScript technology for rendering of 3D objects is explained. In detail, only a web browser (e.g., Google Chrome) rather than plenty of software, plugin, applet, activeX, etc. is needed for visualization of 3D objects. It could be employed for many fields, e.g., scientific visualization for 3D objects in computer sciences, 3D city modeling for urban planning, 3D BIM (Building Information Modeling) for architecture, 3D animation for multi-media design, visualization of 3D skeleton for medical simulation, etc. However, this paper focuses on learning/teaching of 3D visualization in Computer Graphics and GIS (Geographic Information System) using HTML5/WebGL technology.

2 Related Work

As mentioned above, Computer Graphics is widely used in medical, film/animation, 3D architectural modeling and so on. The most popular software or package in traditional learning/teaching for Computer Graphics is OpenGL (https://www.opengl.org/) [1]. OpenSceneGraph (http://www.openscenegraph.org/) [2], as an advanced packages for 3D visualization, is based on OpenGL. The core contents of OpenGL covers representations of geometrical primitives (e.g., GL_POINT, GL_LINE, GL_POLYGON), 'camera-objects' mode using perspective or orthographic projection, 3D realistic augmentation (e.g., hidden-line removal, materials or textures, particle effects). However, OpenGL or OpenSceneGraph are built using C++ (typical C++ files for OpenGL are glut.h, glut.lib, glut.dll) where standalone client is required. Up to now, most applications based on OpenGL are built on either Client/Server architecture or not pure Browser/Server architecture where additional web plugins, Java applet or ActiveX controls must be installed or configured in advance.

Here, related works for visualization of 3D objects are firstly reviewed. (Zhuang Yueting et al. 2003) designed 3D facial modeling for e-learning using C++ and Matlab [3] while (Tang, Jing-Yao et al. 2009) [4] and (Huang, Kuang-Min et al. 2009) [5] designed standalone 2D and 3D adventure game-based assessment clients respectively. The above research are all based on Client/Server architecture which is not convenient for e-learning beginners. Meanwhile, in (Tai, Ying-Shan, et al. 2004) [6], CAD is modeled only in 2D rather than 3D while 3D VRML scenes in (Yun, Ru-wei et al. 2006) [7] and 3D geometrical objects in (Tam, Gary, et al. 2008) [8] are both represented as 2D images. In (Magnenat-Thalmann, Nadia, et al. 2008) [9] an additional Adobe 3D viewer plugin needs installation for sharing 3D dancing while limited functions for manipulation of 3D medical objects are provided in (Nicolaescu, Petru, et al. 2015) [10] although exploratory research work using HTML5 X3D is given.

In this paper, HTML5-based 3D visualization environment for learning/teaching online is provided where no additional plugin, ActiveX control, applet needs installation or configuration. It is purely based on WebGL technology elaborated below.

3 HTML5-Based 3D Rendering

As shown in Fig. 1, OpenGL, as the industry's standard for high performance graphics, was maintained by Khronos Group since 2006. And OpenGL ES (OpenGL Embedded Systems) is OpenGL APIs designed for embedded devices. Meanwhile, HTML5, as the fifth version of HTML, was published in 2008. And the core 'canvas' element in HTML5 supports visualization of 3D objects in the web browser.

WebGL (Web Graphics Library) comes from combination of OpenGL ES and HTML5. In detail, OpenGL ES 2.0-based WebGL is 3D drawing protocol which provides 3D accelerated rendering using HTML 5 Canvas. The ThreeJS JavaScript package (https://threejs.org/) [11] elaborated in Sect. 3.1 for e-learners of Computer Graphics is based on WebGL.

From the perspective of technology, modeling of 3D targets in Computer Graphics mostly depends on local X/Y/Z whose axis obey right-hand rule (also called Three-dimensional Cartesian), and it can be applied to 3D modeling of a single building, a car, or a simple objects, etc. However, when it is applied to scenes which cover a larger area, global elliptic terrain rather than local planar ground should be taken into account, and GIS (Geographic Information System) is the solution for location-related applications. In detail, OpenLayers (https://openlayers.org/) [12], as a typical 2D GIS JavaScript library, is employed here for transforming 3D real world to 2D digital plane. Meanwhile, CesiumJS (https://cesium.com/cesiumjs/) [13], as a typical 3D GIS JavaScript library, is used here for transforming 3D real world to 3D digital globe.

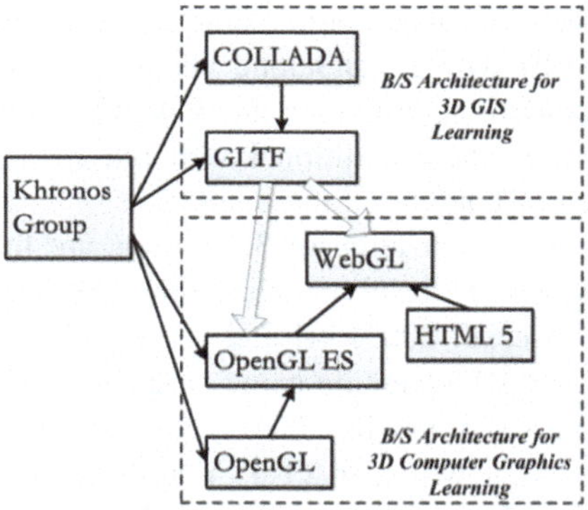

Fig. 1. OpenGL maintained by Khronos Group and GLTF proposed by Khronos Group.

3.1 HTML5-Based Learning for 3D Computer Graphics

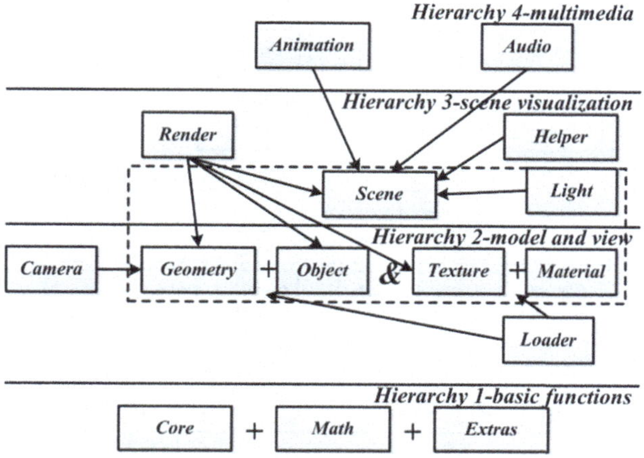

Fig. 2. Hierarchies and core classes in 3D ThreeJS JavaScript Library.

Figure 2 gives hierarchies and core classes of 3D ThreeJS JavaScript library where 'Geometry' and 'Object' in 'Scene' are rendered using 'Camera' while 'Light', 'Texture' and 'Material', 'Animation' and 'Audio' are used for 3D realistic augmentation. In ThreeJS 3D environment, online learners could intuitively see the effective of 3D manipulating operation purely using the web browser. And not only operating including 3D pan, 3D rotation, 3D scaling, 3D reflection but also advanced functions (e.g., 3D collision detection, accurate picking from thousands of 3D targets by mouse in Fig. 3) are provided.

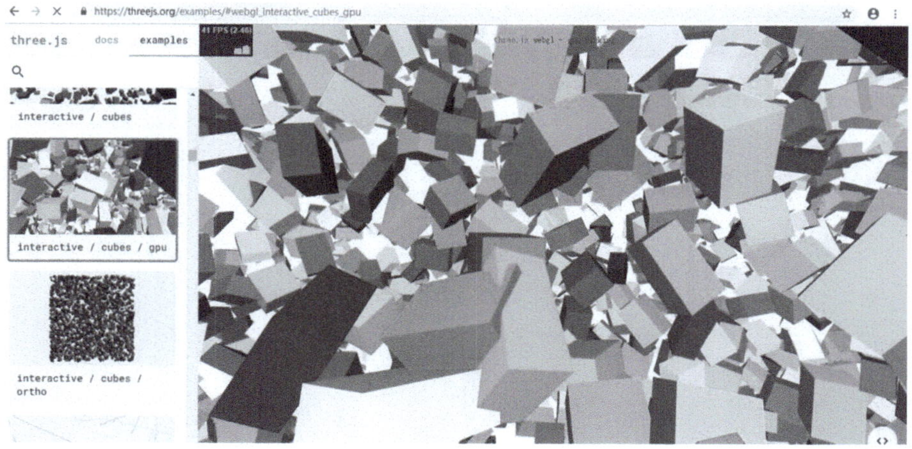

Fig. 3. An example using 3D ThreeJS (accurate pick using mouse from thousands of cuboids).

3.2 HTML5-Based Learning for 2D/3D GIS

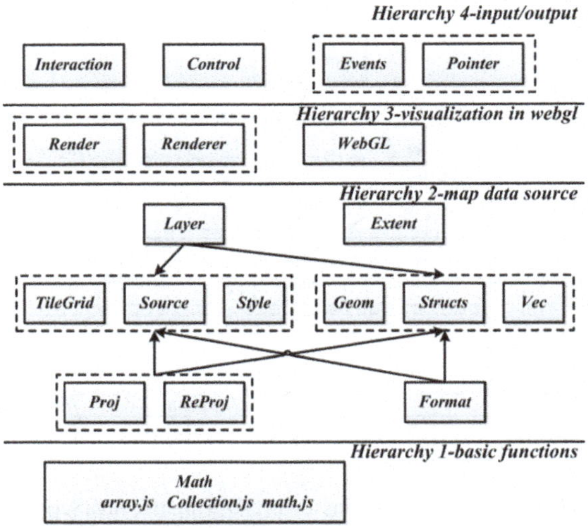

Fig. 4. Hierarchies and core classes in 2D OpenLayers JavaScript Library.

E-learning for 2D GIS. Figure 4 gives hierarchies and core classes in OpenLayers JavaScript library where different data source (e.g. OpenStreeMap, Microsoft Bing-Map, Google Map) as 'Layer' could be organized in the form of 'TileGrid' while different 'Proj' or 'ReProj' (e.g., Mecator projection, UTM projection) are employed for mapping 3D real world to 2D digital plane. And e-learners could view and manipulate the 2D map (see Fig. 5) using the web browser.

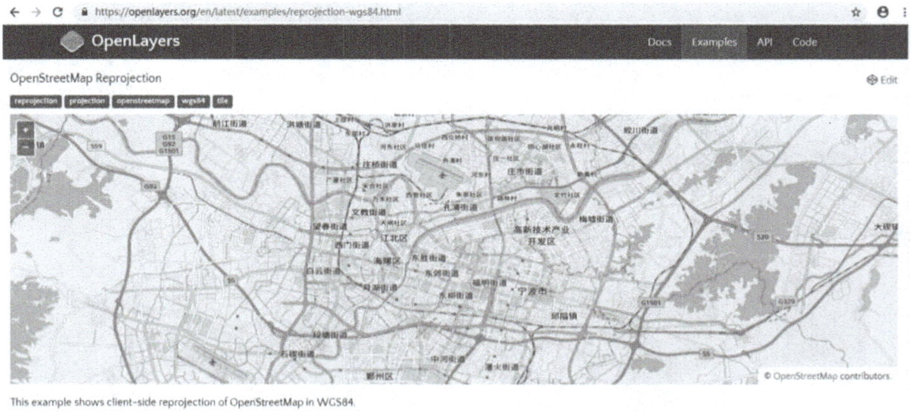

Fig. 5. An example using 2D OpenLayers (a map showing roads, rivers, text annotations from OpenStreetMap data source in Ningbo city, Zhejiang Province, China).

E-learning for 3D GIS. Figure 7 gives hierarchies and core in CesiumJS JavaScript library where GLTF (Graphics Language Transmission Format) and 3D Tiles play important roles. The core class in GLTF in json format (see Fig. 6) is 'scene' and 'node', and 'mesh' and 'bufferView' under 'node' is viewed by 'camera' while 'material' and 'texture' could also be employed for 3D realistic augmentation. The above 3D coordinate systems are mostly based on WGS (World Geodetic System) 84.

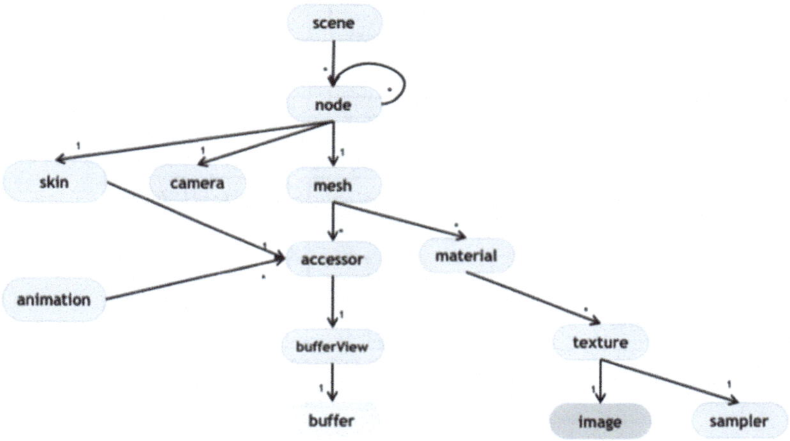

Fig. 6. Logical hierarchical classes of GLTF file in json format.

3D Tiles (similar to 2D Tiles in OpenLayers), as implementation of GLTF, could be employed to fast accessing of 3D objects on the web browser. Data formats for 3D Tiles can be Batched3DModel3DTile(.b3dm), Instanced3DModel3DTile(.i3dm), Vector3DTile(.vctr), PointCloud3DTile(.pnts), Composite3DTile(.cmpt). The above GLTF and its implementation 3D Tiles both proposed by Khronos Group aim to be the transmission standard for 3D data (similar to JPEG standard for 2D data).

Figure 8 gives an example of CesiumJS where 1.1 million buildings in New York city are loaded whose data source files are over 700 MB, however, e-learners could view 3D building group and manipulate them (e.g., different heights of buildings are rendered by distinguished colors) over the web browser freely and fluently without additional installation or plugins.

The design principle using the pure JavaScript library is beneficial to integrating OpenLayers and CesiumJS with other JavaScript library seamlessly, e.g., terrain representation using 2D delaunay triangulation (see. Fig. 9).

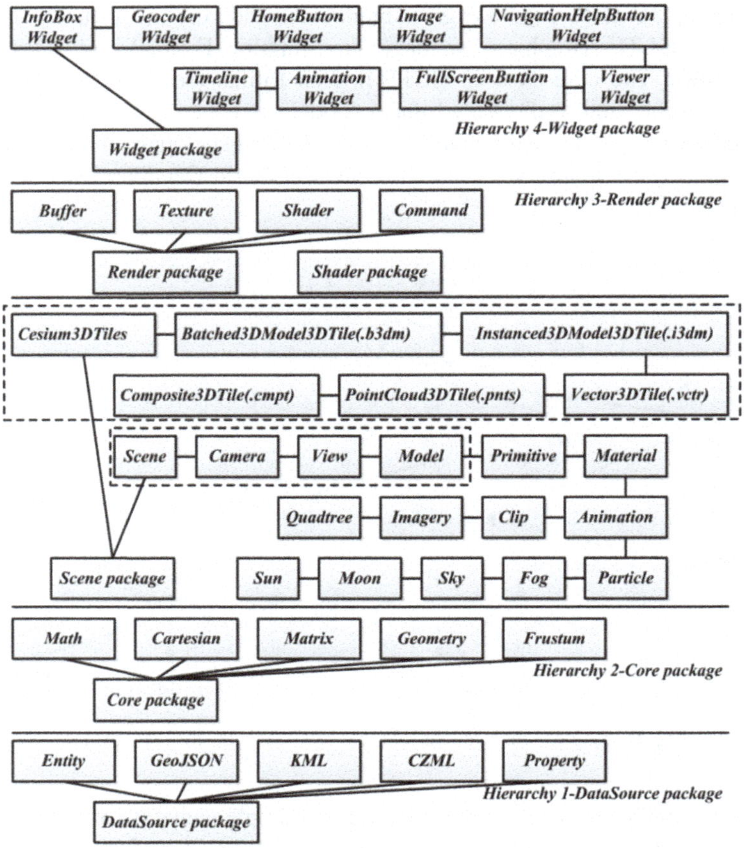

Fig. 7. Hierarchies and core classes in 3D CesiumJS JavaScript Library.

Fig. 8. An example using 3D CesiumJS (loading 1.1 Million 3D buildings in New York City). (Color figure online)

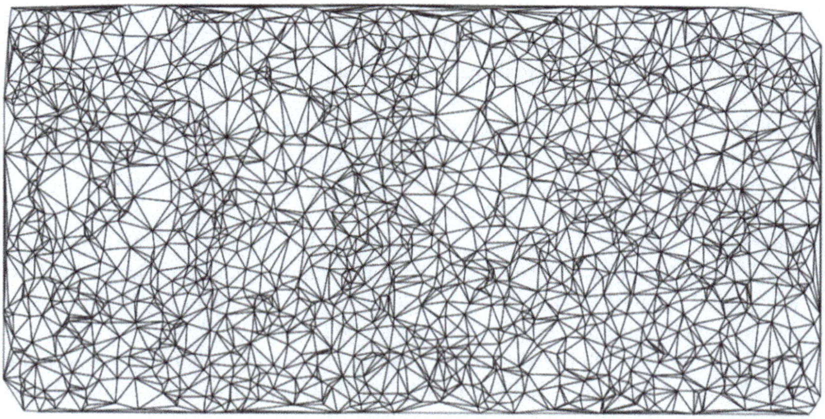

Fig. 9. 2D delaunay triangulation based on JavaScript could be seamlessly embedded into OpenLayers and CesiumJS.

4 Conclusions and Future Work

In this paper, e-learners for Computer Graphics or GIS-related fields could employ ThreeJS, OpenLayers and CesiumJS libraries to view and manipulate 3D targets in 3D web browser purely using JavaScript, and no additional applet, plugin, or ActiveX control needs installation or configuration. And the future work will be extend the current JavaScript libraries. Taking extension of CesiumJS library for example, integrating current popular JavaScript libraries for table showing (e.g., Jquery-EasyUI, LayUI) and libraries for graph showing (e.g., GoogleCharts, Echarts, D3) with CesiumJS would be necessary, because besides visualization of 3D objects intuitive demonstration of line graph, bar graph, pie graph, radar graph, etc. is also helpful to e-learners. Meanwhile, backends using Java/Java Servlet as web services for remote calling would also be employed further where much more advanced functions (e.g., complicated 3D spatial analysis using other attribute data stored in the sql or nosql databases) could be offered to e-learners.

Acknowledge. The Project Supported by the Open Fund of Key Laboratory of Urban Land Resources Monitoring and Simulation, Ministry of Natural Resources (No. KF-2019-04-017), Zhejiang Provincial Natural Science Foundation of China under Grant No. LQ19D010005, Ningbo Natural Science Foundation (Grant No. 2018A610121), also funded by Open Research Fund Program of Key Laboratory of Digital Mapping and Land Information Application Engineering, NASG (Grant No. GCWD201801).

References

1. OpenGL. https://www.opengl.org/. Accessed 26 July 2020
2. OpenSceneGraph. http://www.openscenegraph.org/. Accessed 26 July 2020

3. Zhuang, Y., Su, C., Huang, L., Wu, F.: Subdivision feedback based 3D facial modeling for E-learning. In: Zhou, W., Nicholson, P., Corbitt, B., Fong, J. (eds.) ICWL 2003. LNCS, vol. 2783, pp. 218–229. Springer, Heidelberg (2003). https://doi.org/10.1007/978-3-540-45200-3_21
4. Tang, J.-Y., Chen, J.-H., Ku, D.T., Chao, L.R., Shih, T.K., Weng, M.M.: Constructing the 2D adventure game-based assessment system. In: Spaniol, M., Li, Q., Klamma, R., Lau, R. W.H. (eds.) ICWL 2009. LNCS, vol. 5686, pp. 376–385. Springer, Heidelberg (2009). https://doi.org/10.1007/978-3-642-03426-8_45
5. Huang, K.-M., Chen, J.-H., Shen, C.-Y., Chao, L.R., Shih, T.K.: Developing the 3D adventure game-based assessment system with Wii remote interaction. In: Spaniol, M., Li, Q., Klamma, R., Lau, R.W.H. (eds.) ICWL 2009. LNCS, vol. 5686, pp. 192–195. Springer, Heidelberg (2009). https://doi.org/10.1007/978-3-642-03426-8_24
6. Tai, Y.S., Liu, W.: A web-based CAD system for learning and laboratory purpose. In: Liu, W., Shi, Y., Li, Q. (eds.) ICWL 2004. LNCS, vol. 3143, pp. 309–316. Springer, Heidelberg (2004). https://doi.org/10.1007/978-3-540-27859-7_40
7. Yun, R., Chen, G., Li, Y.: Using VRML and JAVA to build virtual game-based learning environment for addition and subtraction operation. In: Liu, W., Li, Q., W.H. Lau, R. (eds.) ICWL 2006. LNCS, vol. 4181, pp. 146–153. Springer, Heidelberg (2006). https://doi.org/10.1007/11925293_14
8. Tam, G.K.L., Lau, R.W.H., Zhao, J.: A 3D geometry search engine in support of learning. In: Leung, H., Li, F., Lau, R., Li, Q. (eds.) ICWL 2007. LNCS, vol. 4823, pp. 404–415. Springer, Heidelberg (2008). https://doi.org/10.1007/978-3-540-78139-4_36
9. Magnenat-Thalmann, N., Protopsaltou, D., Kavakli, E.: Learning how to dance using a web 3D platform. In: Leung, H., Li, F., Lau, R., Li, Q. (eds.) ICWL 2007. LNCS, vol. 4823, pp. 1–12. Springer, Heidelberg (2008). https://doi.org/10.1007/978-3-540-78139-4_1
10. Nicolaescu, P., Toubekis, G., Klamma, R.: A microservice approach for near real-time collaborative 3D objects annotation on the web. In: Li, F.W.B., Klamma, R., Laanpere, M., Zhang, J., Manjón, B.F., Lau, R.W.H. (eds.) ICWL 2015. LNCS, vol. 9412, pp. 187–196. Springer, Cham (2015). https://doi.org/10.1007/978-3-319-25515-6_17
11. ThreeJS. https://threejs.org/. Accessed 26 July 2020
12. OpenLayers. https://openlayers.org/. Accessed 26 July 2020
13. CesiumJS. https://cesium.com/cesiumjs/. Accessed 26 July 2020

An Analysis on Learners' Word Reading and Writing in an English Reading-to-Write Task System

Min Kang[1(✉)], Koichi Kawamura[2], Harumi Kashiwagi[3], and Kazuhiro Ohtsuki[1]

[1] Graduate School of Intercultural Studies, Kobe University, Kobe 657-8501, Japan
kang@kobe-u.ac.jp
[2] ICCRC, Kobe University Hospital, Kobe 650-0047, Japan
[3] Institute for Promotion of Higher Education, Kobe University, Kobe 657-8501, Japan

Abstract. Writing is a difficult activity for learners in foreign language learning. Besides the organization and clarity in content, the accurate use of lexical and syntactic knowledge, is needed. Considerable effort has been put in teaching writing to enhance learners' writing proficiency. Reading is regarded as one of the important strategies of enhancing writing proficiency. However, few studies have reported the linguistic knowledge that learners pay attention to and how they use the knowledge of web-based learning in their writings. In this paper, we analyze learners' word reading and writing in an English reading-to-write task system. We focus the words that are read in reading and are written in writing tasks by learners on verbs. Based on the analysis, we assume that word reading may not influence word writing if there are no special Focus on Form approaches to force students' attention on the words and the development of Focus on form approaches for web-based learning environments is needed.

Keywords: Word reading and writing · Reading-to-write task · Focus on form approach · Web-based EFL writing

1 Introduction

Writing is considered to be a difficult activity for learners in foreign language learning [1]. Besides the organization and clarity in content, the accurate use of lexical and syntactic knowledge, is needed [2]. Considerable effort has been put in teaching writing to enhance learners' writing proficiency [3–7]. Reading is regarded as one of the important strategies of enhancing writing proficiency. Reading and writing are interdependent and writers' linguistic skills, contextual awareness, and strategies, etc., are influenced by the information in source texts, in writers' prior experiences and learning etc. [3]. It has been stated that learners' writing skills concerning content, organization, vocabulary, and language use are associated with their reading skills [8]. Many studies have focused on the relationship between reading-to-write and writing [9–12]. Most claimed that reading-to-write strongly influences writing. Furthermore, a recent

C. Pang et al. (Eds.): SETE 2020/ICWL 2020, LNCS 12511, pp. 47–53, 2021.
https://doi.org/10.1007/978-3-030-66906-5_5

research, while the subjects are children of native English speakers, has reported a longitudinal examination about reading-writing relations and indicated that reading-writing relations are stronger at the lexical than at the discourse level [12]. Concerning learners of English, it is still an important issue to investigate how reading influences writing in detail not only in face-to-face classrooms but also in web-based reading-writing environments.

On the other hand, as Focus on Form has been emphasized as an effective approach to language teaching, text-enhancement techniques in web-based language learning have been developed to help learners enhance lexical and syntactic knowledge. In reading, learners tend to comprehend the meaning of the text and may pay little attention to the linguistic form. A visual-syntactic text formatting technology that visualizes syntactic structures has been experimentally used on reading to enhance syntactic awareness. The experimental results clarified that the technology raised students' awareness of syntactic structures, and the written conventions and writing strategies of low-proficiency students were significantly influenced by the technology [13].

We have developed an English reading-to-write task system in our previous work to observe how syntactic structures in reading influence learners' writing [14]. In this paper, we analyze learners' word reading and writing by use of the system. We focus the words that are read in reading materials and are written in writing tasks by learners on the verbs. Based on the analysis, we aim to clarify the relation between word reading and word writing in a web-based learning environment and discuss the necessity to develop Focus on form web-based systems on verb-enhancement.

In the next section, we explain the details of the reading-to-write system. We propose the analysis results in Sect. 3, and then give the conclusion in Sect. 4.

2 Reading-to-Write Task System

2.1 System

We developed a web-based reading-to-write task system. The system consists of three steps: measuring learners' levels of word recognition, reading-to-write tasks and a recognition test of words in reading materials.

In the step of measuring levels of word recognition, 8 words are randomly chosen from the JACET8000 list at the Level 3000 and the Level 4000 and shown to ask learners the meanings and word familiarity [15].

Two webpages were designed to provide two reading-to-write tasks. Each page included a paragraph essay on the top followed by two related questions. The first question asks if the essay has been read. The second asks to write a response essay in relation to the essay's topic. An input space for writing is given below the second question. Meanwhile, in order to clarify if learners are sensitive to salient syntactic structures, we emphasized visually the present tense verbs in third person singular, which appear on the second page, by coloring the verbs red or green, and using bold-faced type [16].

In order to confirm the effectiveness of the emphasis, we designed a recognition test in the last step. We choose 4 verbs that appear in the first essay (We call the essay Reading_A) in the first reading-to-write task, 4 verbs that appear in the second essay (We call the essay Reading_B) in the second reading-to-write task, 2 verbs that are used in both essays to ask in which essay the words appeared.

2.2 Reading Materials

As the study focuses on how word reading influences word writing, easy-to-read materials were used to reduce comprehension difficulties and errors in writing. Two paragraphs, Reading_A and Reading_B, were chosen from a text book for the freshmen of Kobe University. Reading_A and Reading_B consist of 156 words (13 sentences) and 152 words (15 sentences) with the topics focusing on bosses in offices and future jobs, respectively. The questions related to the response essays on the first page were as follows:

- Question 1: Have you read the paragraph before?
- Question 2: Please write a short essay on your boss.

2.3 Participants and Procedure

There were 12 participants consisting of second-year, third-year, and senior students of Kobe University, with a major in global culture who used the system. Empirically, we consider that the reading materials are easy to read for them.

The students were required to log in to the system, and then answer the questions at each step of the system. In the two reading-to-write tasks, no dictionary was allowed. To avoid losing participants' attention, the essays were limited to 5 sentences or 70 words.

3 Analysis and Results

We collected 35 sentences and 34 sentences in the two tasks, respectively. Here, simple sentences, complex sentences, and complicated sentences are included.

First, we summarized students' responses at the step of measuring levels of word recognition and confirmed that most of students are familiar to the words at the Level3 of JACET8000. This means that the students had sufficient reading proficiency and writing proficiency to comprehend the essays and complete the writing tasks with little grammatical errors.

3.1 Students' Word Reading and Writing

Because a predicate is the basic part of a sentence and one cannot omit predicates in writing, we primarily focused the analysis on predicates. We used Stanford Parser to parse the essays and students' writings and extracted all predicates in the essays [17]. The percentages of the predicates that appear in students' writings are calculated. Figure 1 and Fig. 2 show the percentages related to Reading_A and Reading_B respectively. In each figure, the predicates used in each reading material are given along the horizontal axis. The vertical axis represents the ratio of a predicate frequency to the number of predicates in students' writings.

We noted that the students prefer to use their own words instead of reusing the words they read in reading. Figure 1 shows that there are only about 40%of the predicates (i.e., 9 verbs) in Reading_A that appear in the students' writings. In additions, the predicates are not directly related to the topic. The predicates related directly to the topic almost have not been used in writings. In Fig. 2, only about 20% of the predicates (i.e., 5 verbs) in Reading_B appear in the students' writings. The reused percentages are lower. The words "be", "is", "think", "likes" and "make" totally do not reflect the meanings of the essay. Moreover, in both figures, the values of the percentages of the reused word is low. It seems that word reading does not influence word writing by use of the reading-to-write task system. This suggests that systems with special Focus on form approaches to force students' attention on the words is necessary.

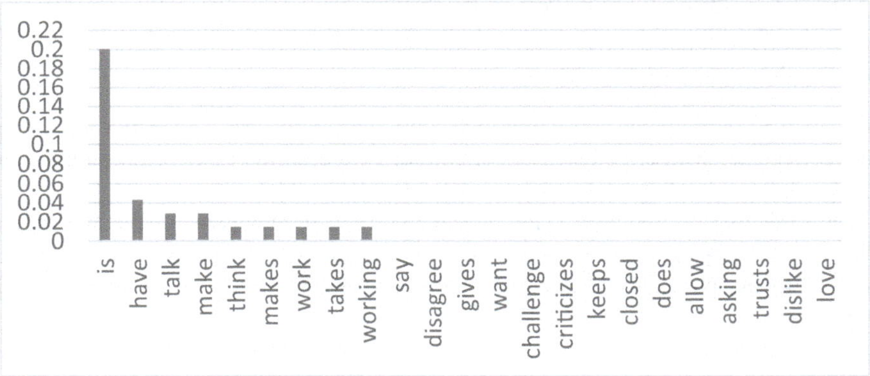

Fig. 1. The word percentages that reused by the students in the first task

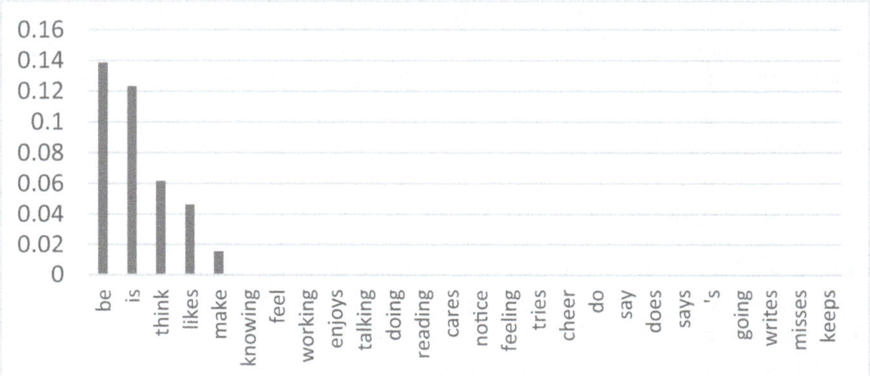

Fig. 2. The word percentages that reused by the students in the second task

3.2 Results of the Recognition Test

In Fig. 3, we give the correct percentages in the last step that measure if the students could correctly memorize in which essay the words appeared. The red bar means that the word "misses" was colored red. The green bars correspond to the words colored green. The black bar represents that the word "keeps" was displayed in bold-faced type. The other words were not emphasized visually.

Figure 3 shows that the approach of word coloring did not help students a lot to recognize words. In comparison with uncolored words, the correct percentages of all colored words are under 50%. Conversely, the uncolored words, "encourages", "notice", "challenge" and "trusts", were recognized correctly by over 50% students. It is noticed that the uncolored words are strongly related to the meanings of the reading materials. In other words, the students paid more attention to the meanings rather than the linguistic form even the coloring approach was used.

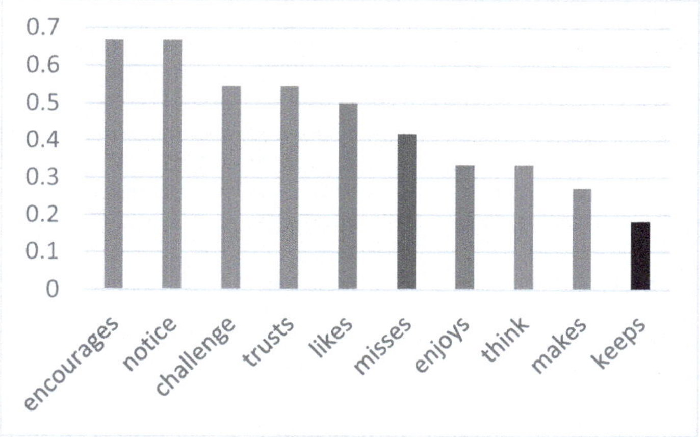

Fig. 3. The correct percentages of the words in the recognition test (Color figure online)

Consequently, as the literature [13] mentioned that low-proficiency students were significantly influenced by the salient syntactic structures, in this paper, the writing proficiency may result in the students paid little attention to the linguistic form. The reading-to-write tasks were considerably easy for the students. It is assumed that other approaches are necessary for high-proficiency students.

4 Conclusion

In this paper, we analyzed learners' word reading and writing in an English reading-to-write task system. We focused the words on the verbs. Two essays were used in reading and writing tasks related to the topics of the essays were requested. By use of Stanford Parser, the predicates in the essays were extracted and reused percentages in learners' writings were calculated. Furthermore, the effectiveness of word coloring as a Focus on form approach was examined. It seems that word reading does not influence word writing in the reading-to-write task system. Leaners tend to pay more attention to the meaning of a word rather than linguistic form in systems without special Focus on form approaches. In addition, the approach of word coloring did not help students a lot to recognize words. Further Focus on form approaches for word reading and writing in web-based learning environments should be developed.

Acknowledgement. This work is supported by JSPS KAKENHI Grant Number JP17K01081.

References

1. Kroll, B.: Second Language Writing. Cambridge University Press, Cambridge (1990)
2. Kimura, H., Kimura, T., Shiki, O.: Theory and Practice in Reading and Writing: Nurturing Independent Learning. TAISHUKAN Publishing Co., Ltd., Tokyo (2010). (in Japanese)
3. Ackerman, J.M.: Reading, writing, and knowing: the role of disciplinary knowledge in comprehension and composing. Res. Teach. Engl. **25**(2), 133–178 (1991)
4. Truscott, J.: The case against grammar correction in L2 writing classes. Lang. Learn. **46**(2), 327–369 (1996)
5. Chandler, J.: The efficacy of various kinds of error feedback for improvement in the accuracy and fluency of L2 student writing. J. Second Lang. Writ. **12**(3), 267–296 (2003)
6. Leacock, C., Chodorow, M., Gamon, M., Tetreault, J.: Automated Grammatical Error Detection for Language Learners, 2nd edn. Morgan & Claypool Publisher, San Rafael (2014)
7. Kawamura, K., Kashiwagi, H., Kang, M.: An approach toward automatic error detection in learners' English writing based on the source language. In: 10th International Proceedings on Mobile, Hybrid, and On-line Learning, pp. 62–65. IARIA, Roma (2018)
8. Ito, F.: L2 reading–writing correlation in Japanese EFL high school students. Lang. Teach. **35**(5), 23–29 (2011)
9. Asencion Delaney, Y.: Investigating the reading-to-write construct. J. Engl. Acad. Purp. **7**, 140–150 (2008)
10. Weigle, S.C., Parker, K.: Source text borrowing in an integrated reading/writing assessment. J. Second Lang. Writ. **21**, 118–133 (2012)
11. Liu, H., Brantmeier, C., Strube, M.: EFL test preparation in China: the multidimensionality of the reading-writing relationship. Read. Foreign Lang. **31**, 44–61 (2019)
12. Kim, Y.G., Petscher, Y., Wanzek, J., Al Otaiba, S.: Relations between reading and writing: a longitudinal examination from grades 3 to 6. Read. Writ. **31**, 1591–1618 (2018). https://doi.org/10.1007/s11145-018-9855-4
13. Park, Y., Warschauer, M.: Syntactic enhancement and second language literacy: an experimental study. Lang. Learn. Technol. **20**(3), 180–199 (2016)
14. Kang, M., Kawamura, K., Shao, S., Kashiwagi, H., Ohtsuki, K.: Measuring similarity to observe learners' syntactic awareness in web-based writing environments. In: Herzog, M.A., Kubincová, Z., Han, P., Temperini, M. (eds.) ICWL 2019. LNCS, vol. 11841, pp. 100–107. Springer, Cham (2019). https://doi.org/10.1007/978-3-030-35758-0_10

15. JACET Vocabulary (n.d.). http://j-varg.sakura.ne.jp/publications/in-dex.html
16. Kuhbandner, C., Pekrun, R.: Joint effects of emotion and color on memory. Emotion **13**(2), 375–379 (2013)
17. Klein, D., Manning, C.D.: Accurate unlexicalized parsing. In: Proceedings of the 41st Meeting of the Association for Computational Linguistics, pp. 423–430 (2003)

Learning Analytics

Evaluation of Approaches for Automatic E-Assessment Item Annotation with Levels of Bloom's Taxonomy

Roy Meissner[1(✉)], Daniel Jenatschke[2(✉)], and Andreas Thor[3(✉)]

[1] Leipzig University, Leipzig, Germany
roy.meissner@uni-leipzig.de
[2] Leipzig University of Telecommunications, Leipzig, Germany
d.jenatschke@gmx.de
[3] Leipzig University for Applied Sciences, Leipzig, Germany
andreas.thor@htwk-leipzig.de

Abstract. The classification of e-assessment items with levels of Bloom's taxonomy is an important aspect of effective e-assessment. Such annotations enable the automatic generation of parallel tests with the same competence profile as well as a competence-oriented analysis of the students' exam results. Unfortunately, manual annotation by item creators is rarely done, either because the used e-learning systems do not provide the functionality or because teachers shy away from the manual workload. In this paper we present an approach for the automatic classification of items according to Bloom's taxonomy and the results of their evaluation. We use natural language processing techniques for pre-processing from four different NLP libraries, calculate 19 item features with and without stemming and stop word removal, employ six classification algorithms and evaluate the results of all these factors by using two real world data sets. Our results show that 1) the selection of the classification algorithm and item features are most impactful on the F1 scores, 2) automatic classification can achieve F1 scores of up to 90% and is thus well suited for a recommender system supporting item creators, and 3) some algorithms and features are worth using and should be considered in future studies.

Keywords: E-assessment · Items · Annotation · Bloom's taxonomy · Data mining · Machine learning systems · Performance levels · Knowledge based systems

1 Introduction

E-Assessment is an integral part of e-learning and many learning management systems support the creation of items and the execution of online tests or online

This work was supported by the German Federal Ministry of Education and Research for the tech4comp project under grant No 16DHB2102.

C. Pang et al. (Eds.): SETE 2020/ICWL 2020, LNCS 12511, pp. 57–69, 2021.
https://doi.org/10.1007/978-3-030-66906-5_6

exams. Despite this technical support, item creation is still a time-consuming process for teachers whose main goal is to prepare an online test in time. Additional work, like the annotation of items with metadata, often falls by the wayside, as its added value is only apparent after a certain timespan. An example for such annotation of items are levels of Bloom's taxonomy [5]. This annotation has no influence on the conduct of the online test and is not visible to the students at all. For large item pools, such annotations support teachers in the long run, since they can, for example, compare exams if they are competence-equivalent or create tests specifically for individual student groups.

Automatic item annotation is therefore a promising way to ensure high and comprehensive data quality (as many items as possible are correctly annotated) with low resource input (teachers check only difficult cases if necessary). Content and wording of the problem context as well as the actual question of an item usually contain all the information that a human being needs to classify an item into one of Bloom's levels. On the other hand, automatic annotation is difficult because domain experts have multi-layered background knowledge against which they assign an item to a certain level.

This paper presents a comprehensive evaluation of approaches for automatic item annotation with levels of Bloom's taxonomy. Using two real-world item pools from the fields of Computer Science and Educational Sciences, both machine-learning-based and rule-based methods are evaluated. The individual parts are systematically varied, e.g. pre-processing by means of NLP techniques or the selected machine learning method. As a result, the evaluation discusses the main factors influencing effective and efficient automatic item annotation.

The paper is structured as follows: In Sect. 2 we introduce two generic approaches to automatic item annotation. Both are based on a preprocessing using typical NLP techniques and use a machine learning method or a rule-based knowledge base for classification. Section 3 presents the main contribution of this paper and discusses the results of the comprehensive evaluation. Different parameter configurations are systematically evaluated with two datasets and their results are interpreted. Section 4 discusses related work before we end with a short summary and an outlook on future work in Sect. 5.

2 Automatic Item Classification

Bloom defined six classes to structure cognitive learning outcomes [4]. He associated them with several inclusion conditions, based on encountered words, like verbs such as *arrange*, *define* or *describe* for the first taxonomy level *knowledge*. In this section we briefly describe two approaches for automatic item classification using Bloom's taxonomy levels. The first approach, rule-based, utilizes a set of manually curated rules that look for the aforementioned keywords and assign corresponding weights to the levels. The second approach employs machine learning and thus requires test and training data, i.e., items have already been manually classified.

First of all, all items are preprocessed in a uniform way as shown in Fig. 1. Each item is first converted into a document that contains the item's context

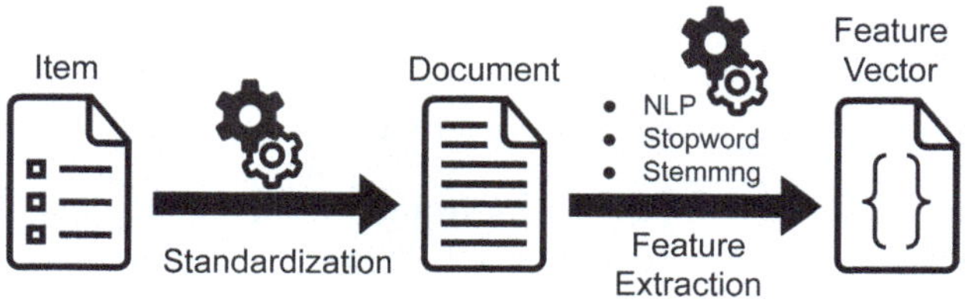

Fig. 1. Common item preprocessing pipeline

description and the question. Item type-specific information such as answer options for single- and multiple-choice questions, which are naturally missing for free-text questions, are not considered. Then, item features are extracted from the document using Natural Language Processing (NLP) techniques. All 19 features from Table 3 are calculated for each item and form specific feature vectors. For the extraction of the items we use standard NLP libraries. In addition, the removal of stop words and stemming is done. Thus, items are transformed into feature vectors, which are used for both, the rule-based approach and the machine learning-based approach.

For the rule-based approach we follow the work of [12]. Bloom listed keywords (primarily verbs) and their assigned levels in [4], which we converted to rules. The assignment is done with a rule weight to characterize their relevance. Using the created rule set for item classification is straight forward: The obtained feature vector is searched sequentially for the keywords that occur in the rule set and the weights are added per level. One exemplary rule is: if the term *arrange* is found within the vector, increment the level *knowedge*. The item is lastly classified into the level with the maximum weight sum.

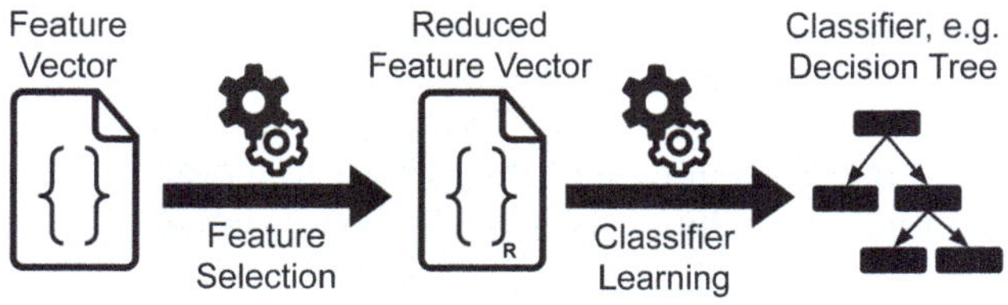

Fig. 2. Process for classifier learning

Figure 2 shows the procedure for the machine learning-based approach. Here the feature vectors of the training data are first optionally reduced, so that not all information has to be included in the creation of the classifier (we evaluate the influence of item features in Subsect. 3.2). The reduced item vectors are then passed to a machine learning algorithm (e.g. a decision tree algorithm) so that an executable classifier can be calculated. We will evaluate the quality in the following section.

3 Evaluation

3.1 Data Sets, Configuration and Metrics

For the evaluation we use two data sets from two different domains: Computer Science and Educational Sciences. Table 1 shows key statistics on the number of items and their size. Each item was independently classified by three domain experts and the majority opinion was adopted for the item, as for some items the domain experts ended up with different classifications. Since item classification is a highly context sensitive task it might be worth allowing more than one class per item. We leave this multi-class classification as subject for future work.

Table 1. Description of the data set used in the evaluation. The length of an item is defined as the number of characters of its context description and question.

Data Set	Domain	#Items	Avg. Item Length in Characters
A	Computer Sc.	83	141.6
B	Educational Sc.	292	283.5

As so often, real world data sets show imbalances in their characteristics. For data set A, for example, the number of items in the level *apply* is about three times as large as for *knowledge*. In contrast, for data set B the class *knowledge* is three times as large as *comprehension*.

Items are used as both, training and test data sets for the machine learning approach and we used a 10-fold cross-validation method[1]. Partitioning, training and testing is not needed for the rule-based approach, as the rules were already verbalized [3].

In our experiments we varied the following parameters:

- NLP Libraries: Pattern, Open NLP, TreeTagger and Stanford CoreNLP
- Text Preprocessing: Stemming and Stop Word Removal
- Machine Learning Algorithms: Six different classifiers (see Table 2), which are all supported by the used library scikit-learn[2].
- Item Features: Up to four out of 19 item features (see Table 3)

We ran the evaluation for all possible parameter combinations and report the minimum, maximum and average F1 scores, as well as the standard deviation of the F1 scores.

[1] https://scikit-learn.org/stable/modules/cross_validation.html.
[2] https://scikit-learn.org/stable/index.html.

Table 2. Description of the six employed machine learning-based classifiers including their configuration.

Classifier	Configuration
DTC: Decision Tree Classifier	max_depth = 5
GNB: Gaussian Naive Bayes	none
KNN: k-nearest-Neighbor	#neigbors = 5
SVM: Support Vector Machine	gamma = 2, C = 1 cache_size = 7000
RFC: Random Forest Classifier	max_depth = 5 max_trees = 10
QDA: Quadr. Discriminant Analysis	none

Table 3. List of item features

Code	Attribute
W	Bag of Words
V	Bag of Words but verbs only
N	Bag of Words but nouns only
CS	Number of sentences
CT	Number of tokens
CV	Number of verbs
CN	Number of nouns
TPS	Number of tokens per sentence (CT/CS)
NPS	Number of nouns per sentence (CN/CS)
NPT	Number of nouns per token (CN/CT)
VPS	Number of verbs per sentence (CV/CS)
VPT	Number of verbs per token (CV/CT)
KM	Number of keywords
KMR	Number of keywords per token (KM/CT)
PD	Number of Part of Speech (POS) Tags
PRD	Number of POS Tags per token (PD/CT)
PCD	Number of POS classes
PCRD	Number of POS classes per token (PCD/CT)
IT	Item type

3.2 Machine Learning-Based Approaches

Figure 3 illustrates the F1 scores grouped by the NLP library. Obviously the employed library has only minor influence on the results and in general it seems irrelevant which NLP library is used. Maximum values differ only up to 0.38% and average values differ up to 3.9%, which corresponds with a differing in standard deviation of up to 2.4%. Only the library Pattern shows a small advantage due to its low standard deviation and high average score. Of course, the library used can have a large influence on the runtime, but this is not part of this evaluation.

Figure 4 illustrates the results by the used classifier. Obviusly, 5 out of 6 algorithms deliver comparable results, from which the Decision Tree Classifier (DTC) algorithm performs best. It delivered the highest maximum F1 score of 0.907, the highest minimum F1 score, performs 0.074 points better then the second best algorithm on average and has the second best standard deviation of 0.078. Second best algorithms are GNB, and SVN, but GNB has a much larger dispersion than SVN but only slightly better average and maximum values. This classifies the algorithms DCT and SVN as candidates for future work, e.g. for algorithm tuning. On the other hand, GNB is a parameter-free method and thus a suitable candidate for first exploratory studies. The same conclusions can be

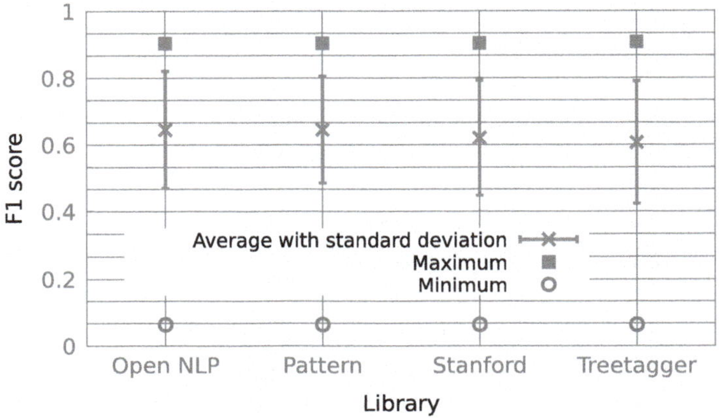

Fig. 3. F1 scores for different NLP libraries (Data set A)

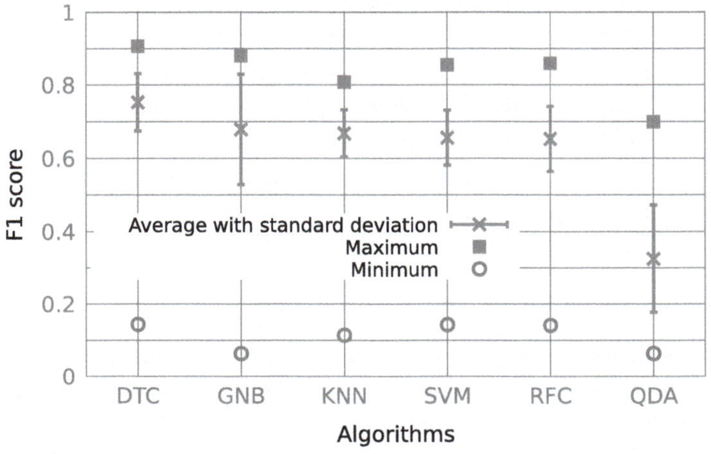

Fig. 4. F1 scores for the different classifiers (Data set A)

Table 4. F1 scores for stemming (Stem) and stop word removal (Stop) for data set A. A check mark indicates usage of the algorithm, a cross mark that the algorithm was not used.

Stem	Stop	Avg	Min	Max	σ
✗	✓	0.608	0.0632	0.872	0.178
✓	✓	0.607	0.0632	0.888	0.177
✗	✗	0.638	0.0632	0.907	0.170
✓	✗	0.635	0.0632	0.883	0.168

drawn from data set B whose results show a similar distribution with higher minimum and lower maximum values, but about the same average and standard deviation values. Interestingly the QDA algorithm performs much better for data set B (avg. F1 score is 0.62), but is also outperformed by the first four algorithms from Fig. 4.

In our third experiment we investigate in the influence of stop word removal and stemming. The results for these pre-processing steps are shown in Table 4. They reveal that stemming has nearly no influence on the average and minimum results, even though there is a mixed difference in the maximum values. In contrast, stop word removal decreases the average F1 scores by 3.75% on average and increases the standard deviation by 0.9% on average. Data set B shows the same characteristics, with the only difference that the usage of stemming increases the maximum results by 1.4% on average. Regarding these mixed results from both data sets and the higher tendency of decreasing the F1 scores, we can not recommend the usage of either of these algorithms for classification performance improvements.

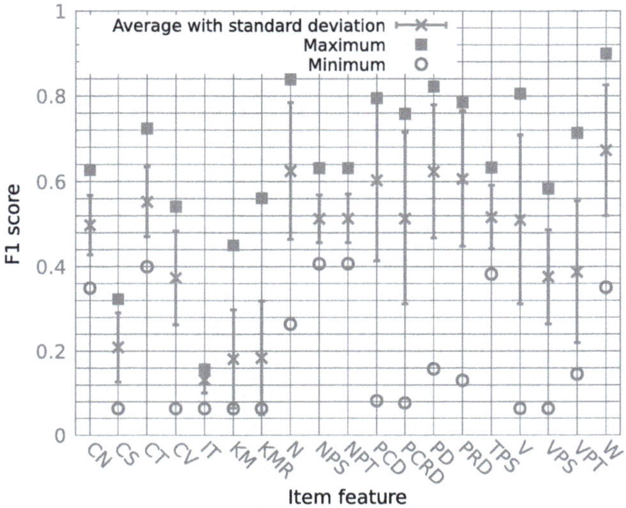

Fig. 5. F1 scores for the different data attributes for data set A. See Table 3 for their abbreviations.

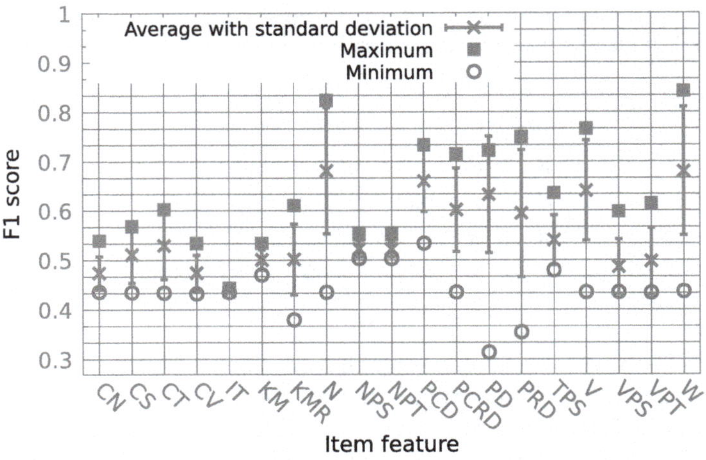

Fig. 6. F1 scores for the different data attributes for data set and B. See Table 3 for their abbreviations.

Figure 5 evaluates the individual influence of the 19 extracted item features for data set A. Only four features achieve a maximum F1 score of more than 0.8 (N, PD, V, W). These features (with the exception of V) also achieve the best average F1 scores, but their standard deviation is among the eight highest ones. In particular, item feature W (Bag of Words) achieves the best overall scores but has a poor standard deviation. The results for Bag of Words might explain why the maximum results of the different NLP libraries are so close. A simple item processing with Bag of Words, without stemming and stop word removal already achieves an F1 score of 0.88 using the Decision Tree Classifier. Such a simple workflow does not require any of the advanced preprocessing techniques and is supported by all NLP named libraries.

A second observation is that all features related to Part of Speech tagging (PD, PRD, PCD, and PCRD) show comparitivly good results, which indicates that analysing the word classes of items gives information about the performance level classification. This corresponds with the high results for verbs (V) and nouns (N). It seems that nouns are more valuable for determining the performance level of items, even though we would have anticipated that verbs are more valuable, as they prompt to do something. These results are contrary to the list of words Bloom defined in [4], which are mostly verbs and which are used for the rule-based approach. A possible explanation is that nouns appear more often within items (about 32% of words are nouns, about 10% are verbs in our data sets). At the same time, the significance of nouns and verbs provides a possible explanation why stemming and stop word removal do not have a significant influence on the results (see Table 4).

Figure 6 shows the individual influence of the 19 item features for data set B. Again all the minimum values are higher and all the maximum values are lower than for data set A. Apart of this fact the results show a comparable impact

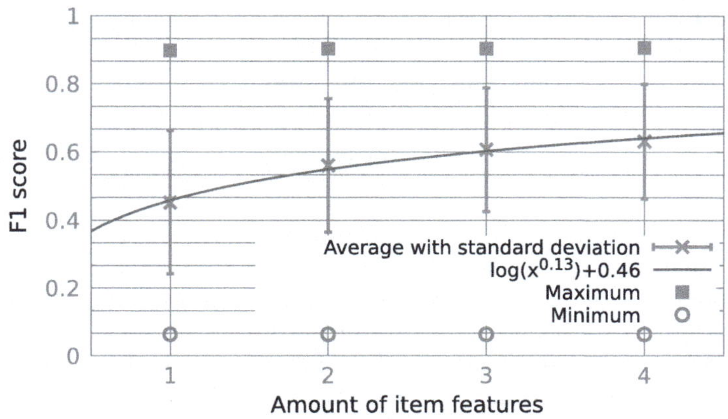

Fig. 7. F1 scores for combinations of item features for data set A. The data represents averages for all runs with x item features.

of the above named features from data set A. Of particular intrest are the bad performing features from Fig. 5 (top) (CS, IT, KM, KMR), which perform much better for data set B. A detailed analysis of the data set regarding these features revealed that they perform better because of data set characteristics, like the usage of specific item types for specific performance levels. There is also much less difference between features V and N which stems from different phrasing of items.

To further increase the quality of the classification, up to four random item features were used in a final experiment. Figure 7 shows the results grouped by the number of features used for data set A. A first observation shows that both the maximum and minimum F1 scores are almost not improved by adding more features. However, the increase of the average F1 score as well as the reduction of the standard deviation is striking, i.e., the classification results become more robust on average. Data set B is also showing the same behaviour with slightly different scores and again lower maximum and higher minimum values.

To summarize the results, the biggest influence factors on the performance of the machine learning-based classification are the chosen machine learning algorithm, as well as the employed item features. Using different NLP libraries has nearly no impact on the maximum values, even though a slight difference in average values is observable. Stop word removal and stemming can be omitted. Combining different item features is not delivering significantly higher maximum F1 scores but increases the average F1 score. In general, the achieved F1 scores (maximums are 0.907 and 0.84 for data set A and B, respectively) may allow for an automatic item annotation but are in any case suitable for a recommender system.

3.3 Rule Based System

In contrast to the machine learning-based approaches, the rule-based approach does not require any training data or parameters to be specified. We thus

Table 5. F1 scores for the rule-based systems for both data sets.

Data Set	Avg	Min	Max	σ
A	0.485	0.18	0.79	0.305
B	0.378	0.08	0.63	0.289

calculated precision, recall and F1 score for each of Blooms classes (i.e., levels) and computed the average, minimum, and maximum value. So in Table 5 the average F1 score is the average of the F1 scores of all classes; minimum and maximum values correspond to the minimum and maximum F1 scores in the distribution of the target classes. Thus these last two values are *not* comparable to Subsect. 3.2 where minimum and maximum were calculated over all parameter configurations.

As visible in Table 5 the average F1 scores are low, differ a lot between the two data sets and have a high standard deviation. Especially the minimum F1 scores are low and show that this approach performs bad for selected target classes. These results are not surprising as we already introduced in Subsect. 3.1 that the mapping of items to target classes is context sensitive. The implemented rule base is not context sensitive, even though the rating vector has been initialized with values that corresponds to the target class distribution within the data set.

As seen in Subsect. 3.2 and Fig. 5 the item features that correspond to the ones used for the rule-based system, namely Number of Keywords (KM) and Item Type (IT), perform bad in comparison to other item features. Especially KM coincides with the results presented in this subsection and reaches with an F1 score of 0.45 about the same dimension as the rule-based system for data set A. Furthermore the results show that the item type alone is no satisfying indicator for performance levels, as attribute IT only reaches a maximum F1 score of 0.16 and 0.44 for data sets A and B. This might stem from the fact that IT provides more information about which performance levels are not applicable than which are.

An analysis of the available items show that the keywords used for the rule based approach only appear in about 12.1% of all items for data set A and 29.6% of all items of data set B. There are two possible interpretations: 1) the item authors did not use the phrasing Bloom proposed for items. They should stick to standard phrasing in order to achieve better classification and have clear items for students. 2) The inclusion rule set, gathered by Bloom, is too small and misses typical keywords and phrases. As discussed for items above, the rule set and thus the amount of keywords and phrases might be too small and should be extended by experts. The focus of the keywords on verbs seems limited, as our results revealed that nouns are used much more often and seem to have a higher impact on the classification results.

4 Related Work

Itemy may be automatically classified during their creation process from existing knowledge sources, like ontologies [2], Linked Open Data [8] or text [9]. This process involves item templates, like sentence patterns or parameterized SPARQL queries. Used templates are annotated with Blooms taxonomy and as a result every created item is also annotated with the respective data [7]. A downside of these approaches is that they only considor the currently created item and are context-free, thus ignoring information the classification might benefit from or depend on, like similar items and available learning material.

There are rule-based, machine learning-based and neural network-based approaches to classify existing items, as well as combinations of the former ones. Rule-based approaches either focus on rules that have been created by experts, or they on creating rules from existing data sets. Chang et al. used the expert approach and obtained comparable results to our study [6]. Haris et al. showed a much better average F1 score of 0.77 for their automatic rule creation system, which in addition analysis the semantic structure of items [10]. Jayakodi et al. use a rule creation system, but did not measure the classification performance in detail. They concentrated on assessing the benefits of available verbs of items, which they summarize to be a limited approach [11].

Machine learning-based approaches for item classification typically employ only a small number of features (e.g. up to four features in [1,15]). Stemming, stop word removal, and various NLP techniques are used by many authors without testing their performance impact, which we have done in our evaluation. A recent survey shows that most approaches focus on support vector machines, k-nearest Neighbor and Naive Bayes as algorithms [14]. The results of our study indicate that decision tree and random forest classifiers show comparable results. Neural Networks are used by Yusof et al., whom focused on convergence time and classification precision, but not on recall and F1 scores [15]. Interestingly they rated the attribute document frequency (DF) as a valuable attribute for precise classification results.

Lastly it is possible to combine different classifiers in voting and ensemble systems to improve the quality of the classification [1,13]. For example Osadi et al. reports an average F1 score of 0.79 for all of Blooms classes, which is comparable to our maximum F1 scores [13].

In general it is noticeable that there is no generic data set available for providing comparable results and that F1 scores, if available at all, differ a lot between different research groups and used data sets, as also shown in the literature review by Sangodiah et al. [14]. Additionally most articles are not analysing their results with respect to external factors and typically do not focus on more than four item features, which, in contrast, we did within this paper.

5 Summary and Future Work

In this paper we presented a comprehensive evaluation of machine learning and rule-based systems for the automatic classification of items with performance levels defined by Bloom. The results show that the machine learning approaches outperform the rule based approach that primarily addresses key verbs. The actual classifier and the employed item features have the biggest influence of the F1 score of the classification.

In future work we will extend the rule-based approach so that not only verbs and the item type are considered. Our evaluation results show that nouns and information from POS tagging should also be considered. Another field of research is the parameter optimization of the algorithms, especially in view of the fact that the results sometimes produced large standard deviations, as well as the consideration of neural networks. There are also a lot of experimental input features available, which have not been tested for the assessment topic so far. Finally, the provision of standardized training and test data sets would also be a valuable contribution to the community. Results, scripts, and programs used for this study are provided at https://gitlab.com/Tech4Comp/automatic-item-annotation.

References

1. Abduljabbar, D.A., Omar, N.: Exam questions classification based on bloom's taxonomy cognitive level using classifiers combination. J. Theoret. Appl. Inf. Technol. **78**(3), 447 (2015)
2. Alsubait, T., Parsia, B., Sattler, U.: Generating multiple choice questions from ontologies: lessons learnt. In: Proceedings of the 11th International Workshop on OWL: Experiences and Directions (OWLED 2014), pp. 73–84 (2014)
3. Beierle, C.: Methoden wissensbasierter Systeme : Grundlagen, Algorithmen, Anwendungen. Vieweg + Teubner, Wiesbaden, 4, verb. aufl. ed. (2008)
4. Bloom, B.S.: Taxonomie von Lernzielen im kognitiven Bereich. Beltz-Studienbuch. Beltz, Weinheim u.a., 3. aufl. ed. (1973)
5. Bloom, B.S., et al.: Taxonomy of educational objectives. In: Cognitive Domain, vol. 1, pp. 20–24. McKay, New York (1956)
6. Chang, W.-C., Chung, M.-S.: Automatic applying Bloom's taxonomy to classify and analysis the cognition level of English question items. In: 2009 Joint Conferences on Pervasive Computing (JCPC), pp. 727–734. IEEE (2009)
7. Cubric, M., Tosic, M.: Towards automatic generation of e-assessment using semantic web technologies. Int. J. e-Assess. (2011)
8. Foulonneau, M.: Generating educational assessment items from linked open data: the case of dbpedia. In: Garcia-Castro, R., Fensel, D., Antoniou, G. (eds.) Proceedings of the Semantic Web ESWC 2011 Workshops (2011)
9. Gutl, C., Lankmayr, K., Weinhofer, J., Hofler, M.: Enhanced automatic question creator-EAQC: concept, development and evaluation of an automatic test item creation tool to foster modern e-education. Electron. J. e-Learn. **9**(1), 23–38 (2011)
10. Haris, S.S., Omar, N.: A rule-based approach in Bloom's taxonomy question classification through natural language processing. In: 2012 7th International Conference on Computing and Convergence Technology (ICCCT), pp. 410–414. IEEE (2012)
11. Jayakodi, K., Bandara, M., Perera, I.: An automatic classifier for exam questions in engineering: a process for Bloom's taxonomy. In: 2015 IEEE International Conference on Teaching, Assessment, and Learning for Engineering (TALE), pp. 195–202. IEEE (2015)

12. Omar, N., et al.: Automated analysis of exam questions according to Bloom's taxonomy. Procedia-Soc. Behav. Sci. **59**, 297–303 (2012)
13. Osadi, K., Fernando, M., Welgama, W., et al.: Ensemble classifier based approach for classification of examination questions into Bloom's taxonomy cognitive levels. Int. J. Comput. Appl. **162**(4), 76–92 (2017)
14. Sangodiah, A., Muniandy, M., Heng, L.E.: Question classification using statistical approach: a complete review. J. Theoret. Appl. Inf. Technol. **71**(3) (2015)
15. Yusof, N., Hui, C.J.: Determination of Bloom's cognitive level of question items using artificial neural network. In 2010 10th International Conference on Intelligent Systems Design and Applications, pp. 866–870. IEEE (2010)

Competence of an ICT Teacher Concerning Didactic and Methodological Support in Teaching ICT at Primary School

Irena Loudova(✉)

University of Hradec Kralove, Rokitanskeho 62,
500 03 Hradec Kralove, Czech Republic
irena.loudova@uhk.cz

Abstract. The widely accepted rhetoric is that our industrial society has been changing to an information society that manages huge amounts of information, which can be disclosed and processed with the help of information and communication technology (ICT). This fact has been reflected in the revised school curricula, and subsequently in the growing role and importance of both, ICT implementation into non-informatics subjects, and the computer science teachers in schools of all levels. These teachers should master the new competencies related to the changes brought by the Educational Policy 2030 + in the Czech Republic, which clearly defines the implementation of computer science in non-informatics subjects. Our research aimed to find out, through a field survey, the real situation in the field of professional readiness of ICT teachers in selected primary schools - with special regard to the perception of the situation by professional teachers. The results of our survey showed that especially computer science teachers in primary schools are not sufficiently professionally or technically prepared for the implementation of ICT in the teaching of non-informatics subjects. They also encounter several problems that prevent them from successfully applying the new educational strategy. Our findings are important in terms of current developments when the COVID-19 pandemic completely paralyzed full-time teaching and underlined the importance of the educational environment to be prepared for such exceptional situations while accelerating the implementation of computer science in non-informatics subjects.

Keywords: Information and communication technology (ICT) · Implementation of ICT · Non-informatics subjects · ICT competences · Professional readiness

1 Introduction

Schools across the globe have already closed in the wake of the coronavirus pandemic. As a result, teachers are suddenly faced with the challenge of how to continue their students' education. While this might seem a daunting task, there are several ways teachers can utilise the technology and resources already available to support online learning and ensure students still receive a quality education. In light of the coronavirus,

C. Pang et al. (Eds.): SETE 2020/ICWL 2020, LNCS 12511, pp. 70–81, 2021.
https://doi.org/10.1007/978-3-030-66906-5_7

companies such as Microsoft and Google have opened up the availability of their remote learning tools to schools and are offering support to teachers and students to help them use their resources. One of the biggest concerns for teachers is to be prepared to provide technology-supported learning opportunities for their students. Being prepared to use technology and knowing how that technology can support student learning have become integral skills in every teacher's professional repertoire. Teachers need to be prepared to empower students with the advantages technology can bring. Schools and classrooms, both real and virtual, must have teachers who are equipped with technology resources and skills and who can effectively teach the necessary subject matter content while incorporating technology concepts and skills. Interactive computer simulations, digital and open educational resources, and sophisticated data-gathering and analysis tools are only a few of the resources that enable teachers to provide previously unimaginable opportunities for conceptual understanding.

However, the Czech School Inspectorate found out that only 9.5 per cent of large primary schools meet digital education standards. Within small primary schools, it is only less than 5 per cent. The research aims to find out, through a field survey, the real situation in the field of professional readiness in the teaching of ICT subjects in primary schools - with special regard to the perception of the situation by professional teachers.

2 Theoretical Basis

The skills acquired in the educational area of Information and Communication Technologies allow pupils to apply computer technology using a wide range of educational software and information sources in all areas of their basic education. This application level goes beyond the content of the educational area of Information and Communication Technologies and becomes part of all educational areas of basic education.

The involvement and use of modern technologies in teaching modifies the teaching style, the approach of teachers and pupils to the instruction. It subsequently changes the image of the school, climates at school, processes in teaching, and cooperation of teachers.

2.1 ICT Competency

"Successful integration of ICT into teaching and learning requires rethinking the role of teachers in planning and applying ICT to enhance and transform learning. Education systems need to regularly update and reform teacher preparation and professional development according, ensuring that all teachers can harness technology for education [1]. The UNESCO ICT Competency Standards for Teachers focus on teachers in primary and secondary schools. However, these approaches apply to all levels of education.

New technologies require new teacher roles, new pedagogies, and new approaches to teacher [2]. Teacher professional development is a crucial component of the educational improvement and thus the framework can be used to localize or tailor a teacher competency program as illustrated in Fig. 1.

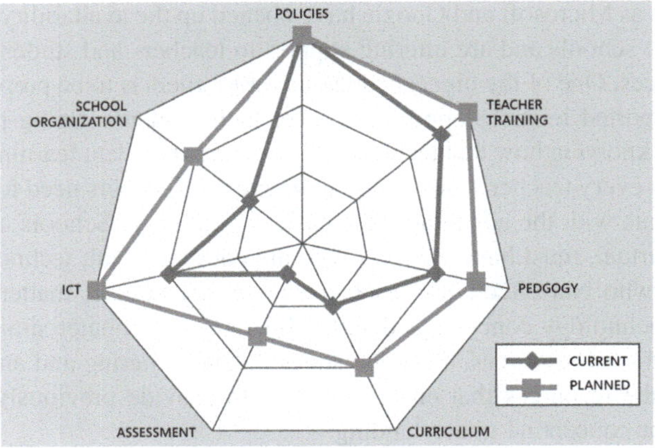

Fig. 1. Example of a teacher competency program.

2.2 Teacher's Professional Readiness in Teaching ICT Subjects

There is a shortage of computer science (ICT) teachers at all levels of schools in the Czech Republic, especially the first stage of primary school (elementary school), as there is no such field at pedagogical faculties. The study is only for the second stage of primary or secondary school (secondary school). In any case, ICT teachers at primary schools in the Hradec Kralove region are ready to develop digital literacy of pupils from the didactic and methodological point of view. As part of the new education policy, they are placed in a new role: to train teachers of other subjects to use ICT technologies in teaching. At present, several subject didactics talk about the information and communication concept of their field, which emphasizes the growing importance of informatics as a science, as a subject and as a necessary basis for the effective use of digital technologies at school and in practical life [3].

Many research studies with a particular focus on ICT (e.g. Hennessy et al., 2005; John, 2005; Watson, 2001) have indicated an apparent conflict concerning whether to use ICT in order to facilitate subject learning, or whether the emphasis should be on demonstrating ways in which ICT can be used and on teaching technical skills [4]. Research findings also emphasize the crucial importance of teachers' awareness about the type of support that ICT can provide, with the purpose of implementing digital resources to enhance pedagogy and pupils' capacity to learn within and across subject domains [5, 6].

Examples from primary classrooms demonstrate that using ICT without a clear method does not bring about progresses and that digital resources need to be matched to the pedagogic intention of the teacher [7]. These outcomes have underlined the necessity to further investigate teachers' integration of ICT in a primary school practice. This need is currently further supported by the situation brought about by the COVID-19 pandemic and the consequent reduction in full-time teaching in schools.

Digital technology is nowadays one of the tools that provides possibilities and limitations in teaching and learning in schools. Teachers should have an available range of methodologies and forms from which they can choose ones that suit their personality, characteristics and professional orientation. A quality teacher then acquires several selected methods and forms of teaching and then chooses them for the teaching according to the situation, the specific composition of pupils in the classroom and other criteria [8].

2.3 Strategy of Digital Education

Concerning the new role of ICT teachers, which is to train teachers of other subjects to use ICT technologies in the instruction, it is also important to mention the situation in terms of the new Digital Education Strategy. According to the criteria of the Union of School Teachers of Informatics in the Czech Republic, the schools will only be able to teach with ICT, if they have an updated ICT strategy, ICT administrator, make a computer or other device available to more than 50% of teachers, renew computers for pupils after 7 years at the latest, and have sufficient coverage (at least 60% of classrooms) by the internal network for connecting computers.

According to the latest data published by the Ministry of Education, only 14.3% of primary schools meet the minimum criteria for quality teaching and use of ICT in schools. In 33.5% of primary schools do less than half of teachers have a computer, and 9% of primary schools are equipped with the new computers (under five years old) in the classrooms. In a questionnaire survey conducted by the Czech School Inspectorate in 2019 in 5 316 primary schools, teachers mentioned the following reasons that prevent them from implementing ICT in teaching: insufficient equipment 45,8%); lack of time (29,7%); little knowledge of computer operation (28,5%); problems in the organization of the instruction (26,3%); ICT concerns and lack of self-confidence (20,5%); difficulties in linking ICT and learning curriculum (14,1%), negative attitude towards ICT in teaching (12,2%), and previous bad experiences (7,7%), see Fig. 2 [9].

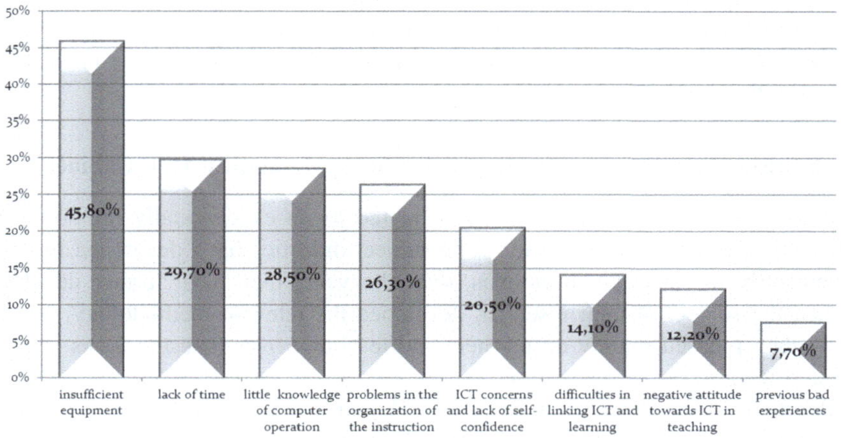

Fig. 2. Reasons that prevent teachers from implementing ICT in teaching.

2.4 The National ICT Curriculum

In connection with the above-mentioned facts, the significant innovation of the national curriculum is needed, namely - to take into account the development of new technologies, its impact on developments in disciplines and related requirements for equipping primary and secondary school graduates with new competencies. The whole general curriculum in the framework of educational programs should be revised and the changes could be divided into three areas:

1. Inclusion of ICT topics in the binding general curriculum. The aim is not to educate IT professionals, it is to develop computational thinking and provide basic insight into the issues of data, information and their processing with the expectation that a certain part of such educated students will be interested in the field of informatics and will continue their studies.
2. The innovation of other educational fields according to how the development of digital technologies has influenced the development in their parent fields and various areas of human activities in general.
3. The condition for the sustainable development of pupils' digital competences means the use of digital technologies by pupils in everyday school work. The new concept is therefore based on the innovations in educational fields, and the expected level of digital competencies is determined by the field problems that the student should be able to solve with the help of technology.

Experience shows that incorporating digital technologies into traditional teaching strategies and working with them using traditional methods does not lead to more effective teaching with better results for pupils [10]. Often the opposite is true. For schools and teachers to prepare pupils for life and work in the world to come, they must work in a different way than usual.

The integration of digital technologies into education and school life is changing established practices. It is a controlled process of transformation that has the same rules and characteristics as introducing any other innovation. Most teachers and schools perceive the need and benefits of using digital technologies and are interested in integrating them into the instruction.

2.5 Subject Didactics as a Starting Point for a Methodical ICT Cabinet

The field or subject didactics can be considered as an exceptionally important area influencing the quality of education. The subject didactics have developed relatively independently as scientific disciplines, over the years they have further developed, specified the subject of their research, developed the relevant methodology, etc. Previously they focused mainly on teachers and teaching, over time the focus is on the pupil and learning in a broader context.

Although subject didactics in the Czech Republic is a relatively young discipline, there are efforts to develop this area. One of the main challenges of subject didactics was the curricular reform, which has been taking place since 2005 and a revision of the framework of the educational programs is currently planned. Subject didactics thus have the opportunity to react and form a new philosophy of subjects and to comment on the transformation of the curriculum itself [11]. For this reason, it is appropriate to coordinate activities of methodical ICT cabinets with subject didactics and reflect their suggestions to the level of individual subjects, or wider educational areas, as is typical for the concept of transdisciplinary didactics, see the suggested concept of the development of digital and IT competences of pupils, Fig. 3.

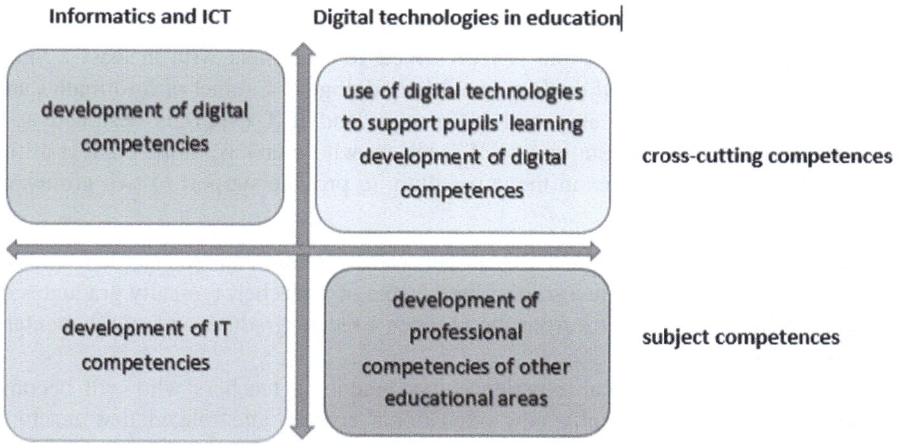

Fig. 3. Concept the development of digital and IT competences of pupils.

In the presented paper, we were focusing on the situation in the field of professional readiness in teaching ICT subjects at primary schools. The emphasis was placed on the perception of the situation by the professional teachers. We detected didactic resources and the overall readiness of ICT teachers for the planned introduction of computer science into non-ICT subjects. The research question was: *How do teachers of ICT subjects at primary schools perceive the specifics of the new role and its management*?[1]

3 Methodology

Applied pedagogical research was carried out at primary schools in the Hradec Kralove region, focused on the detection of didactic aids and the overall readiness of computer science teachers for the planned introduction of computer science into non-ICT subjects. Research and data collection was realized directly in the field. The research method we applied was a structured questionnaire. The questionnaire contained two research parts, the first part was focused on exploring a new role (new competencies, subjective satisfaction), and the second part was aimed at the reflection on experience

[1] Note: The basis of the new role is the transmission of competencies of the use of ICT in the education of teachers of other subjects.

(behavioral questions focused on empiricism in transmission, examples of good practice, problem formulation and solutions).

The research group consisted of ICT teachers from the Hradec Kralove region. Inclusive criteria for inclusion in the research group were as follows: ICT teacher with at least 2 years of experience; the teacher was either a member of the Regional Methodological Cabinet of Informatics and ICT (N = 150) or an ICT teacher in selected regional primary school (N = 150, standard research mortality was calculated). Research data were obtained from N = 300 respondents.

Group 1, for details see Table 1, represented ICT teachers with at least 2 years teaching practice, members of the Regional Methodological Cabinet of Informatics and ICT. The Methodological Cabinet of Informatics and ICT (MCI) is designed as a structured professional community of ICT teachers, whose task is, in the context of the mentioned significant changes in the curriculum, to provide support to two groups of teachers:

- Teachers without pedagogical experience (or with minimal pedagogical experience), i.e. those who are just starting in the role of a teacher, typically graduates of teaching studies, or experts from the practice - recent graduates of supplementary pedagogical studies.
- Teachers with pedagogical experience, i.e. practicing teachers who will become acquainted with significantly new educational content and related new teaching methods, forms and strategies.

150 members of this Cabinet took part in our research, three from each district of the Hradec Kralove Region, while each district has its own specifics, such as a large number of so-called small classes, or a lot of pupils from socially excluded localities.

Table 1. Basic input data of the research Group 1.

Gender		Age range			ICT proficiency		Teaching experience (years)		Previous degree		
	N	30s	40s	50s	Qualified	Non-qualified	2	3/more	Bc.	MA/Msc	Ph.D.
F	53	17	23	13	53	0	15	38		50	3
M	97	32	48	17	97	0	29	68		92	5
$\sum$	150	49	71	30	150	0	44	106		142	8

Group 2, see Table 2, was composed of ICT teachers who have at least two years of teaching experience. From each district (the Hradec Kralove region includes 5 districts), 150 teachers from primary schools, recommended by the Czech School Inspectorate, were selected. Criteria of the Czech School Inspectorate assessed the quality of teaching and the corresponding technical equipment at selected primary schools in the region.

Table 2. Basic input data of the research Group 2.

Gender		Age range			ICT proficiency		Teaching experience (years)		Previous degree		
	N	30s	40s	50s	Qualified	Non-qualified	2	3/more	Bc.	MA/Msc	Ph.D.
F	61	21	34	6	54	7	18	43	7	53	1
M	89	29	46	14	78	11	26	63	11	76	2
$\sum$	150	50	80	20	132	18	44	106	18	129	3

The obtained data were processed by quantitative analysis procedures: tests to verify the psychometric properties of the questionnaire, descriptive statistics (non-parametric test to compare two or more files - comparison of files concerning monitored variables: length of practice, overall evaluation of experience, etc.). Behavioral questions in the form of open items were administered through procedures of qualitative analysis (categorical analysis) and computational linguistic analysis (keyword detection, frequency analysis).

4 Research Results

4.1 Research in the Field of a New Role of ICT Teachers

In the first part of our research we focused on the exploration of a new role of ICT teachers as mediators in the triad teacher – pupil – ICT in a non-IT subject. When creating the structured questionnaire, we took an example from the ICT competence framework of UNESCO and focused on five competencies concerning the new role of ICT teacher at school (mediator in the implementation of ICT in non-IT subjects): 1) Communication 2) Digital Competency 3) Connecting theory and practice 4) Adaptation to the use of a new role 5) Motivation. The structured questionnaire contained a total of 12 questions, which included the key concepts: professional readiness, the available offer of methods and forms of teaching, support from school management, opportunities for further education, linking digital technologies and curricula, technical readiness of the school, fear of losing authority over colleagues, technical school readiness.

The non-parametric U-test of Mann and Whitney was used to evaluate given answers. The Likert scale, where 1 meant "definitely no" and 5 meant "definitely yes" was applied. Based on the determination of the p-value from the test, the null hypothesis (for every question) was accepted or rejected. The null hypothesis was set as follows: the answer of Group 1 is identical or similar to the answer of Group 2. Therefore, if a statistically significant difference was detected, the p-value was printed in bold, see Table 3. If there was no statistically significant difference between the answers of the groups (the null hypothesis was not rejected), the p-value in Table 3 was indicated in a basic font.

Table 3. Comparison of responses of Group 1 and Group 2.

Question	p-value
Question 1	0,279
Question 2	**0,006**
Question 3	**0,0002**
Question 4	0,810
Question 5	**0,023**
Question 6	**0,003**
Question 7	**0,253**
Question 8	**0,0002**
Question 9	**0,023**
Question 10	**0,006**
Question 11	**0,003**
Question 12	**0,003**

The results of the first part of the research clearly showed that ICT teachers who are members of the Methodological Cabinet (Group 1) are in all respects much better prepared for the planned changes in the position of ICT teachers at schools. Only in the first and fourth questions of the questionnaire survey there was a statistical agreement. It was a question of professional qualification (Are you professionally prepared for teaching ICT?) and support from school management (Do you have the support of the school management to get prepared for the new role?).

4.2 Qualitative Analysis (Categorical Analysis) and Computational Linguistic Analysis (Keyword Detection, Frequency Analysis)

In the second part of our research we aimed at behavioral questions (empiricism in transmission, examples of good practice, problem formulation and solutions). Figure 4 and Fig. 5 show the key words which appeared in Group 1 (Fig. 4) and Group 2 (Fig. 5) in the open answers most frequently.

During the interview with ICT teachers in Group I (members of the Regional Methodological Cabinet of Informatics) and Group 2 (ICT teachers at selected primary schools), we monitored the occurrence of keywords and their frequency as teachers answered the submitted behavioral questions. The procedures of qualitative (categorical) and computational linguistic analysis (keyword detection, frequency analysis of lemmas) were used.

The results of the survey showed that most often in both groups there are concerns about the insufficient quality of technical equipment of schools (group 1–92%, and Group 2–90%) and also about the implementation of modern ICT technologies in the teaching process (Group 1–82%, Group 2–80%). On the other hand, there has been a fundamental difference in the approach to communication skills (Group 1–30%, Group 2–80%), where ICT teachers from Group 2 were concerned that the lack of qualified communication competencies would prevent them from cooperating smoothly with teachers of other subjects. This can consequently, in the opinion of the respondents, cause a decline in professional prestige among colleagues. This output is consistent with the results of similar research in the world, e.g. K. Hakkarainen from Finland, who pointed to the existing discrepancy between ICT teachers' pedagogical principles that commonly emphasized active construction of knowledge, and their self-reported pedagogical practices [12].

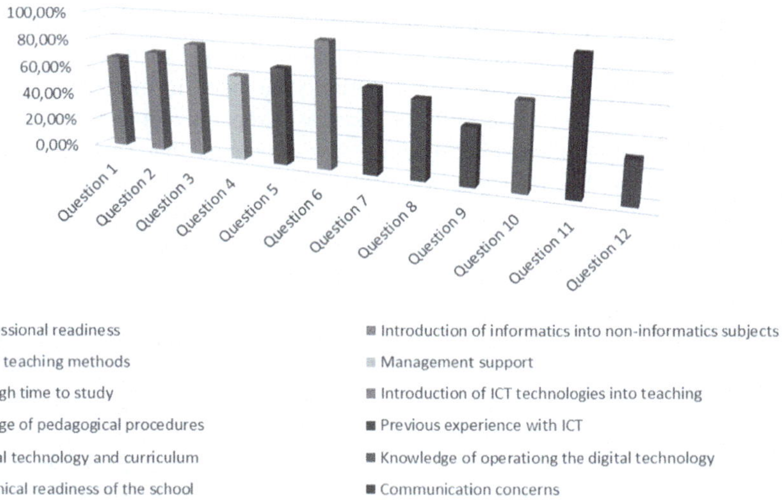

Fig. 4. Frequency of keywords in Group 1 (members of the Regional Methodological Cabinet).

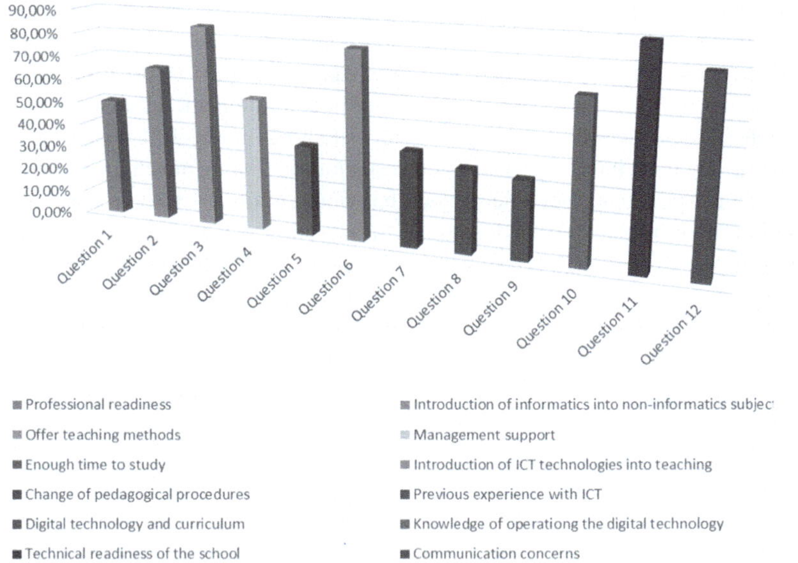

Fig. 5. Frequency of keyword in Group 2 (ICT teachers at selected primary schools).

5 Implications and Conclusions

The implementation of ICT in non-IT subjects is becoming an inevitable step towards the smooth provision of teaching, as we could see at the time of the coronavirus pandemic and the subsequent closure of schools. The failure of full-time instruction and the transition to teaching in the cyber environment has proven the growing importance of the role of the computer science teacher as a methodologist and a mentor in the implementation of ICT to other subjects. The question remains whether ICT teachers are professionally prepared for their new role. It is clear that at least in the Czech Republic, the preparation of teachers for the implementation of ICT in subjects is not yet fully ensured. Although this issue is partially included in the teaching of subject didactics and methodology, it is still not taken as a separate subject.

Our field research pointed to several other problems that schools of all levels gradually encountered, whether it was insufficient technical equipment of schools, absence of quality ICT methodology, fear of teachers in the application of ICT in teaching, legislative unpreparedness for distance learning, lack of quality educational materials in the field of support for distance education and subject didactics, etc. As the research of the Faculty of Education in the Hradec Kralove region (conducted in spring 2020) showed, 30–40% of children did not participate in online education during the COVID-19 pandemic in the Czech Republic. These children skipped lessons not only because of the absence of a computer or good internet connection but simply because this method of teaching does not suit them and their parents did not motivate them to work in an online environment. This issue, however, was not the subject of our research, although it is certainly interesting and would merit a more detailed study.

The results of our research confirmed that most teachers are still not methodically and didactically sufficiently prepared for the new role of ICT teacher, which carries the elements of a mentor. Unfortunately, there are currently not enough quality training programs to help existing ICT teachers complement their communication and mentoring training. Teachers working in Methodological Cabinets have a better starting position, because they are informed well in advance about the steps of the Ministry of Education and they also have sufficient training in communication due to their position. One of their main activities is mapping the situation in the ICT teaching and networking of teachers interested in closer cooperation. Future teachers at pedagogical faculties do not yet have a specific subject that would allow them to prepare for the role assigned to them by the 2030+Educational Strategy in the Czech Republic. Should the situation, urged by the COVID-19 pandemic, changed and the pedagogical programs for future teachers would incorporate the communication competency and mentoring aimed at knowledge sharing skills, the authors believe that the ICT implementation into non-informatics subjects will improve. It is important as ICT have become the transversal axis of all teaching activities in which they almost always have three functions: they are an instrument in learning processes, a tool for information processing and implicit learning content.

Acknowledgement. This paper is supported by Specific research 2020 at the Faculty of Education University of Hradec Kralove. Special thanks go to the dean's office of the Faculty of Education for support, students Jana Novotna and Andrea Cernikova, for assistance in data collection in the field.

References

1. ICT Competency Framework for Teachers harnessing Open Educational Resources. https://shar.es/aHT9dN. Accessed 02 June 2020
2. Makrakis, V.: Training teachers for new roles in the new era: experiences from the United Arab Emirates ICT program (2005)
3. Friesen, N.: Continuing the dialogue: curriculum, didaktik and theories of knowledge. J. Curric. Stud. **50**(6), 724–732 (2018). https://doi.org/10.1080/00220272.2018.1537377
4. Hennessy, S., et al.: Pedagogical approaches for technology-integrated science teaching. Comput. Educ. **48**(1), 137–152 (2007)
5. Engeness, I.: Teacher facilitating of group learning in science with digital technology and insights into students' agency in learning to learn. Res. Sci. Technol. Educ. **38**(1), 42–62 (2020). https://doi.org/10.1080/02635143.2019.1576604
6. Tay, L.Y., Lim, S.K., Lim, C.P., Koh, J.H.L.: Pedagogical approaches for ICT integration into primary school English and mathematics: a Singapore case study. Aust. J. Educ. Technol. **28**(4), 4 (2012). https://doi.org/10.14742/ajet.838
7. Kervin, L., Danby, S., Mantei, J.: A cautionary tale: digital resources in literacy classrooms. Learn. Media Technol. **44**(4), 443–456 (2019). https://doi.org/10.1080/17439884.2019.1620769
8. França, R.P., Iano, Y., Monteiro, A.C.B., Arthur, R: Improvement of the transmission of information for ICT techniques through CBEDE methodology. In: Utilizing Educational Data Mining Techniques for Improved Learning: Emerging Research and Opportunities, pp. 13–34. IGI Global (2020)
9. Czech School Inspectorate. https://www.csicr.cz/getattachment/4e7af154-d761-47d5-8ef6-10645fab4a61/Shrnuti-Vyuzivani-digitalnich-technologii-v-MS,-ZS,-SS-a-VOS.pdf
10. Lemley, P.S.M.M.A.: Intellectual property in the new technological age: 2019. Diss. University of California at Berkeley (2019)
11. Creelman, A., et al.: Didactical use of information communication technology (ICT) and modern teaching practices (MTPs) in higher education. In: 2nd International Scientific Conference Teaching Methods for Economics and Business Sciences (2019)
12. Hakkarainen, K., Muukonen, H., Lipponen, L., Ilomäki, L., Rahikainen, M., Lehtinen, E.: Teachers' information and communication technology (ICT) skills and practices of using ICT. J. Technol. Teach. Educ. **9**(2), 181–197 (2001)

Integrating MOOCs in Blended-Learning Courses: Perspectives of Teachers and Students

Hui Bao[1(✉)] and Xin Zhao[2(✉)]

[1] West Anhui Health Vocational College, Lu'an, China
bh1118@163.com
[2] University of Sheffield, Sheffield, UK
xin.zhao@sheffield.ac.uk

Abstract. Being recognized by Chinese universities, the MOOC has been currently applied in higher education institutions, and has been widely believed as an important opportunity for educational practices. However, only a few teachers have quickly adapted toward MOOC teaching in China. This research highlights the experiences perceived by students' blended learning in their computer courses, as well as the specific challenges that teachers encounter in blended teaching. This study adopts a mixed-method approach. Relevant data include those generated from email interviews with university teachers (n = 12) and from online questionnaires with students (n = 96) in a Chinese university. The results suggest that both from teachers and students' perspectives, the effect of application of blended teaching is satisfactory, as it improves classroom teaching in general. Students' learning interests and their self-study and analytical abilities are likewise improved. Moreover, this research highlights the challenges of teachers based on several aspects – educational technology, teaching content design, students' participation, and integration of online and offline course contents. The paper concludes with pertinent suggestions for blended teaching, such as finding a balance amongst the three aspects of technology, science, and humanities amidst a constant improvement of the teaching methods.

Keywords: Blended-learning · MOOC courses · Education technology

1 Introduction

1.1 Overview of MOOC Development

Since the extensive adoption of MOOC in higher education in 2012, students' learning has transformed [1]. Initially, MOOC was committed to imparting knowledge to the students through short films, reading materials, assignments and exam-related forms. The only interaction between the users is through the BBS, a primitive collaboration style that is named the MOOC 1.0 era [2]. Today, the MOOC has made considerable progress; its most popular concepts in online education are collaborative and blended learning, which is now called the MOOC's 2.0 era [3].

C. Pang et al. (Eds.): SETE 2020/ICWL 2020, LNCS 12511, pp. 82–93, 2021.
https://doi.org/10.1007/978-3-030-66906-5_8

1.2 MOOC in Chinese Higher Education

Alongside a widespread adoption of large-scale online open courses (MOOC) on a global scale, an increasing number of higher education institutions in China have adopted MOOCs in their teaching curricula. In 2017, the Ministry of Education of China launched the first batch of its national quality online open course recognition work, and selected 490 from more than 3,000 courses as the national online open courses [4]. By 2020, it plans to launch another 3,000 online open courses in the country. However, studies suggest that certain challenges still remain in this process, including how to use a network platform to design different types of curriculum contents, learning scenarios and learning evaluation models. Meanwhile, various teaching curriculum changes have also posed challenges to teachers' teaching design methods.

The MOOC has enabled learners to break through the constraints of time and space in the process of acquiring knowledge, thus opening an effective path for active knowledge acquisition [5]. Indeed, it has an advantage that traditional teaching cannot match. However, the relatively independent MOOC has its shortcomings, hence it needs to integrate its learning concepts with the excellent resources in traditional classrooms in order to create a new learning method [6]. Therefore, blended teaching that combines face-to-face classroom interactions with online learning has become a trend in higher education. On this note, this research has a certain significance with regard to the current blended teaching approach based on MOOC.

1.3 Research Gap and Research Questions

Current research on MOOC focuses on four aspects. The first aspect revolves on MOOC's application model and explored the respective problems in teaching and learning from them. The second aspect entails research on MOOC platform and curriculum development. In Yulianto et al.'s [7] analysis of curriculum resource development and platform application, they believe that MOOC's success depends on the courses being offered on the platforms. The third aspect refers to the study of learners' behaviors in MOOCs, including analyzing students' learning behaviors and motivations [8]. The fourth aspect comprises the impact of MOOC on education. De Waard et al. [9] believe that there are several factors that can promote the interaction between MOOC and the learners; mobile network technology is one of them. However, so far, relatively very little analysis has been made on MOOC-based teaching methods, particularly on which methods are effective or ineffective. Furthermore, the influence of MOOC-based blended teaching in the universities' computer courses remains unknown. In this case, we need to capture the experiences perceived by students in learning computer courses, as well as the specific challenges that teachers encounter in teaching design and implementation.

This research introduces MOOC-based blended teaching in Chinese college curriculum and takes the West Anhui Health Vocational College's Course of University Computer Application Foundation as an example. Its aim is to explore teachers' and students' perspectives in integrating MOOC in blended-learning courses. In particular, the following research questions will be answered:

1. What are the challenges that teachers may face in integrating MOOC in their teaching?
2. What are the impacts of blended learning approach that utilizes MOOC courses on student learning?
3. What suggestions can be provided for the educators who wish to adopt a similar teaching approach?

2 Literature Review

2.1 MOOC Platforms and Courses

As a product of highly-developed information technology, the MOOC platform is a new type of online course resource. Its course production, resource dissemination, discussion and examination all reflect the full benefits of automatic feedback and big data technology means, with the aim to achieve learning objectives and evaluations [10]. Presently, there are several MOOC platform providers, business platforms and university-based MOOC platforms that go hand-in-hand. They have provided vital resources for the development of MOOC and have brought much convenience to online learners [11].

However, currently, there are very few MOOC courses that are built by college teachers themselves, who are basically quoting such courses from other platforms. The introduction of excellent MOOC course resources may reduce the durations that college teachers spend in designing online courses, as they will have more energy for classroom activities. However, the MOOC platform has always faced high dropout rates, and many scholars have been focusing on how to keep the learners interested in their selected courses and to ensure learning quality and efficiency [12]. Usually, the learners who choose courses on the platform are definitely interested in such courses, and what they care most about is the quality of those courses [13]. They have mentioned that whether the videos in an MOOC course are well prepared has little impact on the improvement of their knowledge levels. Indeed, they pay more attention to the quality of course contents rather than form.

A survey found that students have varied needs and expectations in MOOC platforms, and a quantitative analysis showed that perceived ease of use (PEOU) and perceived usefulness (PU) have a significant impact on learning willingness in MOOCs [14]. The researchers further pointed out that improving existing MOOC platforms and meeting students' needs can increase both PEOU and PU. Arguably, in order to improve the MOOC platforms' retention rates, they should cater to the MOOC learners by providing appropriate course selections, considering both course quality and student demands.

2.2 Blended Learning Approach Using MOOC Courses

For learners, blended learning is considered from the perspective of "learning"; in contrast, for teachers, blended teaching is considered from the perspective of "teaching". The former means that in order to achieve certain teaching and learning objectives, certain media are employed for communication, while corresponding learning methods are combined to attain teacher-led and student-centered teaching situations [15]. Blended learning programs may vary according to their teaching objectives, classroom proportions, online learning content, and content delivery [16]. As such, several of today's blended learning programs are built around traditional courses, enriching online content and taking advantage of the positive effects of blended learning in both teaching and learning [16]. Osguthorpe and Graham [17] believe that blended teaching includes a mixture of different teaching theories, teaching media, classroom and online environments, and teacher-led and student-centered subjects.

Blended teaching is not only a teaching form, but also a teaching strategy. The online learning approach of MOOC frees students from the constraints of time and place, allowing them to repeatedly watch videos within the release cycle of MOOC to enhance their learning [18]. Yapici and Akbayin [19] adopted the blended learning mode for practice in a university's course data structure and algorithm. Through the implementation of an online learning management system (LMS), it was established that the dropout rate of learners was significantly reduced, and the learning effect on learners was relatively ideal. Gerbic [20] made a critical analysis of the current research field and pointed out two development areas of MOOCs that deserve attention: teachers' concepts and beliefs on blended teaching; and, the change in teachers' roles, especially in curriculum design.

2.3 MOOC and Higher Education

Presently, digital courses are more effective than ever before, as MOOCs have influenced the face of higher education. In 2013, 40 public universities provided online course credits through the MOOC2Degree academic partnership. Many elite universities likewise entered the said field, thus making online learning more credible, whilst reducing the number of enrolled graduate students [21]. With the emergence of MOOC platforms for learning analysis and interactive exercises, a critical and controversial question has emerged: are the higher education programs making themselves bankrupt? Stephen Carson, Head of External Relations of the MIT Open Course, answered in the negative. Nonetheless, the best experience for graduate students is to use relevant equipment, design and create things in laboratories, which may not be realized online for now [22].

The MOOC brings both opportunities and certain challenges for higher education. Once the MOOC approach becomes mature and stable, students can earn their diplomas through MOOCs for their much-desired credit. Yet, some scholars ponder whether universities will disappear in the future, with some claiming that around 25% will disappear or merge within 10 to 15 years [23]. In addition, MOOC construction has certain requirements regarding teachers' professional abilities and school network platform support in higher education. For instance, Israel [23] emphasized that integrating online and offline courses will present certain challenges for teachers.

3 Methods

3.1 Design

This research aims to extensively explore the effect of applying and implementing MOOC-based blended teaching on the teachers, study teachers' views on blended teaching and their encountered challenges in terms of facilities, and analyze students' satisfaction and participation in MOOCs. A mixed-method approach was conducted. Email-based interviews with the teachers (n = 12) were analyzed by thematic analysis. Meanwhile, filled-in questionnaire surveys were collected from the students (n = 96) and then analyzed through the SPSS software. The first method highlights the research subject by investigating teachers' reflections in implementing MOOC-based blended teaching. On the other hand, the second method can provide more detailed and in-depth data by analyzing students' satisfaction with the MOOC platform in terms of learning effect, participation, and expectations for blended learning in the future. In this way, the combination of the two above-mentioned research methods can provide more in-depth conclusions.

3.2 Participants

This study takes 100 students from two freshmen classes in China's West Anhui Health Vocational College as the research subjects. It primarily utilizes the course resources in the Chinese Universities MOOC Platform, rather than the local course resources produced by the aforementioned college's teachers. Six computer teachers from the said college were interviewed via email after the questionnaire survey was completed by the students. In selecting the teacher-respondents, the researcher adopted purposive sampling; six teachers from two universities in other provinces and cities in China and with rich MOOC teaching experience were selected for the email interview.

3.3 Data Collection and Analysis

Data collection has two aspects: questionnaires for the students and email interviews for the teachers. Since this research needed to calculate the correlation of various factors in accordance with the Likert Scale, a closed-ended questionnaire was employed. Based on the research aim and questions, the questionnaire was divided into five parts: course and platform, learning effect before class, teaching process participation, learning effect after blended teaching and learning expectations. There were 22 questions in total. The questionnaires were transmitted to 100 students through the online examination platform. Eventually, the number of valid filled-in questionnaires was 96.

Meanwhile, the email interviews were semi-structured in nature, and were administered online and asynchronously. These interviews allowed the participants to be in familiar settings, which might enable them to feel comfortable when responding and had more time to think before answering the questions. A total of six questions were designed based on the following topics: views on the platform courses, difficulties in designing teaching contents, challenges encountered in classrooms, students' participation, preferred teaching method, and suggestions to improve the mixed teaching method.

This study used the SPSS software, a professional statistical tool, to analyze the filled-in questionnaires. Based on the data from the questionnaire's five parts, correlation analysis was carried out, taking each part as independent of each other. This study used thematic analysis to analyze email interview data; as a useful and flexible qualitative research method, it can be applied to any research question [24]. Through manual coding, this study explored the relationships between the topics, and determined the similarities and differences amongst the teachers' opinions and attitudes.

4 Results

4.1 Teachers' Perspectives on Blended Learning

According to teachers, blended teaching improved classroom teaching efficiency, student motivation and autonomy, as well as helping teachers to understand student learning needs.

Blended Teaching Improved Classroom Teaching Efficiency. Most of the 12 respondents indicated that blended teaching had absorbed the advantages of both traditional and online teaching, and teachers had focused on explaining various vital and difficult contents in their classes to solve certain problems, thus enhancing the quality of classroom teaching. In addition, the participants noted that offline teaching in blended teaching made up for the deficiencies of MOOC online teaching, which required real class interactions. Similarly, Güzer and Caner [25] stated that although students prefer a web-based learning environment, they are not willing to fully deviate from the traditional classroom component of the course, which is precisely one of the principles of a blended teaching method.

Blended Teaching Improved Students' Interests. The participants mentioned that MOOC had increased students' interests and promoted their deep learning. Some of them further proposed that blended teaching had a better classroom atmosphere. Furthermore, they noted that while students had more problems in blended classes, teachers saw some significant improvements in their students' analytical skills. However, some of them pointed out inadequacies in blended teaching, including an overemphasis on students' interests, communication and cooperation, leading to an inaccurate positioning of teachers' roles. Notably, the teaching process should guide students' active learning and knowledge construction; it is a bilateral process in which teachers' teaching and students' learning should be realized together. Therefore, irrespective of the teaching environment, teachers should be the organizers in the teaching process.

Educational Technology Helped Teachers Understand Students' Needs. Some participants suggested that educational technologies enabled the teachers to change their teaching process and adjust their teaching methods according to students' learning conditions, in order to ensure the effectiveness and concentration of students' learning. As relevant literature pointed out, by combining the neural networks established in the regression analysis with the prior knowledge of the experts, the prediction model of students' learning outcomes had an accurate predictive effect.

MOOC Platforms Promoted Teachers' Ability and Enhanced Students' Autonomy. The participants indicated that blended teaching through the MOOC platform had rich learning resources, and the teachers could draw on the excellent experiences of others. It further increased the teachers' teaching skills. Meanwhile, the study manifested that blended teaching improved the students' independent capability, learning motivation and student satisfaction. Notably, one participant pointed out that although the current MOOC platform was rich in teaching resources, it was not made by the university teachers; to some extent, it lacked academic originality. Arguably, the school needed to focus more on building resources for its local courses.

4.2 Challenges Encountered in Blended Teaching

Presentation of Teaching Content. The interview results suggested that teachers faced some challenges in switching from 'teacher-fronted classes' to 'student-centered classes' in a blended learning approach. The design of face-to-face teaching needed to be based on the MOOC classes in order to avoid redundancy of teaching contents and to keep students' engagement in class. This requires the teachers to master how to integrate online learning and face-to-face teaching when designing their teaching schemes, guiding students in their classes, and solving vital and difficult topics, rather than merely employing repetitive teaching.

"From traditional classroom teaching to MOOC teaching, the teachers have switched from 'authoritative' to 'facilitative'. Teachers are the resource providers and organizers in the teaching process." (T7, associate professor)

Allocation of Teaching Time and Communication with Students. Meanwhile, some teachers put forward the problem of time allocation in their teaching content. They expressed that in a mixed learning environment, time allocation must be considered for them to organize mixed teaching activities, such that synchronous and asynchronous tasks could play the best roles in mixed teaching. A blended learning setup needed to provide online content between 30% and 79%, which would reduce face-to-face interactions.

Furthermore, the participants noted that communicating with students and disseminating knowledge were also challenges for them. Indeed, blended teaching is characterized by a learner-centered and technology-mediated teaching that focuses on knowledge construction and real activities. Blended learning is highly different from traditional teaching, as it has more emphasis on the interactions between the teachers and students.

Lack of Computer Literacy, Technology Support and Preparation Time. All 12 teachers mentioned that technology was the biggest challenge for them, as they needed to master video recording and online teaching design technologies. This is consistent with the findings of Hanson [26], who suggested that the emergence of ICT and the Internet is a major challenge for teachers who have been implementing traditional teaching methods; likewise, the teachers must address this issue in order for them to teach confidently amid a blended environment.

Moreover, the participants proposed that blended teaching tended to increase teachers' workload. In a report that explored the motivations, challenges and teaching processes of MOOC teachers, it was pointed out that the main challenge for teachers was the lack of technology and time to produce MOOC videos [26]. Additionally, the participants expressed that teachers must have basic computer literacy and information security knowledge. In this manner, effective teaching can be carried out to eventually solve the problems encountered in the teaching process.

Low Student Engagement. The low participation of students affected teachers' classroom teaching, as teachers were greatly challenged on how to enable the students to independently complete the MOOC contents and continue to actively participate in classroom interactions. The twelve respondents further pointed out that the MOOC completion rate was low; the main reason was students' procrastination. At the same time, the participants pointed out that for those who had studied through MOOC, they did not focus or listen in face-to-face sessions. In this case, how can the teachers achieve their expected objectives? Studies have shown that teachers need to combine proactive help and encouragement in order to improve course submission and module retention. Additionally, one interviewee mentioned lack of standardization in blended learning course assessments.

Lack of Guidance and Financial Support for Teachers. Colleges and universities hoped to enhance their reputation by providing or introducing MOOC courses, establish professional development opportunities for the teachers, and internationalize location-based courses. However, for most higher learning institutions, management system was not perfect. Teachers had no corresponding financial support and professional equipment, leading to their low enthusiasm.

"Lack of financial support and professional equipment in recording videos... the school lacks a perfect management system and a clear reward mechanism, so teachers are not very active in making MOOCs." (T6, associate professor)

Arguably, the school management needs to rethink the basic processes in the education value chain. Based on the required dimensions of integration, technical finance and administrative support processes should be combined in order to improve daily management.

4.3 Student Survey Results

Courses and Platform. It can be concluded that most students agreed with the MOOC platform of Chinese universities due to its simple and convenient operation, as well as the rich performance of its course resources as a way to promote learning experience; meanwhile, only a few students had a neutral attitude on the situation. However, the results showed that students' satisfaction in the content of the computer course on the university's MOOC platform was not very optimistic, with 21% of the respondents expressing strong dissatisfaction and 12% expressing dissatisfaction.

The students were generally optimistic with the level of interest in the content; around 47.92% expressed such interest. This is consistent with the attitude of willingness to carry out independent learning before class and the interest in learning content. Understandably, if the students are interested in the learning content itself, they will then enhance their intrinsic motivation to learn through the MOOC platform.

Students' Participation in the Classroom. The survey showed that 92.71% of students were willing to take part in interactive discussions in blended teaching classes, while only 7.29% were either uncertain or unwilling. In contrast, in the interview results, the teachers thought that students' participation was a challenge for them. Nonetheless, the implementation of blended teaching was successful, as students' willingness was deemed positive. Therefore, compared with traditional teaching, blended teaching seemed more attractive to the students. In this case, the main problem of the research is that students' online participation is inconsistent with classroom participation; moreover, the dropout rate in the MOOC learning platform exists.

Learning Effect Before and After Class. Usually, for a traditional teaching method, students can learn by themselves through textbooks only if their teachers assign lessons in advance. However, the current MOOC learning environment is more attractive to the students. As illustrated in the results, 91 of the 96 retrieved questionnaires expressed agreement that it was easier to invest in pre-class learning using the MOOC platform, and as many as 93 respondents agreed that it was better to use the platform for learning. This is consistent with Prensky's [27] study, which mentioned that millennials in the current Internet context prefer videos as a rich curriculum content for independent learning. In addition, satisfaction in pre-class assistant teaching on the MOOC platform reached 94.79%, with neutrality accounting for the remaining 5.21%. Therefore, blended teaching indicated an optimistic learning effect and satisfaction before class for the students.

In addition, among the 96 questionnaires retrieved, 90% respondents thought that blended teaching would achieve a better learning effect, while only one respondent disagreed. Moreover, 85 people agreed that the MOOC platform was well integrated in terms of online and classroom learning. As for the learning process meeting individual needs, 98.96% agreed, with only one respondent disagreed.

The Learning Effect of Blended Teaching was Related to Each Part. This research used the Cronbach's Alpha reliability analysis to determine the questionnaire's reliability. The results illustrated that the scale of each dimension was reliable, with the reliability coefficient exceeding 0.8. A Cronbach's alpha between 0.7 and 0.8 is deemed acceptable; if greater than 0.8, it is considered both accurate and effective [24]. Therefore, the analysis of the variables in this study was reasonable.

According to the data analysis, each variable is positively correlated. All of them have a significant level of 0.01. Therefore, blended teaching effect has a significant positive correlation with the participation of students in the teaching process ($r = 0.571$, $P < 0.001$), as well as in the courses and platforms ($r = 0.631$, $P < 0.001$). Similarly, it has a positive correlation with the students' pre-class learning effect ($r = 0.451$, $P < 0.001$).

Students' Learning Expectations. Around 88.57% of the students agreed that blended learning had stimulated their interest in the course content. Conversely, 21.88% were not willing to study independently through the MOOC platforms in the future. This result is consistent with the problem in the questionnaire regarding the platform and course. Indeed, there were problems in the course's video content, which had not reached the expected learning effects of the students; in response, the students expressed unwillingness to carry out independent learning in the future.

5 Discussion and Conclusions

5.1 The Impacts of Blended Learning Approach that Utilizes MOOC Courses on Student Learning

The teachers generally think that blended teaching improves their students' interest, enhances the classroom teaching atmosphere, and improves classroom teaching effect. Millennials prefer the MOOC environment, as MOOC platforms provide a strong appeal. The students are satisfied with the Chinese university's MOOC platform and are generally interested in the online video contents. They are also willing to take part in discussions and ask more questions during classroom interactions. Their self-study and analytical abilities are improved in the process.

This study found that the students were satisfied with the pre-class effects of blended teaching, as it was positively correlated with the MOOC platform and courses. Likewise, the blended teaching effect on the students was strongly correlated with the course platform, pre-class learning effect and students' participation in the teaching process. A significant correlation likewise existed between the students' learning expectation and blended teaching effect.

5.2 The Challenges that Teachers Face in Integrating MOOC in Teaching

On the one hand, educational technology helps the teachers understand their students, and MOOC platforms promote their abilities as well. However, on the other hand, the main challenge for teachers is technology, especially in terms of designing MOOC videos. Additionally, recording videos tend to increase workload. Teachers also lack support with regard to equipment and funds.

Teachers particularly face a difficult challenge in teaching content design, as well as in integrating online and offline course contents. Students' participation is also a problem for them. Blended teaching emphasizes on classroom interactions too much, which easily leads to teachers neglecting their supposed leading role in teacher-student interactions. Moreover, MOOC learning platforms tend to have dropout rates, and some ways in students' online participation are deemed inconsistent with their classroom participation.

5.3 Recommendations for Educators

Teachers should master teaching content design and information technology in order to make good use of the blended teaching method. Only when students are able participate actively in both online and offline activities can good teaching results be achieved. The MOOC platform and course resources serve as the support of blended teaching; as such, only when schools are able provide sufficient funds, equipment and management support can blended teaching play its best role. Indeed, blended teaching should find a balance within the three aspects of technology, science, and humanities under a constant improvement of the teaching methods.

5.4 Limitations and Future Research Direction

The research subjects in this paper are freshmen students, which are relatively small in number, leading to a lack of universality of the experimental subjects. This study is also inconclusive on whether the other courses are suitable for implementing a blended teaching approach. This research likewise uses the course resources of the Chinese University MOOC platform; hence, the courses lack the characteristics of local course resources, and may not match well with the situation of teachers and students in the college being studies. Three research directions are deemed worthy of further attention: MOOC-based blended teaching for other types of courses; research areas related to educational technology; and, curriculum design of blended teaching approach.

References

1. Yeager, C., Hurley-Dasgupta, B., Bliss, C.A.: CMOOCs and global learning: an authentic alternative. J. Asynchronous Learn. Netw. **17**(2), 133–147 (2013). https://doi.org/10.24059/olj.v17i2.347
2. Salathe, M.: MOOCs 2.0: Scaling one-on-one learning. https://www.wired.com/insights/2014/09/moocs-2-0/. Accessed 06 Aug 2020
3. Soylev, A.: MOOCs 2.0: the social era of education. Turk. Online J. Dist. Educ. **18**(2), 56–67 (2017). https://doi.org/10.17718/tojde.306557
4. Shen, J., Ye, M., Wang, Y., Zhao, Y.: Massive open online course (MOOC) in China: Status quo, opportunities, and challenges. In: 2016 IEEE Global Engineering Education Conference (EDUCON), pp. 1106–1108. IEEE (2016). https://doi.org/10.1109/educon.2016.7474692
5. García-Peñalvo, F.J., Cruz-Benito, J., Borrás-Gené, O., Blanco, Á.F.: Evolution of the conversation and knowledge acquisition in social networks related to a MOOC course. In: Zaphiris, P., Ioannou, A. (eds.) LCT 2015. LNCS, vol. 9192, pp. 470–481. Springer, Cham (2015). https://doi.org/10.1007/978-3-319-20609-7_44
6. Prohorets, E., Plekhanova, M.: Interaction intensity levels in blended learning environment. Procedia-Soc. Behav. Sci. **174**, 3818–3823 (2015). https://doi.org/10.1016/j.sbspro.2015.01.1119
7. Yulianto, B., Prabowo, H., Kosala, R., Hapsara, M.: Implementation of learning analytics in MOOC by using artificial unintelligence. J. Comput. Sci. **14**(3), 317–323 (2018). https://doi.org/10.3844/jcssp.2018.317.323
8. Barak, M., Watted, A., Haick, H.: Motivation to learn in massive open online courses: examining aspects of language and social engagement. Comput. Educ. **94**, 49–60 (2016). https://doi.org/10.1016/j.compedu.2015.11.010

9. De Waard, I., et al.: Using mLearning and MOOCs to understand chaos, emergence, and complexity in education. Int. Rev. Res. Open Distrib. Learn. **12**(7), 94–115 (2011). https://doi.org/10.19173/irrodl.v12i7.1046
10. O'Reilly, U.M., Veeramachaneni, K.: Technology for mining the big data of MOOCs. Res. Pract. Assess. **9**, 29–37 (2014)
11. Fischer, H., Dreisiebner, S., Franken, O., Ebner, M., Kopp, M., Köhler, T.: Revenue vs. costs of MOOC platforms. Discussion of business models for xMOOC providers, based on empirical findings and experiences during implementation of the project iMooX. In: 7th International Conference of Education, Research and Innovation (ICERI 2014), pp. 2991–3000. IATED (2014)
12. De Freitas, S.I., Morgan, J., Gibson, D.: Will MOOCs transform learning and teaching in higher education? Engagement and course retention in online learning provision. Br. J. Educ. Technol. **46**(3), 455–471 (2015). https://doi.org/10.1111/bjet.12268
13. Yousef, A.M.F., Chatti, M.A., Schroeder, U., Wosnitza, M.: What drives a successful MOOC? An empirical examination of criteria to assure design quality of MOOCs. In: 2014 IEEE 14th International Conference on Advanced Learning Technologies, pp. 44–48. IEEE (2014). https://doi.org/10.1109/icalt.2014.23
14. Aharony, N., Bar-Ilan, J.: Students' perceptions on MOOCs: an exploratory study. Interdisc. J. e-Skills Life Long Learn. **12**, 145–162 (2016). https://doi.org/10.28945/3540
15. Garrison, D.R., Vaughan, N.D.: Blended Learning in Higher Education: Framework, Principles, and Guidelines. Wiley, Hoboken (2008)
16. Gilbert, J.A., Flores-Zambada, R.: Development and implementation of a "blended" teaching course environment. MERLOT J. Online Learn. Teach. **7**(2), 244–260 (2011)
17. Osguthorpe, R.T., Graham, C.R.: Blended learning environments: definitions and directions. Q. Rev. Dist. Educ. **4**(3), 227–233 (2003)
18. Horn, M.B., Staker, H.: Blended: Using Disruptive Innovation to Improve Schools. Wiley, Hoboken (2014). https://doi.org/10.1177/0047239515570584
19. Yapici, I.U., Akbayin, H.: The effect of blended learning model on high school students' biology achievement and on their attitudes towards the internet. Turk. Online J. Educ. Technol.-TOJET **11**(2), 228–237 (2012)
20. Gerbic, P.: Teaching using a blended approach–what does the literature tell us? Educ. Media Int. **48**(3), 221–234 (2011). https://doi.org/10.1080/09523987.2011.615159
21. Long, C.: The changing face of higher education: the future of the traditional university experience. Kennedy School Rev. **13**, 58 (2013)
22. Sharrock, G.: Making sense of the MOOCs debate. J. High. Educ. Pol. Manag. **37**(5), 597–609 (2015). https://doi.org/10.1080/1360080x.2015.1079399
23. Israel, M.J.: Effectiveness of integrating MOOCs in traditional classrooms for undergraduate students. Int. Rev. Res. Open Distrib. Learn. **16**(5), 102–118 (2015). https://doi.org/10.19173/irrodl.v16i5.2222
24. Braun, V., Clarke, V.: Using thematic analysis in psychology. Qual. Res. Psychol. **3**(2), 77–101 (2006). https://doi.org/10.1191/1478088706qp063oa
25. Güzer, B., Caner, H.: The past, present and future of blended learning: an in depth analysis of literature. Procedia-Soc. Behav. Sci. **116**, 4596–4603 (2014). https://doi.org/10.1016/j.sbspro.2014.01.992
26. Hanson, J.: Displaced but not replaced: the impact of e-learning on academic identities in higher education. Teach. High. Educ. **14**(5), 553–564 (2009). https://doi.org/10.1080/13562510903186774
27. Prensky, M.: Digital natives, digital immigrants. On the Horizon. http://www.marcprensky.com/writing. https://doi.org/10.1108/10748120110424816. Accessed 06 Aug 2020

Current Political News Translation Model Based on Attention Mechanism

Xixi Luo[1], Jiaqi Yan[1], Xinyu Chen[1], Yingjiang Wu[2], Ke Wu[1], Meili Lu[1], and Liang Cai[1](✉)

[1] Ningbo Tech University, Ningbo, China
1065109325@qq.com, 1170580691@qq.com, 857654755@qq.com, 459386416@qq.com, 623106592@qq.com, 719675711@qq.com
[2] Ningbo University, Ningbo, China
380586242@qq.com

Abstract. China's foreign exchanges are becoming increasingly frequent, among which current political news, as the recent or ongoing relevant reports of facts in the national political life, plays a great role in the field of information transmission. However, there are a large number of proper nouns as well as long and complex sentences in current political news, so the news translation through traditional machine translation tends to have low accuracy and poor usability. Based on this situation, this paper proposes a translation model of Chinese current political news based on Attention. It uses the classic Long Short Term Memory (LSTM) model and introduces the Attention Mechanism to improve the traditional Encoder-Decoder framework. Through the training of parallel corpora, constraints are established for the proper nouns of current political news, thereby improving the overall translation accuracy. The experiment shows that the translation model used in this paper has higher accuracy than neural machine translation (NMT).

Keywords: Proper noun · Machine translation · LSTM model · Attention mechanism

1 Introduction

Under the important strategic environment of the Belt and Road Initiatives, China's foreign exchanges are becoming more frequent, and foreign friends also hope to learn about China through current political news. However, due to the excessive number and rigorous use of proper nouns in political news, the accuracy of traditional translation tools is not high. With the gradual development of deep learning technology, Neural Machine Translation (NMT) has emerged, which uses neural networks to map source language to target language, and it can significantly improve the quality of machine translation [1].

In recent years, many research scholars have carried out a variety of researches and analyses on the Attention Mechanism and neural network model in the NMT framework, which has promoted the rapid development of machine translation [2]. Kim uses

C. Pang et al. (Eds.): SETE 2020/ICWL 2020, LNCS 12511, pp. 94–102, 2021.
https://doi.org/10.1007/978-3-030-66906-5_9

Convolutional Neural Networks (CNN) to solve topic classification tasks [3]. Gehring proposed an encoder-decoder model based on convolutional neural network, which surpassed Google's model in accuracy and greatly improved the translation speed. As a more common framework in deep learning tasks, Encoder-Decoder has successfully applied in various fields. However, as a general framework model, it is not designed for a specific field, resulting in its limited long-range memory ability.

In 2013, Nal Kalchbrenner created a network model called Dynamic Convolutional Neural Network (DCNN), which was based on the neural network translation method. In terms of semantic construction, this model extracts important semantic information from the sentence by analyzing the combination of words, so that even the distant chunks in the sentence can produce a certain semantic connection, which showed great application potential [4]. Under the premise of making minimum assumptions about the sequence structure, Sutskeve used the LSTM model to present an end-to-end sequence learning method (End-to-end approach), and achieved great results. In a translation task from English to French, the BLUE value of the translation generated by LSTM in the entire test set is 34.8, which is higher than the 33.3 obtained by the SMT system [5]. Some scholars [6–9] also proposed a relationship extraction method that combines the LSTM model with the Attention Mechanism, and tested it on the Sem Eval2010 task 8 corpus. Experimental results showed that the accuracy and stability of this method are further improved compared with traditional deep learning methods.

The Attention Mechanism that was first used in image processing is gradually being used in the field of language processing. Bahdanau and his team are the first to apply this mechanism in the field of natural language processing (NLP) [10]. They used a mechanism similar to Attention Mechanism to translate and align simultaneously on machine translation tasks. Then similar extensions of the RNN model based on the Attention Mechanism began to be applied to various NLP tasks [11]. Shi Yan et al. adopted the self-attention model and compared it with other translation models through the Tensorflow framework. The experimental results showed that the model had better translation effect [12].

Based on the research of this kind of model, scholars have extended it to new application fields, such as the translation of place names [13]. Cai Liang et al. took encyclopedia entries as corpus and established characteristic relations through semantic relations, demonstrating the mechanism of language generation [14]. In the field of Chinese news, there are also relevant studies, such as the Chinese news model based on LSTM- Attention proposed by Lan Wenfei et al. [15]. Based on it, the current political news translation model constructed in this paper will be further discussed in detail.

In the field of Chinese current political news translation, there are only researches on words, phrases, and sentences, and lack of real-time translation of complete texts. Therefore, in the process of translation, the accuracy of text classification is low, which may easily lead to the loss or misuse of semantics and affect readers' understanding of the original text. This paper uses a divide-and-conquer strategy to classify and study the new words, phrases and common semantic structures in current political news, and proposes a Chinese current political news translation model based on LSTM-Attention.

2 Method

2.1 Encoder-Decoder Model

Encoder-Decoder model is the basic method of machine translation by neural network, also known as Seq2Seq model, as shown in Fig. 1 [16]. The operating process of Encoder-Decoder model is to imitate the cognitive process of human beings. The general cognitive process of human beings is divided into three parts: when confronted with new knowledge, people will take the initiative to use their past experience and knowledge to try to understand this knowledge; when fully understood, this knowledge is stored as a memory point in the brain; when faced with real problems that need to be solved, people will try to recall and practice what they have learnt in mind. Encoder, Intermediate encoding and Decoder correspond to these three processes respectively.

Encoder encodes the input sentence into a context vector C, and then decodes C into an output sentence by Decoder, which can effectively solve the problem of unequal length of the input sequence and the output sequence. Given the input word sequence X = [x1, x2, …, xn], the model could find a set of output word sequence Y = [y1, y2, …, yn] to maximize the conditional probability distribution P(Y | X) [17]. Encoder and Decoder both use Recurrent Neural Network (RNN) to model the sentence coding. This algorithm emerged in the early 21st century as the most commonly used network structure in deep learning. It is a kind of neural network for processing sequential data and has a broad prospect.

In practical applications, when encoder encodes the input sequence as the input hidden layer vector H = [H = h1, h2, …, hn], formula (1) is usually used.

$$h_i^x = RNN(x_i, h_{i-1}^x) \tag{1}$$

In the formula, xi is the input word at time i, and hx i − 1 is the input hidden layer at time i − 1. Given the context vector C and the output sequence y1, y2, …, yn − 1, the decoder calculates the output probability of the next yt by softmaxing hy i, as shown in Eq. (2), (3).

$$P(y_n|c, y_1, \cdots, y_{n-1}) = P(y_n|h_{n-1}^y, y_{n-1}) \tag{2}$$

$$P(Y|X) = \prod_{\mathrm{i}=1}^{n} P(\mathrm{y_i}|\mathrm{c}, \mathrm{y}_1, \cdots y_{i-1}) \tag{3}$$

The RNN network is suitable for dealing with time-varying sequence problems, but cannot solve the problem of long-term dependence. LSTM is a special type of RNN, which has two transmission states compared to the traditional RNN mode, breaking through only one memory overlay mode and performing better during training. In practical applications, LSTM is usually used to solve the problems of gradient disappearance and explosion in the long sequence training process. Since the context vector C contains all the information in the original sequence, its length becomes the bottleneck that limits the performance of the model [18]. When the translated sentence is long and one C cannot store all the information, the translation accuracy will drop rapidly.

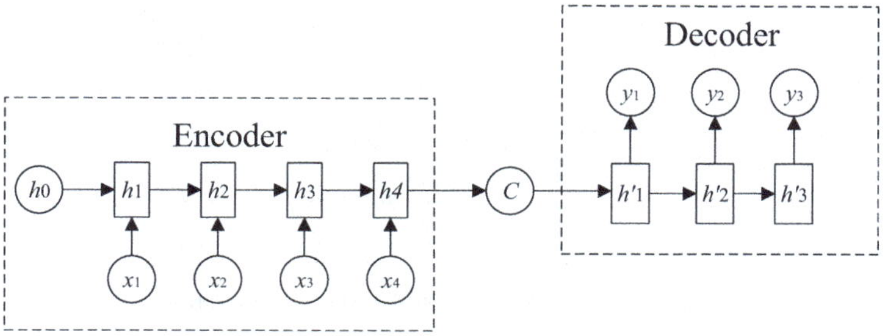

Fig. 1. Encoder–decoder model diagram

2.2 Attention Mechanism

Attention Mechanism can make full use of the information carried by the input sequence. The Encoder working under the Attention mechanism does not need to convert all information into a fixed-length vector, but only needs to encode the input part into a sequence of vectors. At the time of decoding, the Attention Mechanism produces a so-called "attention range", which focuses on the most important part of the input sequence, and then selectively chooses a subset from the vector sequence for processing, to produce an output.

The Attention Mechanism solves the long-sequence dependency problem by inputting different context vectors Ci at each time. Each Ci may correspond to a different probability distribution of the Attention distribution of the source language sentences or words. The model structure is shown in Fig. 2. That is, the fixed context vector C will be replaced with Ci that changes according to the current output word. The calculation formula is shown in formula (4) [19]. Each vector C will select the most suitable context information for the current output sequence y, that is, use aij to calculate the correlation between hj in the j-th stage of the Encoder and the i-th stage in the Decoder. The context vector Ci in the input sequence comes from all hj. The calculation formula for the sum of the weighted values of aij is shown in formula (5)

$$C_{\mathrm{i}} = \sum_{\mathrm{j}=1}^{\mathrm{n}} a_{ij} h_j^n \tag{4}$$

$$a_{ij} = \frac{\exp(e_{ij})}{\sum_{k=1}^{n} \exp(e_{ik})} \tag{5}$$

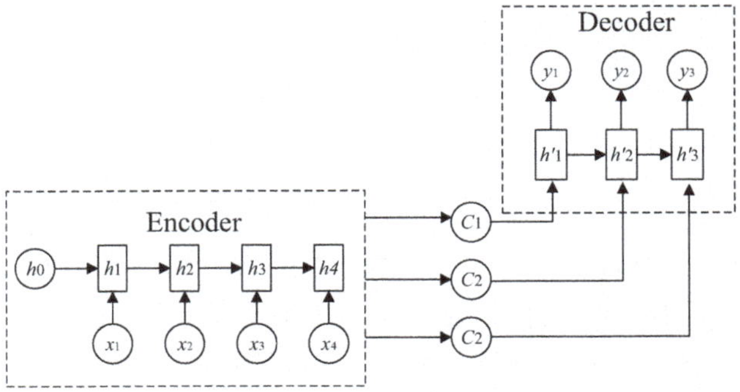

Fig. 2. Attention Mechanism model diagram

3 Model Construction

3.1 Model Framework

Current political news covers a wide range, mainly including the guidelines, policies, and activities of political parties, social groups and social forces in dealing with national life and international relations. Some proper nouns used in news tend to be updated with some important speeches, grand activities and so on. In daily texts, the collection rate of such proper nouns is very low, and they are often not included in the translated glossary in time. Therefore, the symbol <UNK> is used to represent these new words, and the Attention Mechanism is used to filter and replace them. The traditional decoder assigns the same vector to each input value, while the Attention Mechanism can quickly screen out valuable information from a large amount of information, and assign different weights to words according to their differences, so as to achieve the purpose of replacing translated words with professional nouns in the coding process.

The main models in this paper include Encoder, Decoder, the Attention Mechanism and Substitute. When input information, the embedded layer of the encoder converts the information into a fixed-length background vector C and encodes the vector C. The hidden layer outputs the vector in series through the two-way circulating neural network layer of forward coding and reverse coding. The decoder is mainly composed of an embedded layer, hidden layer and Softmax layer. Different from the bidirectional cyclic neural network layer represented by the hidden layer in the encoder, the hidden layer in the decoder is the one-way cyclic neural network layer. The information from the hidden layer is passed on to the Softmax layer, and the word that matches the information most is computed. The Attention Mechanism uses different-background vectors for different time steps of the decoder to reduce the loss of information and retain its original meaning to the greatest extent. The substitute needs to be supported by a database so that each input vector can be interpreted accordingly. Since this paper adopts an open-source parallel corpus in both Chinese and English, its content does not contain the expressions in the latest political news, especially proper nouns. Therefore, although the operation of the model is completely tenable and feasible theoretically, it is necessary to update the corpus in real time to ensure the accuracy of information output.

3.2 LSTM Model

Recurrent Neural Network has a variety of algorithms, such as Simple Recurrent Network (SRN), Gated Algorithm, Depth Algorithm, and Extension Algorithm. Long Short-Term Memory (LSTM) belonging to Gated Algorithm is one of the models adopted in this paper. Due to its special time recursive neural design structure, LSTM can handle long-term dependences, so it is suitable for processing and predicting some important events in time series like relatively long intervals and delays.

In the LSTM model, the cell state plays a core role. Its function is similar to the conveyor belt, running through the whole chain structure to transfer original information and ensure its invariability. In order to control and change information, the concept of "gate" was introduced, consisting of a layer of sigmoid neural network and a point-by-point multiplier. The output range of the Sigmoid layer is between 0 and 1. If the output number 0, the information cannot pass, while the output number 1 makes all the information pass. In this process, three doors have extremely positive effects. When a piece of information is input, a word may continue the previous expression or begin to describe the new content. At this time, the input gate will judge the importance of this word to the whole sentence or a paragraph, while the forget gate will selectively forget the input information of the previous node and discard unimportant information. Next, the output gate determines and outputs useful information contained in hidden layers of the neuron state at the previous moment.

When the encoder and decoder are working, the following two problems are inevitable: 1) The encoding vector cannot effectively store all the information in the source language; 2) Some information is easy to be lost in the encoding and decoding process due to too long information transmission. The Attention Mechanism can to a certain extent deal with these problems by selecting important information directly from the source language, which means it can further solve the problem of low precision.

4 Experiments and Results

4.1 Experiment Procedure

The daily corpus selected for the experiment in this paper comes from Semantic Knowledge Base. In this database, 2000 pieces of political news corpus were randomly selected, and the benchmark test was carried out based on these data. During the test, all data was divided into train, dev, and test, with a ratio of 6:2:2. Before the formal experiment, the data is processed in a pre-trained format: each line has one sentence, with empty lines separating the document so that the source language is aligned with the target language. At the same time, the words in the corpus are expressed in the form of a word number and stored uniformly in the file as a comparison table. Next, it uses the unified standard and the long sentence text in order to make up the short sentence and intercepts the sentence which exceeds the standard length. The preprocessed data will be used as training expectations to initialize the training model.

4.2 Experiment Design

When classifying the current political news, the LSTM model introduced in the Attention Mechanism can achieve the classification prediction more effectively and increase the weight on the important words and sentences, so as to improve the accuracy of the prediction probability of classification. In order to better show the effect of the classification model, Precision, Recall and Accuracy are introduced to measure the classification results. The Precision Rate is based on the prediction results, indicating how many of the prediction samples are positive samples, that is, the proportion of the truly correct ones in all the predictions that are positive. The Recall Rate is based on the original samples, indicating how many positive samples are predicted correctly, more specifically, the proportion of the actual positive and truly correct ones. The Accuracy Rate is the proportion where all the predictions are correct in all the samples.

Besides, in order to more effectively prove the advantages of the LSTM-Attention Model, this paper uses traditional machine learning models to classify them based on the same data and system environment, including the LSTM model, Support Vector Machine (SVM), K-Nearest Neighbor (KNN). At the same time, to ensure that the results have consistent parameter comparisons, word vectors trained by word2vec are adopted as text features in this paper.

4.3 Experimental Results

In this experiment, loss reduction occurred in all models during training. However, compared with other models, the loss value of LSTM-ATTENTION model training has been greatly improved. Specific data are shown in Table 1.

Table 1. Comparison of evaluation standards.

Method	Precision	Accuracy	Recall
SVM	0.880	0.899	0.903
KNN	0.904	0.913	0.919
LSTM	0.937	0.948	0.949
LSTM-ATTENTION	0.978	0.981	0.980

Table 1 shows the difference between The Precision Rate, The Accuracy Rate and The Recall Rate in four different models of SVM, KNN, LSTM and LSTM-ATTENTION. Under the same data, the four models all show good classification effect, among which LSTM-Attention is the best. There are two main reasons for this: on the one hand, word2vec-trained word vectors do play a significant role in describing text features; on the other hand, the Attention Mechanism introduced in this model can flexibly capture the relationship between part of the vocabulary and the whole text in the input text, and reasonably allocate the weight coefficient. Therefore, keywords in current political news, especially some proper nouns, can more appropriately express the semantic meaning of text, and the accuracy of text classification is also improved accordingly.

For example, Google translates "某支特定指数的美国股票" into "U.S. stocks in a specific index", while the result of LSTM-Attention model is "an American stock, a stock that is included in a given index".

5 Conclusion

This paper uses word2vec, a Google's open source word vector calculating tool, to train large-scale Chinese news corpus, and explore the rich semantic relations among knowledge, obtaining the feature representations of text, so as to generate the word vectors of Chinese words. Among the many Language generative paradigms, word2vec turns out to be effective and verifiable vehicles that can automatically establish correlation between semantic information and feature representations vectors of candidate. Therefore, based on the text representation method of word vectors, the LSTM classification model is amended and improved with introduction of Attention Mechanism in the process of translating current political Proper Nouns and long sentences. The calculation of Attention Mechanism is differentiable, which leads to one conclusion that backpropagation is capable of producing the mapping relation between candidates and keyword vector. The other conclusion that can be made is that Attention Mechanism is suitable for the network structure of deep learning in the field of current political news translation. Thus, the LSTM-Attention model is designed. It can also be seen from the experimental results that text classification effect of the LSTM model has improved to a certain extent with the introduction of the Attention mechanism. However, the validity of this model needs further testing in such field as image processing and political news translating.

Acknowledgements. This study is supported by 2019 University Students Science and Technology Innovation Project of Zhejiang Province.

References

1. Sutskever, I. Vinyals, O., Le, V.Q.: Sequence to sequence learning with neural networks. In: Advances in Neural Information Processing Systems, pp. 3104–3112 (2014)
2. Cho, K., Van Merrienboer, B., Gulcehre, C., et al.: Learning phrase representations using rnn encoder-decoder for statistical machine translation. Comput. Sci. (2014). arXiv preprint arXiv:1406.1078
3. Kim, Y.: Convolutional neural networks for sentence classification. arXiv preprint arXiv: 1408.5882 (2014)
4. Kalchbrenner, N., Grefenstette, E., Blunsom, P.: A convolutional neural network for modelling sentences. arXiv preprint arXiv:1404.2188 (2014)

5. Sutskever, I., Martens, J., Hinton, E.G.: Generating text with recurrent neural networks. In: Proceedings of International Conference on Machine Learning. Bellevue, Washington: DBLP, pp. 1017–1024 (2016)
6. Bawden, R., Sennrich, R., Birch, A., et al.: Evaluating discourse phenomena in neural machine translation. In: Conference of the North American Chapter of the Association for Computational Linguistics: Human Language Technologies, pp. 1304–1313 (2018)
7. Zhang, J., Luan, H., Sun, M., et al.: Improving the transformer translation model with document-level context, pp. 533–542 (2018)
8. Sabour, S., Frosst, N., Hinton, G.E.: Dynamic routing between capsules, pp. 3856–3866 (2017)
9. Kuang, S., Xiong, D., Luo, W., et al.: Modeling coherence for neural machine translation with dynamic and topic caches, pp. 596–606 (2017)
10. Bahdanau, D. Kyunghyun, C., Bengio, Y.: Neural machine translation by jointly learning to align and translate. arXiv preprint arXiv:1409.0473 (2014)
11. Zhou, C., Bai, J., Song, J., et al.: An attention-based user behavior modeling framework for recommendation. arXiv preprint arXiv:1711.06632 (2017)
12. Shi, Y., Wang, Y., Wu, S.Q.: Machine translation system based on self-attention model. Comput. Mod. (07), 9–14 (2019)
13. Mao, X. Yan, W., Ma, W.J. Yin, H.M.: Machine translation of English place-names for attention mechanisms. Sci. Surv. Mapp. **44**(06), 296–300+316 (2019)
14. Cai, L., Hu, Z.W.: Construction scheme of network language evolution mechanism based on semantic knowledge base. Audio-Vis. Teach. Foreign Lang. (03), 8 (2018)
15. Lan, W.F., Xu, W., Wang, D.Z., Pan, P.C.: Chinese news text classification based on LSTM-ATTENTION. J. Cent. South Univ. Natl. (Nat. Sci. Edn.) **37**(03), 129–133 (2016)
16. Cui, Y., Liu, T., Chen, Z., et al.: Consensus attention-based neural networks for chinese reading comprehension. arXiv preprint arXiv:1607.02250 (2016)
17. Li, Q., Zhang, C., Woodland, P.C.: Integrating source-channel and attention-based sequence-to-sequence models for speech recognition. arXiv preprint arXiv:1909.06614 (2019)
18. Qiu, Y., Ma, Y., Jin, Y., et al.: Chinese dialects identification using attention-based deep neural networks (2017)
19. Wang, S., Jiang, J.: Machine comprehension using match-LSTM and answer pointer. arXiv preprint arXiv:1608.07905 (2016)

Influencing Factors of Students' Online Learning Satisfaction During the COVID-19 Outbreak: An Empirical Study Based on Random Forest Algorithm

Yuzhuo Zhang[1], Po Zhang[2], Hongji Yang[1], Kangrui Zhao[1], and Chunmei Han[1](✉)

[1] People's Public Security University of China, Beijing, China
hcm426@sina.com

[2] Key Laboratory of Dairy Science, Ministry of Education, Northeast Agricultural University, Harbin, China

Abstract. New coronavirus is wreaking havoc around the world and has a profound impact on the international community, especially on higher education. Online teaching provides an effective path for higher education to avoid the risk of cross-contagion in traditional classroom education under epidemic condition. In order to ensure the quality of online teaching during the epidemic, this study takes the students' satisfaction of online education learning as a measurement object. 1120 online learners from 126 colleges and universities in 26 provinces were investigated through 50 questions survey in 10 dimensions. First, the chi-square test is used to pre-process all the characteristics of the factors, and 30 influencing factors with the highest feature correlation are selected. Random forest algorithm is used to establish a satisfaction classification model in the training set. The accuracy in the test set is 0.72. Through the ranking of feature contribution, the influential factors with higher weight are obtained. The results show that in online learning, the attractiveness of teachers' teaching methods is the most influential factor, while curriculum arrangement and learning environment rank second and third. Finally, according to the research results, this paper puts forward some suggestions and countermeasures.

Keywords: Online learning · Online teaching · Learning satisfaction · Machine learning · Random forest algorithm · COVID-19

1 Introduction

As the Coronavirus continues to spread around the world, many countries started initiating relevant strategies, such as the closure of schools. How can the education system adapt and subversive the impact of the epidemic? When all the scenes of education seem to be moved online, how can instructors build a more perfect and flexible system to guide the future, so as to reshape the educational technology landscape? How will they analyze the data and indicators to measure the different educational responses to crisis management?

C. Pang et al. (Eds.): SETE 2020/ICWL 2020, LNCS 12511, pp. 103–114, 2021.
https://doi.org/10.1007/978-3-030-66906-5_10

It is not the time to stop education process because of this unexpected school closure as it is only the education that can produce man power and professionals to match the demand of COVID-19 pandemic time. Rather, it is the time to embrace different and flexible methods of teaching, and to comply with social distancing and self-isolation by extension learning. Online teaching has emerged as the current mainstream teaching method and a solution to minimizing disruption to education until life returns to normal [1]. The Chinese Ministry of Education has launched an initiative entitled "Disrupted Classes, Undisrupted Learning" to provide flexible online learning to hundreds of millions of students from their homes [2]. It differs from the traditional forms of teaching as it can break the limitation of time or space [3, 4]. In the case of passive large-scale adoption of online teaching, ensuring the quality of students' learning is an issue that higher education must pay attention to.

As the object of online teaching, students' satisfaction is the key factor to evaluating the success or failure of teaching. The transformation to the conventional classroom teaching justifies the essential of introducing scientific instruments in order to analyze the level of learner satisfaction in online teaching environments [5]. A generally accepted connotation of LS refers to the "perception of approval and accomplishment that learners develop in learning environments" [6]. Learners' online learning satisfaction is affected by multiple factors. Researchers summarized the influencing factors of learning satisfaction into six aspects: learner, teacher, curriculum, technology, design and environment [7]. Bolliger emphasized that learning satisfaction should be understood from five dimensions of communication: technology, curriculum management, curriculum website, curriculum interactivity and general information [8]. Analyzing the influence degree of various factors on learner satisfaction is conducive to the further development of online curricula.

At present, the methods of studying learners' online satisfaction are mainly through questionnaire surveys [9–11] interviews, [12] big data, [13, 14] etc. Through Regression Analysis, [9, 15] Decision Tree, [16] Grey Relation Analysis, [17] Confirmatory Factor Analysis [18] and other algorithms for scientific calculation, and finally show the importance of influencing factors through numerical values.

In the current studies, several scholars found that the core element that affects students' satisfaction is the interaction between teachers and students [10, 19, 20]. And some researchers believe that the quality and ability of teachers play a decisive role in the learning effect of students [21–23]. They consider that although online learning relies on information technology, learners' online learning satisfaction does not entirely depend on the perception of technology, while teachers are the most essential factor. Besides, Webster and Hackley (1997) stated that teachers' mastery of technology, style of teaching, and degree of technical approval affect the learning outcomes [22]. Moreover, the stability of learning tools and the reliability of the Internet are of great significance to the effect of online learning [24, 25]. Online learners have different psychological expectations in different periods. Under different conditions, learners will have different needs. Therefore, satisfaction of online learners is a continuous work to meet the psychological needs of learners in various periods.

Although online learning can break the time and space constraints and provide a convenient interactive environment, it should also face the loss of the advantages of face-to-face interaction between teachers and students [26, 27]. In order to make up for the real predicament of online learning, it is necessary to have a deep understanding of the attitude of the target audience and carry out targeted design [10]. Hence, sorting many influencing factors through scientific measurement to find the most relevant elements will help to provide decision-making basis and curriculum improvement path for colleges and universities [11].

2 Method

2.1 Questionnaire Design and Data Collection

In order to ensure the scientificity of the research and cover all influencing factors as much as possible, the questionnaire was designed from ten dimensions (teacher, curriculum schedule, curriculum content, homework, online platform, interaction, learning environment, assessment score, class size, online classroom environment). There are five related questions in each dimension. The above ten dimensions (50 questions) basically cover all links of the current online curriculum, which affect the online learners' perception of satisfaction.

This data collection uses the form of online questionnaire. Internet questionnaires can efficiently collect and organize data. Through the use of self-media for forwarding and appealing by teachers and students of various colleges and universities, it is possible to break the geographical restrictions and collect comprehensive data from as many schools as possible. A total of 1120 valid questionnaires were received from three education levels: undergraduate, master and doctor.

2.2 Participants

The survey instrument was sent out to a total of 1120 individuals including 612 males and 508 females. The sample was constitutive of 951 undergraduates, 163 postgraduates and 6 PhD students from 125 universities, covering 26 provincial-level administrative regions in China. From the perspective of data distribution, the data collected is highly representative. And all respondents are active in online learning during the outbreak of COVID-19 and voluntary to do this survey. The online survey was conducted for one week. All responses were valid with no incomplete response or missing value. Therefore, the learners participating in this survey can reflect the overall situation of the national online curricula.

2.3 Analyzing Instrument

This study uses Random Forest model for analysis. In the research of online learner satisfaction measurements, logistic regression and decision tree are commonly used algorithm, but the logistic regression method is not effective enough for calculating nonlinear data and highly related features. The Random Forest model in machine learning can be used to solve the problems of highly relevant features, and it also can exert good performance on unbalanced data sets. With this algorithm, the calculation result has less errors, which helps to accurately analyze the importance of features. The index framework in this study involves a large number of features and a high correlation between features. In addition, the data classification is not balanced enough. Hence, the use of Random Forest model can obtain higher robustness, which is more suitable for this study.

2.4 Feature Selection and Principal Component Analysis

Because of all features with classes that have natural ordering, we need to transform the ordinal classes into numerical values which maintain the notion of ordering for use in machine learning.

A chi-square ($\chi 2$) statistic was calculated between each feature and the target vector (students' online classroom satisfaction) for removing irrelevant features with classification. We keep 30 features by SelectKBest in Scikit-Learn.

$$\chi^2 = \sum\nolimits_{i=1}^{n} \frac{(Oi - Ei)^2}{E_i} \quad (1)$$

In formula 1, n is the number of target vector class; Oi is the number of observations in class i; Ei is the number of observations in class i. We expect if there is no relationship between the feature and target vector.

Using PCA, we can find the first two principal components, and visualize the data in this new two-dimensional space with a single scatter plot colored by class.

2.5 Model Evaluation

The "hold-out" method divides a data set into two mutually exclusive collections. In order to maintain the category proportion of satisfaction, stratified sampling was adopted to select 75% of individuals as the training set and 25% as the test set. The training set is used to select parameters and build the model, and the test set is used to test the generalization ability of the model.

$$accuracy = \frac{TP + TN}{TP + FP + FN + TN} \quad (2)$$

$$precision = \frac{TP}{TP + FP} \quad (3)$$

$$recall = \frac{TP}{TP + FN} \quad (4)$$

$$f_1 = 2\frac{precision.recall}{precision + recall} \quad (5)$$

Evaluate the model through these four indicators. In the formula (2–5), FP is false positive prediction. FN is false negative prediction. TP is true positive prediction. TN is true negative prediction.

2.6 Modeling

Random Forest is essentially a collection of decision trees, where each tree is slightly different from the others to reduce the amount of overfitting by averaging their results. The Random Forest model was implemented in the Scikit-Learn. There are two important parameters. The n_estimators parameter decides on the number of trees to build. The max_features parameter controls the number of features that are randomly selected in each decision tree node. The algorithm searches the best possible test involving one of these features.

The Scikit-Learn provides the Grid Search CV class. We try the values 100, 200, 300, 400, 500 and 600 for the parameter n_estimators, and 5, 15, 20, 25 and 30 for max_features to conduct a grid search. The mean cross-validation accuracy performed on the training set was used to select parameter setting.

Splitting features with a larger average reduction in Gini impurity (formula 6) are more important features. The random forest provides feature contribution, which are computed by aggregating the feature importances over the trees in the forest. Typically, the feature importance provided by the random forest are more reliable than the ones provided by a single tree.

$$Gini(t) = 1 - \sum\nolimits_{k=1}^{n} p_k^2 \tag{6}$$

N is the number of target vector class. P_k is proportion of class k.

3 Results

3.1 Principle Component Analysis

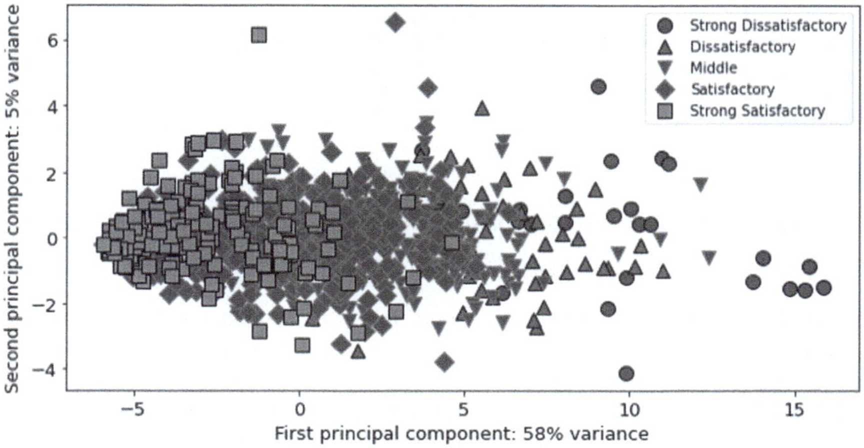

Fig. 1. Two-dimensional scatter plot of the online classroom satisfaction dataset using the first two principal components.

For the Fig. 1 shown here, we plotted the two-dimensional scatter plot, and then used the class information to color the points. The first 2 components contain approximately 63% of the variance. As first principal component increases, classroom satisfaction changes from "strong satisfaction" to "strong dissatisfaction" (Fig. 2).

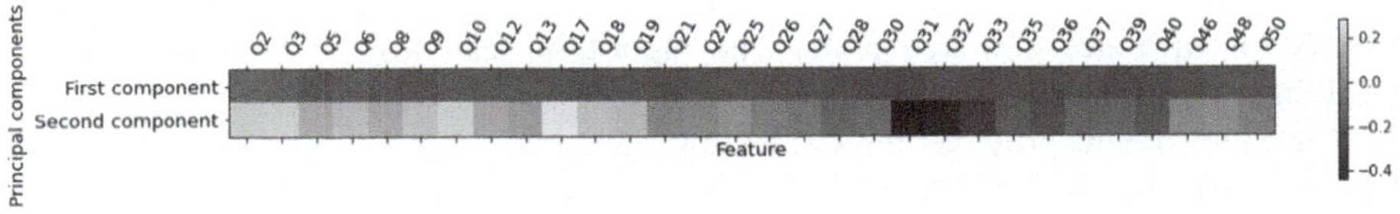

Fig. 2. Heat map of the first two principal components.

In the first component, all features have the same negative sign because of a general correlation among all features, but it doesn't influence which direction the arrow points to. The deeper the color is, the greater contribution to the principal component. Q5, Q8, Q9, Q21 and Q30 contribute a lot to first principal component (Fig. 3).

3.2 Random Forest

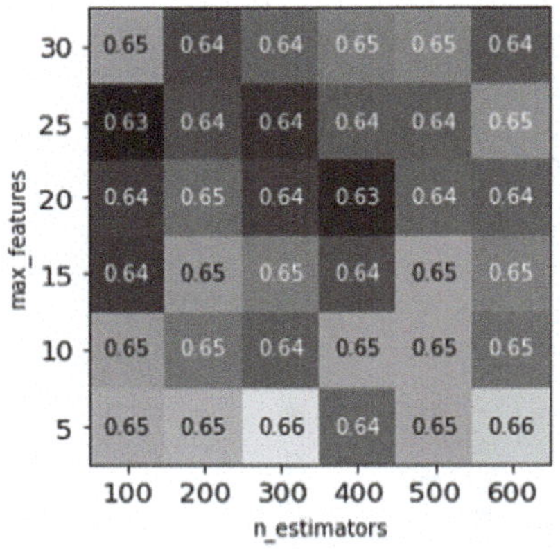

Fig. 3. Heat map of mean cross-validation score as a function of max_features and n_estimators.

Each point in the heat map corresponds to one run of cross-validation, with a particular parameter setting. The color encodes the cross-validation accuracy, with light colors meaning high accuracy and dark colors meaning low accuracy. The panel shows changes in both max_features and n_estimators. The optimum parameter setting is at n_estimators and max_features equal to 300 and 5 respectively. At this parameter setting, the accuracy of cross-validation is 0.66 (Fig. 4).

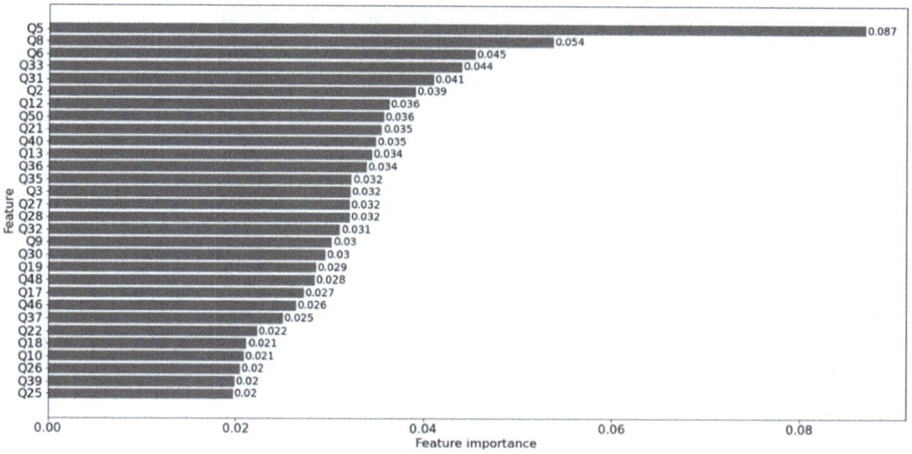

Fig. 4. The importance of features evaluated by Random Forest model.

There is a number between 0 and 1 for each feature, and 0 means "not used at all" and 1 means "perfectly predicts the target". The importance degree of the features always sum to 1. Here we see that the feature Q5 is the most important feature (Fig. 5).

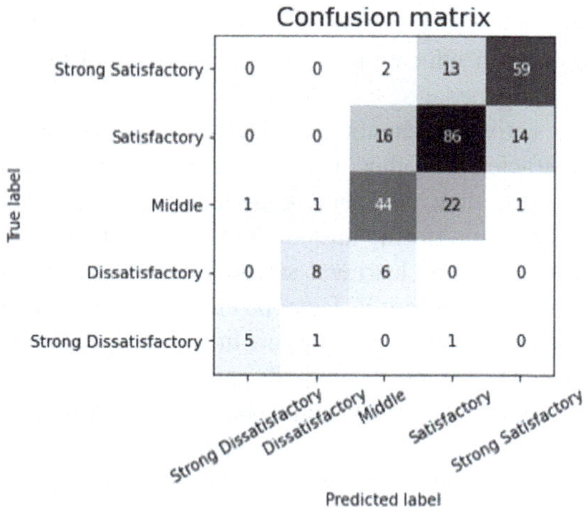

Fig. 5. Confusion matrix for classroom satisfaction classification with random forests.

The test set contains a total of 280 samples. The accuracy of the model in the test set is 0.72. The main reason for misjudgment is that many "Satisfactory" classes were misjudged into "Middle" and "Strong Satisfactory" classes. The number of samples in the "dissatisfied" and "strongly dissatisfied" are small, resulting in poor generalization ability (Table 1).

Table 1. The main evaluation index of the model.

Index	Precision	Recall	f1	Support
Strong dissatisfactory	0.83	0.71	0.77	7
Dissatisfactory	0.80	0.57	0.67	14
Middle	0.65	0.64	0.64	69
Satisfactory	0.70	0.74	0.72	116
Strong satisfactory	0.80	0.80	0.80	74

The support of each class means the number of samples in test set in this class. "Satisfactory" and "Strong Satisfactory" are well predicted. Three indicators of "Strong Satisfactory" are 0.80.

4 Discussion

In this study, a total of 50 feature variables were set, and 20 feature variables with the lowest correlation were eliminated by chi-square test. Among the remaining 30 features, five factors that have the greatest impact on online learner satisfaction (feature importance above 0.040, average value 0.033) are analyzed specifically. These 5 factors are distributed in 3 dimensions (Teacher, Curriculum, Learning environment).

Among all the influencing factors, teaching methods have the highest importance (0.087), and whether they can attract students becomes the most critical factor. The integration of technology in education has not only changed how students learn but also changed the teaching pedagogy by promoting collaborative activities [28]. Students' home learning activities cannot be effectively known by teachers, and their attention is easily distracted by other activities, which lead to a decline in learning effectiveness. At the same time, teachers' knowledge reserve, teaching design level, teaching activity organization form, attitude towards students and interaction with students will greatly affect online learners' satisfaction with online open courses. That is why Yunus is assertive that before ICT can be effectively integrated, lecturers should be provided with adequate training and support in ICT and pedagogy [29]. Therefore, the attractiveness of teachers' teaching methods can attract the attention of learners, and narrow the psychological distance between teachers and students to a greater extent, so as to supplement the spatial distance. Moreover, staff motivation needs to be considered as an important factor for the successful assimilation of technology. Yuen and Ma strongly recommend the need to empower teaching staff and build their confidence so that they are able to implement ICT integrated teaching [30].

Curriculum quality is an important factor that needs to be considered in assessing online learning effectively. Effectiveness of online learning depends on curriculum quality. Just in the e-learning system field, information quality is considered to be a key issue. Roca, Chiu, and Martínez found that information quality had significant effects on user satisfaction, which directly affected the user's continuing intention to use e-learning systems [31]. The curriculum is the core component in the teaching activities, and all the components are connected, such as organizing teaching in the curriculum, reviewing, learning new textbooks, consolidating, and arranging extracurricular homework. Multimedia contents are digital instructional materials that combine text, graphics, audio and animations. Teachers tend to liven up classroom lessons by using these contents to better demonstrate and explain difficult concepts that cannot be easily explained by using text alone [32]. The Study indicated that the appropriate use of multimedia-enhanced content in educational context provides several benefits. Reasonable curriculum can scientifically construct the process for learners to receive curriculum content, which is convenient for learners to master the learning content systematically. F. Mohammadi, Abrizah, and Nazari (2015) explored teachers' perceptions of information quality in Farsi Web-based Learning Resources and extended the list of characteristics to 14 indicators that emerged from focus groups and interviews: student engagement; content accessibility; multimedia interactivity; collaborative resource development; reusability; factual accuracy; stylistic accuracy; authority; currency; target audience; adequacy; selective exposure; active links; and website accessibility [33].

The third factor is the rationality of course schedule (importance of features 0.045). Online course schedule is more flexible than the traditional. The more reasonable the curriculum distribution is, the more learners there will be who can receive information in a learning activity that combines leniency and strictness, which can also fully mobilize the enthusiasm and focus of the learners. In practice, there is an unstable schedule of curricula. The amount of courses for learners varies greatly every day, which seriously affects the enthusiasm of learners in learning. On the other hand, it is common to adjust the course charts according to the teachers' personal wishes.

The fourth factor (0.044) is similar to the fifth factor (0.041), which is the effect of learning environment on learner satisfaction. The learning environment in this study mainly refers to the external environment where learners study. During the epidemic, the number of learner family members is different from usual, and most family members take the initiative to isolate themselves at home. Due to the increase in family population density, family members are fulfilling self-satisfaction in enclosed spaces. TV programs and self-entertainment activities are flooded in the family environment, making online learners' learning environment more complicated. The family environment and the learning environment are intertwined, so that the concentration and satisfaction of online learners will also be negatively affected. Learning environment is not conducive because all family members stay at home and there is limited space to do revision and perform course assessment task [34].

In addition, the acquisition of curriculum content and materials, the function of the learning platform, methods and preparation time of assessment, and online classroom order are also important factors that affect learner satisfaction.

However, the study found that all factors under the three dimensions of homework, classroom interaction, and teaching scale have little effect on learner satisfaction.

5 Conclusion

Across the globe, the spread of COVID-19 has led to profound changes in education. In China, moving to online learning at scale raises profound equity concerns. However, transitioning to online learning at scale is a very difficult and highly complex undertaking for education systems, even in the best of circumstances.

Many existing researches show that students' satisfaction with online learning affects the effect of online learning [35]. In order to provide guidance and reference for the expanding application of online learning, this study uses Random Forest algorithm to make an in-depth empirical analysis of the influencing factors of students' online learning satisfaction. Based on the above study results, it can be concluded that online learning is not only helpful in the midst of COVID-19 pandemic but also spots the light on the teaching method, curriculum content, course schedule, and learning environment. In terms of satisfaction issue, the learners hope that lecturers make use of facilities and multiple and flexible teaching methods as the guarantee. Organizing digital educational content to align with existing curricula can be critical in providing learners and teachers with a way to ensure the learning opportunities. For the implementation, they hope that material and assignment should be preceded by explanation [35]. It implied that the material and instruction implemented by the lecturer in the online learning were not easy to use.

However, because most of the curriculum tasks of Chinese college students are formulated by the University in a unified way, and they have less opportunities for choice and decision of the curriculum, this study focuses on the external factors such as teachers, schools and families when designing questionnaires. And we do not consider the students' own factors for learning satisfaction and the influence of teaching effect, which is a limitation of this study.

In conclusion, with the coming of the post epidemic era, online teaching will gradually become an important development trend in the reform of education methods. How to enhance and improve both home school cooperation and the effect of online learning of students is an important issue that educators should consider and study further.

References

1. Al-Fraihat, D., Joy, M., Masa'Deh, R., Sinclair, J.: Evaluating E-learning systems success: an empirical study. Comput. Hum. Behav. **102** (2019). https://doi.org/10.1016/j.chb.2019.08.004
2. Huang, R., et al.: Handbook on facilitating flexible learning during educational disruption: the chinese experience in maintaining undisrupted learning in COVID-19 outbreak (2020)
3. Yu, C., Chang, H., Chen, K.: Developing a performance evaluation matrix to enhance the learner satisfaction of an e-learning system. Total Qual. Manag. Bus. 1–19 (2016). https://doi.org/10.1080/14783363.2016.1233809
4. Al-Samarraie, H., Selim, H., Teo, T., Zaqout, F.: Isolation and distinctiveness in the design of e-learning systems influence user preferences. Interact. Learn. Envir. **25**(1–4), 452–466 (2017). https://doi.org/10.1080/10494820.2016.1138313

5. Gonzalez-Gomez, F., Guardiola, J., Rodriguez, O.M., Alonso, M.A.M.: Gender differences in e-learning satisfaction. Comput. Educ. **58**(1), 283–290 (2012). https://doi.org/10.1016/j.compedu.2011.08.017
6. Sweeney, J.C., Ingram, D.: A comparison of traditional and web-based tutorials in marketing education: an exploratory study. J. Mark. Educ. **23**(1), 55–62 (2001). https://doi.org/10.1177/0273475301231007
7. Sun, P.C., Tsai, R.J., Finger, G., Chen, Y.Y., Yeh, D.: What drives a successful e-learning? An empirical investigation of the critical factors influencing learner satisfaction. Comput. Educ. **50**(4), 1202 (2008). https://doi.org/10.1016/j.compedu.2006.11.007
8. Bolliger, D., Martindale, T.: Key factors for determining student satisfaction in online courses. Int. J. E-Learn. **3**, 61–67 (2004)
9. Udo, G.J., Bagchi, K.K., Kirs, P.J.: Using SERVQUAL to assess the quality of e-learning experience. Comput. Hum. Behav. **27**(3), 1272–1283 (2011). https://doi.org/10.1016/j.chb.2011.01.009
10. Liaw, S.S.: Investigating students' perceived satisfaction, behavioral intention, and effectiveness of e-learning: a case study of the blackboard system. Comput. Educ. **51**(2), 864–873 (2008). https://doi.org/10.1016/j.compedu.2007.09.005
11. Wu, J.H., Tennyson, R.D., Hsia, T.L.: A study of student satisfaction in a blended e-learning system environment. Comput. Educ. **55**(1), 155–164 (2010). https://doi.org/10.1016/j.compedu.2009.12.012
12. Hong, K.-S.: Relationships between students' and instructional variables with satisfaction and learning from a Web-based course. Internet High. Educ. (2002). http://dx.doi.org/10.1016/S1096-7516(02)00105-7
13. Harrati, N., Bouchrika, I., Tari, A., Ladjailia, A.: Exploring user satisfaction for e-learning systems via usage-based metrics and system usability scale analysis. Comput. Hum. Behav. **61**, 463–471 (2016). https://doi.org/10.1016/j.chb.2016.03.051
14. Elia, G., Solazzo, G., Lorenzo, G., Passiante, G.: Assessing learners' satisfaction in collaborative online courses through a big data approach. Comput. Hum. Behav. **92**(MAR), 589–599 (2019). https://doi.org/10.1016/j.chb.2018.04.033
15. Wang, C., et al.: Need satisfaction and need dissatisfaction: a comparative study of online and face-to-face learning contexts. Comput. Hum. Behav. **95**, 114–125 (2019). https://doi.org/10.1016/j.chb.2019.01.034
16. Thomas, E.H.G.N.: What satisfies students?: Mining student-opinion data with regression and decision tree analysis. Res. High. Educ. **45** (2004). https://doi.org/10.1023/b:rihe.0000019589.79439.6e
17. Chen, C.M., Chen, M.C.: Mobile formative assessment tool based on data mining techniques for supporting web-based learning. Comput. Educ. **52**(1), 256–273 (2009). https://doi.org/10.1016/j.compedu.2008.08.005
18. Jöreskog, K., Sorbom, D.: LISREL 8.8 for Windows [Computer software] (2006)
19. Swan, K., Pickett, A., Pelz, W., Maher, G., Shea, P., Fredericksen, E.: Building knowledge building communities: consistency, contact and communication in the virtual classroom. J. Educ. Comput. Res. **23**(4), 359–383 (2000). https://doi.org/10.2190/XKLM-3A96-2LAV-CB3L
20. Swan, K.: Virtual interaction: design factors affecting student satisfaction and perceived learning in asynchronous online courses. Dist. Educ. **22**(2), 306–331 (2001)
21. Hiltz, S.R.: The Virtual Classroom: Learning Without Limits via Computer Networks. Ablex Publishing Corp. (1994). https://doi.org/10.1016/S0040-1625(96)90017-7
22. Hackley, J.W.A.P.: Teaching effectiveness in technology-mediated distance learning. Acad. Manage. J. **40**(6), 1282–1309 (1997). https://doi.org/10.2307/257034

23. Selim, H.M.: Critical success factors for e-learning acceptance: confirmatory factor models. Comput. Educ. **49**(2), 396–413 (2007). https://doi.org/10.1016/j.compedu.2005.09.004
24. DE Leidner, Jarvenpaa, S.L.: The information age confronts education: case studies on electronic classrooms. Inf. Syst. Res. **1**(4), 24–54 (1993)
25. Jarvenpaa, D.E.L.A.: Special issue on IS curricula and pedagogy || The use of information technology to enhance management school education: a theoretical view. MIS Quart. **19**(3), 265–291 (1995). https://doi.org/10.2307/249596
26. Kinshuk, D., Yang, A.: Web-based Asynchronous Synchronous Environment for Online Learning. United States Dist. Educ. Assoc. J. **17**, 5–17 (2003)
27. Wu, J.H., Tennyson, R.D., Hsia, T.L., Liao, Y.W.: Analysis of E-learning innovation and core capability using a hypercube model. Comput. Hum. Behav. **24**(5), 1851–1866 (2008). https://doi.org/10.1016/j.chb.2008.02.008
28. Haddad, W.D.: Is instructional technology a must for learning? Techknowlogia (1) (2003)
29. Yunus, M.M.: Malaysian ESL Teachers' Use of ICT in Their Classrooms: Expectations and Realities. Cambridge University Press (2007). https://doi.org/10.1017/S0958344007000614
30. Yuen, A.H.K., Ma, W.W.K.: Gender differences in teacher computer acceptance. J. Technol. Teach. Educ. **10**(3), 365–382 (2002)
31. Martínez, J.C.R.A.: Understanding e-learning continuance intention: an extension of the technology acceptance model. Int. J. Hum.-Comput. ST (2006). https://doi.org/10.1016/j.ijhcs.2006.01.003
32. Olabode, T.O., Omoniyi, I.O.: Effectiveness of animation and multimedia teaching on students' performance in science subjects. Br. J. Educ. Soc. Behav. Sci. **2**(4), 201–210 (2014). https://doi.org/10.9734/BJESBS/2014/3340
33. Mohammadi, F., Abrizah, A., Nazari: is the information fit for use? Exploring teachers perceived information quality indicators for Farsi web-based learning resources. Malays. J. Libr. Inf. SC **20** (2014)
34. Ali, W.: Online and Remote learning in higher education institutes: a necessity in light of COVID-19 Pandemic. High. Educ. Stud. **10**, 16 (2020). https://doi.org/10.5539/hes.v10n3p16
35. Girik Allo, M.: Is the online learning good in the midst of Covid-19 Pandemic? Case EFL Learn. **10**, 1–10 (2020)

Granular Computing in Education

Attribute Reduction and Rule Acquisition of Formal Decision Context Based on Dual Concept Lattice

Qian Hu[1(✉)] and Keyun Qin[2]

[1] The School of Information Science and Technology, Southwest Jiaotong University, Chengdu 611756, China
15227193972@163.com

[2] School of Mathematics, Southwest Jiaotong University, Chengdu 611756, China

Abstract. Concept lattice theory is a powerful tool for analyzing and extracting information from data sets. Rule acquisition and attribute reduction are hot research topics in formal concept analysis. This paper mainly proposes three kinds of rules based on formal concepts and dual concepts. In addition, the methods of rule acquisition for different kinds of rules are presented. Finally, the attribute reduction approaches to preserve different kinds of rules are given by using discernibility matrix.

Keywords: Formal concept analysis · Object-oriented and property-oriented concept lattice · Dual concept lattice · Rule acquisition · Attribute reduction

1 Introduction

Formal Concept Analysis (FCA) was proposed by Wille in 1982 [22]. Formal concept analysis is discussed based on formal context, which can be expressed as Boolean two-dimensional data table, where rows represent objects and columns represent attributes. The relationship between an object with an attribute has two situations: "yes" (represented by "1") and "no" (represented by "0"). Concept lattice theory is an effective data analysis tool. At present, concept lattice has been widely used in knowledge discovery, software engineering, digital library, document retrieval and social network, etc. [4–7, 12, 13, 18, 25].

Inspired by the formal concept lattice, Duntsch and Gediga used a pair of approximate modal operators to define the property-oriented concept [2]. It provided a supplementary data view of the derivation operator for the classical form concept analysis. Yao presented the definition of object-oriented concept and compared different kinds of concept lattices in data analysis [24]. Wan proposed some approximate concept acquisition methods based on attribute set and object set [19]. Yao defined the dual concept lattice based on the dual form of

Supported by the National Natural Science Foundation of China (Grant No. 61976130).

C. Pang et al. (Eds.): SETE 2020/ICWL 2020, LNCS 12511, pp. 117–129, 2021.
https://doi.org/10.1007/978-3-030-66906-5_11

the formal concept, and studied their properties and the relationship between them [1].

As one of the most important research directions in FCA, rule acquisition and attribute reduction for formal context have been widely studied by lots of scholars [8,9,14–17,20,21,23,26]. Attribute reduction can not only obtain more concise knowledge, but also reveal the dependency between attributes. Ganter et al. put forward the notions of consistent sub-context, reducible attributes and reducible objects by the perspective of reducing rows and columns, and systematically analyzed the characteristics of related attributes [3]. Zhang et al. gave the judgment theorem of consistent set, put forward the notions of discernibility attribute set and discernibility matrix of formal context, and gave the calculation method of reduction by using Boolean logic formula transformation [27]. Li et al. put forward the notions of the implication relationship between decision rules, redundant decision rules, non-redundant decision rules and systematically studied the reduction problem based on decision rules for formal decision context [9–11]. Based on the property-oriented and object-oriented concept lattice, Qin et al. discussed the rule acquisition of non-redundant decision rules and the attribute reduction for formal decision context [14]. Through the previous research, we note that under the formal decision context, we can get different forms of rules when the rule premise and conclusion are different forms of concepts.

In this paper, we mainly study the rules in which premise and conclusion are the dual (formal) concept and the formal (dual) concept respectively. Then semantic interpretation of each rule are given. Finally we propose methods of rule acquisition and attribute reduction in formal decision context based on these rules.

2 Preliminaries

In this section, the involved notions of FCA are introduced. Please refer to [22] and [24] for details.

Definition 1 [22]. *A formal context is a triple $\mathcal{K} = (G, M, I)$, where G is a finite nonempty set of objects, M represents a finite nonempty set of attributes and $I \subseteq G \times M$ is called an incidence relation, $(x, m) \in I$ indicates that the object x has the attribute m.*

For any $X \subseteq G$ and $B \subseteq M$, we define: $X^* = \{m \in M | \forall x \in X((x, m) \in I)\}$, $B^* = \{x \in G | \forall m \in B((x, m) \in I)\}$. That is, X^* is the set of attributes shared by all the objects in X, B^* is the set of objects that has all attributes in B. A pair (X, B) is called a formal concept (or a concept for short), if $X^* = B$ and $B^* = X$, then X and B are regarded as the extent and the intent of the formal concept (X, B) respectively. All the concepts of (G, M, I) form a complete lattice, which is called the concept lattice of (G, M, I) and denoted by $L(G, M, I)$ (or $L(\mathcal{K})$ for simple), which possesses a partial order relation $(X_1, B_1) \leq (X_2, B_2) \Leftrightarrow X_1 \subseteq$

$X_2 (\Leftrightarrow B_2 \subseteq B_1)$. In addition, if $(X_1, B_1) < (X_2, B_2)$ and $\nexists (X_3, B_3) \in L(\mathcal{K})$ such that $(X_1, B_1) < (X_3, B_3) < (X_2, B_2)$, then we call $(X_1, B_1) \prec (X_2, B_2)$.

For any $x \in G$ and $m \in M$, $\{x\}^*$ and $\{m\}^*$ can be replaced by x^* and m^* respectively for simple. We denote by $ExtL(G, M, I)$ the set of all extents of formal concepts in $L(G, M, I)$.

Yao proposed dual concept lattice based on duality [24].

Definition 2 [24]. *Let $\mathcal{K} = (G, M, I)$ be a formal context. For any $X \subseteq G$ and $B \subseteq M$, we define:*

$$X^\sharp = \{m \in M | \exists x \in X^c ((x, m) \notin I)\},$$

$$B^\sharp = \{x \in G | \exists m \in B^c ((x, m) \notin I)\}.$$

A pair (X, B) is called a dual formal concept, if $X^\sharp = B$ and $B^\sharp = X$, then X and B are regarded as the extent and the intent of the dual formal concept (X, B) respectively. All the dual concepts of (G, M, I) form a complete lattice, which is called the dual concept lattice of (G, M, I) and denoted by $L_d(G, M, I)$(or $L_d(\mathcal{K})$ for simple), which possesses a partial order relation $(X_1, B_1) \leq (X_2, B_2) \Leftrightarrow X_1 \subseteq X_2 (\Leftrightarrow B_2 \subseteq B_1)$. We denote by $ExtL_d(G, M, I)$ the set of all extents of dual formal concepts in $L_d(G, M, I)$.

Some properties of concept lattice and dual concept lattice can be found in [22] and [24], we will not go into details.

Li et al. [9] presented decision rule for formal decision context.

Definition 3. *A formal decision context is a quintuple $\mathcal{F} = (G, M, I, T, J)$, where (G, M, I) and (G, T, J) are two formal contexts, and M and T are regarded as the conditional attribute set and the decision attribute set respectively with $M \cap T = \emptyset$. For any $(X, B) \in L(G, M, I)$ and $(Y, C) \in L(G, T, J)$, if $X \subseteq Y$, and X, B, Y and C are nonempty, then we say that (Y, C) can be implied by (X, B) and $(X, B) \to (Y, C)$ is a decision rule.*

The semantic interpretation of a decision rule $(X, B) \to (Y, C)$ is that if a object x has all attributes in conditional attribute set B, then it must possesses all attributes in decision attribute set C.

3 Rule Acquisition and Attribute Reduction Based on DW-Decision Rules

Decision rules defined by Li [9] have been extensively studied. Clearly, it is established based on formal concept lattices. In order to enrich the rule acquisition theory in formal decision context, we consider dual formal concept for the premise or the conclusion of decision rule as follows.

3.1 Rule Acquisition Method for DW-Decision Rules

Definition 4. *Let $\mathcal{F} = (G, M, I, T, J)$ be a formal decision context. For any $(X, B) \in L_d(G, M, I)$ and $(Y, C) \in L(G, T, J)$, if $X \subseteq Y$, $X \neq \emptyset$ and $Y \neq G$, then $(X, B) \to (Y, C)$ is called a DW-decision rule. We denote by $\mathcal{R}_{DW}(M, T)$ the set of all DW-decision rules for formal decision context $\mathcal{F}$.*

The semantic interpretation of a DW-decision rule $(X, B) \to (Y, C)$ is that if a object x doesn't hold at least one attribute of conditional attribute set B^c then it must possess all attributes in decision attribute set C.

Definition 5. *For two DW-decision rules $(X_1, B_1) \to (Y_1, C_1), (X_2, B_2) \to (Y_2, C_2) \in \mathcal{R}_{DW}(M, T)$, if $X_2 \subseteq X_1 \subseteq Y_1 \subseteq Y_2$, then we call that $(X_2, B_2) \to (Y_2, C_2)$ can be impled by $(X_1, B_1) \to (Y_1, C_1)$, which is denoted by $(X_1, B_1) \to (Y_1, C_1) \Rightarrow (X_2, B_2) \to (Y_2, C_2)$. Furthermore for $(X, B) \to (Y, C) \in \mathcal{R}_{DW}(M, T)$, if there exists $(X', B') \to (Y', C') \in \mathcal{R}_{DW}(M, T) - \{(X, B) \to (Y, C)\}$ such that $(X', B') \to (Y', C') \Rightarrow (X, B) \to (Y, C)$, then we say that $(X, B) \to (Y, C)$ is redundant in $\mathcal{R}_{DW}(M, T)$. Otherwise, $(X, B) \to (Y, C)$ is said to be non-redundant in $\mathcal{R}_{DW}(M, T)$. All non-redundant DW-decision rules in $\mathcal{R}_{DW}(M, T)$ is denoted by $\overline{\mathcal{R}}_{DW}(M, T)$.*

We are more interested in non-redundant rules because redundant rules can be implied by non-redundant rules. In order to acquire non-redundant rules, we give an equivalence relation $R(M, T)$ on $ExtL_d(G, M, I)$ as follow: $R(M, T) = \{(X, Y) \in ExtL_d(G, M, I) \times ExtL_d(G, M, I) | X^{*_T} = Y^{*_T}\}$. We denote by $[X]_{R(M,T)} = \{Y \in Ext_d(G, M, I) | (X, Y) \in R(M, T)\}$ the equivalence class of X.

Lemma 1. *Let $\mathcal{F} = (G, M, I, T, J)$ be a formal decision context. There exists maximum element in $[X]_{R(M,T)}$.*

Proof. Suppose $X_1, X_2 \in [X]_{R(M,T)}$, then there exists B_1, B_2 such that $(X_1, B_1), (X_2, B_2) \in L_d(G, M, I)$. It follows that $(X_1, B_1) \vee (X_2, B_2) = (X_1 \cup X_2, (B_1 \cap B_2)^{\sharp_M \sharp_M}) \in L_d(G, M, I)$, then we have $X_1 \cup X_2 \in ExtL_d(G, M, I)$. Since $(X_1 \cup X_2)^{*_T} = X_1^{*_T} \cap X_2^{*_T} = X^{*_T}$, then we know $X_1 \cup X_2 \in [X]_{R(M,T)}$. Similarly we can conclude that $\bigcup [X]_{R(M,T)} \in [X]_{R(M,T)}$. Obviously, we have $\bigcup [X]_{R(M,T)}$ is the maximum element in $[X]_{R(M,T)}$.

Theorem 1. *Let $\mathcal{F} = (G, M, I, T, J)$ be a formal decision context, then $\overline{\mathcal{R}}_{DW}(M, T) = \{(Max[X]_{R(M,T)}, (Max[X]_{R(M,T)})^{\sharp_M}) \to (X^{*_T *_T}, X^{*_T}) | (X \in ExtL_d(G, M, I)) \bigwedge (X \neq \emptyset) \bigwedge (X^{*_T *_T} \neq G)\}$.*

Proof. Let $\mathcal{R}_{DW} = \{(Max[X]_{R(M,T)}, (Max[X]_{R(M,T)})^{\sharp_M}) \to (X^{*_T *_T}, X^{*_T}) | (X \in ExtL_d(G, M, I)) \bigwedge (X \neq \emptyset) \bigwedge (X^{*_T *_T} \neq G)\}$. From $Max[X]_{R(M,T)} \neq \emptyset$, $(X^{*_T *_T} \neq G)$ and $Max[X]_{R(M,T)} \subseteq (Max[X]_{R(M,T)})^{*_T *_T} = X^{*_T *_T}$, we have $(Max[X]_{R(M,T)}, (Max[X]_{R(M,T)})^{\sharp_M}) \to (X^{*_T *_T}, X^{*_T})$ is a DW-decision rule. Suppose that there exists $(X_1, B_1) \to (Y_1, C_1) \in \overline{\mathcal{R}}_{DW}(M, T)$ such that $(X_1, B_1) \to (Y_1, C_1) \Rightarrow (Max[X]_{R(M,T)}, (Max[X]_{R(M,T)})^{\sharp_M}) \to (X^{*_T *_T}, X^{*_T})$.

By $Max[X]_{R(M,T)} \subseteq X_1 \subseteq Y_1 \subseteq X^{*T*T}$, then we have $X^{*T*T} = (Max[X]_{R(M,T)})^{*T*T} \subseteq X_1^{*T*T} \subseteq Y_1^{*T*T} \subseteq X^{*T*T*T*T} = X^{*T*T}$. Consequently $(Y_1, C_1) = (X^{*T*T}, X^{*T})$ since $Y_1 = Y_1^{*T*T} = X^{*T*T}$. Hence $X_1^{*T} = X_1^{*T*T*T} = Y_1^{*T*T*T} = X^{*T}$, then we have $X_1 \in [X]_{R(M,T)}$. It follows that $X_1 = Max[X]_{R(M,T)}$ since $Max[X]_{R(M,T)} \subseteq X_1$. So we have $(X_1, B_1) = (Max[X]_{R(M,T)}, (Max[X]_{R(M,T)})^{\sharp_M})$, then we know $(Max[X]_{R(M,T)}, (Max[X]_{R(M,T)})^{\sharp_M}) \to (X^{*T*T}, X^{*T}) \in \overline{\mathcal{R}}_{DW}(M,T)$, it follows that $\mathcal{R}_{DW} \subseteq \overline{\mathcal{R}}_{DW}(M,T)$. For any $(X,B) \to (Y,C) \in \overline{\mathcal{R}}_{DW}(M,T)$, we have $X^{*T*T} \subseteq Y^{*T*T} = Y \subset G$ by $X \subseteq Y$. It follows that $(X,B) \to (X^{*T*T}, X^{*T}) \in \mathcal{R}_{DW}(M,T)$ and $(X,B) \to (X^{*T*T}, X^{*T}) \Rightarrow (X,B) \to (Y,C)$. We have $(X^{*T*T}, X^{*T}) = (Y,C)$ since that $(X,B) \to (Y,C)$ is non-redundant. If $X \neq Max[X]_{R(M,T)}$, then there exists $X_1 = Max[X]_{R(M,T)}$ such that $X \subset X_1$. It follows that $X_1 \subseteq X_1^{*T*T} = X^{*T*T}$, then $(X_1, X_1^{\sharp_M}) \to (X^{*T*T}, X^{*T}) \in \mathcal{R}_{DW}(M,T)$ and $(X_1, X_1^{\sharp_M}) \to (X^{*T*T}, X^{*T}) \Rightarrow (X,B) \to (X^{*T*T}, X^{*T})$. Which is contrary to the assumption that $(X,B) \to (Y,C) \in \overline{\mathcal{R}}_{DW}(M,T)$. Consequently we have $\overline{\mathcal{R}}_{DW}(M,T) \subseteq \mathcal{R}_{DW}$.

In summary we have $\overline{\mathcal{R}}_{DW}(M,T) = \{(Max[X]_{R(M,T)}, (Max[X]_{R(M,T)})^{\sharp_M}) \to (X^{*T*T}, X^{*T}) | (X \in ExtL_d(G,M,I)) \bigwedge (X \neq \emptyset) \bigwedge (X^{*T*T} \neq G)\}$.

Table 1. A formal decision context

	a_1	a_2	a_3	a_4	a_5	a_6	d_1	d_2	d_3
x_1	×						×		×
x_2		×		×			×	×	
x_3	×		×		×				×
x_4		×		×		×	×	×	
x_5	×	×	×					×	

Example 1. A formal decision context $\mathcal{F} = (G, M, I, T, J)$ is given by Table 1, where $G = \{x_1, x_2, x_3, x_4, x_5\}$ is the object set , $A = \{a_1, a_2, a_3, a_4, a_5, a_6\}$ is the set of conditional attributes, and $D = \{d_1, d_2, d_3\}$ is the set of decision attributes.

We can compute all the non-redundant DW-decision rules by *Theorem* 1:

$(r_1) : (x_1x_3, a_1a_3a_4a_5a_6) \to (x_1x_3, d_3)$

$(r_2) : (x_2x_4, a_2a_3a_4a_5a_6) \to (x_2x_4, d_1d_2)$

$(r_3) : (x_1x_2x_4, a_2a_4a_5a_6) \to (x_1x_2x_4, d_1)$.

3.2 Attribute Reduction Based on DW-Decision Rules

Next we consider attribute reduction of formal decision context based on DW-decision rules.

Definition 6. *Let $\mathcal{F} = (G, M, I, T, J)$ be a formal decision context, $E, F \subseteq M$. If for any $(X, B) \to (Y, C) \in \mathcal{R}_{DW}(E, T)$, there exists $(X', B') \to (Y', C') \in \mathcal{R}_{DW}(F, T)$ such that $(X', B') \to (Y', C') \Rightarrow (X, B) \to (Y, C)$, then we call $\mathcal{R}_{DW}(F, T) \Rightarrow \mathcal{R}_{DW}(E, T)$.*

Lemma 2. *Let $\mathcal{K} = (G, M, I)$ be a formal context, $E \subseteq M$, then we have $ExtL(G, E, I_E) \subseteq ExtL(G, M, I)$.*

Theorem 2. *Let $\mathcal{F} = (G, M, I, T, J)$ be a formal decision context, $E \subseteq M$, then we have $\mathcal{R}_{DW}(M, T) \Rightarrow \mathcal{R}_{DW}(E, T)$.*

Proof. The proof is straightforward and obvious.

We can define consistent set and attribute reduction for formal decision context based on DW-decision rules as follows.

Definition 7. *Let $\mathcal{F} = (G, M, I, T, J)$ be a formal decision context, $E \subseteq F$, E is called a DW-consistent set of $\mathcal{F}$ if $\mathcal{R}_{DW}(E, T) \Rightarrow \mathcal{R}_{DW}(M, T)$. Furthermore, if E is a DW-consistent set and any proper subset $H \subset E$ is not a DW-consistent set of $\mathcal{F}$, then we call E is a DW-reduction of $\mathcal{F}$.*

Let $\mathcal{U}_{DW}(M, T) = \{X \in ExtL_d(G, M, I) | \exists (Y, B) \in L(G, T, J)((X, X^{\sharp_M}) \to (Y, B)) \in \overline{\mathcal{R}}_{DW}(M, T))\}$. We can conclude that $\mathcal{U}_{DW}(M, T) = \{Max[X]_{R(M,T)} | (X \in ExtL_d(G, M, I)) \bigwedge (X \neq \emptyset) \bigwedge (X^{*_T *_T} \neq G)\}$ by *Theorem* 1.

Theorem 3. *Let $\mathcal{F} = (G, M, I, T, J)$ be a formal decision context. E is a DW-consistent set of $\mathcal{F}$ if and only if $\mathcal{U}_{DW}(M, T) \subseteq \mathcal{U}_{DW}(E, T)$.*

Proof. Suppose that E is a DW-consistent set. It follows that for any $Y \in \mathcal{U}_{DW}(M, T)$, there exists $X \in ExtL_d(G, M, I)$ such that $Y = Max[X]_{R(M,T)}$ and $(Y, Y^{\sharp_M}) \to (Y^{*_T *_T}, Y^{*_T}) \in \overline{\mathcal{R}}_{DW}(M, T)$. Since E is a DW-consistent set, we have there exists $Z \in ExtL_d(G, E, I_E)$ such that $Y \subseteq Z \subseteq Y^{*_T *_T}$, it follows that $Z^{*_T} = Y^{*_T} = X^{*_T}$. On account of $Y = Max[X]_{R(M,T)}$, $Z \in [X]_{R(M,T)}$ and $Y \subseteq Z$, then we have $Y = Z$ and hence $Y \in ExtL_d(G, E, I_E)$. Assume $Y \neq Max[Y]_{R(E,T)}$, then there exists $Y_1 = Max[Y]_{R(E,T)}$ such that $Y \subset Y_1$. We can conclude that $Y_1 \in ExtL_d(G, E, I_E) \subseteq ExtL_d(G, M, I)$ and $Y_1^{*_T} = Y^{*_T} = X^{*_T}$. It follows that $Y_1 \in [X]_{R(M,T)}$ and $Y \subset Y_1$, which is contrary to $Y = Max[X]_{R(M,T)}$. Then $Y = Max[Y]_{R(E,T)}$, hence we have $\mathcal{U}_{DW}(M, T) \subseteq \mathcal{U}_{DW}(E, T)$.

Conversely, assume that $\mathcal{U}_{DW}(M, T) \subseteq \mathcal{U}_{DW}(E, T)$. Then for any $(X, B) \to (Y, C) \in \mathcal{R}_{DW}(M, T)$ there exists $(X_1, B_1) \to (Y_1, C_1) \in \overline{\mathcal{R}}_{DW}(M, T)$ such that $(X_1, B_1) \to (Y_1, C_1) \Rightarrow (X, B) \to (Y, C)$. Since $\mathcal{U}_{DW}(M, T) \subseteq \mathcal{U}_{DW}(E, T)$ we have $X_1 \in \mathcal{U}_{DW}(M, T) \subseteq \mathcal{U}_{DW}(E, T)$ and $(X_1, X_1^{\sharp_E}) \to (Y_1, B_1) \in \overline{\mathcal{R}}_{DW}(E, T)$. It follows that $\mathcal{R}_{DW}(E, T) \Rightarrow \mathcal{R}_{DW}(M, T)$, that is to say E is a DW-consistent set of $\mathcal{F}$.

For any $(X_1, B_1), (X_2, B_2) \in L_d(G, M, I)$, let $\gamma_{DW}((X_1, B_1), (X_2, B_2))$ be the condition $X_1 \in \mathcal{U}_{DW}(M, T) \bigwedge (X_2, B_2) \prec (X_1, B_1)$ or $X_2 \in \mathcal{U}_{DW}(M, T) \bigwedge (X_1, B_1) \prec (X_2, B_2)$. We can define:

$$D_{DW}((X_1, B_1), (X_2, B_2)) = \begin{cases} B_1 \cup B_2 - B_1 \cap B_2, & if \ \gamma_{DW}((X_1, B_1), (X_2, B_2)) \\ \emptyset, & otherwise \end{cases}$$

Then the DW-discernibility attribute set of (X_1, B_1) and (X_2, B_2) is defined by $D_{DW}((X_1, B_1), (X_2, B_2))$. The DW-discernibility matrix of $\mathcal{F}$ is denoted by $\mathcal{D}_{DW} = D_{DW}((X_1, B_1), (X_2, B_2))$, where $(X_1, B_1), (X_2, B_2) \in L_d(G, M, I)$.

Theorem 4. *Let $\mathcal{F} = (G, M, I, T, J)$ be a formal decision context. E is a DW-consistent set of $\mathcal{F}$ if and only if for any $(X_1, B_1), (X_2, B_2) \in L_d(G, M, I)$, if $D_d((X_1, B_1), (X_2, B_2)) \neq \emptyset$, then $E \cap D_{DW}((X_1, B_1), (X_2, B_2)) \neq \emptyset$ must be hold.*

Proof. Assume that E is a DW-consistent set, $(X_1, B_1), (X_2, B_2) \in L_d(G, M, I)$ and $D_{DW}((X_1, B_1), (X_2, B_2)) \neq \emptyset$. Then we have $X_1 \in \mathcal{U}_{DW}(M, T) \bigwedge (X_2, B_2) \prec (X_1, B_1)$ or $X_2 \in \mathcal{U}_{DW}(M, T) \bigwedge (X_1, B_1) \prec (X_2, B_2)$. Without loss of generality, we suppose that $X_1 \in \mathcal{U}_{DW}(M, T) \bigwedge (X_2, B_2) \prec (X_1, B_1)$. From $X_1 \in \mathcal{U}_{DW}(M, T) \subseteq \mathcal{U}_{DW}(E, T) \subseteq ExtL_d(G, E, I_E)$, then we have $X_1 = X_1^{\sharp_E \sharp_E} = (E \cap X_1^{\sharp_M})^{\sharp_E} = (E \cap B_1)^{\sharp_E}$. Suppose $E \cap D_{DW}((X_1, B_1), (X_2, B_2)) = \emptyset$, it follows that $E \cap (B_1 \cup B_2 - B_1 \cap B_2) = E \cap (B_1 \cup B_2) \cap (B_1 \cap B_2)^c = (E \cap B_1 \cap B_2^c) \cup (E \cap B_2 \cap B_1^c) = \emptyset$, then we have $(E \cap B_1 \cap B_2^c) = \emptyset \Rightarrow E \cap B_1 \subseteq B_2 \Rightarrow E \cap B_1 \subseteq E \cap B_2$ and $(E \cap B_2 \cap B_1^c) = \emptyset \Rightarrow E \cap B_2 \subseteq B_1 \Rightarrow E \cap B_2 \subseteq E \cap B_1$. It follows that $E \cap B_1 = E \cap B_2$. Then we have $X_1 = (E \cap B_1)^{\sharp_E} = (E \cap B_2)^{\sharp_E} \subseteq B_2^{\sharp_M} = X_2$, which is conflicted with $(X_2, B_2) \prec (X_1, B_1)$. Hence the assumption is not true, that is to say $E \cap D_{DW}((X_1, B_1), (X_2, B_2)) \neq \emptyset$.

Conversely, assume for any $(X_1, B_1), (X_2, B_2) \in L_d(G, M, I)$, if $D_d((X_1, B_1), (X_2, B_2)) \neq \emptyset$, then $E \cap D_{DW}((X_1, B_1), (X_2, B_2)) \neq \emptyset$ must be hold. For any $X \in \mathcal{U}_{DW}(M, T)$, let $(X, B) \in L_d(G, M, I)$. Suppose that $(B \cap E)^{\sharp_E} \neq X$. Since $(B \cap E)^{\sharp_E} \subseteq B^{\sharp_M} = X$ and $((B \cap E)^{\sharp_E \sharp_M \sharp_M}, (B \cap E)^{\sharp_E \sharp_M}) \in L_d(G, M, I)$, then we have $(B \cap E)^{\sharp_E \sharp_M \sharp_M} \subseteq (B \cap E)^{\sharp_E} \subset X$ and $((B \cap E)^{\sharp_E \sharp_M \sharp_M}, (B \cap E)^{\sharp_E \sharp_M}) < (X, B)$, hence there exists $(X_1, B_1) \in L_d(G, M, I)$ such that $((B \cap E)^{\sharp_E \sharp_M \sharp_M}, (B \cap E)^{\sharp_E \sharp_M}) \leq (X_1, B_1) \prec (X, B)$. It follows that $B \subset B_1 \subseteq (B \cap E)^{\sharp_E \sharp_M}$ and $B_1 - B \subseteq (B \cap E)^{\sharp_E \sharp_M} - B$. From the assumption, we have $E \cap (B_1 - B) \neq \emptyset$ and $E \cap ((B \cap E)^{\sharp_E \sharp_M} - B) \neq \emptyset$. Hence there exists $e \in E$ such that $e \in (B \cap E)^{\sharp_E \sharp_M}$ and $e \notin B$. It follows that there exists $x \in (B \cap E)^{\sharp_{E} c}$ such that $(x, e) \notin I$ from $e \in (B \cap E)^{\sharp_E \sharp_M}$. On account of $(B \cap E)^{\sharp_{E} c} = (B \cap E)^{c_E *_E} = (E - B)^{*_E}$, we can conclude $x \in (E - B)^{*_E}$ and $(x, e) \notin I$. Then we have $e \notin E - B$. By $e \in E$, we know $e \in E \cap B$, which is conflicted with $e \notin B$. Hence the assumption is not true, that is to say $(B \cap E)^{\sharp_E} = X$. Since $X^{\sharp_E} = B \cap E$, we have $(X, X^{\sharp_E}) \in L_d(G, E, I_E)$. Suppose $X \neq Max[X]_{R(E,T)}$, then there exists $X_1 = Max[X]_{R(E,T)}$ such that $X \subset X_1$. By $X_1 \in ExtL_d(G, E, I_E) \subseteq ExtL_d(G, M, I)$ and $X_1^{*_T} = X^{*_T}$, we have $X_1 \in [X]_{R(M,T)}$. Which is conflicted with $X = Max[X]_{R(M,T)}$. Hence the assumption is not true, that is to say $X = Max[X]_{R(E,T)}$, then we have $X \in \mathcal{U}_d(E,T)$. Consequently, $\mathcal{U}_{DW}(M,T) \subseteq \mathcal{U}_{DW}(E,T)$. Then we can conclude E is a DW-consistent set by *Theorem* 3.

Definition 8. *Let $\mathcal{F} = (G,M,I,T,J)$ be a formal decision context. For any $(X_1,B_1),(X_2,B_2) \in L_d(G,M,I)$,*

$$f_{DW} = \bigwedge_{D_{DW}((X_1,B_1),(X_2,B_2))\neq\emptyset} \bigvee D_{DW}((X_1,B_1),(X_2,B_2))$$

is called the DW-discernibility function of $\mathcal{F}$.

Theorem 5. *For a formal decision context $\mathcal{F} = (G,M,I,T,J)$, let $g = \vee_{t=1}^{k} \wedge_{s=1}^{r_t} a_s$ be the minimal disjunctive normal form of the DW-discernibility function of $\mathcal{F}$, then $\{E_t|1 \leq t \leq k\}$ is the set of all DW-reductions of $\mathcal{F}$, where $E_t = \wedge_{s=1}^{r_t} a_s$ is all the prime implicants of the DW-discernibility function g.*

Theorem 5 provides an approach to compute all the DW-reductions of a formal decision context.

Example 2. We consider the formal decision context $\mathcal{F}$ presented in Example 1. By direct computation, we get $\mathcal{U}_{DW}(M,T) = \{x_1x_3, x_2x_4, x_1x_2x_4\}$. Then the DW-reductions of formal decision context $\mathcal{F}$ is $\{a_1, a_2, a_3\}$.

4 Attribute Reduction Based on WD-Decision Rules

Definition 9. *Let $\mathcal{F} = (G,M,I,T,J)$ be a formal decision context. For any $(X,B) \in L(G,M,I)$ and $(Y,C) \in L_d(G,T,J)$, if $X \subseteq Y$, $X \neq \emptyset$ and $Y \neq G$, then $(X,B) \to (Y,C)$ is called a WD-decision rule. We denote by $\mathcal{R}_{WD}(M,T)$ the set of all WD-decision rules for formal decision context $\mathcal{F}$.*

The semantic interpretation of a WD-decision rule $(X,B) \to (Y,C)$ is that if a object x holds all attributes in conditional attribute set B then it doesn't possess at least one attribute in decision attribute set C^c.

Definition 10. *The redundant and non-redundant rules in $\mathcal{R}_{WD}(M,T)$ can be defined similar to the above. All non-redundant WD-decision rules in $\mathcal{R}_{WD}(M,T)$ is denoted by $\overline{\mathcal{R}}_{WD}(M,T)$. Let $\mathcal{U}_{WD}(M,T) = \{X \in ExtL(G,M,I)|\exists(X,X^*) \to (Y,B)) \in \overline{\mathcal{R}}_{WD}(M,T))\}$.*

Definition 11. *Let $\mathcal{F} = (G,M,I,T,J)$ be a formal decision context, $E \subseteq F$, E is called a WD-consistent set of $\mathcal{F}$ if $\mathcal{R}_{WD}(E,T) \Rightarrow \mathcal{R}_{WD}(M,T)$. Furthermore, if E is a WD-consistent set and any proper subset $H \subset E$ is not a WD-consistent set of $\mathcal{F}$, then we call E is a WD-reduction of $\mathcal{F}$.*

Theorem 6. *Let $\mathcal{F} = (G,M,I,T,J)$ be a formal decision context. E is a WD-consistent set of $\mathcal{F}$ if and only if $\mathcal{U}_{WD}(M,T) \subseteq \mathcal{U}_{WD}(E,T)$.*

Proof. The proof is straightforward and obvious.

For any $(X_1, B_1), (X_2, B_2) \in L_d(G, M, I)$, let $\gamma_{WD}((X_1, B_1), (X_2, B_2))$ be the condition $X_1 \in \mathcal{U}_{WD}(M,T) \bigwedge (X_1, B_1) \prec (X_2, B_2)$ or $X_2 \in \mathcal{U}_{WD}(M,T) \bigwedge (X_2, B_2) \prec (X_1, B_1)$. Then we can define:

$$D_{WD}((X_1, B_1), (X_2, B_2)) = \begin{cases} B_1 \cup B_2 - B_1 \cap B_2, \ if \ \gamma_{WD}((X_1, B_1), (X_2, B_2)) \\ \emptyset, \ otherwise \end{cases}$$

Theorem 7. *Let $\mathcal{F} = (G, M, I, T, J)$ be a formal decision context. E is a WD-consistent set of $\mathcal{F}$ if and only if for any $(X_1, B_1), (X_2, B_2) \in L_d(G, M, I)$, if $D_d((X_1, B_1), (X_2, B_2)) \neq \emptyset$, then $E \cap D_{WD}((X_1, B_1), (X_2, B_2)) \neq \emptyset$ must be hold.*

Proof. Assume that E is a WD-consistent set, $(X_1, B_1), (X_2, B_2) \in L(G, M, I)$ and $D_{WD}((X_1, B_1), (X_2, B_2)) \neq \emptyset$. Then we have $X_1 \in \mathcal{U}_{DW}(M,T) \bigwedge (X_1, B_1) \prec (X_2, B_2)$ or $X_2 \in \mathcal{U}_{WD}(M,T) \bigwedge (X_2, B_2) \prec (X_1, B_1)$. Without loss of generality, we suppose that $X_1 \in \mathcal{U}_{WD}(M,T) \bigwedge (X_1, B_1) \prec (X_2, B_2)$. From $X_1 \in \mathcal{U}_{WD}(M,T) \subseteq \mathcal{U}_{WD}(E,T) \subseteq ExtL(G, E, I_E)$, we have $X_1 = X_1^{*_E *_E} = (E \cap X_1^{*_M})^{*_E} = (E \cap B_1)^{*_E}$. Suppose $E \cap D_{DW}((X_1, B_1), (X_2, B_2)) = \emptyset$, it follows that $E \cap B_1 = E \cap B_2$. Then we have $X_1 = (E \cap B_1)^{*_E} = (E \cap B_2)^{*_E} = (E \cap B_2)^{*_M} \supseteq B_2^{*_M} = X_2$, which is conflicted with $(X_1, B_1) \prec (X_2, B_2)$. Hence the assumption is not true, that is to say $E \cap D_{DW}((X_1, B_1), (X_2, B_2)) \neq \emptyset$.

Conversely, suppose for any $(X_1, B_1), (X_2, B_2) \in L(G, M, I)$, if $D_{WD}((X_1, B_1), (X_2, B_2)) \neq \emptyset$, then $E \cap D_{WD}((X_1, B_1), (X_2, B_2)) \neq \emptyset$ must be hold. For any $X \in \mathcal{U}_{WD}(M,T)$, let $(X, B) \in L(G, M, I)$. Suppose that $(B \cap E)^{*_E} \neq X$. Since $(B \cap E)^{*_E} \supseteq B^{*_M} = X$ and $((B \cap E)^{*_E *_M *_M}, (B \cap E)^{*_E *_M}) \in L_d(G, M, I)$, then we have $(B \cap E)^{*_E *_M *_M} \supseteq (B \cap E)^{*_E} \supset X$ and $(X, B) < ((B \cap E)^{*_E *_M *_M}, (B \cap E)^{*_E *_M})$, hence there exists $(X_1, B_1) \in L(G, M, I)$ such that $(X, B) \prec (X_1, B_1) \leq ((B \cap E)^{*_E *_M *_M}, (B \cap E)^{*_E *_M})$. It follows that $(B \cap E)^{*_E *_M} \subseteq B_1 \subset B$ and $B - B_1 \subseteq B - (B \cap E)^{*_E *_M}$. From the assumption, we have $E \cap (B - B_1) \neq \emptyset$ and $E \cap (B - (B \cap E)^{*_E *_M}) \neq \emptyset$. Hence there exists $e \in E$ such that $e \in B$ and $e \notin (B \cap E)^{*_E *_M}$. Since $e \in E$ and $e \in B$, we have $e \in E \cap B$. It follows that $e \in E \cap B \subseteq (B \cap E)^{*_E *_M}$, which is conflicted with $e \notin (B \cap E)^{*_E *_M}$. Hence the assumption is not true, that is to say $(B \cap E)^{*_E} = X$. Since $X^{*_E} = B \cap E$, we have $(X, X^{*_E}) \in L(G, E, I_E)$. Suppose $X \notin \mathcal{U}_{WD}(E,T)$, then there exists $X_1 \in \mathcal{U}_{WD}(E,T)$ such that $X \subset X_1$ and $(X_1, B_1) \to (Y, C) \in \overline{\mathcal{R}}_{WD}(E,T)$. Since $X_1 \in ExtL(G, E, I_E) \subseteq ExtL(G, M, I)$, we have $(X_1, B_1) \to (Y, C) \in \mathcal{R}_{WD}(M,T)$. Then we can conclude that $(X_1, B_1) \to (Y, C) \Rightarrow (X, B) \to (Y, C)$. Which is conflicted with $X \in \mathcal{U}_{WD}(M,T)$. Hence the assumption is not true, that is to say $X \in \mathcal{U}_{WD}(E,T)$. Consequently, $\mathcal{U}_{WD}(M,T) \subseteq \mathcal{U}_{WD}(E,T)$. Then we can conclude E is a DW-consistent set by *Theorem* 6.

By discernibility matrix and the judgment theorem of consistent set (Theorem 7), we can get an approach to compute all the WD-reductions of a formal decision context.

Example 3. We consider the formal decision context $\mathcal{F}$ presented in Example 1. We can get $\mathcal{U}_{WD}(M,T) = \{x_3, x_3x_5, x_1x_3x_5, x_2x_4x_5\}$, then we have the WD-reduction of the formal decision context $\mathcal{F}$ is $\{a_1, a_2, a_3, a_5\}$.

5 Rule Acquisition and Attribute Reduction Based on DD-Decision Rules

Finally, we discuss the premise and the conclusion of decision rule are both dual concepts.

5.1 Rule Acquisition Methods for DD-Decision Rules

Definition 12. *Let $\mathcal{F} = (G, M, I, T, J)$ be a formal decision context, for any $(X, B) \in L_d(G, M, I)$ and $(Y, C) \in L_d(G, T, J)$, if $X \subseteq Y$, $X \neq \emptyset$ and $Y \neq G$, then $(X, B) \to (Y, C)$ is called a DD-decision rule. We denote by $\mathcal{R}_{DD}(M,T)$ the set of all DD-decision rules for formal decision context $\mathcal{F}$.*

The semantic interpretation of a DD-decision rule $(X, B) \to (Y, C)$ is that if a object x doesn't hold at least one attribute in conditional attribute set B^c then it doesn't possess at least one attribute in decision attribute set C^c.

Definition 13. *The redundant and non-redundant rules in $\mathcal{R}_{DD}(M,T)$ can be defined similar to the above. All non-redundant DD-decision rules in $\mathcal{R}_{DD}(M,T)$ is denoted by $\overline{\mathcal{R}}_{DD}(M,T)$. Let $\mathcal{U}_{DD}(M,T) = \{X \in ExtL_d(G,M,I) | (X, X^{\sharp_M}) \to (Y, B) \in \overline{\mathcal{R}}_{DD}(M,T))\}$.*

We are more interested in non-redundant rules because redundant rules can be implied by non-redundant rules. In order to acquire non-redundant rules, we give a equivalence relation $R(T,M)$ on $ExtL_d(G,T,J)$ as follow: $R(T,M) = \{(X,Y) \in ExtL_d(G,T,J) \times ExtL_d(G,T,J) | X^{\sharp_M} = Y^{\sharp_M}\}$. We denote by $[X]_{R(T,M)} = \{Y \in Ext_d(G,T,J) | (X,Y) \in R(T,M)\}$ the equivalence class of X. Let $Min[X]_{R(T,M)}$ be the set of all minimal elements in $[X]_{R(T,M)}$.

Theorem 8. *Let $\mathcal{F} = (G, M, I, T, J)$ be a formal decision context, then $\overline{\mathcal{R}}_{DD}(M,T) = \{(X^{\sharp_M \sharp_M}, X^{\sharp_M}) \to (Y, Y^{\sharp_T}) | (X \in ExtL_d(G,T,J)) \bigwedge (X \neq \emptyset) \bigwedge (X^{*_T *_T} \neq G) \bigwedge (Y \in Min[X]_{R(T,M)})\}$.*

Proof. Similar to *Theorem* 1.

Example 4. Let $\mathcal{F} = (F, M, I, T, J)$ be a formal decision context presented in Example 1. By direct computation, we can compute all the non-redundant DD-decision rules by *Theorem* 8:

$(r_1) : (x_1x_3, a_1a_3a_4a_5a_6) \to (x_1x_3, d_1d_3)$
$(r_2) : (x_2x_4, a_2a_3a_4a_5a_6) \to (x_2x_4x_5, d_1d_2)$
$(r_3) : (x_1x_3x_5, a_1a_3a_5a_6) \to (x_1x_3x_5, d_3)$.

5.2 Attribute Reduction Based on DD-Decision Rules

Next we consider attribute reduction of formal decision context based on DD-decision rules.

Definition 14. *Let* $\mathcal{F} = (G, M, I, T, J)$ *be a formal decision context,* $E, F \subseteq M$*. If for any* $(X, B) \to (Y, C) \in \mathcal{R}_{DD}(E, T)$*, there exists* $(X', B') \to (Y', C') \in \mathcal{R}_{DD}(F, T)$*, then we call* $\mathcal{R}_{DD}(F, T) \Rightarrow \mathcal{R}_{DD}(E, T)$*.*

Theorem 9. *Let* $\mathcal{F} = (G, M, I, T, J)$ *be a formal decision context,* $E \subseteq M$*, then we have* $\mathcal{R}_{DD}(M, T) \Rightarrow \mathcal{R}_{DD}(E, T)$*.*

Then definitions of consistent set and attribute reduction of formal decision context based on DD-decision rules are given as follows.

Definition 15. *Let* $\mathcal{F} = (G, M, I, T, J)$ *be a formal decision context,* $E \subseteq F$*,* E *is called a DD-consistent set of* $\mathcal{F}$ *if* $\mathcal{R}_{DD}(E, T) \Rightarrow \mathcal{R}_{DD}(M, T)$*. Furthermore, if* E *is a DD-consistent set and any proper subset* $H \subset E$ *is not a DD-consistent set of* $\mathcal{F}$*, then we call* E *is a DD-reduction of* $\mathcal{F}$*.*

Theorem 10. *Let* $\mathcal{F} = (G, M, I, T, J)$ *be a formal decision context.* E *is a DD-consistent set of* $\mathcal{F}$ *if and only if* $\mathcal{U}_{DD}(M, T) \subseteq \mathcal{U}_{DD}(E, T)$*.*

Proof. Similar to *Theorem* 6.

For any $(X_1, B_1), (X_2, B_2) \in L_d(G, M, I)$, let $\gamma_{DD}((X_1, B_1), (X_2, B_2))$ be the condition $X_1 \in \mathcal{U}_{DD}(M, T) \bigwedge (X_2, B_2) \prec (X_1, B_1)$ or $X_2 \in \mathcal{U}_{DD}(M, T) \bigwedge (X_1, B_1) \prec (X_2, B_2)$. We can define:

$$D_{DD}((X_1, B_1), (X_2, B_2)) = \begin{cases} B_1 \cup B_2 - B_1 \cap B_2, \ if \ \gamma_{DD}((X_1, B_1), (X_2, B_2)) \\ \emptyset, \ otherwise \end{cases}$$

Then the DD-discernibility attribute set of (X_1, B_1) and (X_2, B_2) is defined by $D_{DD}((X_1, B_1), (X_2, B_2))$. The DD-discernibility matrix of $\mathcal{F}$ is denoted by $\mathcal{D}_{DD} = D_{DD}((X_1, B_1), (X_2, B_2))$, where $(X_1, B_1), (X_2, B_2) \in L_d(M, T)$.

Theorem 11. *Let* $\mathcal{F} = (G, M, I, T, J)$ *be a formal decision context.* E *is a DD-consistent set of* $\mathcal{F}$ *if and only if for any* $(X_1, B_1), (X_2, B_2) \in L_d(G, M, I)$*, if* $D_{DD}((X_1, B_1), (X_2, B_2)) \neq \emptyset$*, then* $E \cap D_{DD}((X_1, B_1), (X_2, B_2)) \neq \emptyset$ *must be hold.*

Proof. Similar to *Theorem* 4.

By discernibility matrix and the judgment theorem of consistent set (Theorem 11), we can get an approach to compute all the DD-reductions of a formal decision context.

Example 5. Let $\mathcal{F} = (G, M, I, T, J)$ be a formal decision context presented in *Example* 1. From *Example* 4, we get $\mathcal{U}_{DD}(M, T) = \{x_1x_3, x_2x_4, x_1x_3x_5\}$. Then the DD-reductions of a formal decision context $\mathcal{F}$ is $\{a_1, a_2, a_4\}$.

6 Conclusion

In this paper, we mainly propose three kinds of rule acquisition and attribute reduction for formal decision context based on dual concept and formal concept. Then rule acquisition methods, consistent set judgment theorems and methods of attribute reduction are proposed. In the further research, we will try to develop more effective rule acquisition and attribute reduction algorithms. In addition, the application of these decision rules in real database is worthy of further study.

Acknowledgement. This work has been partially supported by the National Natural Science Foundation of China (Grant No. 61976130).

References

1. Chen, Y., Yao, Y.: A multiview approach for intelligent data analysis based on data operators. Inf. Sci. **178**(1), 1–20 (2008)
2. Dntsch, I., Gediga, G.: Modal-style operators in qualitative data analysis. In: IEEE International Conference on Data Mining (2002)
3. Ganter, B., Stumme, G., Wille, R.: Formal Concept Analysis: Foundations and Applications (2005)
4. Ganter, B., Wille, R.: Formal Concept Analysis (1999)
5. Godin, R., Missaoui, R., Alaoui, H.: Incremental concept formation algorithms based on galois (concept) lattices. Comput. Intell. **11**(2), 246–267 (2010)
6. Hao, F., Pang, G., Pei, Z., Qin, K.Y., Zhang, Y., Wang, X.: Virtual machines scheduling in mobile edge computing: a formal concept analysis approach. IEEE Trans. Sustain. Comput. (2019)
7. Hao, F., Park, D.S., Min, G., Jeong, Y.S., Park, J.H.: k-cliques mining in dynamic social networks based on triadic formal concept analysis. Neurocomputing **209**, 57–66 (2016)
8. Li, J., Mei, C., Lv, Y.: A heuristic knowledge-reduction method for decision formal contexts. Comput. Math. Appl. **61**(4), 1096–1106 (2011)
9. Li, J., Mei, C., Lv, Y.: Knowledge reduction in decision formal contexts. Knowl.-Based Syst. **24**(5), 709–715 (2011)
10. Li, J., Mei, C., Wang, J., Zhang, X.: Rule-preserved object compression in formal decision contexts using concept lattices. Knowl.-Based Syst. **71**, 435–445 (2014)
11. Liaab, J.: Knowledge reduction in real decision formal contexts. Inf. Sci. **189**(7), 191–207 (2012)
12. Neuss, C., Kent, R.E.: Conceptual analysis of resource meta-information. Comput. Netw. ISDN Syst. **27**(6), 973–984 (1995)
13. Nguyen, P.H.P., Dan, C.: A basic mathematical framework for conceptual graphs. IEEE Trans. Knowl. Data Eng. **18**(2), 261–271 (2006)
14. Qin, K., Li, B., Pei, Z.: Attribute reduction and rule acquisition of formal decision context based on object (property) oriented concept lattices (2019)
15. Shao, M., Leung, Y., Wang, X., Wu, W.: Granular reducts of formal fuzzy contexts. Knowl.-Based Syst. **114**, 156–166 (2016)
16. Shao, M., Leung, Y., Wu, W.: Rule acquisition and complexity reduction in formal decision contexts. Int. J. Approx. Reason. **55**(1), 259–274 (2014)
17. Shao, M., Yang, H., Wu, W.: Knowledge reduction in formal fuzzy contexts. Knowl.-Based Syst. **73**, 265–275 (2015)

18. Sheng, Q., Qian, Q.: Mining materials knowledge with concept lattice algorithm. Mater. Today Commun. **22**, 100726 (2020)
19. Wan, Q., Wei, L.: Approximate concepts acquisition based on formal contexts. Knowl.-Based Syst. **75**, 78–86 (2015)
20. Wei, L.: Attribute reduction theory of concept lattice based on decision formal contexts. Sci. China **51**(7), 910–923 (2008)
21. Wei, L., Liu, L., Qin, J., Qian, T.: Rules acquisition of formal decision contexts based on three-way concept lattices. Inf. Ences **516**, 529–544 (2020)
22. Wille, R.: Restructuring lattice theory: an approach based on hierarchies of concepts. Orderd Sets D Reidel **83**, 314–339 (1982)
23. Wu, W., Leung, Y., Mi, J.: Granular computing and knowledge reduction in formal contexts. IEEE Trans. Knowl. Data Eng. **21**(10), 1461–1474 (2009)
24. Yao, Y.: A comparative study of formal concept analysis and rough set theory in data analysis. In: Tsumoto, S., Słowiński, R., Komorowski, J., Grzymała-Busse, J.W. (eds.) RSCTC 2004. LNCS (LNAI), vol. 3066, pp. 59–68. Springer, Heidelberg (2004). https://doi.org/10.1007/978-3-540-25929-9_6
25. Zhang, W., Ma, J., Fan, S.: Variable threshold concept lattices. Inf. Sci. **177**(22), 4883–4892 (2007)
26. Zhang, W.-X., Wei, L., Qi, J.-J.: Attribute reduction in concept lattice based on discernibility matrix. In: Ślęzak, D., Yao, J.T., Peters, J.F., Ziarko, W., Hu, X. (eds.) RSFDGrC 2005. LNCS (LNAI), vol. 3642, pp. 157–165. Springer, Heidelberg (2005). https://doi.org/10.1007/11548706_17
27. Zhang, W., Wei, L., Qi, J.: Attribute reduction theory and approach to concept lattice. Sci. China **48**(6), 713–726 (2005)

Study on Sample Reduction Method Based on Neighborhood Granulation

Fachao Li, Yunfeng Guo(✉), and Chenxia Jin

School of Economics and Management, Hebei University of Science and Technology, Shijiazhuang, China
lifachao@tsinghua.org.cn, Guoyunfeng0329@qq.com, Jinchenxia2005@126.com

Abstract. In the process of classifier learning, not all the samples contribute to the classifier classification. It has become an important research topic to reduce the samples under the premise of ensuring the classification accuracy of the classifier. In this paper, for the numerical sample reduction problem, the method of random forest computing attribute importance is introduced to reduce the redundant attributes in the sample and reduce the sample dimension. The sample similarity is constructed by measuring the distance between samples. By granulating the neighborhood and reducing the redundant samples in the neighborhood, a sample reduction model based on neighborhood granulation (ng-sr) is proposed The model is compared with two classification methods. Theoretical analysis and experimental results show that ng-sr model is effective, and has a good effect on data preprocessing, and has strong practicability for the actual sample reduction problem.

Keywords: Sample reduction · Neighborhood granulation · Classification method

1 Introduction

With the increasing updating of data acquisition, storage and other equipment, The simple feature data set develops to the complex data set with large storage and high dimension. Sample reduction of complex numerical data has become an urgent problem.

Many scholars at home and abroad have proposed some sample reduction algorithms based on different theories. A support vector data description data reduction method based on information entropy is proposed [1]. By calculating the distribution information entropy of each data, all data whose entropy value is lower than the threshold are excluded. A reduction method of Boolean matrix decomposition samples based on formal concept analysis is proposed [2]. A method is proposed to reduce a given sample set by multiplying the defined absolute density with the defined local density [3]. In [4], a method of surface point sample reduction based on Hausdorff distance error measurement of principal curvature vector is proposed. A sample reduction method based on k-median clustering algorithm is proposed [5]. [6] proposed that the sample redundancy caused by the spatial similarity of aggregate point clouds

C. Pang et al. (Eds.): SETE 2020/ICWL 2020, LNCS 12511, pp. 130–137, 2021.
https://doi.org/10.1007/978-3-030-66906-5_12

can be eliminated. The reduction of preserving approximate samples is similar to the sample reduction method of parametric reduction [7]. A clustering method based on sample stability is proposed [8]. [9] studies an integration method of greatly reducing the transmission of multimedia data. [10] quantifies the extent to which any type of information is discarded when public data reduction is performed. A hierarchical sample reduction mechanism is proposed to reduce the influence of noise [11]. A framework for early data reduction on the client side is proposed, and a business model for end-to-end data reduction in enterprise applications is proposed [12]. In [13], an improved clustering method is proposed, which can reduce the overhead in clustering and message exchange. In [14], a distributed data prediction model, (DDPM), is proposed to prolong the network lifetime by reducing the energy consumption of sensor nodes. Sample reduction is carried out based on high-order singular value decomposition (HOSVD) technique [15].

As can be seen from the above literature, the most current methods of sample reduction, from the viewpoint of reducing the size, ignore the increase of computational complexity caused by the increase of sample dimensions, and do not take into account the similar distance between samples. The main work of this paper is as follows: 1) the redundant attributes in the sample are reduced by the random forest attribute importance algorithm, and the dimension of the sample set is reduced. 2) based on the similarity of the distance between samples, a sample reduction model based on neighborhood granulation is proposed. 3) the performance of the algorithm is analyzed through several commonly used UCI data sets.

2 Preliminary Knowledge

2.1 Overview of Classification Methods

1) C4.5 classification method. C4.5 is the most commonly used decision tree algorithm. It inherits all the advantages of ID3 algorithm and improves ID3 algorithm. The implementation process of C4.5 is divided into four steps. The first step is to calculate the entropy of all samples in the training set. The second step is to calculate the information gain rate of each feature. In the third step, the feature with the largest information gain rate is selected to classify and get the sub nodes. The fourth step iterates the second step and the third step among the unselected features until there is no feature separable or the information gain rate can not meet the required standard.

2) SVM classification method. SVM is a machine learning method. Its basic idea is to construct the optimal hyperplane based on the principle of structural risk minimization, so as to maximize the distance between samples and hyperplane. Given a sample set $E = \{(x_1, y_1), (x_2, y_2), \ldots(x_i, y_i)\}$, $x_i \in R^N$, $y_i \in \{-1, 1\}$, is x_i class marker, the equation of the classification line is $\omega^T \cdot x + b = 0$, ω and x are n dimensional column vectors, ω is the normal vector on the plane, and b is the distance from the classification line to the origin. The classification interval $2/\|\omega\|$ corresponding to the optimal classification line should be the maximum, the maximum $2/\|\omega\|$ is converted to the minimum $\|\omega\|^2/2$, the relaxation variable ξ_i and the penalty factor C are introduced, and the Lagrange multiplier method is used to solve the problem.

$$\begin{cases} \min \frac{\|w\|^2}{2} + C\sum_{i=1}^{n} \xi_i \\ \text{s.t.} y_i(w^T \cdot x_i + b) \geq 1 - \xi_i,\ \xi_i \geq 0, i = 1,\ 2, \cdots,\ n \end{cases} \tag{1}$$

3 Numerical Sample Reduction Model Based on Neighborhood Granulation

From the above analysis, we can see that when reducing numerical samples, we need to consider two factors: sample dimension and sample size. Therefore, from these two angles, this paper first considers the dimensionality reduction of the sample, through the random forest attribute importance algorithm, preserves the attributes with high attribute importance in the sample, and then reduces the sample after dimensionality reduction. In view of the similarity of numerical samples, a sample reduction model based on neighborhood granulation is constructed to eliminate similar samples.

In order to construct the sample neighborhood reduction model, the neighborhood should be determined first. Here, the neighborhood is defined according to the maximum distance from the neighborhood center point to the boundary at a certain degree. Before the sample reduction, we enter a parameter radius δ, then calculate the similarity between the two samples, and select a sample as the reduction initial sample, as long as its similarity with the surrounding sample is less than the given parameter δ, it is considered that the samples in the δ range are similar, reduce other samples in the neighborhood of the sample, retain the sample to complete a neighborhood reduction, and then find the next sample to repeat the process. Until the whole sample reduction is completed. From this, we can find that there are two difficulties in this process: 1) how to select the initial sample of sample reduction; 2) the extension of sample reduction set.

For the convenience, we assume that in the following: 1) Given a sample set $\Omega = \{x_1, x_2, \ldots x_n\}$, x_i is a sample. Sample x_i is represented by the n-dimensional vector $x_i = (a_1, a_2, \ldots, a_n)$. a_i is the attribute of the sample. 2) Euclidean distance is used as the similarity measure of the two samples, then

$$d(x_i, x_j) = \sqrt{(a_{j1} - a_{i1})^2 + (a_{j2} - a_{i2})^2 + \cdots + (a_{jn} - a_{jn})^2} \tag{2}$$

3) B represents a sample set, $B \subset \Omega$, and the distance from sample x_i to B is

$$d(x_i, B) = \min\{d(x_i, y) | y \in B\} \tag{3}$$

4) The sample set after sample reduction is $U = (x_1, x_2, \ldots, x_m)$, $m < n$, we call

$$r(\Omega, U) = (1 - \frac{|U|}{|\Omega|}) \tag{4}$$

the sample reduction rate of $\Omega \rightarrow U$. 5) The classification accuracy and error of samples before reduction is a, and after reduction is b.

3.1 Selection Method of Initial Sample

There are many ways to choose the initial sample in sample reduction, but different initial sample has great influence on the final result. Random initial point of sample reduction is very likely to encounter outliers, we will use random optimization to select the initial sample. The steps are stated as follows:

1) Calculate the Euclidean distance between all samples, divide the maximum Euclidean distance into 5 parts on average, and select the second, third and fourth distances as neighborhood radii to get three neighborhood radius.
2) Randomly select a sample from the sample after attribute reduction, and observe the sample value of the sample under the three different neighborhood radius (the number of samples in the neighborhood of the sample).
3) Each sample takes the mean value of the sample value under the three different neighborhood radius, arranges the 10 mean according to its size, and takes the sample corresponding to the maximum mean as the initial sample of sample reduction.

3.2 Extension of Neighborhood Center Set Based on Minimum Interval

After the initial sample is determined, it is regarded as the neighborhood center which is used to participate in the following reduction process. The samples within its neighborhood radius are regarded as a kind of reduction, but the extension of the neighborhood center is a key problem. In this paper, the extension of neighborhood center set based on minimum interval is proposed, and its steps are as follows:

1) Suppose that B is an expanding set of sample neighborhood centers. After the first neighborhood center y_1 is determined, the samples within the range of y_1 are subtracted approximately, and y_1 is added to B. At this time, the neighborhood center set is $B = \{y_1\}$;
2) From the rest of the sample point selection conforms to the point B, and this point is the second neighborhood center y_2, approximately subtract the samples in the range of y_2, and add y_2 to B, and the neighborhood center set $B = \{y_1, y_2\}$;
3) Repeat 1) and 2) until the neighborhood center of the entire sample set is determined. And then finally, the sample in B is going to be the sample set U that we reduced.

3.3 Selection of Neighborhood Radius Method Based on Sample Reduction

Before the sample reduction, we give the parameter radius δ, because the parameter radius is artificially given, and the sample reduction rate and classification accuracy corresponding to δ are unknown, so we find δ by giving a model to solve the optimal neighborhood radius.

$$\max \delta \\ \text{s.t}\begin{cases} \alpha(k) - 0.05 \le \alpha(\delta) \le \alpha(k) + 0.05 \\ p \le r(\delta) \le l \end{cases} \tag{5}$$

Where, δ represents the neighborhood radius, $\alpha(\delta)$ represents the classification accuracy of the classifier for the reduced data, $\alpha(k)$ represents the classification accuracy of the data after attribute reduction, $r(\delta)$ represents the reduction rate of the data under a given δ, p represents the minimum reduction rate of the data, and l represents the maximum reduction rate of the data. And p is usually 10%, and l is usually 90%.

For the convenience of description, we call the model of sample reduction based on neighborhood granulation as NG-SR.

4 NG-SR Model Solving Algorithm

4.1 NG-SR Algorithm Steps

Input: Sample sets $\Omega = \{x_1, x_2, ...x_n\}$ and δ.

Output: reduced sample set $U = (x_1, x_2, ..., x_m)$, $m < n$, and $r(\Omega, U)$.

Step 1. Through the random forest attribute importance algorithm, the attributes with low attribute importance in the sample set are eliminated;
Step 2. The sample set is normalized and the Euclidean distance $d\ (xi, xj)$ between all samples is calculated;
Step 3. Random optimization to find the initial reduction sample;
Step 4. The initial reduction samples of the optimal sample set are randomly selected, and the samples within the neighborhood radius are classified into one category, the neighborhood center samples are retained, and the other samples are reduced;
Step 5. The second neighborhood center selects samples in accordance with d (x_i, B), repeats the operation of step4 until the nth neighborhood center is selected, and finally gets the reduced sample set U;.

4.2 Analysis of Time Complexity of NG-SR Algorithm

The time complexity of the first step is $O(mnlogn^2)$. The time complexity of the second step is $O(n^2)$. The time complexity of the third step is $O(1)$. The time complexity of the fourth step is $O(mn)$. The time complexity of the fifth step is $O(mn)$. Where n is the number of samples, m is the number of categories, and the number of samples is much larger than the number of categories. So the time complexity of the algorithm is $O(n^2)$.

5 Performance Analysis of NG-SR Algorithm

In order to test the effectiveness of the algorithm, we will use four sets of UCI data to make experiments, and the details are stated as Table 1.

Table 1. Description of sample set.

	Sample	Number	Number of conditional attributes
1	Wine	178	13
2	Sonar	208	60
3	ILPD	583	10
4	Transfusion	748	4

Table 2 gives the results after attribute reduction. The reason why the attributes of transfusion sample set are not reduced is that the importance of these four attributes is very high, so they are preserved (see Table 2).

Table 2. Sample set after attribute reduction of random forest.

	Sample	Number	Number of conditional attributes
1	Wine	178	10
2	Sonar	208	53
3	ILPD	583	8
4	Transfusion	10 point, italic	4

Compared with ng-sr algorithm, the two algorithms reduce the sample through C4.5 classifier, and observe their classification accuracy (see Table 3 and 4).

Table 3. The classification accuracy and error under the initial sample are randomly selected.

Sample	Radius of neighborhood	$r(\Omega, U)$	a	b
Wine	0.4708	0.7697	0.9731 ± 0.0204	0.9153 ± 0.0685
Sonar	0.8	0.2692	0.9239 ± 0.0286	0.8556 ± 0.0778
ILPD	0.05	0.1986	0.8120 ± 0.0832	0.8103 ± 0.0517
Transfusion	0.009	0.2874	0.7750 ± 0.0676	0.6758 ± 0.2077

Table 4. Classification accuracy and error of NG-SR algorithm.

Sample	Radius of neighborhood	$r(\Omega, U)$	a	b
Wine	0.4708	0.7753	0.9813 ± 0.0249	0.9815 ± 0.0247
Sonar	0.8	0.2740	0.9331 ± 0.0307	0.9523 ± 0.0023
ILPD	0.05	0.1986	0.8198 ± 0.0654	0.8286 ± 0.0286
Transfusion	0.009	0.2874	0.6872 ± 0.1774	0.7231 ± 0.2115

The classification accuracy errors in Table 4 are all within 5%, indicating that the sample set obtained by ng-sr algorithm is more reasonable. The optimal neighborhood radius is wine: 0.2282–0.6277, sonar: 0.8–1.0923, ilpd: 0.0355–0.2836, fusion: 0.0102–0.0271.

The original data and the reduced data can get two classification line equations through linear support vector opportunity. The difference between the two classification lines is compared, which is called classification difference (see Table 5 and Table 6).

Table 5. Take the classification difference of the initial sample at random.

Sample	Radius of neighborhood	$r(\Omega, U)$	Classification difference
Wine	0.4708	0.7753	4.0406
Sonar	0.8	0.2740	2.0170
ILPD	0.05	0.1986	0.0011
Transfusion	0.009	0.2874	12.6189

Table 6. Classification difference of NG-SR algorithm.

Sample	Radius of neighborhood	$r(\Omega, U)$	Classification difference
Wine	0.4708	0.7753	3.8704
Sonar	0.8	0.2740	1.5293
ILPD	0.05	0.1986	0.0023
Transfusion	0.009	0.2874	10.5476

Table 5 and Table 6 showed that the samples of ILPD were close to each other, and the differences of the other three groups were smaller than that of NG-SR. It also shows that NG-SR selecting initial samples is reasonable.

6 Conclusion

In this paper, in order to reduce the classification accuracy of the classifier with redundant samples, based on the analysis of the characteristics of the existing sample reduction methods, through the idea of neighborhood granulation, a NG-SR sample reduction model is proposed, and the performance of the model is tested. Theoretical analysis and experiments show that NG-SR has good interpretability and structural characteristics, and has little effect on the classification accuracy and classification difference of samples after sample reduction, which complements the existing sample reduction methods, and has a wide application prospect in many fields, such as data system reduction, text classification, information security and so on.

Acknowledgements. This work is supported by the National Natural Science Foundation of China (71771078, 71371064).

References

1. Li, D., Wang, Z., Cao, C., Liu, Y.: Information entropy based sample reduction for support vector data description. Appl. Soft Comput. **71**, 1153–1160 (2018)
2. Trnecka, M., Trneckova, M.: Data reduction for Boolean matrix factorization algorithms based on formal concept analysis. Knowl.-Based Syst. **158**(15), 75–80 (2018)
3. Wang, J., Yue, S., Yu, X., Wang, Y.: An efficient data reduction method and its application to cluster analysis. Neurocomputing **238**(17), 234–244 (2017)
4. Ma, X., Cripps, R.J.: Shape preserving data reduction for 3D surface points. Comput.-Aided Design **43**(8), 902–909 (2011)
5. Ougiaroglou, S., Diamantaras, K.I., Evangelidis, G.: Exploring the effect of data reduction on neural network and support vector machine classification. Neurocomputing **280**, 101–110 (2018)
6. Li, Y., Liu, H., Tao, Y., Liao, J.: Reasoning mechanism: an effective data reduction algorithm for on-line point cloud selective sampling of sculptured surfaces. Comput. Aided Design **113**, 48–61 (2019)
7. Abu-Khzam, F.N., Bazgan, C., Chopin, M., Fernau, H.: Data reductions and combinatorial bounds for improved approximation algorithms. J. Comput. Syst. Sci. **82**(3), 503–520 (2016)
8. Li, F., Qian, Y., Wang, J., Dang, C., Jing, L.: Clustering ensemble based on sample's stability. Artif. Intell. **273**, 37–55 (2019)
9. Mostefaoui, A., Noura, H., Fawaz, Z.: An integrated multimedia data reduction and content confidentiality approach for limited networked devices. Ad Hoc Netw. **32**, 81–97 (2015)
10. Booth, B.G., Keijsers, N.L.W., Sijbers, J., Huysmans, T.: An assessment of the information lost when applying data reduction techniques to dynamic plantar pressure measurements. J. Biomech. **87**, 161–166 (2019)
11. Tang, X.-Y., Zeng, W.-M., Wang, N.-Z., Shi, Y.-H., Zhao, L.: A novel layered data reduction mechanism for clustering fMRI data. Biomed. Signal Process. Control **33**, 48–65 (2017)
12. ur Rehman, M.H., Chang, V., Batool, A., Wah, T.Y.: Big data reduction framework for value creation in sustainable enterprises. Int. J. Inf. Manage. **36**(6), 917–928 (2016)
13. Radhika, S., Rangarajan, P.: On improving the lifespan of wireless sensor networks with fuzzy based clustering and machine learning based data reduction. Appl. Soft Comput. **83**, 105610 (2019)
14. El-Sayed, W.M., El-Bakry, H.M., El-Sayed, S.M.: Integrated data reduction model in wireless sensor networks. Appl. Comput. Inf. (2019)
15. García-Magariño, A., Sor, S., Velazquezc, A.: Data reduction method for droplet deformation experiments based on high order singular value decomposition. Exp. Ther. Fluid Sci. **79**, 13–24 (2016)

Research on Attribute Reduction Method Based on Local Dependency

Xiaozheng Yang[1], Yexing Ren[2(✉)], and Fachao Li[1,2]

[1] School of Economics and Management, Hebei University of Science and Technology, Shijiazhuang 050018, China
654716439@qq.com, lifachao@tsinghua.org.cn

[2] College of Science, Hebei University of Science and Technology, Shijiazhuang 050018, China
x960312@yeah.net

Abstract. Attribute reduction is one of the research hotspots in the field of data mining. Although the result of attribute reduction algorithm based on single attribute identification matrix is better, it is still not efficient enough to deal with large-scale information system problems. In this paper, the concept of sub matrix of single attribute identification matrix is proposed. Based on the sub matrix, the calculation method of local dependency degree is given, and an attribute reduction algorithm based on local dependency degree is designed. If the equivalence class of information system is regarded as basic knowledge granules, this algorithm first finds an attribute set to separate the first particle from other particles, and then adds attributes to the attribute set in order to separate the second particle from other particles. Repeat the above operation until all particles are distinguished, and the resulting attribute set is called reduction set. This algorithm reduces the time and space complexity of reduction algorithm to a certain extent. The effectiveness of this method is verified by UCI data set, which provides a method for attribute reduction.

Keywords: Attribute reduction · Local dependency · Submatrix · Distinguishing ability · Core attributes

1 Introduction

In 1982, Polish mathematician Pawlak Z proposed rough set theory, which is a new mathematical tool to deal with fuzzy and uncertain knowledge [1, 2]. Attribute reduction is one of the important contents in the research of rough set theory. Scholars always hope to find the minimum reduction of information system. However, literature [3] has confirmed that solving the minimum reduction of information system is a NP hard problem. The general method to solve this kind of problem is to use heuristic algorithm to solve reduction.

The basic idea of heuristic reduction algorithm is to take the core attributes as the starting point, and take the importance of non-core attributes as the standard to add them to the reduction set one by one until a reduction is obtained. How to achieve attribute reduction through some heuristic algorithm is a hot research content in this

C. Pang et al. (Eds.): SETE 2020/ICWL 2020, LNCS 12511, pp. 138–147, 2021.
https://doi.org/10.1007/978-3-030-66906-5_13

field, many scholars have carried on many beneficial discussions. In Ref. [4], an attribute reduction algorithm based on rough set and information entropy is proposed. The change of mutual information caused by an attribute is added to the decision table to reflect the importance of the attribute, and the relative reduction is obtained. In Ref. [5], the basic concepts and main operations of rough set theory are analyzed and discussed from the viewpoint of information theory, and a reduction algorithm of decision table is proposed based on conditional information entropy. A heuristic algorithm based on the importance of attributes is proposed in Ref. [6]. The algorithm starts from the core attribute, takes the attribute gain as the importance, and adds the attribute with the maximum attribute gain into the reduction set until the attribute set becomes the harmonious set. In Ref. [7], an attribute reduction algorithm based on multi criteria and sample selection is proposed. Firstly, the cluster center is extracted from the data, and then the samples far away from the cluster center are selected. This step completes the process of sample selection to reduce the amount of data. Secondly, the attribute reduction based on multi criteria is designed, and the heuristic algorithm is used to reduce the multi criteria on the selected samples. A reduction construction method based on discernibility matrix simplification is proposed in Ref. [8]. Aiming at the problem of attribute ranking, a heuristic reduction algorithm based on attribute measure and user directly is proposed. In Ref. [9], a complete heuristic reduction algorithm based on discernibility matrix absorption law is proposed by improving discernibility matrix and measuring attribute saliency, which effectively reduces the spatial complexity of discernibility matrix reduction algorithm. The Boolean matrix representation of concepts and operations in rough set theory is given in Ref. [10], and it is proved that attribute reduction is equivalent under two different representations of Boolean matrix and algebra. Reference [11] discussed the method based on rough set and Boolean reasoning and its application in pattern recognition, machine learning, data mining and conflict analysis. In Ref. [12], an attribute reduction algorithm based on kernel and conditional distinguishing ability and weighted conditional distinguishing ability is given on the basis of Boolean matrix. Reference [13] combines the importance of attributes with the improved conditional distinguishing ability, and proposes an attribute reduction algorithm based on the core and improved conditional distinguishing ability, and ensures the completeness of the reduction set with the help of reverse deletion. Reference [14] introduces the Boolean matrix of conditional attribute and Boolean matrix of decision attribute, and proposes a new attribute reduction model based on Boolean operation according to the concept of neighborhood rough set. Reference [15] extended the attribute reduction method of concentrated Boolean matrix on equivalence relation to dominance relation, proposed the concept of concentrated Boolean matrix for dominance matrix, and established corresponding efficient reduction method. Reference [16] studied the relationship between attribute reduction in rough set and minimum vertex cover of graph, and pointed out that finding minimum vertex cover of graph is equivalent to attribute reduction of information system derived from graph, and proposes an attribute reduction algorithm for rough sets based on graph theory. The above studies are all successful. Although the single attribute identification matrix solves the problems of large storage space and low generation efficiency, the calculation of single attribute identification matrix still has the problem of high computational complexity, which is not suitable for large-scale information system. To solve the above problems, this paper proposes the concept of sub matrix of single attribute identification matrix, and designs an attribute reduction algorithm based on local dependence. This algorithm greatly reduces the running time of the algorithm and improves the efficiency of the attribute reduction algorithm when the reduction result is not bad.

Based on the information system, this article mainly does the following work: 1) The concept of sub matrix of single attribute identification matrix is given. 2) The definition of the local dependency of the attribute set on the attribute is given, its properties are analyzed and the calculation method is given. 3) An attribute reduction algorithm based on local dependency is designed. 4) The effectiveness of this method is verified by the commonly used UCI database.

2 Preliminaries

This part mainly reviews some basic concepts of rough set and dependency. For the convenience of description, we suppose that: 1) (U, A, F_A) is an information system $U = \{x_1, x_2, \cdots, x_n\}$ is object set, $A = \{a_1, a_2, \cdots, a_m\}$ is a set of attributes, $F_A = \{f_a : U \to V_a | \ a \in A\}$ is the set of relationships between U and A, V_a is the range of a); 2) For (U, A, F_A) and $B \subseteq A, [x_i]_B = \{x_j \in U | \ (x_i, x_j) \in R_B\}$ is the R_B equivalent of x_i, $U/B = \{[x_i]_B | x_i \in U\}$; 3) M^T is the transposition of matrix M.

Definition 1 ([2]). Let (U, A, F_A) be an information system, $B \subseteq A$, $a \in A$. 1) If $R_B = R_A$, then B is a partition coordination set of A; 2) If B is a partition coordination set of A, and any true subset of B is not a partition coordination set of A, then B is a partition reduction set of A; 3) If any partition reduction set B of A by $a \in B$ is constant, then a is a partition core attribute of A.

Definition 2 ([17]). Let (U, A, F_A) be an information system, $B \subseteq A$, $F_B^* = \{f_a^* : U/B \to V_a | f_a^*([x]_B) = f_a(x), \ a \in B\}$. 1) (U, B, F_B) is called a subsystem of (U, A, F_A); 2) $(U/B, B, F_B^*)$ is called the identification information system derived from (U, B, F_B).

Definition 3 ([17]). Let (U, A, F_A) be an information system, $(U/A, A, F_A^*)$ is the identification information system derived from (U, A, F_A), $a_i, a_j \in A$, $u, v \in U/A$, $x, y \in U$, $B \subseteq A$, $B \neq \phi$, $C \subseteq A$,

$$\mathrm{sign}(s, t) = \begin{cases} 1, & \text{if } s \neq t, \\ 0, & \text{if } s = t. \end{cases} \tag{1}$$

1) If $\mathrm{sign}(f_{a_i}(x), f_{a_i}(y)) = 1$, then a_i can distinguish x from y; otherwise, a_i cannot distinguish x from y.

2) Let $a_t(u, v) = \mathrm{sign}\left(f_{a_t}^*(u), f_{a_t}^*(v)\right)$,

$$G(a_i) = \sum_{u,v \in U/A} a_i(u,\ v) \tag{2}$$

$$G(B) = \sum_{u,v \in U/A} \max_{a_i \in B} a_i(u,\ v) \tag{3}$$

$$D(a_i|B) = \sum_{u,v \in U/A} a_i(u,v) \cdot \operatorname{sign}\left(\max_{a_k \in B} a_k(u,v),\ a_i(u,v) \right) \tag{4}$$

Then $G(a_i)$ is called the distinguishing ability of a_i, $G(B)$ is called the comprehensive distinguishing ability of attribute set B, and $D(a_i|B)$ is called the dependence of B on a_i.

3 Calculation Method of Local Dependence

Based on the identification information system, this part focuses on the calculation of local dependence. For convenience, we assume that:

1) $\delta(x) = 1$ for any $x > 0$, and $\delta(x) = 0$ for any $x \leq 0$.
2) For $X = (x_1,\ x_2, \cdots, x_n)$, $Y = (y_1,\ y_2, \cdots, y_n)$, $X_i = (x_{i1},\ x_{i2}, \cdots,\ x_{in})$, $i = 1,\ 2, \cdots,\ s$, $S(X) = \sum_{i=1}^{n} x_i$, $\delta(X) = (\delta(x_1),\ \delta(x_2), \cdots,\ \delta(x_n))$, $X - Y = (x_1 - y_1,\ x_2 - y_2, \cdots,\ x_n - y_n)$, $\vee_{i \in I} X_i = (\vee_{i \in I} x_{i1},\ \vee_{i \in I} x_{i2}, \cdots,\ \vee_{i \in I} x_{in})$.
3) If the first row of matrix Q is $(a_1,\ a_2, \cdots, a_m)$, and the elements of other rows are 0 or 1, then Q is called the identity Boolean matrix of $a_1,\ a_2, \cdots, a_m$, denoted as $Q = \begin{pmatrix} a_1 & a_2 & \cdots & a_m \\ P(Q, a_1) & P(Q, a_2) & \cdots & P(Q, a_m) \end{pmatrix}$.

3.1 Submatrix of Single Attribute Identification Matrix

Definition 4. Let $(U,\ A,\ F)$ be an information system, $(U/A,\ A,\ F_A^*)$ the identification information system derived from $(U,\ A,\ F_A)$, $A = \{a_1,\ a_2, \cdots,\ a_m\}$, $U/A = \{U_1,\ U_2, \cdots,\ U_N\}$,

$$r_{ij}^{(k)} = \begin{cases} 1, & \text{if } f_{a_j}^*(U_k) \neq f_{a_j}^*(U_i), \\ 0, & \text{if } f_{a_j}^*(U_k) = f_{a_j}^*(U_i), \end{cases} \tag{5}$$

$M(U_k) = (r_{ij}^{(k)})_{(N-k) \times m}$ is a matrix of order $(N-k) \times m$ with $r_{ij}^{(k)}$ as its element, there $k = 1,\ 2,\ \cdots,\ N-1$, $i = k+1,\ k+2,\ \cdots,\ N$, $j = 1,\ 2,\ \cdots,\ m$. $\vec{A} = (a_1,\ a_2, \cdots,\ a_m)$,

1) The following $[N(N-1)/2 + 1] \times m$ order matrix:

$$M = \begin{bmatrix} \vec{A} & M(U_1) & M(U_2) & \cdots & M(U_{N-1}) \end{bmatrix}^T \\ \triangleq \begin{bmatrix} a_1 & a_2 & \cdots & a_m \\ P(M,\ a_1) & P(M,\ a_2) & \cdots & P(M,\ a_m) \end{bmatrix} \tag{6}$$

M is the single attribute identification matrix [17] of information system $(U,\ A,\ F)$.
2) The following $[(N-k)+1] \times m$ order matrix:

$$M_{U_k} = \begin{bmatrix} \vec{A} \\ M(U_k) \end{bmatrix} \triangleq \begin{bmatrix} a_1 & a_2 & \cdots & a_m \\ P(M_{U_k},\ a_1) & P(M_{U_k},\ a_2) & \cdots & P(M_{U_k},\ a_m) \end{bmatrix} \tag{7}$$

M_{U_k} is the sub matrix of single attribute identification matrix of information system $(U,\ A,\ F)$.

It is easy to see that M_{U_k} just adds an attribute line $\vec{A} = (a_1,\ a_2,\cdots,a_m)$ on the basis of $M(U_k)$. If the R_A equivalence class of $(U,\ A,\ F_A)$ is regarded as a basic knowledge particle, then: 1) M_{U_k} is the 0–1 comprehensive description of whether each attribute a_j $(j = 1,\ 2,\cdots,m)$ can distinguish between particles U_k and particles U_i; 2) The attributes in the column where 1 is located in row t $(t = 1,\ 2,\cdots,N-k)$ of M_{U_k} can distinguish U_k and U_{k+t}. This indicates that the number of 1 in this row is the number of attributes in A that can distinguish U_k from U_{k+t} (in particular, if there is only one 1 in this row, then the attribute in the column where the 1 is located must be the core attributes). 3) The number of sub matrices of single attribute identification matrix is related to the number of equivalence classes of information system.

3.2 Properties and Calculation Method of Local Dependence

We assume that: 1) $M_{U_k} \ominus \{a_k\}$ is the new matrix formed by deleting the row in which the column element of attribute a_k is 1 and deleting the column in which a_k is located in matrix M_{U_k}; 2) Let $B \subset A$, $M_{U_k} \ominus B$ be the matrix formed by deleting the row corresponding to element 1 in $\vee_{a_t \in B} P(M_{U_k},\ a_t)$ and the column corresponding to attribute in B in matrix M_{U_k}; 3) $M \ominus \phi = M$.

Definition 5. Let $(U,\ A,\ F)$ be an information system, $A = \{a_1,\ a_2,\cdots,\ a_m\}$, $U/A = \{U_1,\ U_2,\cdots,\ U_N\}$, $(U/A,\ A,\ F_A^*)$ be the identification information system derived from $(U,\ A,\ F_A)$, $A = \{a_1,\ a_2,\ \cdots,\ a_m\}$, M_{U_k} be the sub matrix of the single attribute identification matrix of $(U,\ A,\ F_A)$, $B \subset A$, Then when distinguish particle U_k and particle $U_i (i = k+1,\ k+2,\cdots,N)$, the local dependency of attribute set B on attribute a_j is

$$D(a_j|B)_{U_k} = 2 \cdot S(\delta(P(M_{U_k},\ a_j) - \vee_{a_t \in B} P(M_{U_k},\ a_t))). \tag{8}$$

It can be seen from Definition 5 that the local dependency $D(a_j|B)_{U_k}$ is the number of times that particle U_k and particle U_i cannot be distinguished by B but can be distinguished by a_j.

Theorem 1. Let (U, A, F) be an information system, $A = \{a_1, a_2, \cdots, a_m\}$, $U/A = \{U_1, U_2, \cdots, U_N\}$, $(U/A, A, F_A^*)$ be the identification information system derived from (U, A, F_A), $A = \{a_1, a_2, \cdots, a_m\}$, M_{U_k} be the sub matrix of the single attribute identification matrix of (U, A, F_A), $B \subset A$, $a_i, a_j \in A$. Then:

1) $G(a_i) = 2 \cdot \sum_{k=1}^{N-1} S(P(M_{U_k}, a_i), G(B) = 2 \cdot \sum_{k=1}^{N-1} S(\vee_{a_t \in B} P(M_{U_k}, a_t));$

2) $D(a_j|B) = 2 \cdot \sum_{k=1}^{N-1} D(a_j|B)_{U_k};$

3) The necessary and sufficient condition for B to be a partition coordination set of A is $G(B) = N(N-1)$ (that is, $\vee_{a_t \in B} P(M, a_t)$ is a vector whose components are all 1).

4 Algorithm Design

The attribute reduction algorithm based on core and conditional distinguishing ability in reference [12] (hereinafter referred to as algorithm 1) and the reverse deletion attribute reduction algorithm based on core and improved conditional discriminant ability (hereinafter referred to as algorithm 2) in reference [13] are established based on the dependency of attribute set on single attribute, and the execution process of Algorithm 1 is as follows:

Input: Information system $S = (U, A, F_A)$;

Output: A reduction attribute set B of the information system;

Step 1: Construct the corresponding single attribute identification matrix M according to information system S. Let $B = \phi$;

Step 2: Calculate the sum of the elements of each row in M. If there is a row whose sum is 1, add the attribute a_i corresponding to 1 in the row to B, Set $B = B \cup \{a_i\}$ and update the corresponding $M \ominus B$ at the same time. Otherwise, go to step3;

Step 3: Calculate the column sum of each column in $M \ominus B$, select the corresponding attribute a_i with the largest result and add it to the reduction attribute set B (if there are multiple attributes at the same time, take any one of them), then $B = B \cup \{a_i\}$, and update the corresponding $M \ominus B$ at the same time;

Step 4: If $M \ominus B$ is not a row vector, go to step 3; Otherwise, stop and output B as the reduced set of properties.

It is worth noting that step 3 of algorithm 1 is arbitrary when adding the attribute with the largest sum of elements in each column. Since the number of 1 in the row of single attribute identification matrix is less, the attribute corresponding to 1 is more important. Therefore, algorithm 2 designs a reverse deletion attribute reduction algorithm with core and improved conditional discrimination ability. When there are multiple attributes with the largest sum of column elements at the same time, the attribute with the smallest row sum and the highest frequency is selected. The implementation process is as follows:

Input: Information system $S = (U, A, F_A)$;

Output: A reduction attribute set B of the information system;

Step 1: Construct the corresponding single attribute identification matrix M according to information system S. Let $B = \phi$;

Step 2: Calculate the sum of the elements of each row in M. If there is a row whose sum is 1, add the attribute a_i corresponding to 1 in the row to B, Set $B = B \cup \{a_i\}$ and update the corresponding $M \ominus B$ at the same time. Otherwise, go to step3;

Step 3: Calculate the column sum of each column in $M \ominus B$, select the corresponding attribute a_i with the largest result and add it to the reduction attribute set B (If there are multiple attributes with the largest result at the same time, the attribute with the smallest row sum and the most frequent occurrence is selected to add to the reduction set), then $B = B \cup \{a_i\}$, and update the corresponding $M \ominus B$ at the same time;

Step 4: If $M \ominus B$ is not a row vector, go to step 3; Otherwise, go to step 5;

Step 5: For the initial reduction set B, B is used as a new attribute set to construct a single attribute identification matrix. If a new zero row appears in the new matrix after removing the column of an attribute, the attribute should be retained; otherwise, the attribute is deleted and the reduction set B is output.

The idea of this algorithm is to first add the core attributes to the reduction set B, update the matrix $M \ominus B$; judge whether the matrix $M \ominus B$ is a row vector, and if so, output the reduction set B; otherwise, add the attribute with the greatest dependency on the reduction set (If there are more than one attribute, the attribute with the highest frequency of row sum minimum occurrence is selected), update the matrix $M \ominus B$. Finally, the initial reduction set B is obtained by reverse deleting redundant attributes to get the final reduction set. This shows that algorithm 2 is an improvement of algorithm 1, but the time and space complexity of the two algorithms are relatively high, and they are not suitable for dealing with the problem of large-scale information system. In order to solve this problem, this paper designs an attribute reduction algorithm based on local dependency degree (hereinafter referred to as algorithm 3), its execution process is as follows:

Input: Information system $S = (U, A, F_A)$;

Output: A reduction attribute set B of the information system;

Step 1: Using equivalence relation classification, an object in each class is selected as a new object set, let $k = 1$ $B = \phi$;

Step 2: Calculate the sub matrix M_{U_k} of the single attribute identification matrix and update the matrix $M_{U_k} \ominus B$;

Step 3: Calculate the sum of the elements of each row in $M_{U_k} \ominus B$. If there is a row whose sum is 1, add the attribute a_i corresponding to 1 in the row to B, Set $B = B \cup \{a_i\}$ and update the corresponding $M \ominus B$ at the same time. Otherwise, go to step5;

Step 4: Calculate the local dependence of attribute set B on other attributes, select the corresponding attribute a_i with the largest result to add to B, then $B = B \cup \{a_i\}$, and update the matrix $M_{U_k} \ominus B$ at the same time;

Step 5: If $M \ominus B$ is not a row vector, go to step 4; Otherwise, go to step 6;

Step6: Let $k = k + 1$, if $k = N$, output B; otherwise, go to step 2.

The basic idea of this algorithm is that if the equivalence class of information system is regarded as basic knowledge granules, an attribute set B is first found to distinguish the first particle from other particles, and then attributes are added to the attribute set to distinguish the second particle from other particles until all particles are distinguished. The attribute set B is the reduction set.

The computational complexity of the above algorithms is analyzed as follows: Assuming the number of equivalence classes of information system is n, the time complexity of calculating single attribute identification matrix is $O(n^2 \times |A|)$. The time and space complexity of algorithm 1 and algorithm 2 are $O(n^2 \times |A|)$. The time complexity of algorithm 3 to calculate the sub matrix of single attribute identification matrix is $O(n \times |A|)$. Because this algorithm needs to cycle $n - 1$, the time complexity of algorithm 3 is $O(n^2 \times |A|)$ and the space complexity is $O(n \times |A|)$. Therefore, although the time complexity of algorithm 3 is the same as the other two algorithms, the spatial complexity of algorithm 3 is lower than that of algorithm 1 and algorithm 2.

5 Experimental Simulation and Comparative Analysis

In this part, nine commonly used UCI datasets are selected to compare the performance of the above three algorithms. Since the algorithm in this article is designed for information systems, if the database processed in the experiment is a decision table, only the conditional attribute set is selected for processing. The results are shown in Table 1 (where |U| is the number of samples, |A| is the number of attributes, |B| is the number of attributes in the reduction set, T is the average running time (seconds) and * indicates that the running is out of memory).

Table 1. Reduction results and running time

UCI database	|U|	|A|	Algorithm 1		Algorithm 2		Algorithm 3	
			|B|	T	|B|	T	|B|	T
Zoo	101	16	11	2.98	11	3.55	11	4.07
Soybean (Large)	307	35	12	5.97	12	7.23	14	3.48
abalone	4177	8	3	1448.47	3	2089.31	4	148.05
Tic-tac-toe	958	9	8	24.65	8	44.10	8	6.24
Ionosphere	351	34	7	7.31	7	7.76	10	4.06
Hayes-roth	132	5	5	3.21	5	3.64	5	3.22
Primary	132	18	14	3.12	14	4.17	14	3.62
Chess	3196	36	29	*	29	*	31	682.80
Car Evaluation	1728	6	6	84.81	6	170.50	6	15.18

From the above experimental results, it can be seen that the reduction results of the above three algorithms are not much different. Although algorithm 3 is slightly inferior to algorithm 1 and algorithm 2 in reduction results, algorithm 3 is obviously faster than the other two algorithms in terms of running time, and this advantage is more obvious with the increase of data set. This shows that algorithm 3 has higher efficiency of attribute reduction, which is an improvement of algorithm 1 and algorithm 2. The experimental results are consistent with the theoretical analysis, which verifies the effectiveness of the algorithm.

6 Conclusion

Although the single attribute identification matrix can solve the shortcomings of large storage space and low generation efficiency of identification matrix to a certain extent, the calculation of single attribute identification matrix also has the problem of high computational complexity, which is not suitable for dealing with large-scale information system. In order to solve this problem, this paper proposes the concept of sub matrix of single attribute identification matrix, designs an attribute reduction algorithm based on local dependence, and verifies the effectiveness of this algorithm by combining UCI data set. Compared with the existing attribute reduction algorithm of single attribute identification matrix representation, the algorithm designed in this paper is more efficient, and it is used in attribute reduction of decision table, and has strong application value in data processing, data mining and other fields.

Acknowledgements. This work is supported by the National Natural Science Foundation of China (71771078, 71371064).

References

1. Zhang, W., Qiu, G.: Uncertain Decision Making Based on Rough Sets. Tsinghua University Press, Beijing (2005)
2. Pawlak, Z.: Rough sets. Int. J. Comput. Inf. Sci. **11**(5), 341–356 (1982)
3. Wang, S.K.M., Ziarko, W.: On optimal decision rules in decision tables. Bull. Polish Acad. Sci. **33**(11–12), 693–696 (1985)
4. Wu, S., Gou, P.: Attribute reduction algorithm on rough set and information entropy and its application. Comput. Eng. **37**(7), 56–58 (2011)
5. Wang, G., Yu, H., Yang, D.: Decision table reduction based on conditional information entropy. Chinese J. Comput. **25**(7), 759–766 (2002)
6. Jelonek, J., Krawiec, K., Slowinski, R.: Rough set reduction of attributes and their domains for neural networks. Int. J. Comput. Intell. **11**(2), 339–347 (1995)
7. Gao, Y., Chen, X., Yang, X., et al.: Neighborhood attribute reduction: a ulticriterion strategy based on sample selection. Information **9**(11), 282 (2018)
8. Yao, Y., Zhao, Y.: Discernibility matrix simplification for constructing attribute reducts. Inf. Sci. **179**(7), 867–882 (2009)
9. Wang, J., Gao, C.: Improved algorithm for attribute reduction based on discernihility matrix. Comput. Eng. **35**(3), 66–68 (2009)
10. Li, L., Yun, S., Yang, B.: Attribute reduction heuristic algorithm in rough sets based on Boolean matrix. Comput. Eng. **33**(10), 205–206 (2007)
11. Pawlak, Z., Skowron, A.: Rough sets and Boolean reasoning. Inf. Sci. **177**(1), 41–73 (2007)
12. Tong, X., Yun, S., Zhang, Y.: Attribute reduction algorithm of rough set based on Boolean matrix representation. J. Luoyang Inst. Sci. Technol. (Nat. Sci. Edn.) **19**(1), 69–71 (2009)
13. Feng, W., Zhang, M.: Converse delete attribute reduction algorithm based on core and improved condition distinguishing ability. Comput. Appl. Softw. **033**(005), 252–255 (2016)
14. Shi, Y., Huang, Y., Wang, C., et al.: Attribute reduction based on the Boolean matrix. Granular Comput. **4**(3), 313–322 (2019)

15. Li, Y., Guo, N., Wu, T., et al.: Attribute reduction based on concentration Boolean matrix under dominance relations. Comput. Sci. **45**(10), 229–234 (2018)
16. Mi, J., Chen, J.: Graph-based approaches for attribute reduction in rough sets. J. Northwest Univ. (Nat. Sci. Edn.) **49**(4), 508–516 (2019)
17. Li, F., Ren, Y., Jin, C.: Research on attribute reduction method based on dual attribute comprehensive dependency. J. Shanxi Univ. (Nat. Sci. Edn.) (2020, to be accepted)

Research on Uncertain Prediction Method Based on Credibility Distribution

Yan Li[1(✉)], Chenxia Jin[2], and Ying Wang[1]

[1] College of Science, Hebei University of Science and Technology, Shijiazhuang 050018, China
byxliyan@163.com

[2] School of Economics and Management, Hebei University of Science and Technology, Shijiazhuang 050018, China
jinchenxia2005@126.com

Abstract. With the rapid development of information society, traditional data mining is unable to meet the actual needs. In this paper, an uncertain prediction method based on credibility distribution (RDP) is proposed. Firstly, the implementation mechanism of the uncertain prediction based on credibility distribution in sampling is given. Secondly, combining with the law of large Numbers, the convergence characteristics of test credibility of the decision attribute corresponding to the value of a conditional attribute in sampling are analyzed. Finally, the validity of RDP is verified through Simulation experiment of UCI database. Theoretical analysis and simulation results show that RDP is feasible in interpretability and operability.

Keywords: Prediction · Credibility distribution · Big data · Sampling with replacement

1 Introduction

With the development of science and technology, the prediction based on data (regression and classification) becomes more important. Many scholars have done relevant research around classification prediction.

The commonly used classification algorithms are Genetic Algorithm [1], Bayesian [2], Decision Tree [3], and [4] Rough Sets. Das and Sengupta [5] proposed a group incremental feature selection algorithm based on rough set theory, the method is good in time and classification accuracy; Ricardo and Marcio et al. [6] proposed to use genetic algorithms to induce hierarchical and multi-label rules, and the multi-label classification rules of different levels are obtained; Wu and Pan et al. [7] proposed a new Artificial Immune System (AIS) based Adaptive Attribute-Weighted Naive Bayesian classification method (AISWNB), which was significantly better than NB in classification accuracy; Karabadj and Khelf et al. [8] proposed a method to generate the optimal decision tree by selecting the optimal training samples and attribute subsets; Zhai and Wang et al. [9] proposed a tolerance rough fuzzy decision tree algorithm (TRFDT) based on tolerance rough fuzzy set for the problem of loss of information caused by the fuzziness of conditional attributes; Trabelsi and Elouedi et al. [10]

C. Pang et al. (Eds.): SETE 2020/ICWL 2020, LNCS 12511, pp. 148–159, 2021.
https://doi.org/10.1007/978-3-030-66906-5_14

proposed the gain ratio belief decision tree (gr-bdt) and the diffusion ratio belief decision tree (dr-bdt) to deal with uncertain data of attribute values and class labels due to the traditional decision tree was not capable of processing incomplete information; Obregon and Kim et al. [11] proposed a new method, Rule COSI, which combined the output of the binary decision tree set and simplified it into a set of rules generation. All the above studies have accelerated the development of classification and prediction to some extent, but they also have the following two problems: 1) Predictions based on rules cannot directly give the prediction results for the conditional attributes of specific values; 2) When the credibility of the rule does not reach the credibility threshold, the data in the information system involved in the rule cannot provide the prediction results.

This paper will propose a prediction method based on the credibility of uncertain distribution (RDP), aiming at providing the value distribution of the decision attribute. The structure of this paper is as follows: Sect. 2 gives some preliminaries; In Sect. 3, the implementation mechanism of RDP is given by combining with sampling with replacement. In Sect. 4, the test credibility convergence characteristics of various possible values of decision attribute corresponding to a conditional attribute value is analyzed. In Sect. 5, the validity of RDP is verified through UCI database. Combined with the theoretical analysis and the experimental results, we know that RDP has good theoretical interpretability and operability in practical applications.

2 Preliminaries

In this paper, we suppose, 1) $\Delta = (\Omega, A, d, V, V_d)$ represents a decision information system (here, Ω represents the example set, $A = \{a_1, a_2, \cdots, a_n\}$ represents the conditional attribute set, d represents the decision attribute, $V = V_1 \times V_2 \times \cdots \times V_n$ represents the range of $(a_1, a_2, \cdots, a_n)$, and V_d represents the range of d); 2) for $\Delta = (\Omega, A, d, V, V_d)$ and $a_i \in A$, $e \in \Omega$, $a_i(e)$ represents the value of e at a_i, $d(e)$ represents the value of e at d($b_i \in V_i$ or $b_i = *$;* represents the various possible values of a_i; i.e., d_k is independent of the value of a_i when $b_i = *$ is true); 3) $\mathrm{B} = (b_1, b_2, \cdots, b_n)$ represents the values of a group of condition attributes in decision information system $\Delta = (\Omega, A, d, V, V_d)$, and $V_d = \{d_1, d_2, \cdots, d_l\}$, we call

$$\mathrm{P}(d_k|\mathrm{B}^\Delta) = \frac{\mathrm{P}(d_k \cap \mathrm{B}^\Delta)}{\mathrm{P}(\mathrm{B}^\Delta)} = \frac{N(d_k \cap \mathrm{B}^\Delta)}{N(\mathrm{B}^\Delta)} \tag{1}$$

as the credibility of when the value of the condition attribute is B, the decision attribute is d_k. (here, $N(\mathrm{B})$ represents the number of examples in which the value of condition attribute satisfy Bin $\Delta = (\Omega, A, d, V, V_d)$, $N(d_k \cap \mathrm{B}^\Delta)$ represents the number of examples in which the value of condition attribute satisfy B and the value of decision attribute satisfy D). 4) in $\Delta = (\Omega, A, d, V, V_d)$, the probability distribution column (Table 1) of the decision attribute of all examples whose value of the conditional attribute satisfies B, so that the decision-maker can make prediction according to the probability distribution of the value of the decision attribute.

Table 1. Distribution columns of d.

d	d_1	d_2	$\cdots$	d_l
P	$P(d_1\|B^{\Delta})$	$P(d_2\|B^{\Delta})$	$\cdots$	$P(d_l\|B^{\Delta})$

3 The Implementation Mechanism of RDP

In practice, decision makers usually believe in decisions with certain credibility. Therefore, this part studies how to determine the test credibility based on the values of decision attributes corresponding to condition attributes.

RDP is an uncertain information prediction method based on Credibility under the background of big data. This method will have a certain impact in the research field and practical application. The implementation steps include:

Step 1: Input decision information system $\Delta = (\Omega, A, d, V, V_d)$, condition attribute value $B = (b_1, b_2, \cdots, b_n)$, matching capacity m, sampling times n;

Step 2: The samples in $\Delta = (\Omega, A, d, V, V_d)$ are extracted by sampling with replacing until the number of the conditional attributes of the extracted samples meets the requirement of $B = (b_1, b_2, \cdots, b_n)$ and reaches the matching capacity m;

Step 3: Take the decision attribute value d in Ω^* as a random variable, and calculate the distribution column of d in Ω^*, $V_d = \{d_1, d_2, \cdots, d_l\}$. Then the credibility of d_k is $\bar{P}(d_k|\ B^{\Delta_i})$. The distribution column of d_k is shown in Table 2;

Table 2. Distribution columns of d under the ith sampling condition.

D	d_1	d_2	$\cdots$	d_l
P	$P(d_1\|\ B^{\Delta_i})$	$P(d_2\|\ B^{\Delta_i})$	$\cdots$	$P(d_l\|\ B^{\Delta_i})$

If $i<n$, turn to step 2; if not, turn to step 4;

Step 4: According to the following formula (Where $\bar{P}(d_k|\ B^{\Delta_i})$ is taken as the test credibility of the conditional attribute B and the decision attribute is d_k), the distribution columns of decision attribute d whose conditional attribute satisfies B are determined, as shown in Table 3:

Table 3. Credibility of the condition attribute satisfies B and the decision attribute takes d_k

d	d_1	d_2	$\cdots$	d_l
P	$\bar{P}(d_1\|\ B^{\Delta})$	$\bar{P}(d_2\|\ B^{\Delta})$	$\cdots$	$\bar{P}(d_l\|\ B^{\Delta})$

Note 4: The matching capacity m in step 1 should be large enough; and the sampling number n should not be too large (see part 4 for its theoretical analysis).

4 The Credibility Convergence of Classification Rules in Sampling Background

This part will combine the law of large Numbers to discuss the statistical law of distribution columns of decision attributes satisfying a certain conditional attribute in the sampling background. These will provide theoretical basis for our prediction method.

Lemma 1. Let X be a discrete random variable, its probability distribution is $P(\{X = x_i\}) = p_i$, $i = 1, 2, \cdots$, and $x_i \in [0, 1]$ is constant for any $i \in \{1, 2, \cdots\}$, $a \in (0, 1]$. Then we have the following conclusions:

1) $E(X)$ and $Var(X)$ all exist, and $E(\zeta)$, $Var(X) \in [0, 1]$;
2) For any $\varepsilon \in (0, a/2)$ and $\delta \in (0, \varepsilon]$, $|E(X) - a| \leq 2\varepsilon$ holds if $P(\{|X - a| \leq \varepsilon\}) \geq 1 - \delta$.

Proof: 1) Obvious;

2) According to the definition of mathematical expectation.

$$E(X) = \sum_{|x_i - a| \leq \varepsilon} x_i p_i + \sum_{|x_i - a| > \varepsilon} x_i p_i \leq \sum_{|x_i - a| \leq \varepsilon} (a + \varepsilon) p_i + \sum_{|x_i - a| > \varepsilon} 1 \cdot p_i \leq a + 2\varepsilon$$

$$E(X) = \sum_{|x_i - a| \leq \varepsilon} x_i p_i + \sum_{|x_i - a| > \varepsilon} x_i p_i \geq \sum_{|x_i - a| \leq \varepsilon} (a - \varepsilon) p_i + \sum_{|x_i - a| > \varepsilon} 0 \cdot p_i \geq a - 2\varepsilon$$

So, $|E(X) - a| \leq 2\varepsilon$.

Theorem 2. Let $\mathrm{B} = (b_1, b_2, \cdots, b_n)$ be a classification rule based on decision information system (Ω, A, d, V, V_d), $N(\mathrm{B}^{\Delta})$ be the number of examples in which the value of condition attribute satisfy B in $\Delta = (\Omega, A, d, V, V_d)$, $N(d_k \cap \mathrm{B}^{\Delta})$ be the number of examples in which the value of condition attribute satisfy B and the value of decision attribute satisfy d_k in $\Delta = (\Omega, A, d, V, V_d)$; $(\Omega_m^n, A, d, V, V_d)$ represents the decision information system that the condition attribute satisfy B with capacity of m obtained from Ω through sampling with replacement in the nth time, $N(d_k \cap \mathrm{B}^{\Delta_m^n})$ represents the number of examples in which the value of decision attribute satisfy d_k in $(\Omega_m^n, A, d, V, V_d)$, $X_{m;k}^n = N(d_k \cap \mathrm{B}^{\Delta_m^n})/m$, then for any natural number n,

$$\frac{1}{n}\sum_{i=1}^{n} X_{m;k}^{i} \xrightarrow[m \to \infty]{P} \frac{N(d_k \cap \mathrm{B}^{\Delta})}{N(\mathrm{B}^{\Delta})} \tag{2}$$

2) For any natural number m, there is

$$\frac{1}{n}\sum_{i=1}^{n} X_{m;k}^{i} \xrightarrow[n \to \infty]{[}]Pa_m \tag{3}$$

3) For any $\varepsilon \in (0,\ s/(2t)\ ]$, $\delta \in (0,\ \varepsilon]$ and natural number n, there must be a natural number M, when $m \geq M$, there is

$$P\left(\left\{\left|\frac{1}{n}\sum_{i=1}^{n} X_{m;\,k}^{i} - \frac{N(d_k \cap \mathrm{B}^{\Delta})}{N(\mathrm{B}^{\Delta})}\right| \leq 3\varepsilon\right\}\right) \geq (1-\delta)^2 \tag{4}$$

Where, $a_m = E\left(X_{m;\,k}^{1}\right)$, $\sigma_m^2 = Var\left(X_{m;\,k}^{1}\right)$.

Prove: 1) Let Ω^* represent the example set where the value of conditional attribute in Ω satisfy B, Ω_k^* represents the examples in Ω^* where the value of decision attribute satisfy d_k, $\Omega_{m\,;\,k}^{n}$ represents the number of examples in Ω_m^n where the value of decision attribute satisfy d_k, then any natural number n, if the acquisition process of $\Omega_{m\,;\,k}^{n}$ from Ω_m^n is understood as m times of independent sample extraction test, then the probability of each sample extracted is $N(d_k \cap \mathrm{B}^{\Delta})/N(\mathrm{B}^{\Delta})$ in Ω_k^*. From this and theorem 1: i) If $\tau_{k,r}$ represents the number of examples where the value of decision attribute of the rth extraction satisfies d_k, it obeys $b(1,\ N(d_k \cap \mathrm{B}^{\Delta})/N(\mathrm{B}^{\Delta}))$ and $E(\tau_{k,r}) = N(d_k \cap \mathrm{B}^{\Delta})/N(\mathrm{B}^{\Delta})$;

ii) $X_{m;k}^{n} = \frac{1}{m}N\left(d_k \cap \mathrm{B}^{\Delta_m^n}\right) = \frac{1}{m}\sum_{r=1}^{m}\tau_{k,r} \xrightarrow[m\to\infty]{[}]PN\left(d_k \cap \mathrm{B}^{\Delta}\right)/N\left(\mathrm{B}^{\Delta}\right)$;

iii) $\frac{1}{n}\sum_{i=1}^{n} X_{m;k}^{i} \xrightarrow[m\to\infty]{[}]PN\left(d_k \cap \mathrm{B}^{\Delta}\right)/N\left(\mathrm{B}^{\Delta}\right)$.

2) For any given nature number m, $N(d_k \cap \mathrm{B}^{\Delta_m^n})$ obeys two distributions $b\left(m,\ N(d_k \cap \mathrm{B}^{\Delta})/N(\mathrm{B}^{\Delta})\right)$, $N(d_k \cap \mathrm{B}^{\Delta_m^n}) \leq m$, that means $\{X_{m;\,k}^{n}\}_{n=1}^{\infty}$ is an independent, identically distributed sequence of random variables with both expected value $E(X_{m;\,k}^{n}) \triangleq a_m$ and variance $Var(X_{m;\,k}^{n}) \triangleq \sigma_m^2$, from this and Theorem 1 we know that $\frac{1}{n}\sum_{k=1}^{n} X_{m;k}^{i} \xrightarrow[n\to\infty]{P} a_m$.

3) It can be proven from the proof process of 1), 2) and Lemma 1 that:

i) $X_{m;\,k}^{1} \xrightarrow[m\to\infty]{P} N(d_k \cap \mathrm{B}^{\Delta})/N(\mathrm{B}^{\Delta})$, there exists natural number M, if $m \geq M$, then $P\left\{|X_{m;\,k}^{1} - N(d_k \cap \mathrm{B}^{\Delta})/N(\mathrm{B}^{\Delta})| \leq \varepsilon\right\} \geq 1-\delta$ is true;

ii) $\left\{|X_{m;\,k}^{1} - N(d_k \cap \mathrm{B}^{\Delta})/N(\mathrm{B}^{\Delta})| \leq \varepsilon\right\}$ and $\left\{|\frac{1}{n-1}\sum_{k=2}^{n} X_{m;\,k}^{i} - a_m| \leq \varepsilon\right\}$ are independent of each other, $P\left\{|\frac{1}{n-1}\sum_{k=2}^{n} X_{m;\,k}^{i} - a_m| \leq \varepsilon\right\} \geq 1-\delta$;

iii) When $\left\{|\frac{1}{n-1}\sum_{k=2}^{n} X_{m;\,k}^{i} - a_m| \leq \varepsilon\right\}$, $\left\{|X_{m;\,k}^{1} - N(d_k \cap \mathrm{B}^{\Delta})/N(\mathrm{B}^{\Delta})| \leq \varepsilon\right\}$ occur simultaneously,

$$\left|\frac{1}{n}\sum_{i=1}^{n} X_{m;\,k}^{i} - \frac{N(d_k \cap \mathrm{B}^{\Delta})}{N(\mathrm{B}^{\Delta})}\right| \leq \frac{1}{n}\left|X_{m;\,k}^{1} - \frac{N(d_k \cap \mathrm{B}^{\Delta})}{N(\mathrm{B}^{\Delta})}\right| + \frac{n-1}{n}\left|\frac{1}{n-1}\sum_{i=2}^{n} X_{m;\,k}^{i} - \frac{N(d_k \cap \mathrm{B}^{\Delta})}{N(\mathrm{B}^{\Delta})}\right| \leq 3\varepsilon \left\{|\frac{1}{n}\sum_{i=1}^{n} X_{m;\,k}^{i} - a_m| \leq \varepsilon\right\} \cap \left\{|X_{m;\,k}^{1} - N(d_k \cap \mathrm{B}^{\Delta})/N(\mathrm{B}^{\Delta})| \leq \varepsilon\right\} \subset$$

$\left\{|\frac{1}{n}\sum_{i=1}^{n} X_{m;\,k}^{i} - N(d_k \cap \mathrm{B}^{\Delta})/N(\mathrm{B}^{\Delta})| \leq 3\varepsilon\right\}$ this shows that when $m \geq M$, formula (5) is valid.

Theorem 2 describes the change law of $\frac{1}{n}\sum_{i=1}^{n} X_{m;\,k}^{i}$ from different aspects, the basic characteristics are: i) $X_{m;\,k}^{i}$ is an estimate of $N(d_k \cap \mathrm{B}^{\Delta})/N(\mathrm{B}^{\Delta})$ in sampling with replacement (when $|\Omega|$ is large enough, we can approximately treat the sampling without replacement as sampling with replacement); ii) As m increases, $\frac{1}{n}\sum_{i=1}^{n} X_{m;\,k}^{i}$ is the consistent estimator of $N(d_k \cap \mathrm{B}^{\Delta})/N(\mathrm{B}^{\Delta})$ (only when m is large enough, can $\frac{1}{n}\sum_{i=1}^{n} X_{m;\,k}^{i}$ approach $N(d_k \cap \mathrm{B}^{\Delta})/N(\mathrm{B}^{\Delta})$ with greater credibility); iii) No matter how big n is, there is no approximation between $\frac{1}{n}\sum_{i=1}^{n} X_{m;\,k}^{i}$ and $N(d_k \cap \mathrm{B}^{\Delta})/N(\mathrm{B}^{\Delta})$ when m is not big enough.

5 Simulation Experiment

This part will make systematic simulation on the six decision tables in Table 4. For convenience, we suppose that: $\bar{P}(d_k|\ \mathrm{B}_i^{\Delta})$ represents the test credibility of all possible values d_k of the decision attribute corresponding to the conditional attribute B obtained by sampling in decision information system Δ; $\bar{P}(d_k|\ \mathrm{B}_i^{\Delta})$ is to consider the test credibility of all possible values d_k of the decision attribute corresponding to the conditional attribute B obtained from all samples.

Table 4. Basic information for several UCI databases

Database number	Database name	Data size	Number of conditional attributes	Number of values of decision attributes
U_1	Qualitative_ Bankruptcy	250	6	2
U_2	Ecoli	336	7	8
U_3	Balance Scale	625	4	3
U_4	Tic-Tac-Toe End game	958	9	2
U_5	Yeast	1484	8	10
U_6	Car Evaluation	1728	6	4

This part carries out simulation tests on the selected conditional attributes of each database in Table 4. This simulation mainly observes the approximation performance of decision attribute value distribution $\bar{P}(d_k|\ \mathrm{B}^{\Delta})$ determined by each attribute under the conditions of different matching capacity m and sampling times n. Where: 1) arbitrarily Select the attribute value B_i in the Uiarbitrarily, $\mathrm{B}_1 = (2, 2, *, *, *, *)$, $\mathrm{B}_2 = (1, *, 2, 2, 2, *, *)$, $\mathrm{B}_3 = (4, 5, *, *)$, $\mathrm{B}_5 = (*, *, 2, *, 2, *, *, *)$, $\mathrm{B}_6 = (*, 2, 2, *, *, *)$; 2) The distribution of various possible values of the decision attribute corresponding to B_i is simulated by means of sampling with replacement (Where, the matching capacity m is 50, 200, 400, 800 and 1600 respectively, and the sampling number n is 1, 5 and 10 respectively). The results are shown in Table 5. All samples are considered to simulate the distribution of various possible values corresponding to B_i, and the results are shown in Table 6.

Table 5. Simulation of the decision attribute values distribution for B under all samples

$P(d_k \mid B_i^{\Delta})$		i					
		1	2	3	4	5	6
k	1	0.2432	0.7474	0.8800	0.3500	0.0303	0.7778
	2	0.7568	0	0.0800	0.6500	0	0.2037
	3		0	0.0400		0	0
	4		0			0.1515	0.0185
	5		0.2316			0	
	6		0			0.0606	
	7		0			0.7273	
	8		0.0211			0.0303	
	9					0	
	10					0	

Table 6. Simulation of the decision attribute values distribution for B under sampling

Δ and B	Sampling number	$\overline{P}(d_k \mid B_i^{\Delta})$	Matching capacity m				
			50	200	400	800	1600
$\Delta = U_1$ B_1	1	$\overline{P}(d_1 \mid B_1^{\Delta})$	0.3000	0.2250	0.2450	0.2388	0.2492
		$\overline{P}(d_2 \mid B_1^{\Delta})$	0.7000	0.7750	0.7550	0.7613	0.7508
		time/s	0.0814	0.0850	0.1014	0.1317	0.1855
	5	$\overline{P}(d_1 \mid B_1^{\Delta})$	0.2680	0.2360	0.2460	0.2445	0.2425
		$\overline{P}(d_2 \mid B_1^{\Delta})$	0.7320	0.7640	0.7540	0.7555	0.7575
		time/s	0.0930	0.1617	0.2195	0.3454	0.5839
	10	$\overline{P}(d_1 \mid B_1^{\Delta})$	0.2240	0.2490	0.2485	0.2404	0.2403
		$\overline{P}(d_2 \mid B_1^{\Delta})$	0.7760	0.7510	0.7515	0.7596	0.7597
		time/s	0.3855	0.2570	0.3548	0.6013	1.0897
$\Delta = U_2$ B_2	1	$\overline{P}(d_1 \mid B_2^{\Delta})$	0.8200	0.7500	0.7750	0.7513	0.7231
		$\overline{P}(d_2 \mid B_2^{\Delta})$	0	0	0	0	0
		$\overline{P}(d_3 \mid B_2^{\Delta})$	0	0	0	0	0
		$\overline{P}(d_4 \mid B_2^{\Delta})$	0	0	0	0	0
		$\overline{P}(d_5 \mid B_2^{\Delta})$	0.1400	0.2500	0.2075	0.2275	0.2575
		$\overline{P}(d_6 \mid B_2^{\Delta})$	0	0	0	0	0
		$\overline{P}(d_7 \mid B_2^{\Delta})$	0	0	0	0	0
		$\overline{P}(d_8 \mid B_2^{\Delta})$	0.0400	0	0.0175	0.0213	0.0194
		time/s	0.5428	0.5000	0.5220	0.5349	0.5709

(continued)

Table 6. (*continued*)

Δ and B	Sampling number	$\overline{P}(d_k \mid B_i^{\Delta})$	Matching capacity *m*				
			50	200	400	800	1600
	5	$\overline{P}(d_1 \mid B_2^{\Delta})$					0.7505
		$\overline{P}(d_2 \mid B_2^{\Delta})$	0	0	0	0	0
		$\overline{P}(d_3 \mid B_2^{\Delta})$	0	0	0	0	0
		$\overline{P}(d_4 \mid B_2^{\Delta})$	0	0	0	0	0
		$\overline{P}(d_5 \mid B_2^{\Delta})$	0.2240	0.2420	0.2420	0.2303	0.2290
		$\overline{P}(d_6 \mid B_2^{\Delta})$	0	0	0	0	0
		$\overline{P}(d_7 \mid B_2^{\Delta})$	0	0	0	0	0
		$\overline{P}(d_8 \mid B_2^{\Delta})$	0.0280	0.0245	0.0245	0.0175	0.0205
		time/s	0.5223	0.6124	0.6124	0.7086	0.9583
	10	$\overline{P}(d_1 \mid B_2^{\Delta})$	0.7452	0.7524	0.7484	0.7431	0.7464
		$\overline{P}(d_2 \mid B_2^{\Delta})$	0	0	0	0	0
		$\overline{P}(d_3 \mid B_2^{\Delta})$	0	0	0	0	0
		$\overline{P}(d_4 \mid B_2^{\Delta})$	0	0	0	0	0
		$\overline{P}(d_5 \mid B_2^{\Delta})$	0.2340	0.2255	0.2293	0.2358	0.2319
		$\overline{P}(d_6 \mid B_2^{\Delta})$	0	0	0	0	0
		$\overline{P}(d_7 \mid B_2^{\Delta})$	0	0	0	0	0
		$\overline{P}(d_8 \mid B_2^{\Delta})$	0.0208	0.0221	0.0224	0.0212	0.0217
		time/s	0.6548	1.0749	1.6062	2.7701	5.0945
Δ and B	Sampling number	$\overline{P}(d_k \mid B_i^{\Delta})$	matching capacity*m*				
			50	200	400	800	1600
$\Delta = U_3$ B_3	1	$\overline{P}(d_1 \mid B_3^{\Delta})$	0.9	0.8900	0.8625	0.8858	0.8744
		$\overline{P}(d_2 \mid B_3^{\Delta})$	0.1	0.0700	0.0800	0.0736	0.0744
		$\overline{P}(d_3 \mid B_3^{\Delta})$	0	0.0400	0.0575	0.0406	0.0513
		time/s	0.1137	0.1348	0.1803	0.1787	0.1896
	5	$\overline{P}(d_1 \mid B_3^{\Delta})$	0.8800	0.8820	0.8800	0.8777	0.8814
		$\overline{P}(d_2 \mid B_3^{\Delta})$	0.1040	0.0660	0.0810	0.0802	0.0838
		$\overline{P}(d_3 \mid B_3^{\Delta})$	0.0160	0.0520	0.0390	0.0421	0.0348
		time/s	0.1509	0.3054	0.5176	0.5266	0.5253
	10	$\overline{P}(d_1 \mid B_3^{\Delta})$	0.8740	0.8725	0.8821	0.8793	0.8780
		$\overline{P}(d_2 \mid B_3^{\Delta})$	0.0780	0.0915	0.0770	0.0812	0.0819
		$\overline{P}(d_3 \mid B_3^{\Delta})$	0.0480	0.0360	0.0410	0.0395	0.0401
		time/s	0.2146	0.5282	0.9367	0.9677	0.9545

(*continued*)

Table 6. (*continued*)

Δ and B	Sampling number	$\overline{P}(d_k \mid B_i^{\Delta})$	Matching capacity *m*				
			50	200	400	800	1600
Δ = U_4 B_4	1	$\overline{P}(d_1 \mid B_4^{\Delta})$	0.3000	0.2250	0.2450	0.2388	0.2492
		$\overline{P}(d_2 \mid B_4^{\Delta})$	0.7000	0.7750	0.7550	0.7613	0.7508
		time/s	0.0814	0.0850	0.1014	0.1317	0.1855
	5	$\overline{P}(d_1 \mid B_4^{\Delta})$	0.2680	0.2360	0.2460	0.2445	0.2425
		$\overline{P}(d_2 \mid B_4^{\Delta})$	0.7320	0.7640	0.7540	0.7555	0.7575
		time/s	0.0930	0.1617	0.2195	0.3454	0.5839
	10	$\overline{P}(d_1 \mid B_4^{\Delta})$	0.2240	0.2490	0.2485	0.2404	0.2403
		$\overline{P}(d_2 \mid B_4^{\Delta})$	0.7760	0.7510	0.7515	0.7596	0.7597
		time/s	0.3855	0.2570	0.3548	0.6013	1.0897
Δ = U_5 B_5	1	$\overline{P}(d_1 \mid B_5^{\Delta})$	0.8200	0.7500	0.7750	0.7513	0.7231
		$\overline{P}(d_2 \mid B_5^{\Delta})$	0	0	0	0	0
		$\overline{P}(d_3 \mid B_5^{\Delta})$	0	0	0	0	0
		$\overline{P}(d_4 \mid B_5^{\Delta})$	0	0	0	0	0
		$\overline{P}(d_5 \mid B_5^{\Delta})$	0.1400	0.2500	0.2075	0.2275	0.2575
		$\overline{P}(d_6 \mid B_5^{\Delta})$	0	0	0	0	0
		$\overline{P}(d_7 \mid B_5^{\Delta})$	0	0	0	0	0
		$\overline{P}(d_8 \mid B_5^{\Delta})$	0.0400	0	0.0175	0.0213	0.0194
		$\overline{P}(d_9 \mid B_5^{\Delta})$					
		$\overline{P}(d_{10} \mid B_5^{\Delta})$					
		time/s	0.5428	0.5000	0.5220	0.5349	0.5709
Δ and B	Sampling number	$\overline{P}(d_k \mid B_i^{\Delta})$	matching capacity*m*				
			50	200	400	800	1600
Δ = U_5 B_5	5	$\overline{P}(d_1 \mid B_5^{\Delta})$	0.0240	0.0350	0.0362	0.0358	0.0348
		$\overline{P}(d_2 \mid B_5^{\Delta})$	0	0	0	0	0
		$\overline{P}(d_3 \mid B_5^{\Delta})$	0	0	0	0	0
		$\overline{P}(d_4 \mid B_5^{\Delta})$	0.1560	0.1420	0.1442	0.1526	0.1595
		$\overline{P}(d_5 \mid B_5^{\Delta})$	0	0	0	0	0
		$\overline{P}(d_6 \mid B_5^{\Delta})$	0.0720	0.0650	0.0671	0.0571	0.0455
		$\overline{P}(d_7 \mid B_5^{\Delta})$	0.7280	0.7250	0.7191	0.7249	0.7293
		$\overline{P}(d_8 \mid B_5^{\Delta})$	0.0200	0.0330	0.0335	0.0296	0.0309
		$\overline{P}(d_9 \mid B_5^{\Delta})$	0	0	0	0	0
		$\overline{P}(d_{10} \mid B_5^{\Delta})$	0	0	0	0	0
		time/s	0.3225	0.6004	0.6504	0.6381	0.6301

(*continued*)

Table 6. (*continued*)

Δ and B	Sampling number	$\bar{P}(d_k\| B_i^{\Delta})$	Matching capacity m				
			50	200	400	800	1600
	10	$\bar{P}(d_1\| B_5^{\Delta})$					0.0289
		$\bar{P}(d_2\| B_5^{\Delta})$	0	0	0	0	0
		$\bar{P}(d_3\| B_5^{\Delta})$	0	0	0	0	0
		$\bar{P}(d_4\| B_5^{\Delta})$	0.1300	0.1456	0.1468	0.1429	0.1510
		$\bar{P}(d_5\| B_5^{\Delta})$	0	0	0	0	0
		$\bar{P}(d_6\| B_5^{\Delta})$	0.0660	0.0622	0.0646	0.0665	0.0648
		$\bar{P}(d_7\| B_5^{\Delta})$	0.7340	0.7284	0.7332	0.7322	0.7234
		$\bar{P}(d_8\| B_5^{\Delta})$	0.0360	0.0352	0.0249	0.0252	0.0319
		$\bar{P}(d_9\| B_5^{\Delta})$	0	0	0	0	0
		$\bar{P}(d_{10}\| B_5^{\Delta})$	0	0	0	0	0
		time/s	0.4388	1.0289	1.0838	1.0865	1.0791
$\Delta = U_6$ B_6	1	$\bar{P}(d_1\| B_6^{\Delta})$	0.9	0.8900	0.8625	0.8858	0.8744
		$\bar{P}(d_2\| B_6^{\Delta})$	0.1	0.0700	0.0800	0.0736	0.0744
		$\bar{P}(d_3\| B_6^{\Delta})$	0	0.0400	0.0575	0.0406	0.0513
		$\bar{P}(d_4\| B_6^{\Delta})$					
		time/s	0.1137	0.1348	0.1803	0.1787	0.1896
	5	$\bar{P}(d_1\| B_6^{\Delta})$	0.8800	0.8820	0.8800	0.8777	0.8814
		$\bar{P}(d_2\| B_6^{\Delta})$	0.1040	0.0660	0.0810	0.0802	0.0838
		$\bar{P}(d_3\| B_6^{\Delta})$	0.0160	0.0520	0.0390	0.0421	0.0348
		$\bar{P}(d_4\| B_6^{\Delta})$					
		time/s	0.1509	0.3054	0.5176	0.5266	0.5253
	10	$\bar{P}(d_1\| B_6^{\Delta})$	0.8740	0.8725	0.8821	0.8793	0.8780
		$\bar{P}(d_2\| B_6^{\Delta})$	0.0780	0.0915	0.0770	0.0812	0.0819
		$\bar{P}(d_3\| B_6^{\Delta})$					
		$\bar{P}(d_4\| B_6^{\Delta})$	0.0480	0.0360	0.0410	0.0395	0.0401
		time/s	0.2146	0.5282	0.9367	0.9677	0.9545

As can be seen from Table 6, the difference between $\bar{P}(d_k| B^{\Delta})$ and $P(d_k| B^{\Delta})$ gradually decreases with the increase of matching capacity m (Since one sample (i.e., n = 1) is contingency, we mainly refer to n = 5 and n = 10, for the condition attribute $B_1 = (2, 2, *, *, *, *)$ in database U1, when $m \geq 400$, $|\bar{P}(d_k| B_1^{\Delta}) - P(d_k| B_1^{\Delta})| < 0.5\%$; For the condition attribute $B_2 = (1, *, 2, 2, 2, *, *)$ in database U2, when $m \geq 800$, $|\bar{P}(d_k| B_2^{\Delta})$ - $P(d_k| B_2^{\Delta})| < 0.5\%$; For the condition attribute $B_3 = (4, 5, *, *)$ in U3, when $m \geq 400$, $|\bar{P}(d_k| B_3^{\Delta}) - P(d_k| B_3^{\Delta})| < 0.5\%$; For $B_4 = (*, *, 2, *, 2, *, *, *, 1)$ in database U4, when $m \geq 800$, $|\bar{P}(d_k| B_4^{\Delta}) - P(d_k| B_4^{\Delta})| < 0.5\%$; For the condition attribute $B_5 = (*, *, 2, *, 2, *, *, *)$ in U5, when $m \geq 1600$,

$|\bar{P}(d_k|\ B_5^{\Delta})$ - $P(d_k|\ B_5^{\Delta})|$ <0.5%; For $B_6 = (*, 2, 2, *, *, *)$ in database U6, when $m \geq 800$, $|\bar{P}(d_k|\ B_6^{\Delta})$ - $P(d_k|\ B_6^{\Delta})|$ <0.5%). However, there is no significant dependence on the sampling number n, which is in partial agreement with the theoretical analysis in Sect. 4 (the matching capacity is at least 1000 and the sampling times is 5–10).

6 Conclusion

Aiming at the problem of incomplete prediction of big data with structural characteristics, an uncertainty prediction method based on credibility distribution (RDP) is proposed in this paper, and we have done the following work: 1) Combined with the distribution in the sense of probability, the implementation mechanism of the uncertainty prediction based on credibility is given; 2) Taking the law of large Numbers as the background, the convergence characteristics of the test credibility of various decision attribute values corresponding to a condition attribute value in sampling condition has been analyzed; 3) The validity of RDP has been verified through UCI database by taking the test credibility of the corresponding decision attribute as the observation point when the conditional attribute takes a certain value. Combined with the theoretical analysis of the law of large Numbers and the experimental results of UCI database, it is shown that RDP has good theoretical interpretability and operability in practical applications, which provides a lot of theoretical and experimental support for risk prediction analysis in the current background of big data application.

Acknowledgements. This work is supported by the National Natural Science Foundation of China (71771078, 71371064).

References

1. Kirkpatrick, S., Gelatt, C.D., Vecchi, M.P.: Optimization by simulated annealing. Science **220**(4598), 671–680 (1983)
2. Cooper, G.F., Herskovits, E.A.: A Bayesian method for the induction of probabilistic networks from data. Mach. Learn. **9**(4), 309–347 (1992)
3. Quinlan, J.R.: Induction of decision trees. Mach. Learn. **1**(1), 81–106 (1986)
4. Pawlak, Z.: Rough sets. Int. J. Comput. Inf. Sci. (IJCIS) **11**, 341–356 (1982)
5. Das, A.K., Sengupta, S., Bhattacharyya, S.: A group incremental feature selection for classification using rough set theory based genetic algorithm. Appl. Soft Comput. **65**, 400–411 (2018)
6. Cerri, R., Basgalupp, M.P., Barros, R.C., de Carvalho, A.C.P.L.F.: Inducing hierarchical multi-label classification rules with genetic algorithms. Appl. Soft Comput. **77**, 584–604 (2019)
7. Wu, J., Pan, S., Zhu, X., Cai, Z., Zhang, P., Zhang, C.: Self-adaptive attribute weighting for Naive Bayes classification. Expert Syst. Appl. **42**(3), 1487–1502 (2015)
8. Karabadji, N.E.I., Khelf, I., Seridi, H., Aridhi, S., Remond, D., Dhifli, W.: A data sampling and attribute selection strategy for improving decision tree construction. Expert Syst. Appl. **129**, 84–96 (2019)

9. Zhai, J., Wang, X., Zhang, S., Hou, S.: Tolerance rough fuzzy decision tree. Inform. Sci. **465**, 425–438 (2018)
10. Trabelsi, A., Elouedi, Z., Lefevre, E.: Decision tree classifiers for evidential attribute values and class labels. Fuzzy SetSyst. **366**, 46–62 (2019)
11. Obregon, J., Kim, A., Jung, J.Y.: RuleCOSI: combination and simplification of production rules from boosted decision trees for imbalanced classification. Expert Syst. Appl. **126**, 64–82 (2019)
12. Mao, S., Cheng, Y., Pu, X.: Course of probability theory and mathematical statistics. Higher Education Press, Peking (2004)

Variable Precision Multi-granularity Rough Soft Sets Based on Multiple Thresholds

Xiaomin Wang(✉), Yan Shang, Yang Liu, and Hanxiang He

School of Mathematics and Statistics, Northeastern University at Qinhuangdao, Qinhuangdao 066004, China
wangxiaomin@neuq.edu.cn, 2425228436@qq.com, 617027741@qq.com, liuy960117@163.com

Abstract. In this paper the concepts of variable precision rough soft sets and variable precision multi-granularity rough soft sets are proposed. Based on this, considering set different fault tolerance thresholds for different knowledge granularity, a variable precision multi-granularity rough soft set model based on multiple thresholds is also proposed, and some basic properties are presented. Finally, the application of the model in decision-making is also discussed with an example.

Keywords: Rough soft sets · Multi-granularity rough soft sets · Variable precision rough soft sets · Multiple thresholds

1 Introduction

In the field of classical mathematics, all concepts of definition and all mathematical tools used are accurate. However, with the continuous development of society, especially the rapid development of network technology, the amount of data has increased sharply, and the uncertain or inaccurate information is increasing. Because of the proposition and development of probability theory, fuzzy theory, etc., uncertain and inaccuracy problems have studied, but the priori information they need for data processing is usually difficult to obtain. In 1982, Pawlak [1] proposed rough set theory, which was mainly used to deal with inaccurate and incomplete problems. Compared with the probability theory and fuzzy sets, it does not need to provide any prior knowledge beyond the data that the problem needs to deal with. The theory has been widely used in many fields and has achieved gratifying results. But whether it is the probability theory, fuzzy sets [2], rough sets, their common shortcoming is the lack of parametric tools. In 1999, Molodtsov [3] proposed the soft set theory, which is a new mathematical method to deal with ambiguity, incompleteness information.

Later, people extended the classic rough set at different levels. On the one hand, Ziarko [4] applied the concept of variable precision to rough sets, and added the fault tolerance rate on the basis of classic rough sets. On the other hand, Chinese scholars Qian [5] used attribute set sequences to divide the original universe of rough sets from the perspective of granular computing, and proposed a multi-granularity rough set model. Qian [6] proposed optimistic multi-granularity rough sets and pessimistic multi-granularity rough sets. Considering the rough set theory and the soft set theory are two

C. Pang et al. (Eds.): SETE 2020/ICWL 2020, LNCS 12511, pp. 160–170, 2021.
https://doi.org/10.1007/978-3-030-66906-5_15

different tools for dealing with uncertainty. Feng [7] combined rough sets and soft sets to produce the concept of mixed models such as rough soft sets and soft rough sets. Wang [8] proposed the multi-granularity rough soft set model, which further extended the mixed model of rough sets and soft sets. In addition, many scholars [9–11] made a special study on rough soft sets, put forward rough soft semi-rings theories, and applied rough soft sets to decision-making systems.

In this paper, the rough soft set is extended in two steps. Firstly, based on the concept of the multi-granularity rough soft sets and the variable precision rough soft sets, the concept of variable precision multi-granularity rough soft sets is proposed. Secondly, in the study of variable-precision multi-granularity rough soft sets, it is found that the same fault-tolerance threshold is used for each granularity, that ignores the difference in the importance of different granularity in the study of different problem. Based on this, a variable precision multi-granularity rough soft set model based on multiple thresholds is proposed. This model can flexibly set parameters according to the needs of the problem, making it more suitable for mixed and incomplete information systems.

2 Preliminaries

In this section, we recall definitions of rough set, soft set, rough soft set and some related concepts.

Definition 2.1 [1]. Let U be a nonempty finite set and R be an equivalence relation on U. If X can be written as the union of some equivalence classes, then we say X is definable otherwise it is not definable. If X is not definable, then we can approximate it by two definite subsets called lower and upper approximations of X as the following:

$$\underline{apr}_R(X) = \{X \in U : [X]_R \subseteq X\}, \overline{apr}_R(X) = \{X \in U : [X]_R \cap X \neq \phi\},$$

where $[x]_R$ denotes the equivalence class determined by equivalence relation R containing some element $x \in U$. A rough set is the pair $(\underline{apr}(X), \overline{apr}(X))$.

Clearly $\underline{apr}(X)$ is the greatest definable set contained in X, whereas $\overline{apr}(X)$ is the least definable set containing X.

Definition 2.2 [3]. Let U be an initial universe of objects and E be the set of certain parameters in relation to the objects in U. Let $P(U)$ denote the power set of U. A pair $S = (F, A)$ is called soft set over U, where $A \subseteq E$ and $F : A \to P(U)$ is a set-valued mapping.

In other words, a soft set over U is a parameterized family of subsets of the universe U. For $e \in A, F(e)$ might be considered as the set of $e-$ approximate elements in the soft set $S = (F, A)$. It is worth noting that $F(e)$ might be arbitrary: some of them might be empty, and some might have nonempty inter-section. The absence of any restrictions on the approximate description in soft set theory makes this theory very convenient and easily applicable in practice. Actually, we can use any suitable parameterization with the help of words and sentences, real numbers, functions, mappings, and so on.

Definition 2.3 [7]. Let $K = (U, R)$ be an Pawlak approximation space, where U is a nonempty finite set and R is a equivalence relation on U. Then the pair $S = (F, A)$ is a soft set over U. Based on K, the following two operations were defined:

$$\underline{apr}_R(S) = (\underline{F}, A), \overline{apr}_R(S) = (\overline{F}, A),$$

for $\forall e \in A$,

$$\begin{aligned}\underline{F}(e) = \underline{apr}_R(S) = \{x \in U : [x]_R \subseteq F(e)\}, \overline{F}(e) = \overline{apr}_R(S) \\ = \{x \in U : [x]_R \cap F(e) \neq \varphi\},\end{aligned}$$

assigning to every soft sets S two sets $\underline{apr}_R(S)$ and $\overline{apr}_R(S)$, which are called the rough $R-$ lower approximation and rough $R-$ upper approximation of S, respectively. In general, we refer to $\underline{apr}_R(S)$ and $\overline{apr}_R(S)$ as rough soft approximation of S with respect to R. Moreover, the sets

$$Pos_R(S) = \underline{apr}_R(S), Neg_R(S) = [\overline{apr}_R(S)]^c, Bnd_R(S) = \overline{apr}_R(S) \backslash \underline{apr}_R(S),$$

are called the rough $R-$positive region, the rough $R-$negative region and the rough $R-$boundary region of S, respectively.

If $\overline{apr}_R(S) = \underline{apr}_R(S)$, S is said to be rough definable; otherwise S is called a rough soft set.

Definition 2.4. Let $K = (U, R)$ be an Pawlak approximation space, where U is a nonempty finite set and R is a family of equivalence relations on U, $R_1, R_2, \cdots R_n \in R$. Then the pair $S = (F, A)$ is a soft set over U.

The following four operations were defined:

$$\underline{apr}^O_{\sum_{i=1}^n R_i}(S) = (\underline{F}^O, A), \overline{apr}^O_{\sum_{i=1}^n R_i}(S) = \sim \underline{apr}^O_{\sum_{i=1}^n R_i}(\sim S);$$

$$\underline{apr}^P_{\sum_{i=1}^n R_i}(S) = (\underline{F}^P, A), \overline{apr}^P_{\sum_{i=1}^n R_i}(S) = \sim \underline{apr}^P_{\sum_{i=1}^n R_i}(\sim S),$$

for $\forall e \in A$,

$$\underline{F}^O(e) = \{x \in U : [x]_{R_1} \subseteq F(e) \vee [x]_{R_2} \subseteq F(e) \vee \cdots \vee [x]_{R_n} \subseteq F(e)\};$$

$$\underline{F}^P(e) = \{x \in U : [x]_{R_1} \subseteq F(e) \wedge [x]_{R_2} \subseteq F(e) \wedge \cdots \wedge [x]_{R_n} \in F(e)\}.$$

If $\underline{apr}^O_{\sum_{i=1}^n R_i}(S) = \overline{apr}^O_{\sum_{i=1}^n R_i}(S)$, S is said to be exact set; otherwise S is called a optimistic multi-granularity rough soft set.

If $\underline{apr}^{P}_{\sum_{i=1}^{n} R_i}(S) = \overline{apr}^{P}_{\sum_{i=1}^{n} R_i}(S)$, S is said to be exact set; otherwise S is called a pessimistic multi-granularity rough soft set.

3 Variable Precision Multi-granularity Rough Soft Set Model Based on Multiple Thresholds

Definition 3.1. Let $K = (U, R)$ be an Pawlak approximation space, where U is a nonempty finite set and R is a equivalence relation on U. Then the pair $S = (F, A)$ is a soft set over U. $0 \leq \beta < 0.5$.
The following two operations were defined:

$$\underline{apr}^{\beta}_{R}(S) = (\underline{F}^{\beta}_{R}, A), \overline{apr}^{\beta}_{R}(S) = \sim \underline{apr}^{\beta}_{R}(\sim S),$$

for $\forall e \in A$,

$$\underline{F}^{\beta}_{R}(e) = \{x \in U : c([x]_R, F(e)) \leq \beta\},$$

where

$$c(X, Y) = \begin{cases} 1 - \frac{|X \cap Y|}{|X|}, & |X| > 0 \\ 0, & |X| = 0 \end{cases},$$

$|X|$ represent the cardinal number of the set X. $c(X, Y)$ is called the relative error classification rate of set X with respect to set Y.

If $\underline{apr}^{\beta}_{R}(S) = \overline{apr}^{\beta}_{R}(S)$, S is said to be exact set; otherwise S is called a variable precision rough soft set.

Definition 3.2. Let $K = (U, R)$ be an Pawlak approximation space, where U is a nonempty finite set and R is a family of equivalence relation on U, $R_1, R_2, \cdots R_n \in R$. $0 \leq \beta < 0.5$. Then the pair $S = (F, A)$ is a soft set over U.

The following two operations were defined:

$$\underline{apr}^{\beta}_{\sum_{i=1}^{n} R_i}(S) = (\underline{F}^{\beta}_{\sum_{i=1}^{n} R_i}, A), \overline{apr}^{\beta}_{\sum_{i=1}^{n} R_i}(S) = \sim \underline{apr}^{\beta}_{\sum_{i=1}^{n} R_i}(\sim S),$$

for $\forall e \in A$,

$$\underline{F}^{\beta}_{\sum_{i=1}^{n} R_i}(e) = \left\{x \in U : c([x]_{R_i}, F(e)) \leq \beta\right\}$$

if $\underline{apr}^{\beta}_{\sum_{i=1}^{n} R_i}(S) = \overline{apr}^{\beta}_{\sum_{i=1}^{n} R_i}(S)$, S is said to be exact set; otherwise S is called a variable precision multi-granularity rough soft set.

Definition 3.3. Let $K = (U, R)$ be an Pawlak approximation space, where U is a nonempty finite set and R is a family of equivalence relation on U, $R_1, R_2, \cdots R_n \in R$.

$0 \leq \beta < 0.5$, $B = \{\beta_1, \beta_2, \cdots, \beta_n\}$ is a set of variable precision thresholds set for different knowledge granularity. The pair $S = (F, A)$ is a soft set over U.

The following two operations were defined:

$$\underline{apr}^{(B,O)}_{\sum_{i=1}^{n} R_i}(S) = (\underline{F}^{(B,O)}_{\sum_{i=1}^{n} R_i}, A), \overline{apr}^{(B,O)}_{\sum_{i=1}^{n} R_i}(S) = \sim \underline{apr}^{(B,O)}_{\sum_{i=1}^{n} R_i}(\sim S),$$

for $\forall e \in A$,

$$\underline{F}^{(B,O)}_{\sum_{i=1}^{n} R_i}(e) = \left\{x \in U : c([x]_{R_1}, F(e)) \leq \beta_1 \vee \cdots \vee c([x]_{R_n}, F(e)) \leq \beta_n\right\},$$

if $\underline{apr}^{(B,O)}_{\sum_{i=1}^{n} R_i}(S) = \overline{apr}^{(B,O)}_{\sum_{i=1}^{n} R_i}(S)$, S is said to be exact set; otherwise S is called a variable precision optimistic multi-granularity rough soft set based on multiple thresholds.

Moreover, the positive region, negative region and boundary region of the variable precision optimistic multi-granularity rough soft set model based on multiple thresholds are:

$$\begin{aligned} POS(S) &= \underline{apr}^{(B,O)}_{\sum_{i=1}^{n} R_i}(S), NEG(S) = U - \overline{apr}^{(B,O)}_{\sum_{i=1}^{n} R_i}(S), BND(S) \\ &= \overline{apr}^{(B,O)}_{\sum_{i=1}^{n} R_i}(S) - \underline{apr}^{(B,O)}_{\sum_{i=1}^{n} R_i}(S). \end{aligned}$$

Definition 3.4. Let $K = (U, R)$ be an Pawlak approximation space, where U is a nonempty finite set and R is a family of equivalence relation on U, $R_1, R_2, \cdots R_n \in R$.

$0 \leq \beta < 0.5$, $B = \{\beta_1, \beta_2, \cdots, \beta_n\}$ is a set of variable precision thresholds set for different knowledge granularity. The pair $S = (F, A)$ is a soft set over U.

The following two operations were defined:

$$\underline{apr}^{(B,P)}_{\sum_{i=1}^{n} R_i}(S) = (\underline{F}^{(B,P)}_{\sum_{i=1}^{n} R_i}, A), \overline{apr}^{(B,P)}_{\sum_{i=1}^{n} R_i}(S) = \sim \underline{apr}^{(B,P)}_{\sum_{i=1}^{n} R_i}(\sim S),$$

for $\forall e \in A$,

$$\underline{F}^{(B,P)}_{\sum_{i=1}^{n} R_i}(e) = \left\{x \in U : c([x]_{R_1}, F(e)) \leq \beta_1 \wedge \cdots \wedge c([x]_{R_n}, F(e)) \leq \beta_n\right\},$$

if $\underline{apr}^{(B,P)}_{\sum_{i=1}^{n} R_i}(S) = \overline{apr}^{(B,P)}_{\sum_{i=1}^{n} R_i}(S)$, S is said to be exact set; otherwise S is called a variable precision pessimistic multi-granularity rough soft set based on multiple thresholds.

Moreover, the positive region, negative region and boundary region of the variable precision pessimistic multi-granularity rough soft set model based on multiple thresholds are:

$$POS(S) = \underline{apr}^{(B,P)}_{\sum_{i=1}^{n} R_i}(S), NEG(S) = U - \overline{apr}^{(B,P)}_{\sum_{i=1}^{n} R_i}(S), BND(S) \\ = \overline{apr}^{(B,P)}_{\sum_{i=1}^{n} R_i}(S) - \underline{apr}^{(B,P)}_{\sum_{i=1}^{n} R_i}(S).$$

Theorem 3.5. Let $K = (U, R)$ be an Pawlak approximation space, where U is a nonempty finite set and R is a family of equivalence relation on U, $R_1, R_2, \cdots R_n \in R$. $B = \{\beta_1, \beta_2, \cdots, \beta_n\}$ is a set of variable precision thresholds set for different knowledge granularity, $0 \leq \beta_i < 0.5$. The pair $S = (F, A)$ is a soft set over U.

For $\forall e \in A$,

(1) Let $\beta_{\max} = \max\{\beta_1, \beta_2, \cdots, \beta_n\}$, then

$$\underline{Apr}^{B}_{\sum_{i=1}^{n} R_i}(S) \subseteq \underline{Apr}^{\beta_{\max}}_{\sum_{i=1}^{n} R_i}(S), \overline{Apr}^{B}_{\sum_{i=1}^{n} R_i}(S) \supseteq \overline{Apr}^{\beta_{\max}}_{\sum_{i=1}^{n} R_i}(S)).$$

(2) Let $\beta_{\min} = \min\{\beta_1, \beta_2, \cdots, \beta_n\}$, then

$$\underline{Apr}^{\beta_{\min}}_{\sum_{i=1}^{n} R_i}(S) \subseteq \underline{Apr}^{B}_{\sum_{i=1}^{n} R_i}(S), \overline{Apr}^{\beta_{\min}}_{\sum_{i=1}^{n} R_i}(S) \supseteq \overline{Apr}^{B}_{\sum_{i=1}^{n} R_i}(S)).$$

Proof: (1) If $\forall x \in \underline{apr}^{B}_{\sum_{i=1}^{n} R_i}(S)$, then x satisfy:

$$c([x]_{R_i}, F(e)) \leq \beta_i,$$

and since $\beta_{\max} = \max\{\beta_1, \beta_2, \cdots, \beta_n\}$, thus $\forall i = 1, 2, \cdots, n, \beta_i \leq \beta_{\max}$,

so x meets the condition:

$$c([x]_{R_i}, F(e)) \leq \beta_{\max},$$

then

$$x \in \underline{Apr}^{\beta_{\max}}_{\sum_{i=1}^{n} R_i}(S).$$

hence

$$\underline{apr}^{B}_{\sum_{i=1}^{n} R_i}(S) \subseteq \underline{apr}^{\beta_{\max}}_{\sum_{i=1}^{n} R_i}(S).$$

If $\forall x \in \overline{apr}^{\beta_{\max}}_{\sum_{i=1}^{n} R_i}(S)$, then x satisfy:

$$c([x]_{R_i}, F(e)) < 1 - \beta_{\max},$$

and since $\beta_{\max} = \max\{\beta_1, \beta_2, \cdots, \beta_n\}$, then $\forall i = 1, 2, \cdots, n$, $1-\beta_{\max} \leq 1 - \beta_i$, so x satisfy the condition:

$$c([x]_{R_i}, F(e)) < 1 - \beta_i,$$

then

$$x \in \overline{apr}^{B}_{\sum_{i=1}^{n} R_i}(S),$$

hence

$$\overline{Apr}^{B}_{\sum_{i=1}^{n} R_i}(S) \supseteq \overline{Apr}^{\beta_{\max}}_{\sum_{i=1}^{n} R_i}(S)).$$

(2) The demonstration process is similar to (1).

Proposition 3.6. Let $K = (U, R)$ be an Pawlak approximation space, where U is a nonempty finite set and R is a family of equivalence relation on U,$R_1, R_2, \cdots R_n \in R$, $T \subseteq R$, $0 \leq \beta < 0.5$, $B = \{\beta_1, \beta_2, \cdots, \beta_n\}$ is a set of variable precision thresholds set for different knowledge granularity,

$$S = (\underline{apr}^{(B,O)}_{\sum_{i=1}^{n} R_i}(S), \overline{apr}^{(B,O)}_{\sum_{i=1}^{n} R_i}(S))$$

is a variable precision optimistic multi-granularity rough soft set.

(1) $\underline{apr}^{(B,O)}_{\sum_{i=1}^{n} R_i}(U) = \overline{apr}^{(B,O)}_{\sum_{i=1}^{n} R_i}(U) = U.$

(2) $\underline{apr}^{(B,O)}_{\sum_{i=1}^{n} R_i}(\phi) = \overline{apr}^{(B,O)}_{\sum_{i=1}^{n} R_i}(\phi) = \phi.$

(3) $S_1 \subseteq S_2 \Rightarrow \underline{apr}^{(B,O)}_{\sum_{i=1}^{n} R_i}(S_1) \subseteq \underline{apr}^{(B,O)}_{\sum_{i=1}^{n} R_i}(S_2)$.

(4) $S_1 \subseteq S_2 \Rightarrow \overline{apr}^{(B,O)}_{\sum_{i=1}^{n} R_i}(S_1) \subseteq \overline{apr}^{(B,O)}_{\sum_{i=1}^{n} R_i}(S_2)$.

(5) $\underline{apr}^{(B,O)}_{\sum_{i=1}^{n} R_i}(S_1 \cap S_2) \subseteq \underline{apr}^{(B,O)}_{\sum_{i=1}^{n} R_i}(S_1) \cap \underline{apr}^{(B,O)}_{\sum_{i=1}^{n} R_i}(S_2)$.

(6) $\underline{apr}^{(B,O)}_{\sum_{i=1}^{n} R_i}(S_1 \cup S_2) \supseteq \underline{apr}^{(B,O)}_{\sum_{i=1}^{n} R_i}(S_1) \cup \underline{apr}^{(B,O)}_{\sum_{i=1}^{n} R_i}(S_2)$.

(7) $\overline{apr}^{(B,O)}_{\sum_{i=1}^{n} R_i}(S_1 \cap S_2) \subseteq \overline{apr}^{(B,O)}_{\sum_{i=1}^{n} R_i}(S_1) \cap \overline{apr}^{(B,O)}_{\sum_{i=1}^{n} R_i}(S_2)$.

(8) $\overline{apr}^{(B,O)}_{\sum_{i=1}^{n} R_i}(S_1 \cup S_2) \supseteq \overline{apr}^{(B,O)}_{\sum_{i=1}^{n} R_i}(S_1) \cup \overline{apr}^{(B,O)}_{\sum_{i=1}^{n} R_i}(S_2)$.

The above properties are also satisfied by variable precision pessimistic multi-granularity rough soft sets.

Theorem 3.7. Let $K = (U, R)$ be an Pawlak approximation space, where U is a nonempty finite set and R is a family of equivalence relation on U, $R_1, R_2, \cdots R_n \in R$. $B = \{\beta_1, \beta_2, \cdots, \beta_n\}$ is a set of variable precision thresholds set for different knowledge granularity, and $0 \leq \beta < 0.5$. $S = (F, A)$ is a soft set over U. Define two operators:

$$\underline{apr}^{B}_{\sum_{i=1}^{n} R_i}(S) = (\underline{F}^{B}_{\sum_{i=1}^{n} R_i}, A), \overline{apr}^{B}_{\sum_{i=1}^{n} R_i}(S) = \sim \underline{apr}^{B}_{\sum_{i=1}^{n} R_i}(\sim S).$$

(1) When $\beta_1 = \beta_2 = \cdots = \beta_n = \beta$,

$$\underline{F}^{B}_{\sum_{i=1}^{n} R_i}(e) = \{x \in U : c([x]_{R_1}, F(e)) \leq \beta \vee \cdots \vee c([x]_{R_n}, F(e)) \leq \beta\},$$

if $\underline{apr}^{(B,O)}_{\sum_{i=1}^{n} R_i}(S) \neq \overline{apr}^{(B,O)}_{\sum_{i=1}^{n} R_i}(S)$, S is a variable precision optimistic multi-granularity rough soft set.

(2) When $\beta_1 = \beta_2 = \cdots = \beta_n = \beta$,

$$\underline{F}^{B}_{\sum_{i=1}^{n} R_i}(e) = \{x \in U : c([x]_{R_1}, F(e)) \leq \beta \wedge \cdots \wedge c([x]_{R_n}, F(e)) \leq \beta\},$$

if $\underline{apr}^{(B,P)}_{\sum_{i=1}^{n} R_i}(S) \neq \overline{apr}^{(B,P)}_{\sum_{i=1}^{n} R_i}(S)$, S is a variable precision pessimistic multi-granularity rough soft set.

(2) When $n = 1, \beta = 1$, The model completely degenerates into rough soft set model.

4 Example

The Table 1 shows the results of nine students who took part in the examination for primary school to junior high school. The universe

$$U = \{x_1, x_2, x_3, x_4, x_5, x_6, x_7, x_8, x_9\}$$

indicates that all students who participated in the school's interview. The conditional attribute set is $R = \{R_1, R_2, R_3, R_4\}$, among them are the results of Chinese scores, Math scores, English scores, and Interview scores. The decision set is $D = \{d\}$, indicates whether the student has planned to accept by the school. Let $A = \{e_1, e_2\}$ is an attribute set, where e_1 means planned to be admitted, and e_2 means planned not to be admitted. Then $S = (F, A)$ is a soft set,

$$F(e_1) = \{x_1, x_3, x_5, x_7, x_9\}, F(e_2) = \{x_2, x_4, x_6, x_8\}.$$

Next, we take optimistic multi-granularity as an example to analyze the admission of students. By comparing the multi-granularity rough soft set model with the variable precision multi-granularity rough soft set model based on multiple thresholds, to show that the variable precision multi-granularity rough soft set model based on multiple thresholds is more flexible and feasible in the problem decision analysis.

(a) According to Definition 2.4, The lower approximation and upper approximation of multi-granularity rough soft set are as follows:

$$\underline{apr}^{O}_{\sum_{i=1}^{4} R_i}(S) = \underline{F}^{O}(e_1) = \{x_1, x_3, x_5, x_7, x_9\},$$

$$\overline{apr}^{O}_{\sum_{i=1}^{4} R_i}(S) = \overline{F}^{O}(e_1) = \{x_1, x_2, x_3, x_4, x_5, x_6, x_7, x_8, x_9\};$$

$$POS(S) = \{x_1, x_3, x_5, x_7, x_9\}, NEG(S) = \phi, BND(S) = \{x_2, x_4, x_6, x_8\}.$$

(b) According to Definition 3.3 and the teaching strategies of the school for different disciplines and admission conditions, we can set $B = \{0.4, 0.2, 0.4, 0.4\}$.

The lower approximation and upper approximation of variable precision multi-granularity rough soft set based on multiple thresholds are as follows:

$$\underline{apr}^{(B,O)}_{\sum_{i=1}^{4} R_i}(S) = \underline{F}^{(B,O)}_{\sum_{i=1}^{4} R_i}(e_1) = \{x_1, x_3, x_5, x_7, x_9\},$$

$$\overline{apr}^{(B,O)}_{\sum_{i=1}^{4} R_i}(S) = \overline{F}^{(B,O)}_{\sum_{i=1}^{4} R_i}(e_1) = \{x_1, x_3, x_4, x_5, x_6, x_7, x_9\},$$

$$POS(S) = \{x_1, x_3, x_5, x_7, x_9\}, NEG(S) = \{x_2, x_8\}, BND(S) = \{x_4, x_6\}.$$

According to the calculation results of the lower and upper approximation operators of variable precision optimistic multi-granularity rough soft set based on multiple thresholds, Candidate x_1, x_3, x_5, x_7, x_9 is fully qualified for admission and determines the candidate to be admitted, Candidate x_2, x_8 is not eligible for admission and is determined Candidates who are not accepted, although Candidate x_4, x_6 is not on the proposed admission list, it is the one with the most potential to be admitted.

Compared with the results of (a) and (b), we can find that the results of (b) can provide more decision-making opinions for further formal admission of the school. So adding variable precision and multiple threshold ideas to multi-granularity rough soft sets can increase the flexibility of the model and improve the classification accuracy and quality of the object.

Table 1. Decision information

	R_1	R_2	R_3	R_4	d
x_1	A	A	A	A	Yes
x_2	C	B	C	B	No
x_3	B	A	A	A	Yes
x_4	B	C	C	B	No
x_5	A	C	C	A	Yes
x_6	B	C	C	C	No
x_7	C	A	B	B	Yes
x_8	C	B	C	C	No
x_9	B	A	B	C	Yes

5 Conclusion

In this paper, we have proposed the new concept of variable precision multi-granularity rough soft sets, which can be viewed as a generalized model of rough soft sets model. Base on this, we have also proposed the concept of variable precision multi-granularity rough soft sets on multiple thresholds. Then, we present important properties of variable precision multi-granularity rough soft sets on multiple thresholds and establish the relations of the model and rough soft sets. Finally, we analyze the application of the model in decision making through an example. As future work, in order to improve the practicability of the variable precision multi-granularity rough soft set model based on multiple thresholds, the attribute reduction algorithm of the model could be explored.

References

1. Pawlak, Z.: Rough sets. Int. J. Parallel Program. **11**(5), 341–356 (1982)
2. Zadeh, L.A.: Fuzzy sets. Inf. Control **8**(3), 338–353 (1965)
3. Molodtsov, D.: Soft set theory—first results. Comput. Math. Appl. 19–31 (1999)
4. Ziarko, W.: Variable precision rough set model. J. Comput. Syst. Sci. **46**, 39–59 (1993)
5. Qian, Y., Liang, J., Yao, Y., et al.: MGRS: a multi-granulation rough set. Inf. Sci. **180**(6), 949–970 (2010)
6. Qian, Y.H., Liang, J.Y., Dang, C.Y.: Incomplete multi-granulation rough set. IEEE Trans. Syst. Man Cybern. Part A Syst. Hum. **40**(2), 420–431 (2010)
7. Feng, F., Li, C., Davaz, B., et al.: Soft sets combined with fuzzy sets and rough sets: a tentative approach. Soft Comput. **14**(9), 899–911 (2010)
8. Wang, X., Guo, C., Li, P., et al.: On rough Multi-granularity soft sets and its application in decision making. In: The 29th Chinese Control and Decision Conference, pp. 3046–3050 (2017)
9. Ali, M.: A note on soft sets, rough soft sets and fuzzy soft sets. Appl. Soft Comput. **11**(4), 3329–3332 (2011)
10. Zhan, J.M., Zhu, K.: Reviews on decision making methods based on (fuzzy)soft sets and rough soft sets. J. Intell. Fuzzy Syst. **29**(3), 1169–1176 (2015)
11. Liu, Y., Qin, K., Martinez, L., et al.: Improving decision making approaches based on fuzzy soft sets and rough soft sets. Appl. Soft Comput. 320–332 (2018)
12. Wang, G.Y., Yao, Y.Y., Yu, H.: Summary of Rough Set Theory and Application Research. **32**(7), 1229–1246 (2009)
13. Xu, Y., Li, C.: Variable precision multi-granularity rough set model based on multiple thresholds. Comput. Eng. Sci. **38**(08), 1727–1734 (2016)

Fuzzy N-soft Ordered Semigroups with Application

Ziyue Chen(✉), Jianbo Liu(✉), Yanan Chen, and Yanyan Zhang

Northeastern University at Qinhuangdao, Princeton 060004, HeBei, China
{jbliu,zhangyy}@neuq.edu.cn, ziyuechen1996@163.com, chenyn0515@163.com

Abstract. In this article, we introduce a new hybrid model called fuzzy N-soft ordered semigroup of S by a suitable combination of ordered semigroup S with fuzzy N-soft sets. Some useful operations and properties are given. Moreover, a novel multi-attribute decision making method combined with ordered semigroups is introduced inside we added the definition of grades that provides a finer granularity. This method is more cautious and reasonable.

Keywords: N-soft sets · Fuzzy N-soft sets · Ordered semigroup · Fuzzy N-soft ordered semigroup · Multi-attribute decision making

1 Introduction

The emergence of soft set theory provides us with a parameterized method to solve life uncertainty problems by Molodtsov in [1]. Maji et al. [2] proposed fuzzy soft set by combining the soft set and fuzzy set. And they combined soft sets with other mathematical model and got hesitant fuzzy soft sets, intuitionistic fuzzy soft sets, neutrosophic soft set, interval-valued neutrosophic soft sets and so on [3–6]. The soft set was in a binary environment whose values were 0 or 1. But the data in real life was non-binary essentially. Motivated by these problems, Fatimach et al. [7] create the improved soft set model called N-soft set which is a new research orientation and the importance of hierarchy was emphasized. For flexibility in decision making, a novel hybrid structure fuzzy N-soft sets[8] made up of N-soft sets and fuzzy N-soft set. Kuroki in [9] introduced fuzzy sets in terms of semigroups in 1991. He also studied fuzzy ideals in semigroups [10,11]. Several classes including simple, regular, intra-regular, weakly regular, and many more were introduced. Later, Kehayopulu [12,13] first gave the concept of an ordered semigroup and further related Ideals and Greens relation with ordered semigroup. Jun et al. [14] initiated the concept of soft ordered semigroups and later many researchers worked on the notions of fuzzy soft ordered semigroups, fuzzy soft left (resp. right) ideals and many more [15,16].

The paper is organized as follows. Section 2 provides the relevant theoretical knowledge. In Sect. 3, we introduce a new hybrid model called fuzzy N-soft ordered semigroups based on both fuzzy N-soft set and ordered semigroups.

C. Pang et al. (Eds.): SETE 2020/ICWL 2020, LNCS 12511, pp. 171–178, 2021.
https://doi.org/10.1007/978-3-030-66906-5_16

Basic operations and its properties are got. In Sect. 4, a novel multi-attribute decision making method combined with ordered semigroups is proposed and we added the definition of grades that provides a finer granularity. Finally, we present our conclusion in Sect. 5.

2 Preliminaries

In this section, some basic definitions and examples were recapitulated, which is serviceable and appropriate to understand the new concepts to be proposed. There is a binary operation called multiplication denoted by "$\cdot$". A set S, $S\neq\emptyset$, $(S,\leq)$ is an ordered set and $(S,\cdot)$ is combinative.

Definition 1 [14]. *$(S,\cdot,\leq)$ is represented an ordered semigroup if $\forall$ $(s_1,s_2)\in S$, $s_1\leq s_2$, then $\forall a\in S$, $as_1\leq as_2$, $s_1a\leq s_2a$.*

Definition 2 [15]. *Let S is an ordered semigroup, M is called subsemigroup of S: (1) $M\subseteq S$, $M\neq\emptyset$; (2) $M^2\subseteq M$.*

Definition 3 [1]. *Let U be an universe sets and E be a set of parameters $A\subseteq E$, $A\neq\emptyset$, a pair (f,A) is called a soft set over U, if f is a mapping from A to $P(U)$, f:$A\rightarrow p(U)$, $p(U)$ is the power set of U.*

Definition 4 [16]. *Let S be an ordered semigroup and R be a set of parameters, $E\subseteq R$, $R\neq\emptyset$, a pair (f,E) is called a soft ordered semigroup over S. If f is a mapping from E to $p(s)$, f: $E\rightarrow p(s)$, $p(s)$ is the power set of S.*

Definition 5 [16]. *Let (f,E) be a soft set over S, then (f,E) is represented a soft ordered semigroup if each $e\in E$, $f(e)\neq\emptyset$, $f(e)$ is subsemigroup of S.*

Definition 6 [7]. *Let U be a universe set of objects and R be a set of attributes, $E\subseteq R$. $M=\{0,1,\cdots,N-1\}$ be a set of ordered grades where $N\in\{2,3,\cdots\}$, (f,E,N) is an N-soft set on U if there is a mapping f: $E\rightarrow p(M)$.*

Definition 7 [9]. *Let $(S,\cdot,\leq)$ be an ordered semigroup and R be a set of parameters $E\subseteq R$ of S, $(\bar{f},E)$ is fuzzy soft ordered semigroup of S, there is a mapping $\bar{f}$: $E\rightarrow F(s)$, $F(S)$ is a fuzzy soft set of S, $\forall e\in E$, $\bar{f}(e)\neq\emptyset$, $\bar{f}(e)$ is fuzzy subsemigroups of S.*

3 Fuzzy N-soft Ordered Semigroups

In this subsection, we define our extension of this concept and propose some fundamental operations for the new setting.

Definition 8. *Let $(S,\cdot,\leq)$ be an ordered semigroup and R be a set of parameters $E\subseteq R$ of S, $M=\{0,1,\cdots,N-1\}$ be a set of ordered grades where $N\in\{2,3,\cdots\}$, $(\bar{f}_{\bar{g}},E,N)$ is fuzzy N-soft ordered semigroup of S, if there is a mapping $\bar{f}_{\bar{g}}$: $E\rightarrow F(s)\times p(M)$; $\bar{f}$: $E\rightarrow F(S)$, $F(S)$ is a fuzzy soft set, $\bar{g}\rightarrow p(M)$, $p(M)$ is the power set of M. $\forall e\in E$, $\bar{f}(e)\neq\emptyset$, $\bar{g}(e)\neq\emptyset$, $\bar{f}(e)$ $\bar{g}(e)$ is fuzzy subsemigroups of S.*

Fuzzy N-soft ordered semigroup is represented as $FNSOS$ in this article.

Example 1. Let $U{=}\{u_1, u_2, u_3, u_4\}$ be an ordered semigroup and under the following multiplication table1 and ordered relation represented as follows: $\leq:=\{(u_1, u_1), (u_2, u_2), (u_3, u_3), (u_4, u_4), (u_1, u_2)\}$. The multiplication table is represented by the Table 1.

Table 1. Multiplication table

$\cdot$	u_1	u_2	u_3	u_4
u_1	u_1	u_1	u_1	u_1
u_2	u_1	u_1	u_1	u_1
u_3	u_1	u_1	u_2	u_1
u_4	u_1	u_1	u_2	u_2

When $N{=}5$, $(\bar{f}, E, 6)$ is 5-soft set of S, $M{=}\{0, 1, 2, 3, 4\}$, Let $E{=}\{e_1, e_2\}$ be an attributes set, the fuzzy 5-soft set $(\bar{f}_{\bar{g}}, E)$ is defined by a mapping, $\bar{f}$: $E{\rightarrow}F(S)$, $\bar{g}$: $E{\rightarrow}P(M)$, we have, $\forall\ e_j{\in}E$, $j = 1, 2$.

$$\bar{f}(e) = \begin{bmatrix} \frac{u_1}{0.9} & \frac{u_2}{0.7} & \frac{u_3}{0.6} & \frac{u_4}{0.1} \\ \\ \frac{u_1}{0.7} & \frac{u_2}{0.5} & \frac{u_3}{0.2} & \frac{u_4}{0.3} \end{bmatrix} \tag{1}$$

Thus, these grades are given to u_i $(i = 1 \cdots 4)$ decided by the $\bar{f}(e_j)$. In other words, the selection panel follows this criteria on the basis of qualities of u_i $(i = 1, \cdots, 4)$ as follows:

$$\begin{aligned} 0.0 \leq \bar{f}|_{nu}(e) < 0.2 &\Leftrightarrow \bar{g_u}(e) = 0; \\ 0.2 \leq \bar{f}|_{nu}(e) < 0.4 &\Leftrightarrow \bar{g_u}(e) = 1; \\ 0.4 \leq \bar{f}|_{nu}(e) < 0.6 &\Leftrightarrow \bar{g_u}(e) = 2; \\ 0.6 \leq \bar{f}|_{nu}(e) < 0.8 &\Leftrightarrow \bar{g_u}(e) = 3; \\ 0.8 \leq \bar{f}|_{nu}(e) \leq 1.0 &\Leftrightarrow \bar{g_u}(e) = 4. \end{aligned}$$

Thus, $\forall\ e_j{\in}E$, $j = 1, 2$.

$$\bar{g}(e) = \begin{bmatrix} 4\,3\,3\,0 \\ 3\,2\,1\,1 \end{bmatrix} \tag{2}$$

$\bar{f}_{\bar{g}}(e){\neq}\emptyset$ also $\bar{f}(e)(u)$, $\bar{g}(e)(u)$ are fuzzy ordered subsemigroups of S. Then, $(\bar{f}_{\bar{g}}, E_1, N_1)$ is a $FNSOS$.

$$\bar{f}_{\bar{g}}(e) = \begin{bmatrix} \frac{u_1}{0.9}, 4 & \frac{u_2}{0.7}, 3 & \frac{u_3}{0.6}, 3 & \frac{u_4}{0.1}, 0 \\ \\ \frac{u_1}{0.7}, 3 & \frac{u_2}{0.5}, 2 & \frac{u_3}{0.2}, 1 & \frac{u_4}{0.3}, 1 \end{bmatrix} \tag{3}$$

Definition 9. *Let* $(S, \cdot, \leq)$ *be an ordered semigroup and* $(\bar{f}_{\bar{g}}, E)$ *is* $FNSOS$ *over* S, $\bar{f}$, $\bar{g}$ *be two fuzzy ordered subsemigroup of* S. *Then* $\forall t \in [0,1]$, $l \in \{0,1,2,3,4\}$, *we defined a fuzzy soft level set as follows:* $L(\bar{f}_{\bar{g}}, t, l) = \{u_i \in S | \bar{f}(u_i) \geq t, \bar{g}(u) \geq l\}$.

Theorem 1. $(\bar{f}_{\bar{g}}, E_1)$ *be a* $FNSOS$ *over* S, *if* $E_2 \subseteq E_1$, $E_2 \neq \emptyset$, *then* $(\bar{h}_{\bar{g}}, E_2)$ *is also a* $FNSOS$.

Proof. The proof is straightforward from the definition 13.

Now we proceed to define some basic algebraic operations in the new model that we have introduced.

Definition 10. *Let* $(\bar{f}_{\bar{g}}|_1, E_1, N_1)$ *and* $(\bar{f}_{\bar{g}}|_2, E_2, N_2)$ *be two fuzzy* N*-soft ordered semigroups over* S. *Then their restricted intersection is denoted by* $(\bar{f}_{\bar{g}}|_1, E_1, N_1) \cap_R (\bar{f}_{\bar{g}}|_2, E_2, N_2) = (\bar{k}_{\bar{g}}, E_3, N_3)$, *where (1)* $E_3 = E_1 \cap E_2 \neq \emptyset$; *(2)* $N_3 = N_1 \wedge N_2$; *(3)* $\forall e \in E_3$,

$$\bar{k}_{\bar{g}}(e) = \bar{f}_{\bar{g}}|_1(e) \cap \bar{f}_{\bar{g}}|_2(e) = \begin{cases} \bar{f}_u|_1(e) \wedge \bar{f}_u|_2(e), \\ \bar{g}_u|_1(e) \wedge \bar{g}_u|_2(e). \end{cases} \tag{4}$$

Theorem 2. $(\bar{f}_{\bar{g}}|_1, E_1, N_1)$ *and* $(\bar{f}_{\bar{g}}|_2, E_2, N_2)$ *be two fuzzy* N*-soft ordered semigroups over* S, *their restricted intersection is also fuzzy* N*-soft ordered semigroup over* S.

Proof. From the definition 10, $(\bar{f}_{\bar{g}}|_1, E_1, N_1) \cap_R (\bar{f}_{\bar{g}}|_2, E_2, N_1) = (\bar{k}_{\bar{g}}, E_3, N_3)$. For each $e \in E_3 = E_1 \cap E_2$, so $e \in E_1$ and $e \in E_2$, thus from the equation (3), if $\bar{f}_u|_1(e) \wedge \bar{f}_u|_2(e) = \bar{f}_u|_1(e)$ and from the definition 8, $\bar{g}_u|_1(e) \wedge \bar{g}_u|_2(e) = \bar{g}_u|_1(e)$, so $\bar{k}_{\bar{g}}(e) = \bar{f}_{\bar{g}}|_1(e) \neq \emptyset$, since $\bar{f}_u(e)|_1$, $\bar{g}_u(e)|_1$ are fuzzy subsemigroups of S, so $(\bar{f}_{\bar{g}}|_1, E_1)$ is fuzzy N-soft ordered semigroup, $(\bar{k}_{\bar{g}}, E_3)$ is also fuzzy N-soft ordered semigroup. If $\bar{f}_u|_1(e) \wedge \bar{f}_u|_2(e) = \bar{f}_u|_2(e)$ and $\bar{g}_u|_1(e) \wedge \bar{g}_u|_2(e) = \bar{g}_u|_2(e)$, so $\bar{k}_{\bar{g}}(e) = \bar{f}_{\bar{g}}|_2(e) \neq \emptyset$, $\bar{f}_u(e)|_2$, $\bar{g}_u(e)|_2$ are fuzzy subsemigroup of S, so $(\bar{f}_{\bar{g}}|_2, E_2)$ is fuzzy N-soft ordered semigroup of S, $(\bar{k}_{\bar{g}}, E_3)$ is also fuzzy N-soft ordered semigroup of S.
If $\bar{f}_{\bar{g}}|_1(e) \wedge \bar{f}_{\bar{g}}|_2(e) = \bar{k}_{\bar{g}}(e) \neq \emptyset$, from the Definition 13, $\bar{k}(e) \neq \emptyset$, $\bar{g}(e) \neq \emptyset$, $\bar{k}(e)$ and $\bar{g}(e)$ are fuzzy subsemigroups of S. So $(\bar{k}_{\bar{g}}, E_3, N_3)$ is fuzzy N-soft ordered semigroup of S.

Definition 11. *Let* $(\bar{f}_{\bar{g}}|_1, E_1, N_1)$ *and* $(\bar{f}_{\bar{g}}|_2, E_2, N_2)$ *be two fuzzy* N*-soft ordered semigroups over* S, *Then their extend intersection is denoted by* $(\bar{f}_{\bar{g}}|_1, E_1, N_1) \cap_E (\bar{f}_{\bar{g}}|_2, E_2, N_2) = (\bar{h}_{\bar{g}}, E_3, N_3)$, *where (1)* $E_3 = E_1 \cup E_2$; *(2)* $N_3 = N_1 \vee N_2$; *(3) For each* $e \in E_3$:

$$\bar{h}_{\bar{g}}(e) = \begin{cases} \bar{f}_{\bar{g}}|_1(e), if e \in E_1 - E_2 \\ \bar{f}_{\bar{g}}|_1(e), if e \in E_2 - E_1 \\ \bar{f}_{\bar{g}}|_1(e) \cap \bar{f}_{\bar{g}}|_2(e) = \begin{cases} \bar{f}_u|_1(e) \wedge \bar{f}_u|_2(e) \\ \bar{g}_u|_1(e) \wedge \bar{g}_u|_2(e), if e \in E_1 \cap E_2. \end{cases} \end{cases} \tag{5}$$

Theorem 3. *Let* $(\bar{f}_{\bar{g}}|_1, E_1, N_1)$ *and* $(\bar{f}_{\bar{g}}|_2, E_2, N_2)$ *be two fuzzy* N*-soft ordered semigroups over* S*. If* $E_1 \cap E_2 = \emptyset$*, their extend restriction is also fuzzy* N*-soft ordered semigroups over* S*.*

Proof. From the Definition 11, $(\bar{f}_{\bar{g}}|_1, E_1, N_1) \cap_E (\bar{f}_{\bar{g}}|_2, E_2, N_2) = (\bar{h}_{\bar{g}}, E_3, N_3)$, when for each $e \in E_1/E_2$ or $e \in E_2/E_1$, $\bar{h}_{\bar{g}}(e) = \bar{f}_{\bar{g}}|_1(e)$ or $\bar{f}_{\bar{g}}|_2(e)$, $(\bar{f}_{\bar{g}}|_1, E_1, N_1)$ and $(\bar{f}_{\bar{g}}|_2, E_2, N_2)$ are $FNSOSs$, so $\bar{f}_{\bar{g}}|_1(e)$ and $\bar{f}_{\bar{g}}|_2(e)$ are fuzzy N-soft ordered subsemigroups over S and $\bar{h}_{\bar{g}}(e)$ is fuzzy N-soft ordered subsemigroup over S, we get $(\bar{h}_{\bar{g}}, E_3, N_3)$ is a $FNSOSs$.

Definition 12. *Let* $(\bar{f}_{\bar{g}}|_1, E_1, N_1)$ *and* $(\bar{f}_{\bar{g}}|_2, E_2, N_2)$ *be two fuzzy* N*-soft ordered semigroups over* S*, then their restricted union is represented by* $(\bar{f}_{\bar{g}}|_1, E_1, N_1) \cup_R (\bar{f}_{\bar{g}}|_2, E_2, N_2) = (\bar{q}_{\bar{g}}|_2, E_3, N_3)$*. Where (1)* $E_3 = E_1 \cap E_2$*; (2)* $N_1 \vee N_2$*; (3)* $\forall e \in E_3$ *:*

$$\bar{q}_{\bar{g}}(e) = \bar{f}_{\bar{g}}|_1(e) \cap \bar{f}_{\bar{g}}|_2(e) = \begin{cases} \bar{f}_u|_1(e) \vee \bar{f}_u|_2(e), \\ \bar{g}_u|_1(e) \vee \bar{g}_u|_2(e). \end{cases} \tag{6}$$

Theorem 4. $(\bar{f}_{\bar{g}}|_1, N_1, E_1)$ *and* $(\bar{f}_{\bar{g}}|_2, N_2, E_2)$ *be two fuzzy* N*-soft ordered semigroups over* S*, then their restricted union is also fuzzy* N*-soft ordered semigroups.*

Proof. The proof is similar to Theorem 2.

Definition 13. *Let* $(\bar{f}_{\bar{g}}|_1, E_1, N_1)$ *and* $(\bar{f}_{\bar{g}}|_2, E_2, N_2)$ *be two fuzzy* N*-soft ordered semigroups over* S*, Then their extend union is denoted by* $(\bar{f}_{\bar{g}}|_1, E_1, N_1) \cup_E (\bar{f}_{\bar{g}}|_2, E_2, N_2) = (\bar{r}_{\bar{g}}, E_3, N_3)$*, where (1)* $E_3 = E_1 \cup E_2$*; (2)* $N_3 = N_1 \vee N_2$*; (3) For each* $e \in E_3$ *:*

$$\bar{h}_{\bar{g}}(e) = \begin{cases} \bar{f}_{\bar{g}}|_1(e), if e \in E_1 - E_2 \\ \bar{f}_{\bar{g}}|_1(e), if e \in E_2 - E_1 \\ \bar{f}_{\bar{g}}|_1(e) \cup \bar{f}_{\bar{g}}|_2(e) = \begin{cases} \bar{f}_u|_1(e) \vee \bar{f}_u|_2(e) \\ \bar{g}_u|_1(e) \vee \bar{g}_u|_2(e), if e \in E_1 \cap E_2. \end{cases} \end{cases} \tag{7}$$

Theorem 5. *Let* $(\bar{f}_{\bar{g}}|_1, E_1, N_1)$ *and* $(\bar{f}_{\bar{g}}|_2, E_2, N_2)$ *be two fuzzy* N*-soft ordered semigroups over* S*. If* $E_1 \cap E_2 = \emptyset$*, their extend restriction is also fuzzy* N*-soft ordered semigroups over* S*.*

Proof. The proof is similar to Theorem 3.

Definition 14. *Let* $(\bar{f}_{\bar{g}}|_1, E_1, N_1)$ *and* $(\bar{f}_{\bar{g}}|_2, E_2, N_2)$ *be two fuzzy* N*-soft ordered semigroups over* S*. Then their AND operation is denoted by* $(\bar{f}_{\bar{g}}|_1, E_1, N_1) \bigvee (\bar{f}_{\bar{g}}|_2, E_2, N_2) = (\bar{t}_{\bar{g}}, E_4, N_4)$*, where (1)* $E_4 = E_1 \times E_2$*; (2)* $N_4 = N_1 \wedge N_2$*; (3)* $\forall (a, b) \in E_4$ *:*

$$\bar{t}_{\bar{g}}(a, b) = \bar{f}_{\bar{g}}|_1(a) \cap \bar{f}_{\bar{g}}|_2(b) = \begin{cases} \bar{f}_u|_1(a) \wedge \bar{f}_u|_2(b), \\ \bar{g}_u|_1(a) \wedge \bar{g}_u|_2(b). \end{cases} \tag{8}$$

Theorem 6. *Let* $(\bar{f}_{\bar{g}}|_1, E_1, N_1)$ *and* $(\bar{f}_{\bar{g}}|_2, E_2, N_2)$ *be two fuzzy N-soft ordered semigroups over S, then their AND operation is also fuzzy N-soft ordered semigroups over S.*

Proof. For each $(a,b)\in E_1\times E_2$, $u\in S$, since $(\bar{f}_{\bar{g}}|_1, E_1, N_1)$ and $(\bar{f}_{\bar{g}}|_2, E_2, N_2)$ are $FNSOSs$, we have $\bar{f}_{\bar{g}}|_1(a)\neq\emptyset$, $\bar{f}_{\bar{g}}|_2(b)\neq\emptyset$, so $\bar{f}(a)\neq\emptyset$, $\bar{g}(a)\neq\emptyset$, $\bar{g}(a)\neq\emptyset$ and $\bar{g}(b)\neq\emptyset$, so $\bar{f}_{\bar{g}}|_1(a)$ and $\bar{f}_{\bar{g}}|_2(b)$ are fuzzy N-soft ordered subsemigroups of S.
If $\bar{f}|_1(a)(u)\wedge\bar{f}|_2(b)(u)=\bar{f}|_1(a)(u)$, so $\bar{g}|_1(a)(u)\wedge\bar{g}|_2(b)(u)=\bar{g}|_1(a)(u)$, $\bar{t}_{\bar{g}}(a,b)=\bar{f}|_1(a)\neq\emptyset$, so $\bar{t}_{\bar{g}}(a,b)$ is a fuzzy subsemigroup of S, $(\bar{t}_{\bar{g}}, E_4, N_4)$ is a $FNSOSs$ of S. Similarly, if $\bar{f}|_1(a)(u)\wedge\bar{f}|_2(b)(u)=\bar{f}|_2(b)(u)$, so $\bar{g}|_1(a)(u)\wedge\bar{g}|_2(b)(u)=\bar{g}|_2(b)(u)$, $\bar{t}_{\bar{g}}(a,b)=\bar{f}|_2(b)\neq\emptyset$, so $\bar{t}_{\bar{g}}(a,b)$ is a fuzzy subsemigroup of S, hence $(\bar{t}_{\bar{g}}, E_4, N_4)$ is a fuzzy N-soft ordered semigroup of S.

4 Application

This section is the application of fuzzy N-soft ordered semigroups in multi-attribute decision making. This method is suitable for different experts to get different order relations and then have different fuzzy N-soft sets, which can take into account the recommendations of each expert to obtain a comprehensive ranking result.

Example 2. $T=\{e_1, e_2, e_3, e_4\}=\{$professional skill, ideological and moral cultivation, language competence$\}$. Different attributes depend on the nature of the people employed by different companies. A school pays more attention to these attributes $E=\{$professional skill, ideological and moral cultivation$\}$. After a period of internship, HR will make an order relations based on the candidates' comprehensive performance. Different HR will give different order relation according to different performance of five cadet teachers.

Let $S=\{u_1, u_2, u_3, u_4\}$ be an ordered semigroup and under the following multiplication Table 2 and Table 3. The ordered relations be represented by the HR_n $n=1,2$ respectively. HR_1:$\leq:=\{(u_1,u_1),(u_2,u_2),(u_3,u_3),(u_4,u_4),(u_1,u_2)$; HR_1:$\leq:=\{(u_1,u_1),(u_1,u_3),(u_1,u_4),(u_2,u_2),(u_2,u_4),(u_3,u_3),(u_4,u_4)\}$.

Table 2. Multiplication table from HR_1

$\cdot$	u_1	u_2	u_3	u_4
u_1	u_1	u_1	u_1	u_1
u_2	u_1	u_1	u_1	u_1
u_3	u_1	u_1	u_1	u_2
u_4	u_1	u_1	u_2	u_3

Table 3. Multiplication table HR_2

$\cdot$	u_1	u_2	u_3	u_4
u_1	u_1	u_4	u_1	u_4
u_2	u_1	u_2	u_1	u_4
u_3	u_1	u_4	u_3	u_4
u_4	u_1	u_4	u_1	u_4

When $N=5$, $(\bar{g}, E, 5)$ is 5-soft set of S, $M=\{0,1,2,3,4\}$, the corresponding grades are same as Example 1; the fuzzy 5-soft set $(\bar{f}_{\bar{g}}, E)$ is defined by a mapping, $\bar{f}|_n$: $E\rightarrow F(S)$, $\bar{g}|_n$: $E\rightarrow P(M)$, $n=1,2$. We have, $\forall\ e_j\in E$, $j=1,2,3$.

$$\bar{f}_{\bar{g}}|_1(e_j) = \begin{bmatrix} 0.8,4 & 0.7,4 & 0.6,3 & 0.6,3 \\ 0.7,3 & 0.6,3 & 0.3,1 & 0.2,1 \\ 0.8,4 & 0.6.3 & 0.4,2 & 0.4,2 \end{bmatrix}, \bar{f}_{\bar{g}}|_2(e_j) = \begin{bmatrix} 0.7,3 & 0.6,3 & 0.5,2 & 0.8,4 \\ 0.6,3 & 0.3,1 & 0.6,3 & 0.7,3 \\ 0.4,2 & 0.2,1 & 0.6,3 & 0.9,4 \end{bmatrix}$$

Here we apply AND operation of fuzzy N-soft ordered semigroups, $\bar{h}(e_j)=\bar{f}|_1(e_j)\wedge\bar{g}|_2(e_j)$;

$\bar{h}(e_1,e_1)=\{(\frac{u_1}{0.8\wedge 0.7},4\wedge 3),(\frac{u_2}{0.7\wedge 0.6},4\wedge 3),(\frac{u_3}{0.6\wedge 0.5},3\wedge 2),(\frac{u_4}{0.6\wedge 0.8},3\wedge 4)\}$;
$\bar{h}(e_1,e_2)=\{(\frac{u_1}{0.8\wedge 0.6},4\wedge 3),(\frac{u_2}{0.7\wedge 0.3},4\wedge 1),(\frac{u_3}{0.6\wedge 0.6},3\wedge 3),(\frac{u_4}{0.6\wedge 0.7},3\wedge 3)\}$;
$\bar{h}(e_1,e_3)=\{(\frac{u_1}{0.8\wedge 0.4},4\wedge 2),(\frac{u_2}{0.7\wedge 0.2},4\wedge 1),(\frac{u_3}{0.6\wedge 0.6},3\wedge 3),(\frac{u_4}{0.6\wedge 0.9},3\wedge 4)\}$;
$\bar{h}(e_2,e_1)=\{(\frac{u_1}{0.7\wedge 0.7},3\wedge 3),(\frac{u_2}{0.6\wedge 0.6},3\wedge 3),(\frac{u_3}{0.3\wedge 0.5},1\wedge 2),(\frac{u_4}{0.2\wedge 0.8},1\wedge 4)\}$;
$\bar{h}(e_2,e_2)=\{(\frac{u_1}{0.7\wedge 0.6},3\wedge 3),(\frac{u_2}{0.6\wedge 0.3},3\wedge 1),(\frac{u_3}{0.3\wedge 0.6},1\wedge 3),(\frac{u_4}{0.2\wedge 0.7},1\wedge 3)\}$;
$\bar{h}(e_2,e_3)=\{(\frac{u_1}{0.7\wedge 0.4},3\wedge 2),(\frac{u_2}{0.6\wedge 0.2},3\wedge 1),(\frac{u_3}{0.3\wedge 0.6},1\wedge 3),(\frac{u_4}{0.2\wedge 0.9},1\wedge 4)\}$;
$\bar{h}(e_3,e_1)=\{(\frac{u_1}{0.8\wedge 0.7},4\wedge 3),(\frac{u_2}{0.6\wedge 0.6},3\wedge 3),(\frac{u_3}{0.4\wedge 0.5},2\wedge 2),(\frac{u_4}{0.4\wedge 0.8},2\wedge 4)\}$;
$\bar{h}(e_3,e_2)=\{(\frac{u_1}{0.8\wedge 0.6},4\wedge 3),(\frac{u_2}{0.6\wedge 0.3},3\wedge 1),(\frac{u_3}{0.4\wedge 0.6},2\wedge 3),(\frac{u_4}{0.4\wedge 0.7},2\wedge 3)\}$;
$\bar{h}(e_3,e_3)=\{(\frac{u_1}{0.8\wedge 0.4},4\wedge 2),(\frac{u_2}{0.6\wedge 0.2},3\wedge 1),(\frac{u_3}{0.4\wedge 0.6},2\wedge 3),(\frac{u_4}{0.4\wedge 0.9},2\wedge 3)\}$.

From the Definition 14, the level l=1 and t=0.3, so $\forall\ e_i,e_j \in E\times E$, $i,j = 1,2,3$.

$$\bar{h}_{\bar{g}}|(e_i,e_j) = \begin{bmatrix} 0.7,3 & 0.6,3 & 0.5,2 & 0.6,3 \\ 0.6,3 & 0.3,1 & 0.6,3 & 0.6,3 \\ 0.4,2 & 0.2.1 & 0.6,3 & 0.6,3 \\ 0.7,3 & 0.6,3 & 0.3,1 & 0.2,1 \\ 0.6,3 & 0.3,1 & 0.3,1 & 0.2,1 \\ 0.73 & 0.6,3 & 0.4,2 & 0.4,2 \\ 0.73 & 0.6,3 & 0.4,2 & 0.4,2 \\ 0.63 & 0.3,1 & 0.4,2 & 0.4,2 \\ 0.42 & 0.2,1 & 0.4,2 & 0.4,2 \end{bmatrix}, \bar{g}^l(e_j) = \begin{bmatrix} 3 & 3 & 2 & 3 \\ 3 & \diamond & 3 & 3 \\ 2 & \diamond & 3 & 3 \\ 3 & 3 & \diamond & \diamond \\ 3 & \diamond & \diamond & \diamond \\ 2 & \diamond & \diamond & \diamond \\ 3 & 3 & 2 & 2 \\ 3 & \diamond & 2 & 2 \\ 2 & \diamond & 2 & 2 \end{bmatrix}, \bar{h}^t(e_i,e_j) = \begin{bmatrix} 0.7 & 0.5 & 0.6 \\ 0.6 & 0.6 & 0.6 \\ 0.4 & 0.6 & 0.6 \\ 0.7 & 0.4 & 0.4 \\ 0.6 & 0.4 & 0.4 \\ 0.4 & 0.4 & 0.4 \end{bmatrix}$$

Let's look at the column vectors of the matrix, if the number of "$\diamond$"≥ 4, the alternative will be at the end of the rank, so the u_2 is the last in the position. Other alternatives should go on with discussion. Let's look at the row vectors of the matrix, the number of "$\diamond$"≥ 2, the corresponding attributes (e_2,e_1), (e_2,e_2), (e_2,e_3) will not be considered.

For each $(e_i,e_j)\in E\times E$, the comparison table of $\bar{h}^t(e_i,e_j)$ i=1, 2, j=1, 2, 3 is given by Table 4.

Table 4. Comparison table of $\bar{h}^t(e_i,e_j)$

·	(e_1,e_1)	(e_1,e_2)	(e_1,e_3)	(e_3,e_1)	(e_3,e_2)	(e_3,e_3)
(e_1,e_1)	–	u_1	u_1	u_1	u_1,u_3,u_4	u_1,u_3,u_4
(e_1,e_2)	u_3	–	u_1	u_3,u_4	u_3,u_4	u_1,u_3,u_4
(e_1,e_3)	u_3	–	–	u_3,u_4	u_3,u_4	u_3,u_4
(e_3,e_1)	–	u_1	u_1	–	u_1	u_1
(e_3,e_2)	–	–	u_1	–	–	u_1
(e_3,e_3)	–	–	–	–	–	–

According to the comparison Table 4, We can count the values of u_1, u_3, u_4; u_1=13, u_3=10, u_4=8. Thus the final rank is $u_1 > u_3 > u_4 > u_2$.

5 Conclusion

In this paper, the basic concepts are reviewed and the novel hybrid structure fuzzy N-soft ordered is proposed. Meanwhile, we give relate properties and operators. In terms of application, we combine order relations with grades. Moreover, the level method and the AND operation are integrated into the multi-attribute decision making. This method is suitable for different experts to get different order relations and then have different fuzzy N-soft sets, which can take into account the recommendations of each expert to obtain a comprehensive ranking result. In the future, we will extend our research in the more hybrid models and applications (1) To investigate various classes of ordered semigroups including regular and intra-regular in terms of fuzzy N-soft ideals, fuzzy N-soft interior ideals, fuzzy N-soft bi-ideals and fuzzy N-soft quasi-ideals. (2) To derive the application of interval value N-soft ordered semigroup in decision sciences.

References

1. Molodtsov, D.: Soft set theory-first results. Comput. Math. Appl. **37**(4), 19–31 (1999)
2. Maji, P.K., Biswas, R., Roy, A.R.: Fuzzy soft set. J. Fuzzy Math. **9**, 589–602 (2001)
3. Maji, P.K., Biswas, R., Roy, A.R.: Intuitionistic fuzzy soft sets. Fuzzy Math. **9**, 677–692 (2001)
4. Babitha, K.V., John, S.J.: Hesitant fuzzy soft sets. New Results Sci. **3**, 98–107 (2013)
5. Maji, P.K.: Neutrosophic soft set. Comput. Math. Appl. **45**, 555–562 (2013)
6. Deli, I.: Interval-valued neutrosophic soft sets and its decision making. Int. J. Mach. Learn. Cybern. **8**(2), 665–676 (2015). https://doi.org/10.1007/s13042-015-0461-3
7. Fatimah, F., Rosadi, D., Hakim, R.B.F., Alcantud, J.C.R.: N-soft sets and their decision-making algorithms. Soft Comput. **22**(12), 3829–3842 (2018)
8. Akram, M., Adeel, A., Alcantud, J.C.R.: Fuzzy N-soft sets: a novel model with applications. Intell. Fuzzy Syst. **35**(4), 4757–4771 (2018)
9. Kuroki, N.: On fuzzy semigroups. Inform. Sci. **53**, 203–236 (1991)
10. Kuroki, N.: Fuzzy bi-ideals in semigroups. Comment. Math. Univ. St. Pauli **28**, 17–21 (1979)
11. Kuroki, N.: On fuzzy ideals and fuzzy bi-ideals in semigroups. Fuzzy Sets Syst. **5**, 203–215 (1981)
12. Kehayopulu, N.: Ideals and Greens relations in ordered semigroups. Int. J. Math. **61286**, 1–8 (2006)
13. Kehayopulu, N.: On weakly prime ideals of ordered semigroups. Japanese J. Math. **35**, 1051–1056 (1990)
14. Jun, Y.B., Lee, K.J., Khan, A.: Soft ordered semigroups. J. Math. Logic Q. **56**, 42–50 (2009)
15. Yin, Y., Zhan, J.: The characterization of ordered semigroups in terms of fuzzy soft ideals. Bull. Malaysian Math. Soc. Ser. **24**(4) (2012)
16. Yang, C.F.: Fuzzy soft semigroups and fuzzy soft ideals. Comput. Math. Appl. **61**(4), 255–261 (2011)

Dominance-Based N-soft Rough Sets

Di Zhang, Qian Wu, Sheng-Nan Cao, and Pi-Yu Li(✉)

School of Mathematics and Statistics, Northeastern University at Qinhuangdao, Qinhuangdao 066004, China
lpy91132006@aliyun.com

Abstract. In the present paper, a new approach dominance-based N-soft rough sets is being introduced to study roughness through dominance relation and N-soft sets. Moreover, approximation operators and some useful properties with respect to dominance-based N-soft rough approximation space are introduced. Furthermore, a potential application in matching the corresponding courses according to the students' knowledge structure and knowledge state are presented with decision-making procedures. This algorithm can address the problems of multi-criteria classification in inconsistent ordered information system.

Keywords: Rough sets · N-soft sets · Dominance-based N-soft rough sets · Multi-criteria classification

1 Introduction

Certain problems existing in the field of social science, medical diagnosis, pattern recognition and economics, etc, are closely associated with uncertainty and impreciseness information. However, there is such problem in the side of education - Matching courses to students' performance. Consider a situation like this: Students are facing a teacher, who is probing his knowledge of high school mathematics. The students are freshly arrived from a foreign country, and some mathematical questions must be answered. To which class should the students be assigned? What are their strengths and weaknesses? Should the students take a remedial course in some parts? The teacher will grade the answers given by the students. After this process, a picture of the students' knowledge structure and knowledge state [2] will emerge, the teacher will match corresponding courses to students' performance. Because the students' performance are graded by a teacher, this issue can be seen as a multi-criteria classification, which is a fundamental problem of multi-criteria decision-making (MCDM) [1]. The multi-criteria classification problem can be stated as follows: given a set of objects described by a set of criteria (attributes with preference-ordered domains), assign these objects to some pre-defined decision classes, in such a way that each object is assigned to exactly one class. In recent years, researchers have made efforts to the theories and approaches of multi-criteria classification problem.

C. Pang et al. (Eds.): SETE 2020/ICWL 2020, LNCS 12511, pp. 179–188, 2021.
https://doi.org/10.1007/978-3-030-66906-5_17

In 1982, Z. Pawlak proposed rough sets theory (RST) [22] based on equivalence relations, as a systematic approach for the classification of objects. However, RST fails to taken into account preference ordering in the domain values of the attributes. To address these limitations, in 1999, Greco et al. [16] characterized the dominance-based rough set approach(DRSA) by substituting equivalence relations in RST [22] with dominance relations. Furthermore, Greco et al. [18] discussed multi-criterion classification problem and improved classification performance in ordered information system. Interesting extension models of the dominance-based rough set have been developed by some scholars [10,11,18,19]. Recently, Du et al. established the dominance-based rough fuzzy set model [12], the innovation of fuzzy dominance relation be considered to deal with fuzziness in preference representation. Rehman, Ali et al. proposed variable precision multi decision λ-soft dominance based rough sets [23] based on soft preference and soft dominance relation.

Soft sets was first introduced by D. Molodtsov [20] in 1999, which showed powerful applicabilities for decision problems. Furthermore, a partial relation are introduced in soft sets [8,9] by Alshami et al. In particular, soft rough sets was proposed by Feng et al. [13,14] who discussed the relationships between soft rough sets and Pawlak's rough sets, the approach also achieved the effect of classification successfully. Researchers have studied the relevant theories of soft rough sets [7,26] and applied it to multiple-criteria classification problems [24,27]. In practical application, in order to show the importance of criteria, Fatimah et al. [15] introduced N-soft sets theory, which introduced the parameterized characterizations of the universe of elements that depend on a finite number of grades. Furthermore, various kinds of extensions of N-soft sets were proposed and applied to resolve multi-criteria decision-making problems in the types of fuzzy N-soft set [4], hesitant N-soft sets [5], hesitant fuzzy N-soft sets [6] and intuitionistic fuzzy N-soft rough sets [3], etc, these models account for the possibility of fuzzy or hesitant when the decision makers evaluated the parameters ranks of objects.

Motivated by these concerns, a new approach is constructed by combining rough sets, N-soft sets and dominance relation. The purposes of this paper are to pose a data analysis method for multiple-criteria classification problems and demonstrate its usefulness in the field of education. The advantages of this method are that it can classify the elements in the universe set by approximating a set divided according to a decision attribute, and address the inconsistency problems in ordered information systems.

The organization of this research article is as follows. To facilitate discussion, we first describe some basic concepts of information system in Sect. 2. Moreover, in Sect. 3, we introduce our new hybrid model, basic operations. In Sect. 4, a potential application in the education field is presented with a decision support algorithm.

2 Preliminaries

Definition 1. [2] A knowledge structure $(\mathcal{Q}, \mathcal{K})$ is called a *learning space* if it satisfies the two following conditions.
[1] Learning smoothness. For any two states K, L such that $K \subset L$, there exists a finite chain of states

$$K = K_0 \subset K_1 \subset \cdots \subset K_p = L$$

such that $|K_i \backslash K_{i-1}| = 1$ for $1 \leqslant i \leqslant p$ and so $|L \backslash K| = p$. Intuitively, in pedagogical language: If the state K of the learner is included in some other state L then the learner can reach state L by mastering the missing items one at a time.
[2] Learning consistency. If K, L are two states satisfying $K \subset L$ and q is an item such that $K + q \in K$, then $L \cup q \in K$.

Definition 2. [22] Information systems can be formally expressed as a four-tuple $\mathcal{I} = \langle U, Q, V, \varphi \rangle$, where $U = (u_1, u_2, \ldots, u_n)$ is a non-empty finite set of objects and called the universe, $Q = C \cup D$ is a non-empty finite set of attributes in which C and D are respectively condition attributes and decision attributes and V is the domain of attributes and $\varphi : U \times Q \to V$ assigning each attribute of each object with a value in V. In this article, the values of V are integers.

Definition 3. [17] For a given information system $\mathcal{I} = \langle U, Q, V, \varphi \rangle$, if there is a partial order relation "$\geq_q$" on the value range of an attribute $q \in Q$, then we call q as a criterion, and $u_i \geq_q u_j$ represents that u_i is at least as good as u_j under the criterion q, where $u_i, u_j \in U$, i.e., u_i is superior than u_j on q. When all attributes in the information system are criteria, the information system is called an ordered information system. For a subset of attributes $A \subseteq Q$, $u_i \geq_A u_j$ means that u_i dominates u_j on all the criteria in A.

In general, we denote an ordered information system by $\mathcal{S}_{\overline{D}}^{\succeq} = \{U, Q, V, \varphi\}$, where $Q = C \cup D$. For simplicity, we only consider attributes with increasing preference in the following. In an ordered information system $\mathcal{S}_{\overline{D}}^{\succeq}$, we say that u_i dominates u_j with respect to $A \subseteq C$ if $u_i \succeq_A u_j$, denoted by $u_i R_{\overline{A}}^{\succeq} u_j$, and u_i dominates u_j with respect to D if $u_i \succeq_D u_j$, denoted by $u_i R_{\overline{D}}^{\succeq} u_j$.

Denote

$$R_{\overline{A}}^{\succeq} = \{(u_i, u_j) \in U \times U | u_i \succeq_A u_j\} = \{(u_i, u_j) | \varphi(u_i, q) \geq \varphi(u_j, q), \forall q \in A\},$$
$$R_{\overline{D}}^{\succeq} = \{(u_i, u_j) \in U \times U | u_i \succeq_D u_j\} = \{(u_i, u_j) | \varphi(u_i, d) \geq \varphi(u_j, d), \forall d \in D\},$$

$$[u_i][u_i]_{\overline{A}}^{\succeq} = \{u_j \in U | (u_j, u_i) \in R_{\overline{A}}^{\succeq}\} = \{u_j \in U | \varphi(u_j, q) \geq \varphi(u_i, q), \forall q \in A\},$$
$$[u_i]_{\overline{D}}^{\succeq} = \{u_j \in U | (u_j, u_i) \in R_{\overline{D}}^{\succeq}\} = \{u_j \in U | \varphi(u_j, d) \geq \varphi(u_i, d), \forall d \in D\}.$$

Where $R_{\overline{A}}^{\succeq}$ and $R_{\overline{D}}^{\succeq}$ are called dominance relation and dominance class [16,17] in ordered information system $\mathcal{S}_{\overline{D}}^{\succeq}$.

Definition 4. [25] Let $\mathcal{S}_{\overline{D}}^{\succeq} = \{U, Q, V, \varphi\}$ be an ordered information system, where $Q = C \cup D$. If $R_{\overline{C}}^{\succeq} \nsubseteq R_{\overline{D}}^{\succeq}$, then the ordered information system is an inconsistent ordered information system.

Next, we recall some basic concepts regarding rough sets, N-soft sets and N-soft rough sets.

Definition 5. [22] Let U be a non-empty finite universe and R an equivalence relation on U. The pair (U, R) is called a Pawlak approximation space. Using the equivalence relation R, one can define the following two operations:

$$\underline{X}_R = \{x \in U : [x]_R \subseteq X\},$$

$$\overline{X}_R = \{x \in U : [x]_R \cap X \neq \emptyset\}$$

assigning to every subset $X \subseteq U$, two sets $\underline{X}_R$ and $\overline{X}_R$ called the $R-lower$ and the $R-upper$ rough approximation of X, respectively.

Definition 6. [15] Let U be a universe of objects and A the set of attributes, $A \subseteq Q$. Let $G = \{0, 1, \ldots, N-1\}$ be a set of ordered grades where $N \in \{2, 3, \ldots\}$. A triple (F, A, N) is called a N-soft set on U if $F : A \rightarrow 2^{U \times G}$, with the property that for each $q \in A$, there exists a unique $(u, g_q) \in U \times G$ such that $(u, g_q) \in F(q), u \in U, g_q \in G$.

Let $U = \{u_i, i = 1, 2, \ldots, m\}$ and $A = \{q_j, j = 1, 2, \ldots, n\}$ be finite sets. The N-soft set can be presented by Table 1, where r_{ij} means $(u_i, r_{ij}) \in F(q_j)$ or $F(q_j)(u_i) = r_{ij}$.

Table 1. Table for (F, A, N)

(F, A, N)	q_1	q_2	...	q_n
u_1	r_{11}	r_{12}	...	r_{1n}
u_2	r_{21}	r_{22}	...	r_{2n}
...	...	...	...	...
u_m	r_{m1}	r_{m2}	...	r_{mn}

Definition 7. [21] Let $S = (F, A, N)$ be a N-soft set over U and $0 \leq t < N$ a threshold. For every $q \in A$, the set associated with (F, A, N) and t denoted by $f(q, t^{\geq})$, where $f : q \rightarrow 2^U$, and it is defined by the expression $f(q, t^{\geq}) = \{u \in U | F(q)(u) \geq t\}$.

Definition 8. [21] Let $S = (F, A, N)$ be a N-soft set over U. Then $R = (U, S, N)$ is called a *N-soft rough approximation space.* For each subset $X \subseteq U$, there are two approximate operations

$$\underline{L}_R(X, A, t) = \{u \in U : \exists q \in A, [u \in f(q, t^{\geq}), u \in f(q, t^{\geq}) \subseteq X]\},$$

$$\overline{L}_R(X, A, t) = \{u \in U : \exists q \in A, [u \in f(q, t^{\geq}), u \in f(q, t^{\geq}) \cap X \neq \emptyset]\}.$$

3 Dominance-Based N-soft Rough Approximations and Dominance-Based N-soft Rough Sets

In this section, we introduce dominance-based N-soft rough approximation space and dominance-based N-soft rough sets. Let U be the initial universe and $X \subseteq U$. The complement of X is denoted by $-X$.

Definition 9. Let $S = (F, A, N)$ be a N-soft sets over U and $t \in \{0, 1, \ldots, N-1\}$ a set of ordered grades, where $N \in \{2, 3, \ldots\}$. For $A_i \subseteq A$ and $q \in A_i$, the set A_i associated with (F, A, N) and t denoted by $f(A_i, t^{\geq})$, where $F : q \to 2^{U \times R}$, and it is defined by the expression $f(A_i, t^{\geq}) = \{u \in U | F(q)(u) \geq t, q \in A_i\}$ and $f(A_i, t^{\leq}) = \{u \in U | F(q)(u) \leq t, q \in A_i\}$.

Remark 1. Let $\mathcal{S}_{\overline{D}}^{\succeq} = \{U, Q, V, \varphi\}$ be an ordered information system. For every $q \in A_i \subseteq C$ and $d \in D_i \subseteq D$, we denote that $f(A_i, t_{\overline{A}_i}^{\geq}) = \{u \in U | F(q)(u) \geq t_{A_i}, q \in A_i\}$ and $f(D_i, t_{\overline{D}_i}^{\geq}) = \{u \in U | F(d)(u) \geq t^{\geq}, d \in D_i\}$. Moreover, when there's only one attribute q in A_i, then the dominance-based N-soft rough sets model in degenerates in to the Definition 7 in N-soft rough sets [21].

Definition 10. Let $\mathcal{S}_{\overline{D}}^{\succeq} = \{U, Q, V, \varphi\}$ be an ordered information system, $D_i \subseteq D$ and $S = (F, A, N)$ a N-soft sets over U. Then $R = (U, S, N)$ is called a *dominance-based N-soft rough approximation space.* The lower and the upper approximations of $f(D_i, t_{\overline{D}_i}^{\geq})$ are defined as follows:

$$\underline{L}_R(f(D_i, t_{\overline{D}_i}^{\geq})) = \{u \in U : \forall A_i \subseteq A, f(A_i, t_{\overline{A}_i}^{\geq}) \subseteq f(D_i, t_{\overline{D}_i}^{\geq})\},$$

$$\overline{L}_R(f(D_i, t_{\overline{D}_i}^{\geq})) = \{u \in U : \exists A_i \subseteq A, f(A_i, t_{\overline{A}_i}^{\geq}) \cap f(D_i, t_{\overline{D}_i}^{\geq}) \neq \emptyset\}.$$

The two sets $\underline{L}_R(f(D_i, t_{\overline{D}_i}^{\geq})$ and $\overline{L}_R(f(D_i, t_{\overline{D}_i}^{\geq})$ are called the *dominance-based N-soft rough R-lower approximation* and the *dominance-based N-soft rough R-upper approximation* of X, respectively. Analogously, the lower and the upper approximations of $f(D_i, t_{\overline{D}_i}^{\leq})$ are defined as follows:

$$\underline{L}_R(f(D_i, t_{\overline{D}_i}^{\leq}) = \{u \in U : \forall A_i \subseteq A, f(A_i, t_{\overline{A}_i}^{\leq}) \subseteq f(D_i, t_{\overline{D}_i}^{\leq}\}),$$

$$\overline{L}_R(f(D_i, t_{\overline{D}_i}^{\leq}) = \{u \in U : \exists A_i \subseteq A, f(A_i, t_{\overline{A}_i}^{\leq}) \cap f(D_i, t_{\overline{D}_i}^{\leq}) \neq \emptyset\}.$$

In the following parts of this paper, we discuss the situation of $t^{\geq}$. Moreover, the sets

$$Pos_R(f(D_i, t_{\overline{D}_i}^{\geq})) = \underline{L}_R(f(D_i, t_{\overline{D}_i}^{\geq}))$$

$$Neg_R(f(D_i, t_{\overline{D}_i}^{\geq})) = -\overline{L}_R(f(D_i, t_{\overline{D}_i}^{\geq}))$$

$$Bnd_R(f(D_i, t_{\overline{D}_i}^{\geq})) = \overline{L}_R(f(D_i, t_{\overline{D}_i}^{\geq})) - \underline{L}_R(f(D_i, t_{\overline{D}_i}^{\geq}))$$

are called *dominance-based N-soft rough R-positive region, the dominance-based N-soft rough R-negative region* and the *dominance-based N-soft rough R-boundary region* of X, respectively. If $\underline{L}_R(f(D_i, t_{\overline{D}_i}^{\geq})) = \overline{L}_R(f(D_i, t_{\overline{D}_i}^{\geq}))$, that is

$Bnd_R(f(D_i, t_{D_i}^{\geq})) = \emptyset$, then X is said to be a *dominance-based N-soft rough R-definable set.* If $\underline{L}_R(f(D_i, t_{D_i}^{\geq})) \neq \overline{L}_R(f(D_i, t_{D_i}^{\geq}))$, that is $Bnd_R(f(D_i, t_{D_i}^{\geq})) \neq \emptyset$, then X is said to be a *dominance-based N-soft R-rough set.*

The lower approximation $\underline{L}_R(f(D_i, t_{D_i}^{\geq}))$ includes all actions in the class $f(D_i, t_{D_i}^{\geq}))$ and the upper approximation $\overline{L}_R(f(D_i, t_{D_i}^{\geq}))$ includes all actions possibly included in the class $f(D_i, t_{D_i}^{\geq})$. We can present the interpretation for the positive region, negative region and boundary region in the same way.

Proposition 1. Let $\mathcal{S}_{\overline{D}}^{\succ} = \{U, Q, V, \varphi\}$ be an ordered information system,$D_i \subseteq D$ and $S = (F, A, N)$ a N-soft sets over U. From the above definition, we have the following two approximations.

$$\underline{L}_R(f(D_i, t_{D_i}^{\geq}))) = \bigcap_{A_i \subseteq A} \{u \in U : f(A_i, t_{A_i}^{\geq}) \subseteq f(D_i, t_{D_i}^{\geq})\},$$

$$\overline{L}_R(f(D_i, t_{D_i}^{\geq}))) = \bigcup_{A_i \subseteq A} \{u \in U : f(A_i, t_{A_i}^{\geq}) \cap f(D_i, t_{D_i}^{\geq}) \neq \emptyset\}.$$

Remark 2. Let $\mathcal{S}_{\overline{D}}^{\succ} = \{U, Q, V, \varphi\}$ be an ordered information system, where $Q = C \cup D$. If $\mid D \mid= 1$, i.e., there is only one decision-maker, we denote by $\mathcal{S}_{\overline{d}}^{\succ} = \{U, Q, V, \varphi\}$. In this case, the set $f(d, t_{\overline{d}}^{\geq})$ and $Cl_t^{\geq}$ are equivalent.

In the following parts of this paper, we discuss following propositions in $\mathcal{S}_{\overline{d}}^{\succ} = \{U, Q, V, \varphi\}$.

Proposition 2. Let $\mathcal{S}_{\overline{d}}^{\succ} = \{U, Q, V, \varphi\}$ be a information system, $S = (F, A, N)$ a N-soft sets over U, $0 \leq t_{A_i}, t_{A_j} < N$ a threshold and $R = (U, S, N)$ a N-soft rough approximation space. Based on above definitions, we have

(1) If $A_j \subseteq A_i$, then $f(A_i, t_{A_i}^{\geq}) \subseteq f(A_j, t_{A_j}^{\geq})$;
(2) $\mid f(A_i, t_{A_i}^{\geq}) \mid\geq 1$ for any $0 \leq t_{A_i} < N$;
(3) Supposed that $x_i \in f(A_i, t_{A_i}^{\geq})$ and $x_j \in f(A_j, t_{A_j}^{\geq})$. If $x_j \in f(A_i, t_{A_i}^{\geq})$, then $f(A_j, t_{A_j}^{\geq}) \subseteq f(A_i, t_{A_i}^{\geq})$.

4 A Application of Dominance-Based N-soft Rough Sets and Its Decision Making Procedures

In [2], two key concepts are: the 'knowledge state', a subset of problems that some individual is capable of solving correctly, and the 'knowledge structure', which is a distinguished collection of knowledge states. We believe that the problem involved in any assessment can be chosen in a pool covering the entire curriculum. Both of these concepts are crucial when we get the students' performance and try to solve the matching problem of the students' curriculum. Now let's consider the following application.

Example 1. Let $U = \{u_1, u_2, u_3, \ldots, u_{10}\}$ be a set of students from a foreign country and $A = \{q_1, q_2, q_3, q_4, q_5, q_6, q_7, q_8, d\}$ a basic set of such questions. The set A can be used for assess the students' capability, in which $q_1, q_2, q_3, q_4, q_5, q_6, q_7, q_8, d$ represent "Set theory", "Solving all quadratic equations", "Exponential function", "Logarithmic function", "Trigonometric functions", "Vector quantity", "Plane analytic geometry", "Conic curve" and "Comprehensive quality assessment". According to the high school mathematics knowledge module, we classified $A = \{A_1, A_2, A_3\}$, where $A_1 = \{q_1\}$ stands for the set question, $A_2 = \{q_2, q_3, q_4, q_5\}$ is the function questions, $A_3 = \{q_6, q_7, q_8\}$ is the geometric questions. In addition, the score of "comprehensive quality assessment" is also an essential item for evaluating students. The students' performance are given by a test in Table 2, for the same criteria, the higher the attribute value is, the better the students master such knowledge. Certainly, we also know the knowledge state and knowledge structure of these students. Now, we intend to assign these students to $ClassA$, B and C, students in $Class\ A$ don't need students to make up the missed lessons, students in $Class\ B$ need to make up the missed lessons moderately, and students in $Class\ C$ need a lot of missed lessons.

Table 2. Students' performance

U	q_1	q_2	q_3	q_4	q_5	q_6	q_7	q_8	d
u_1	6	4	1	2	2	3	5	3	5
u_2	4	7	3	3	5	4	6	3	3
u_3	7	5	4	4	2	4	5	5	5
u_4	7	4	3	3	3	4	6	4	3
u_5	5	6	2	2	2	2	4	2	4
u_6	8	8	4	4	4	5	7	4	3
u_7	7	4	4	2	4	4	6	3	2
u_8	6	6	2	2	2	4	6	1	1
u_9	6	4	3	4	3	4	7	4	5
u_{10}	5	4	3	1	2	2	5	4	4

Step 1: Input the original information system.
From the Table 2, we can get

$[u_1]_{\widetilde{A}}^{\succeq} = \{u_1, u_3, u_4, u_6, u_7, u_9\}$; $[u_1]_{\widetilde{d}}^{\succeq} = \{u_1, u_3, u_9\}$;
$[u_5]_{\widetilde{A}}^{\succeq} = \{u_5, u_6\}$; $[u_5]_{\widetilde{d}}^{\succeq} = \{u_1, u_3, u_5, u_9, u_{10}\}$;
$[u_9]_{\widetilde{A}}^{\succeq} = \{u_6, u_9\}$; $[u_9]_{\widetilde{d}}^{\succeq} = \{u_1, u_3, u_9\}$;
$[u_{10}]_{\widetilde{A}}^{\succeq} = \{u_3, u_4, u_6, u_9, u_{10}\}$; $[u_{10}]_{\widetilde{d}}^{\succeq} = \{u_1, u_3, u_5, u_9, u_{10}\}$;

and

$$[u_1]^{\succeq}_{A} \nsubseteq [u_1]^{\succeq}_{d}; \qquad [u_5]^{\succeq}_{A} \nsubseteq [u_5]^{\succeq}_{d}$$
$$[u_9]^{\succeq}_{A} \nsubseteq [u_9]^{\succeq}_{d}; \qquad [u_{10}]^{\succeq}_{A} \nsubseteq [u_{10}]^{\succeq}_{d}$$

Therefore, the information system can be seen as an inconsistent ordered information system.

Step 2: Based on the attribute d, U can be divided into different class by different grade t. We require that the "comprehensive quality assessment" grades of students belonging to $Class\ A$ and $Class\ B$ should be greater than 2. Therefore, we get $f(d, 3^{\geq}) = \{u_1, u_2, u_3, u_4, u_5, u_6, u_9, u_{10}\}$.

Step 3: By Definition 9, we can get

$$f(q_1, 7^{\geq}) = \{u_3, u_4, u_6, u_7\}; \qquad f(q_2, 3^{\geq}) = \{U\};$$
$$f(q_3, 3^{\geq}) = \{u_2, u_3, u_4, u_6, u_7, u_9, u_{10}\}; f(q_4, 3^{\geq}) = \{u_2, u_3, u_4, u_6, u_9\};$$
$$f(q_5, 3^{\geq}) = \{u_2, u_4, u_6, u_7, u_9\}; \qquad f(q_6, 4^{\geq}) = \{u_2, u_3, u_4, u_6, u_7, u_8, u_9\};$$
$$f(q_7, 4^{\geq}) = \{U\}; \qquad f(q_8, 4^{\geq}) = \{u_3, u_4, u_6, u_9, u_{10}\}.$$

and

$$f(A_1, 7^{\geq}) = \{u_3, u_4, u_6, u_7\};$$
$$f(A_2, 3^{\geq}) = \{u_2, u_4, u_6, u_9\};$$
$$f(A_3, 4^{\geq}) = \{u_3, u_4, u_6, u_9\}.$$

Step 4: According to Definition 10, the sets of lower approximation $\underline{L}_R(f(d, t^{\geq}_{d}))$ and the upper approximation $\overline{L}_R(f(d, t^{\geq}_{d}))$, respectively.

$$\underline{L}_R(f(d, t^{\geq}_{d})) = \{u_4, u_6\}$$

and

$$\overline{L}_R(f(d, t^{\geq}_{d})) = \{u_2, u_3, u_4, u_6, u_7, u_9\}.$$

Step 5: Therefore, the distribution of ten students are

$$Class\ A = \{u_4, u_6\};$$
$$Class\ B = \{u_2, u_3, u_7, u_9\};$$
$$Class\ C = U - \overline{L}_R(f(d, t^{\geq}_{d})) = \{u_1, u_5, u_8, u_{10}\}.$$

The above results show that dominance-based N-soft rough sets approximation is a worth considering alternative to the rough sets approximation. In a inconsistent ordered information system, dominance-based N-soft rough sets have more advantages to classify a universe set of objects.

5 Conclusions

In this article, we pose a new concept called dominance-based N-soft rough sets to studied practical problems in ordered information system. We give the dominance-based N-soft rough approximation operators based on dominance-based N-soft rough approximation space. Different from soft rough sets [14], the method of dominance-based N-soft rough sets allow the practitioner to introduce subjectivity in order to account for personal preferences and attribute rank. Moreover, combined with an example from life, we give the specific decision-making procedures. This approach could provide simpler and more efficient assessment, which makes the decision result more scientific and reasonable.

References

1. Wallenius, J., Dyer, J.S., Fishburn, P.C.: Multiple criteria decision making, multi-attribute utility theory: recent accomplishments and what lies ahead. Manag. Sci. **54**(7), 1336–1349 (2008)
2. Falmagne, J.C., Doignon, J.P.: Learning Spaces: Interdisciplinary Applied Mathematics. Springer, Heidelberg (2011)
3. Akram, M., Ali, G., Alcantud, J.C.R.: New decision-making hybrid model: intuitionistic fuzzy N-soft rough sets. Soft. Comput. **23**(20), 9853–9868 (2019)
4. Akram, M., Adeel, A., Alcantud, J.C.R.: Fuzzy N-soft sets: a novel model with applications. J. Intell. Fuzzy Syst. **35**(4), 4757–4771 (2018)
5. Akram, M., Adeel, A., Alcantud, J.C.R.: Group decision-making methods based on hesitant N-soft sets. Expert Syst. Appl. **115**(JAN), 95–105 (2019)
6. Akram, M., Adeel, A., Alcantud, J.C.R.: Hesitant fuzzy N-soft sets: a new model with applications in decision-making. J. Intell. Fuzzy Syst. **36**(6), 6113–6127 (2019)
7. Ali, A., Ali, M., Rehman, N.: Soft dominance based rough sets with applications in information systems. Int. J. Approx. Reason. **171–195** (2019)
8. Al-shami, T.M., El-Shafei, M.E.: Partial belong relation on soft separation axioms and decision-making problem, two birds with one stone. Soft. Comput. **24**(7), 5377–5387 (2019). https://doi.org/10.1007/s00500-019-04295-7
9. Alshami, T.M., Elshafei, M.E., Aboelhamayel, M.: On soft topological ordered spaces. J. King Saud Univ.-Sci. **31**(4), 556–566 (2019)
10. Błaszczyński, J., Greco, S., Słowiński, R., Szeląg, M.: Monotonic variable consistency rough set approaches. Rough Sets Knowl. Technol. **50**(7), 979–999 (2009)
11. Błaszczyński, J., Słowiński, R., Szeląg, M.: Sequential covering rule induction algorithm for variable consistency rough set approaches. Inf. Sci. **181**(5), 987–1002 (2011)
12. Du, W.S., Hu, B.Q.: Dominance-based rough fuzzy set approach and its application to rule induction. Eur. J. Oper. Res. **261**(2), 690–703 (2017)
13. Feng, F., Li, C., Davvaz, B., Ali, M.I.: Soft sets combined with fuzzy sets and rough sets: a tentative approach. Soft. Comput. **14**(9), 899–911 (2010)
14. Feng, F., Liu, X., Leoreanu-Fotea, V., Jun, Y.B.: Soft sets and soft rough sets. Inf. Sci. **181**(6), 1125–1137 (2011)
15. Fatimah, F., Rosadi, D., Hakim, R.F., Alcantud, J.C.R.: N-soft sets and their decision making algorithms. Soft. Comput. **22**(12), 3829–3842 (2018)
16. Greco, S., Matarazzo, B., Słowiński, R.: Rough approximation of a preference relation by dominance relations. Eur. J. Oper. Res. **117**(1), 63–83 (1999)

17. Greco, S., Matarazzo, B., Słowiński, R.: Rough sets theory for multicriteria decision analysis. Eur. J. Oper. Res. **129**(1), 1–47 (2001)
18. Greco, S., Matarazzo, B., Słowiński, R.: Extension of the rough set approach to multicriteria decision support. INFOR: Inf. Syst. Oper. Res. **38**(3), 161–195 (2000)
19. Inuiguchi, M., Yoshioka, Y., Kusunoki, Y.: Variable-precision dominance-based rough set approach and attribute reduction. Int. J. Approx. Reason. **50**(8), 1199–1214 (2009)
20. Molodtsov, D.: Soft set theoryfirst results. Comput. Math. Appl. **37**(4–5), 19–31 (1999)
21. Zhang, D., Li, P.-Y., An, S.: J. Intell. Fuzzy Syst. (2020). https://doi.org/10.3233/JIFS-200338
22. Pawlak, Z.: Rough sets. Int. J. Comput. Inf. Sci. **11**(5), 341–356 (1982)
23. Rehman, N., Ali, A., Shah, S.I., Park, C.: Variable precision multi decision λ-soft dominance based rough sets and their applications in conflict problems. J. Intell. Fuzzy Syst. **36**(6), 5345–5360 (2019)
24. Shaheen, T., Mian, B., Shabir, M., Feng, F.: A novel approach to decision analysis using dominance-based soft rough sets. Int. J. Fuzzy Syst. **21**(3), 954–962 (2019). https://doi.org/10.1007/s40815-019-00612-2
25. Xu, W., Zhang, W.: Methods for knowledge reduction in inconsistent ordered information systems. J. Appl. Math. Comput. **313–323** (2008)
26. Zhan, J., Alcantud, J.C.R.: A novel type of soft rough covering and its application to multicriteria group decision making. Artif. Intell. Rev. **52**(4), 2381–2410 (2018). https://doi.org/10.1007/s10462-018-9617-3
27. Zhang, G., Li, Z., Qin, B.: A method for multi-attribute decision making applying soft rough sets. J. Intell. Fuzzy Syst. **30**(3), 1803–1815 (2016)

Data-Driven Decision

Trend Analysis of At-a-station Hydraulic Geometry Relations on the Loess Plateau of China

Shumei Zhou[1,2](✉) and Qinghong Yan[3]

[1] School of Management and Economics, Hebei University of Science and Technology, Shijiazhuang, China
shumei1984@163.com
[2] Center for Ecological Economy and Sustainable Development of Hebei Province, Shijiazhuang, China
[3] Department of Irrigation and Drainage, China Institute of Water Resources and Hydropower Research, Beijing, China

Abstract. Empirical at-a-station hydraulic geometry (AHG) power function relations have been widely used to depict relationships between flow width (*W*), flow depth (*D*), mean flow velocity (*V*) and discharge rate (*Q*) for natural rivers and streams. However, it has seldom been considered whether AHG relations hold true for describing hydraulic relations at watershed outlet channels. In this study, observed data from three experimental watersheds of different sizes (1–100 km^2) in the past three decades were used to verify stability of AHG at watershed outlet channels on the Loess Plateau in China. Trend analysis via the sequential Mann-Kendall test was used to detect the stability and abrupt changes in AHG during the study course. The results suggested strong stability in AHG exponents and significant decline in coefficients ($P < 0.05$). The abrupt decline for coefficients was found to be influenced by human activities such as land use changes. This study could provide some useful aids for watershed hydrology management.

Keywords: At-a-station hydraulic geometry (AHG) · Power function · Mann-Kendall test · Watershed · Loess Plateau

1 Introduction

Hydraulic geometry (HG) concept was first put forward by Leopold and Maddock (1953) based on statistical analyses of hydrologic data in the Great Plains and Southwest of the United States of America [1]. It was found that the relationships between discharge rate and three hydraulic variables (flow width, flow depth and mean flow velocity) could be similarly expressed by the power function models:

$$W = aQ^b \tag{1a}$$

$$D = cQ^f \tag{1b}$$

C. Pang et al. (Eds.): SETE 2020/ICWL 2020, LNCS 12511, pp. 191–201, 2021.
https://doi.org/10.1007/978-3-030-66906-5_18

$$V = kQ^m \tag{1c}$$

where W and D are flow width and depth; V is mean flow velocity; Q is discharge rate; and a, b, c, f, k and m are numeric constants. The HG relations were termed as "at-a-station hydraulic geometry" (AHG) when considering instantaneous river flow width, flow depth, mean flow velocity and discharge rate at a point. The exponents b, f and m represent the respective magnitudes of change rate in W, D and V, as Q changes, while the constant parameters a, c and k are scale factors that are equivalent to W, D and V for unit Q.

Although AHG has been widely used all over the world since its first establishment in 1950s, due to the failure to accommodate any other influencing factors but only discharge rate, the relations have been questioned by many scholars. In terms of AHG parameters, most scholars focused on exponents, believed that they are stable and have derived site-specific parameter values. For example, Leopold and Maddock found that flow width, depth and flow velocity exponents averaged 0.26, 0.40 and 0.34 [1], while Carlston derived those exponent parameters as 0.46, 0.38 and 0.16, respectively [2]. Some scholars argue that the values of AHG parameters vary due to unique geological, hydrological and climatic conditions. Ferguson introduced the rational explanation by satisfactorily reducing AHG to "hydraulics and geometry". However, Ferguson ignored AHG coefficients and failed to give explicit formulations explaining the interactions between AHG, channel cross-section geometry and hydraulic factors [3]. Dingman derived explicit formulations for exponents and coefficients of AHG power function relations and found that both exponents and coefficients are dependent on other AHG parameters and channel geometry [4]. Despite Ferguson's reduction of AHG to hydraulics and geometry and Dingman's theoretical derivation of exponents and coefficients, there are still many researchers who remain interested in verifying AHG's existence in diverse environments and its temporal variation [5–9].

The Chinese Loess Plateau is characterized by loose and thick loess soil layer, poor vegetation, intense summer storms and thus severe soil and water loss, causing highly sediment-concentrated floods [10]. With increasing global warming, the extreme climate events frequently yielded disastrous results. The ability to understand and predict the form of these hydraulic geometry relationships across watersheds therefore represents a powerful tool for watershed managers. Therefore, the response laws of watershed hydrological processes must be discussed in-depth [9].

In this study, AHG was extended to describe relations of hydraulic variables at outlet channels of small watersheds in the special context of the Loess Plateau region, to increase knowledge and understanding for changes and its driving forces of watershed hydraulic relations that can help for watershed hydrologic modeling. The objectives of this case study were: 1) to verify AHG power function models in the Loess Plateau region, and 2) to try to give scientific explanation for AHG parameter changes at watershed levels in the study area.

2 Materials and Methods

2.1 Study Area

The hilly-gully region on the Loess Plateau is an area with serious soil erosion in China. In this study, Qiaozi-East (1.36 km^2, W1), Lüergou (12.01 km^2, W10) and Luoyugou (72.79 km^2, W100) watersheds were chosen in the loess hilly-gully region of the Loess Plateau near Tianshui City, Gansu Province, China (Fig. 1). The three experimental watersheds could be considered as representative watershed scales of about 1, 10 and 100 km^2, respectively (Table 1). The study area has a dry and continental climate. The indigenous vegetation in the study watersheds mainly comprises warm temperate deciduous forest. The remaining area is farmland mainly planted with wheat (*Triticum aestivum Linn.*), corn (*Zea mays L.*) and potatoes (*Solanum tuberosum L.*). The annual mean precipitation is 496–628 mm, and about 60% of precipitation occurs during summer days (typically from June to September) as rainstorms of high intensity and short duration. Soil and water conservation practices have been implemented including a ban on grazing, constructing check dams and terraces, restricting cropland farming to less steep slopes and vegetation restoration in W1, W10 and W100 since 1980s.

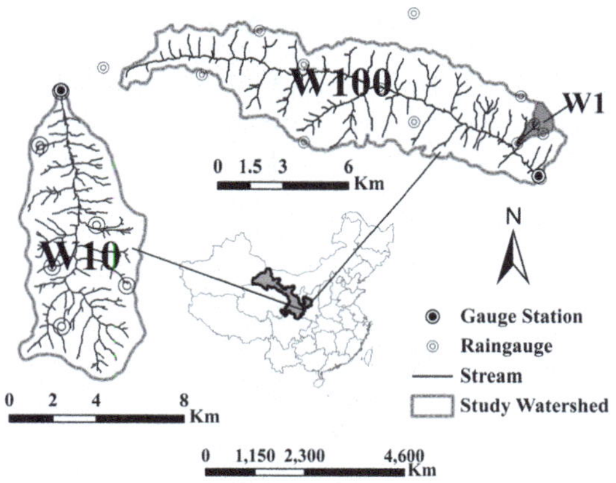

Fig. 1. Locations of three experimental watersheds in Gansu Province, China (W1, Qiaozi-East; W10, Lüergou; W100, Luoyugou).

Table 1. General information of the three experimental watersheds

Watershed name	Area (km^2)	Length (km)	Width (km)	Gully density (km km^{-2})	Mean gully gradient (%)	Number of rain gauges	Number of hydrometric stations
W1	1.4	2	0.7	5.1	8	2	1
W10	12.0	6.6	2.6	3.8	4	6	1
W100	72.8	21.6	3.4	3.8	2.3	24	3

2.2 Data Collection

Both geographic and hydrologic data were collected from the Tianshui Soil and Water Conservation Station of the Yellow River Conservancy Commission (YRCC) in Tianshui City, Gansu Province, China. Discharge rate was derived using flow velocity and cross sectional area by Eq. (2):

$$Q = \alpha VA \tag{2}$$

where A is cross sectional area calculated from the measured water level and channel shape parameters; α is flow velocity adjusting factor, which equals 0.85 and 0.65 (values supplied by the Yellow River Institute of Hydraulic Research, YRCC, Zhengzhou) during level and flood periods, respectively. Flow velocity data were acquired by buoy method [11] for the three study watersheds before 2006. In W100 and W10, the riverbanks at the watershed outlets have been raised due to urbanization however the shapes of the outlets were natural. As AHG coefficients are unit dependent [12], hydrologic data including discharge rate, flow width, flow depth and flow velocity, were transformed into dimensionless data through dividing observed data by the corresponding annual mean values for each year in the experimental watersheds, and the values of AHG exponents would not be changed due to data treatment.

2.3 Mann-Kendall (MK) Trend Detection Test

The nonparametric Mann–Kendall trend test [13, 14] has been widely used to examine the significance of monotonic trends in meteorological, climatological and hydrological data series [15–17]. The MK test statistic S is calculated using following Eqs. (3) and (4):

$$\mathrm{sgn}(X_j - X_i) = \begin{cases} +1\ ; & \text{if } X_j > X_i \\ 0\ ; & \text{if } X_j = X_i \\ -1; & \text{if } X_j < X_i \end{cases} \tag{3}$$

$$S = \sum_{i=1}^{n-1} \sum_{j=i+1}^{n} sgn(X_j > X_i) \tag{4}$$

where X_i and X_j are data values at time i and j in the data series. The notation n indicates the length of data set. Under the assumption that data are independent and identically distributed, the mean and variance of the statistic S in Eq. (4) are given by the following:

$$E(S) = 0 \tag{5}$$

$$Var(S) = \frac{n(n-1)(2n+5) - \sum_{k=1}^{p} t_k(t_k - 1)(2t_k + 5)}{18} \tag{6}$$

In Eq. (6), p is the number of tied groups (there is a tie when $x_j = x_i$), and the summary sign (Σ) indicates the summation over all tied groups. The t_k is the number of data values in the p_{th} group. If there are not tied groups, this summary process can be ignored [18].

The distribution of S tends to normality as the number of observations becomes large [14]. The significance of trends can be checked by comparing standardized variable Z, calculated by Eq. (7) [19], with the standard normal variate at the desired confidence levels ($\alpha = 10\%$, $\alpha = 5\%$, $\alpha = 1\%$). If the calculated $|Z| > |Z_{1-\alpha/2}|$, the null hypothesis (H_0) is invalid, i.e. the trend is statistically significant. A positive value of Z indicates an upward trend, while a negative Z indicates a downward trend. Otherwise, H_0 hypothesis is accepted, that the trend is not statistically significant in the time series. In this study, α was set at 0.05 significant levels unless otherwise stated.

$$Z = \begin{cases} \frac{S-1}{\sqrt{\mathrm{Var}(s)}}; & \text{if } S > 0 \\ 0; & \text{if } S = 0 \\ \frac{S+1}{\sqrt{\mathrm{Var}(s)}}; & \text{if } S < 0 \end{cases} \quad (7)$$

2.4 Mann–Kendall (MK) Abrupt Detection Test

The sequential Mann-Kendall test was used to detect abrupt changes in the data series [20]. Suppose there are samples $X_1, X_2, \ldots, X_n$ from set of random variable X based on rank series of progressive and retrograde rows of the sample. Magnitudes of X_i with $i = 1, \ldots, k$ are compared with X_j, where $j = 1, \ldots, i - 1$. For each comparison, the number of cases $X_i > X_j$ is counted and denoted by r_i in Eq. (8).

$$r_i = \begin{cases} +1; & \text{if } X_i > X_j \\ 0; & \text{else} \end{cases} \quad (8)$$

The test statistic S_k is determined by Eq. (9) as following:

$$S_k = \sum_{i=1}^{k} r_i \; (k = 1, \ldots, n) \quad (9)$$

Under the assumption of stochastic independence of time series, the mean and variance of S_k are determined by Eqs. (10) and (11) as

$$E(S_k) = k(k-1)/4 \quad (10)$$

$$Var(S_k) = k(k-1)(2k+5)/72 \quad (11)$$

The sequential values of the statistic UF_k are calculated as

$$UF_k = \frac{S_k - E(S_k)}{\sqrt{Var(S_k)}} \quad (12)$$

which is the forward sequence, and the backward sequence UB_k is calculated in the same way but in the reverse data series. A positive UF_k means an upward trend while the reverse denotes a downward trend [21]. The occurrence of a trend could be detected from the intersection of the forward (UF_k) and backward (UB_k) curves of the test statistic if it occurs between the critical values [22, 23].

Trends and abrupt changes in the annual AHG parameters for the experimental watersheds were searched for using MK trend test and the sequential version of MK abrupt detection method, respectively. All the graphs were carried out using Grapher 8.0 software, available at http://www.goldensoftware.com.

3 Results and Discussion

Hydrologic data generally showed a good adaption to a power function relationship with average R^2 falling in the range of 0.74–0.96, suggesting strong stability of a power function in typical small watersheds in the Loess Plateau region.

3.1 Annual AHG Coefficient Trend

AHG coefficients fluctuated greatly for different years but the declining trends of AHG coefficients were comparable at both intra-watershed and inter-watershed space scales in the three experimental watersheds. Based on MK trend analysis, AHG coefficients (*a*, *c* and *k*) exhibited a downward trend in the experimental watersheds during the study period ($P < 0.05$), see Table 2.

Table 2. Calculated Z values of at-a-station hydraulic geometry (AHG) parameters in the three experimental watersheds

Watersheds	Coefficients			Exponents		
	a	*c*	*k*	*b*	*f*	*m*
W1	-2.34^{**}	-1.69^{*}	-2.37^{**}	0.91	0.16	−0.16
W10	-2.29^{*}	-2.45^{**}	-2.23^{*}	-3.71^{**}	3.35^{**}	−0.59
W100	-1.79^{*}	-2.20^{*}	-1.85^{*}	0.37	−0.29	0.66

Note: symbols * and ** indicate the trend is significant at $P = 0.05$ and 0.01, respectively.

3.2 Changes of AHG Coefficients

According to Mann-Kendall trend analysis, there is a significant decline in AHG coefficients for the study watersheds ($P < 0.05$). Hence, the sequential Mann–Kendall test was applied to graphically illustrate the abrupt changes of AHG coefficients in 1982–2013 (Fig. 2). The intersection point of *UF* and *UB* curves indicates the starting point of abrupt change for coefficients. Abrupt change in AHG parameter becomes significant at the point where the curves go beyond the dash-dotted lines.

Figure 2-1 illustrates that decline of *a* started in 2003, accelerated in 2007 and became significant in 2009 in W1. The parameter *c* declined significantly in 2009 (Fig. 2-4) and a decrease *k* also started in 2003 as *a*, accelerated in 2007 and got significant in 2011 (Fig. 2-7, $P < 0.05$).

In W10, abrupt decline of coefficients *a* and *k* started in 2007, accelerated and became significant in 2011 (Fig. 2-2 and Fig. 2-8). Coefficient *c* started to decline as early as in 1998, and the decline accelerated in 2007 and became significant in 2010 (Fig. 2-5).

Figure 2-3 shows that decline of coefficient *a* in W100 started in 2007 and became significant in 2012 ($P < 0.05$). Coefficient *c* began to decline in 2001 and declined significantly in 2010 (Fig. 2-6, $P < 0.05$). For coefficient *k*, the decrease happened in 2007, the same year of *a*, and became significant in 2012 ($P < 0.05$), as shown in Fig. 2-9. Although descending pattern is similar to those in W1 and W10, the decline of coefficients in W100 is not as significant as those in the two smaller watersheds ($P < 0.05$). In general, 2007–2012 is the period of abrupt decline in AHG coefficients in the three experimental watersheds.

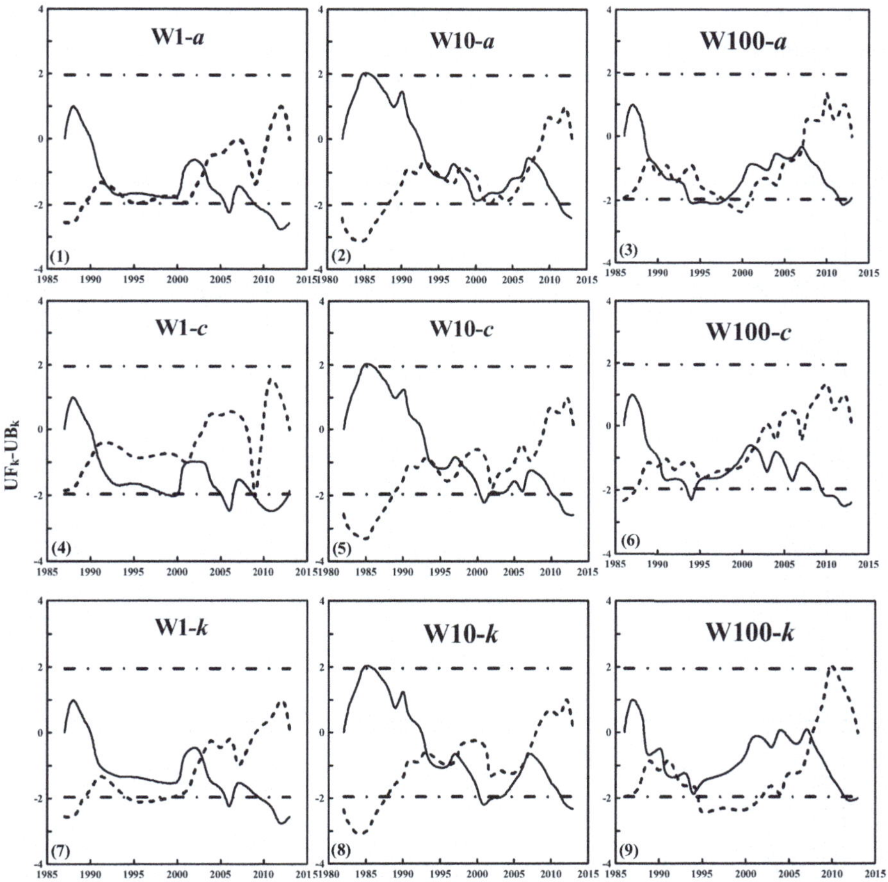

Fig. 2. Sequential Mann–Kendall test for annual at-a-station hydraulic geometry (AHG) coefficients in the three experimental watersheds (W1, W10 and W100) with forward-trend *UF*(*k*)—solid line, and backward-trend *UB*(*k*)—dash-dotted horizontal lines represent critical values corresponding to the 95% confidence interval.

3.3 Annual AHG Exponent Trend

Based on Mann–Kendall analysis, AHG exponents (1982–2013) for W1 and W100 showed sound stability according to Table 2. To be noted, exponent *b* presented a significantly downward trend ($P < 0.01$) while f exhibited a significantly upward trend ($P < 0.01$) in W10, different from those in the other two study watersheds. The exponent m had no significant change ($P < 0.05$).

3.4 Changes of AHG Exponents

AHG exponents b, f and m showed strong stability during the study course in W1 and W100 (Fig. 3). However, the case in W10 was different. Figure 3-2 depicts a significant decline in exponent b; the decline began in 1994, and became significant in 2000 ($P < 0.05$). On the contrary, exponent f in W10 had an increasing trend (Fig. 3-5); the abrupt increase took place in 1998 and became significant in 2006 ($P < 0.05$). For exponent m, there was no significant tendency ($P < 0.05$), see Fig. 3-8.

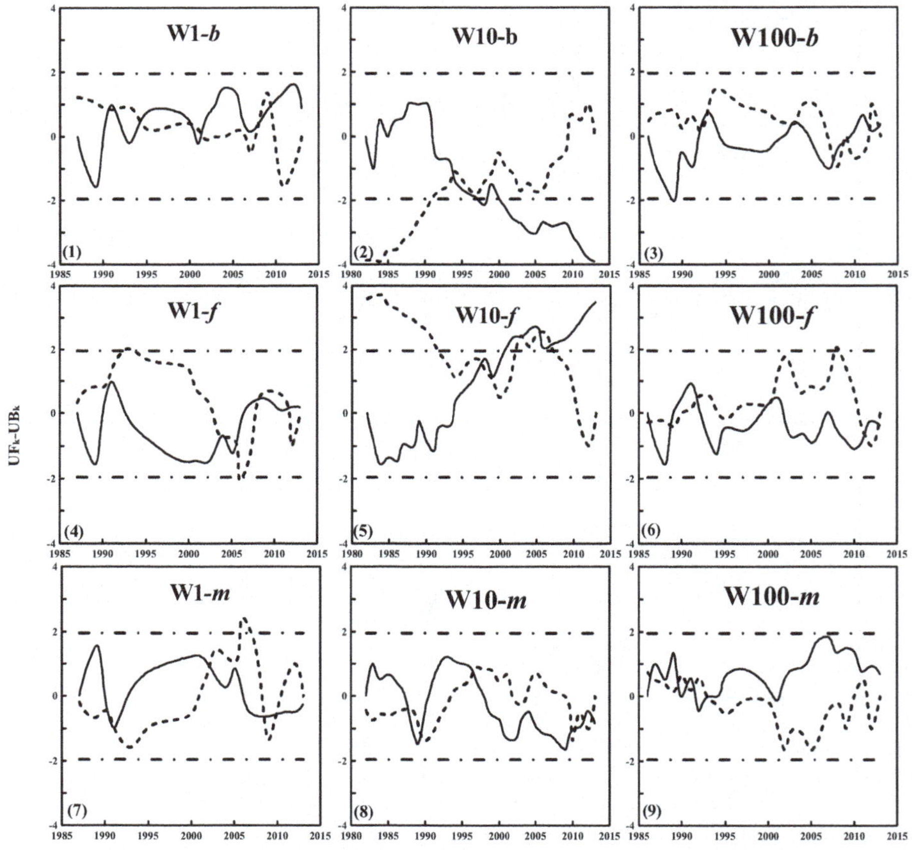

Fig. 3. Sequential Mann–Kendall test for annual at-a-station hydraulic geometry (AHG) exponents in the three experimental watersheds (W1, W10 and W100) with forward-trend $UF(k)$—solid line, and backward-trend $UB(k)$—dash-dotted horizontal lines represent critical values corresponding to the 95% confidence interval.

3.5 Discussion

The maximum and mean annual sediment contents at outlet channels during the study period were measured as 1290 $kg \cdot m^{-3}$ and 254.83 $kg \cdot m^{-3}$ for W1, 1450 $kg \cdot m^{-3}$ and 78.74 $kg \cdot m^{-3}$ for W10, and 1200 $kg \cdot m^{-3}$ and 113.91 $kg \cdot m^{-3}$ for W100, respectively. As AHG relations are directly determined and affected by cross-sectional shape, frictional characteristics and a law between average velocity and friction and depth [3], they would be surely influenced by extremely high sediment content in the floods and short time of confluence, under short-term and high-intensity summer rainstorms in the Loess Plateau region. The hyperconcentrated flow is a Binghamian flow rather than a Newtonian flow, and its physical properties and mechanical behavior differ from those of normal flow [24, 25]. The changes in flow resistance by different sediment concentrations would impose great effects on AHG relations.

Stability of AHG Power Function Relations. Basically, AHG exponents maintained remarkable stability during the study period, which indicated that AHG power function relations generally kept stable at a watershed level on the Loess Plateau. In this study, hydrologic data were regrouped for different sediment contents in each of the three watersheds. AHG power function models were then fitted for each data group. The results showed that coefficient of determination (R^2) values of different groups generally increased as flow sediment contents increased for both intra-watershed and inter-watershed space scales, indicating that AHG power functions perform well in the flows with high sediment content, and even better than in those with lower sediment content on the Loess Plateau of China.

Leopold and Maddock reported that HG relationships existed in stable irrigation canals which neither scour nor aggrade their beds [1], and Ferguson concluded that AHG corresponded to a stable cross-section [3]. In this study, flow measurement flumes were built at W1 outlet, and riverbanks were raised on either side of the outlet channels of W10 and W100. Hence, outlet channels of the study watersheds could be assumed relatively stable and thus the power function relations should be able to accurately describe watershed hydraulic geometry relations, agreeing with the results of our previous studies [26], although the specific values of AHG exponents are different from those in previous studies in the other places in the world [1, 2].

Driving Forces for AHG Coefficient Decline. AHG coefficients were found to be affected in a complex way by channel geometry parameters of cross-sectional shape, slope, and physical characteristics, and more attention should be paid to AHG coefficients [3, 4]. In this study, Mann–Kendall test analysis showed that AHG coefficients in the experimental watersheds declined significantly ($P < 0.05$). Considering the physical meaning as well as data treatment, decline in coefficients a, c and k indicated that flow width, depth and mean flow velocity for annual mean discharge rate decreased, which further suggested watershed annual runoff declined over the study period. This is likely due to soil and water conservation practices in the three study watersheds, which help increase infiltration and reduce runoff [27]. More specifically, the abrupt decrease of a, c and k in the three study watersheds corresponded with period of 2007–2012. This might be because that in W1, huge land use changes by human started in 2006. The area of farmland (sloped cropland and terrace) was decreased dramatically, while the increase of woodland area was the highest (increased by 61.52%), followed by road and unused land. The main changes of farmland to woodland helped to decrease soil erosion during the rainstorms, thus to decrease sediment content of the floods. Therefore, runoff and sediment transport were significantly reduced, which might have some impacts on AHG coefficients in the latter several years. Parker stated that the scale factors a, c and k varied from locality to locality while the exponents b, f and m exhibited a remarkable degree of consistency and stability, which also supported the results of this study [28].

4 Conclusions

In this study, coefficients and exponents of AHG power function relations from three typical experimental watersheds of different sizes were examined by Mann-Kendall trend and abrupt change detection methods, to test stability of AHG relations on the Loess Plateau of China. Mann-Kendall detection analysis indicated that AHG exponents (b, f and m) generally showed sound stability while coefficients (a, c and k) declined dramatically probably due to land use changes, channel bank reconstruction and hydrologic measuring technique improvement in the study watersheds in the last three decades. Therefore, AHG power function relations should be calibrated to derive exact parameter values especially for coefficients before application. AHG at watershed outlet channels could serve as an aid in flood and sediment yield prediction, water resource management in the similar watersheds in Chinese Loess Plateau region and agricultural irrigation regulation systems downstream.

Acknowledgements. This work was financially supported by National Natural Science Foundation of China (41401307), 2020 Hebei Province Social Science Development Research Project (20200203114) and Natural Science Foundation of Hebei Province, China (D2016208100). We are profoundly grateful to editors and anonymous reviewers for carefully reviewing every stage of the manuscript.

References

1. Leopold, L.B., Maddock, T.J.: The Hydraulic Geometry of stream channels and some physiographic implications. United States Government Pringting Office, Washington D.C. (1953)
2. Carlston, C.W.: Downstream variations in the hydraulic geometry of streams; special emphasis on mean velocity. Am. J. Sci. **267**(4), 499–509 (1969)
3. Ferguson, R.I.: Hydraulics and hydraulic geometry. Prog. Phys. Geogr. **10**(1), 1–31 (1986)
4. Dingman, S.L.: Analytical derivation of at-a-station hydraulic-geometry relations. J. Hydrol. **334**, 17–27 (2007)
5. Baki, A.B.M., Zhu, D.Z., Courtice, G.: Hydraulic geometry and resistance to flow in headwater streams in the Northwest Territories, Canada. Can. J. Civil Eng. **39**(12), 1252–1263 (2012)
6. Ran, L.S., Wang, S.J., Lu, X.X.: Hydraulic geometry change of a large river: a case study of the upper Yellow River. Environ. Earth Sci. **66**(4), 1247–1257 (2012)

7. Zhou, S.M., Lei, T.W., Lei, Q.X., Zhang, M.L.: Hydraulic geometry model at small watershed outlets on the Loess Plateau of China. J. Beijing Forestry Univ. **37**(9), 45–52 (2015)
8. Morel, M., et al.: Revisit-ing the drivers of at-a-station hydraulic geometry in stream reaches. Geomorphology **328**, 44–56 (2019)
9. Song, X.L.: Stochastic evolution of hydraulic geometry relations in the lower Yellow River of China under environmental uncertainties. Int. J. Sedim. Res. **35**(4), 328–346 (2020)
10. Zheng, M.G., Cai, Q.G., Cheng, Q.J.: Modelling the runoff-sediment yield relationship using a proportional function in hilly areas of the Loess Plateau. North China. Geomorphol. **93** (3/4), 288–301 (2008)
11. Tsubaki, R., Fujita, I., Tsutsumi, S.: Measurement of the flood discharge of a small-sized river using an existing digital video recording system. J. Hydro-Environ. Res. **5**(4), 313–321 (2011)
12. Shen, C.P., Wang, S.L., Liu, X.F.: Geomorphological significance of at-many-stations hydraulic geometry. Geophys. Res. Lett. **43**(8), 3762–3770 (2016)
13. Mann, H.B.: Nonparametric tests against trend. Econometrica **13**, 245–259 (1945)
14. Kendall, M.G.: Rank Correlation Methods. Griffin, London (1975)
15. Gao, R., Li, F., Wang, X., Liu, T., Du, D., Bai, Y.: Spatiotemporal variations in precipitation across the Chinese Mongolian plateau over the past half century. Atmos. Res. **193**, 204–215 (2017)
16. Kisi, O., Ay, M.: Comparison of Mann-Kendall and innovative trend method for water quality parameters of the Kizilirmak River, Turkey. J. Hydrol. **513**, 362–375 (2014)
17. Hamed, K.H.: Trend detection in hydrologic data: the Mann-Kendall trend test under the scaling hypothesis. J. Hydrol. **349**(3–4), 350–363 (2008)
18. Zhang, A., Zheng, C., Wang, S., Yao, Y.: Analysis of streamflow variations in the Heihe River Basin, northwest China: trends, abrupt changes, driving factors and ecological influences. J. Hydrol. Reg. Stud. **3**, 106–124 (2015)
19. Douglas, E.M., Vogel, R.M., Kroll, C.N.: Trends in floods and low flows in the United States: impact of spatial correlation. J. Hydrol. **240**(1–2), 90–105 (2000)
20. Sneyers, R.: Sur L'analyse statistique des series D'observations, p. 189. OMM, Geneve (1975)
21. Yang, Y., Tian, F.: Abrupt change of runoff and its major driving factors in Haihe River Catchment, China. J. Hydrol. **374**(3–4), 373–383 (2009)
22. Moraes, J.M., Pellegrino, G.Q., Ballester, M.V., Martinelli, L.A., Victoria, R.L., Krusche, A. V.: Trends in hydrological parameters of a southern Brazilian watershed and its relation to human induced changes. Water Resour. Manag. **12**(4), 295–311 (1998)
23. Yu, Y., Meng, X.Y., Zhang, X.: Trends and abruption analysis on the visibility in the urban area of Beijing City during 1980–2011. Res. Environ. Sci. **26**(2), 129–136 (2013)
24. Xu, J.: Hyperconcentrated flows in the slope-channel systems in gullied hilly areas on the Loess Plateau, China. Geografiska Annaler: Ser. A Phys. Geogr. **86**(4), 349–366 (2004)
25. Zheng, W., Guo, Q.C., Lu, Q.: Review of basic theories of hyperconcentrated flows. J. Sediment Res. **2**, 75–80 (2011)
26. Zhou, S.M., Lei, T.W., Warrington, D.N., Lei, Q.X., Zhang, M.L.: Does watershed size affect simple mathematical relationships between flow velocity and discharge rate at watershed outlets on the Loess Plateau of China. J. Hydrol. **444–445**, 1–9 (2012)
27. Yan, Q., et al.: Effects of watershed management practices on the relationships among rainfall, runoff, and sediment delivery in the hilly-gully region of the Loess Plateau in China. Geomorphology **228**, 735–745 (2015)
28. Parker, G.: Hydraulic geometry of active gravel rivers. J. Hydraulics Div. **105**(9), 1185–1201 (1979)

Preparing Primary School Teachers for Teaching Computational Thinking: A Systematic Review

Maria Ausiku(✉) and Machdel Matthee

Department of Informatics, University of Pretoria, Pretoria, South Africa
kamweshi@gmail.com, machdel.matthee@up.ac.za

Abstract. The purpose of this study is to systematically examine the existing literature on the teaching approaches and tools used to prepare primary school teachers to integrate computational thinking (CT) in their offering. In addition, the study considers perceptions of teachers towards teaching approaches for CT. Thirty (30) journal articles and conference proceedings that met the selection criteria were reviewed and thematically analysed. Teaching approaches and tools that have been used to train teachers on teaching CT in primary schools include unplugged computing, robotics, programming and game-based learning activities. Teachers' perceptions toward the teaching approaches for integration of CT were positive and progressive after interventions. To support teachers, most studies introduced modules within teacher-education curriculum or a professional development course for in-service teachers. The results indicated that most of the integrations are done within mathematics and science classes. Also, coding remains a useful way to teach CT. To prepare teachers to teach CT concepts and skills, both pre-service and in-service teachers need to practice teaching those concepts in authentic contexts. The review highlights the need for research that addresses the developing country context and working context of teachers as teaching strategies would differ from the developed countries context which has dominated the current research done.

Keywords: Computational thinking · Teacher education · Primary schools · Teachers' perceptions · Teaching approaches

1 Introduction

Computational thinking (CT) is a term that has emerged widely in the last decade. Although first used in 1980 by Papert (1980), it is only in the last decade that the term computational thinking has gained traction. Selby (2014) has inferred a comprehensive definition of CT and portrayed it as a problem-solving approach that uses decomposition, algorithms, abstractions, evaluation and pattern recognition by incorporating thought processes. In her seminal article, Wing (2006) stated that "computational thinking is a fundamental skill for everyone, not just for computer scientists". "To reading, writing, and arithmetic, we should add computational thinking to every child's analytical ability" (Wing 2006, p. 33). For this to happen, school curricula need redesigning and teachers have to be trained (Collins et al. 2011). Literature reviews on

C. Pang et al. (Eds.): SETE 2020/ICWL 2020, LNCS 12511, pp. 202–213, 2021.
https://doi.org/10.1007/978-3-030-66906-5_19

teachers' training for teaching computing are scarce, especially in primary education (elementary (K–6) education). This study presents a systematic review of existing literature on teaching strategies used to train primary school teachers for teaching computational thinking skills.

2 Background

For computational thinking to be infused in compulsory education, teachers need to be educated of what CT skills are and how they relate to their existing curriculum and what they do already on a day-to-day basis. Unfortunately, most primary school teachers lack the knowledge in content and pedagogy of computer science (CS) (Ng 2017; Rich et al. 2017b; Stanton et al. 2017). CT does not equate CS, but they are related in the sense that CS offers unique opportunities for developing computational thinking and that CT's practices can be applied to different domains besides CS. CT involves a set of problem-solving skills and techniques that software engineers use to write programs that underlie computer applications (Wing 2006). CT education, especially at primary level, is not about creating programmers or computer scientists, but it is enabling learners to solve problems using this powerful strategy and CS concepts (ISTE© - International Society for Technology in Education 2014).

Numerous related studies exist that suggest ways to prepare computational thinking teachers effectively. For example, several literature reviews have been published about teacher professional development (PD) (Desimone et al. 2002; Guskey and Yoon 2009; IAEL 2004), technology integration (e.g., (Ertmer and Ottenbreit-Leftwich 2010; Hew and Brush 2007; Lawless and Pellegrino 2007), computing education (Crick 2017; Garneli et al. 2015; Kallia 2017; Rich et al. 2017a); Waite 2017), and CT in education (Grover and Pea 2013; Heintz et al. 2016); Ilic et al. 2018; Lockwood and Mooney 2018; Lye and Koh 2014).

While wide-ranging research has been published on teachers (PD) and integration of technology in teaching, little research exists on preparing primary school teachers to teach CT skills (Yadav et al. 2019). An exception is the work of Mason and Rich (2019) who did a systematic review of literature describing ways to prepare primary school teachers (in-service and pre-service) to teach computing, coding and computational thinking. The findings suggest that programs that involve the active participation of teachers can improve teachers' computing self-efficacy, attitudes, and knowledge. This study is similar to Mason and Rich (2019)'s work in that it systematically reviews the literature on preparing primary school teachers to teach computational thinking, but the focus is on teachers' perceptions of CT teaching strategies and the teaching approaches and software tools that are used in training teachers to teach CT in primary schools.

3 Methodology

3.1 Research Questions

In an attempt to understand the current state of CT within primary schools in terms of teachers' readiness to teach and teaching approaches, the following questions were used to guide the review:

1. What teaching approaches and tools have been used in training teachers to teach CT in primary school?
2. What are the primary school teachers' perceptions of CT teaching strategies?

3.2 Review Methods

This systematic review followed Kitchenham and Charters (2007)'s guidelines for systematic reviews in software engineering research. Following these guidelines, our research methodology included three main phases: planning the review, conducting the review and reporting the results.

3.3 Data Sources

Conducting the review started with the identification of the relevant primary studies. Databases searched were electronic and concerned about the areas of education and technology/computer science. Overall, our search spanned over six digital libraries that are the most commonly used in similar studies, namely: ACM Digital Library, Springer Link, IEEE Xplore, Science Direct, Taylor & Francis Journals and ERIC. Our study collected data from conference proceedings and journals articles only.

3.4 Search Strategy

The following search string was used to search each database on the 08/03/2020: ("teachers" OR "educators") AND ("computational thinking") AND ("perceptions") AND ("teaching approaches" OR "pedagogy" OR "teaching" OR "teacher education" OR "teacher training"). All searches were made against article title and abstracts. Search filters were used on some databases to align with the screening criteria. Results of the initial search were ACM Digital Library (483), Springer Link (326), IEEE Xplore (16), Science Direct (139), Taylor & Francis Journals (226) and ERIC (25).

3.5 Study Selection

The selection process started by scanning through the title, keywords and abstract to ensure that the article is focused on our study before a paper was downloaded for a full read. The inclusion and exclusion criteria are presented in Table 1.

Table 1. Selection criteria

Inclusion criteria	Exclusion criteria
Studies that focus on teaching/integrating CT, teacher training	Studies that do not focus on the keywords or not written in the English Language
Studies that focus on CT in primary education	Studies that focus on CT at high school or university level
Studies that are published in peer-reviewed journals or conference proceedings	Materials that are not peer-reviewed (audio/video files, PPT, etc.)
Studies that presented teaching approaches or tools for teaching CT at primary level	Studies that did not answer the research questions

3.6 Study Quality Assessment

To assess the quality of the included articles, each was judged on the quality criteria of objectives, methods, results and conclusions with answer scores for the items as No = 0; Partially = 0.5; Yes = 1. A total of 30 papers were retained for the study.

3.7 Data Extraction

The extraction protocol which guided data extraction from the retained articles included research purpose, participant characteristics, description of context or setting, research design used in the research, and key findings. Extracted data were stored in an Excel spreadsheet.

4 Results

4.1 Descriptive Statistics

This section presents the quantitative results regarding the distribution of studies reviewed by publication year, what teaching approaches and tools were used. The majority of studies reviewed were carried out in developed countries with the USA dominating, only one study was done in a developing country which was Colombia and none was done in Africa.

Frequency of Publication. Figure 1 below shows the distribution of studies according to the publication year. While we didn't define a date range, data gathered revealed when publications on teachers training on CT gained attention.

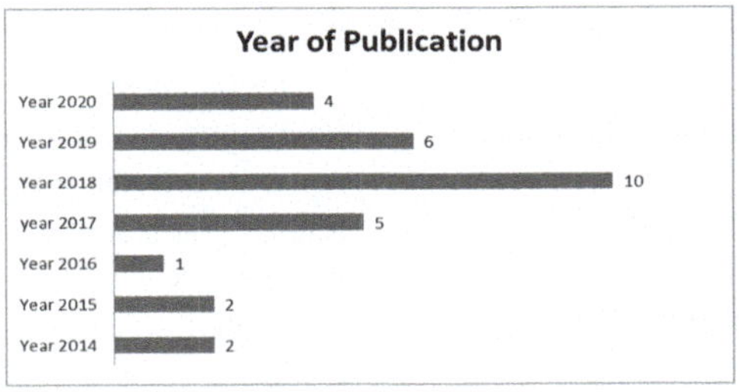

Fig. 1. Distribution of studies by year of publication

Teaching Approach Used. Figure 2 below indicates which teaching approaches were used in the studies to implement CT in the lessons, with programming being the popular approach.

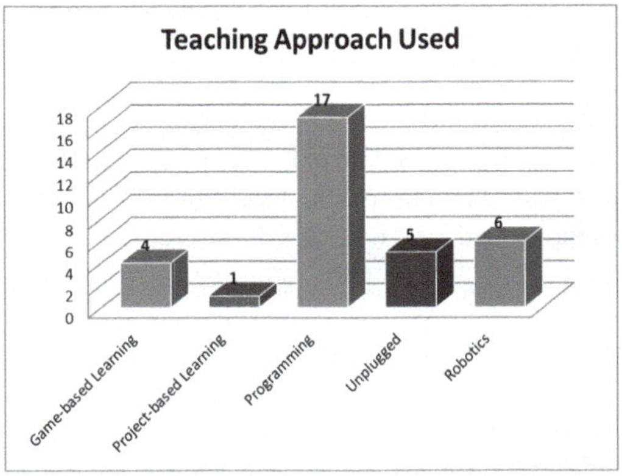

Fig. 2. Teaching approach used

Software Tools Used. Figure 3 below displays the teaching tools used in the studies to teach or integrate CT into the classrooms with Scratch activities being the most popular ones.

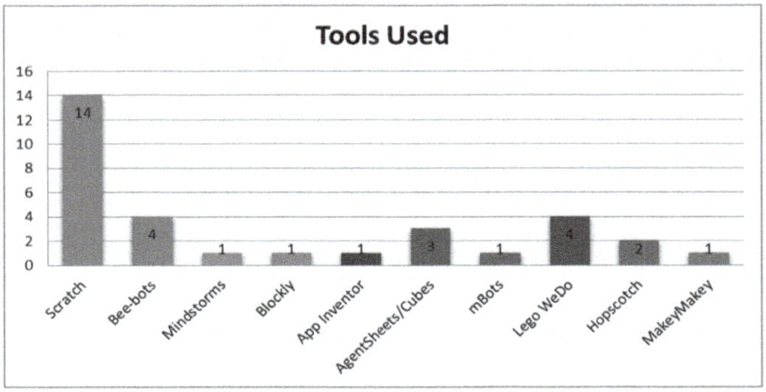

Fig. 3. Tools used

4.2 Thematic Analysis

First, extracted data were examined for patterns or trends across each extracted variable and over time. Next, the Findings or Results, Discussion, and Conclusions sections of the articles were analysed using thematic analysis, a qualitative data reduction method for identifying patterns or themes within data (Braun and Clarke 2006). The themes are grouped according to teaching approaches.

Unplugged Approach: These refer to CT activities without digital devices. Regardless of the teachers' socio-economic status, teachers can apply these concepts in different domains. Ouyang et al. (2018) and Rich et al. (2020) helped teachers to use non-programming/unplugged science and mathematics activities to present CT concepts and improve students' CT skills. Curzon et al. (2014) presented a comprehensive set of workshops for teaching primary school teachers CT concepts through unplugged activities and indicated that it works to build confidence and fill teachers' knowledge gaps about CT. Mannila et al. (2014) showed how unplugged activities and digital story-telling can be used in a variety of approaches.

Game-Based Learning and Robotics: Leonard et al. (2017) and Leonard et al. (2018) learned that game design enabled learners to take familiar concepts in STEM and apply them to a range of complex tasks to create representations and models using LEGO as a tool. Chalmers (2018) examined how primary school teachers infused robotics and coding in their classrooms in Australia using LEGO robot kits and the results determined that teachers built their confidence and knowledge. Jaipal-Jamani and Angeli (2017) and Esteve-mon et al. (2019) studied the understanding of science concepts, self-efficacy, and CT of pre-service teachers as they use robotics in a science methods course using LEGO and MakeyMakey as tools and the findings suggest that the robotics activities have increased the interest in robotics and were an effective teaching strategy to enhance CT skills and increase self-efficacy to instruct with robotics. Nickerson et al. (2015) described a framework for learning different sets of computational thinking concepts and explored how teachers employing a CT curriculum known as Scalable Game Design are using these contexts in their teaching. AgentSheets/Cubes were used as tools and it was shown that CT can be effectively taught if scaffolding is provided for learning specific skills. Game design and robotics afford teachers an opportunity to collaborate with others and engage computer scientists to deepen their understanding of computer science concepts.

Programming as Teaching Approach: A few studies show a positive response of teachers when trained to use Scratch (block-based programming) in CT. These positive responses include a greater understanding and appreciation for its usefulness (Gleasman and Kim 2020; Cetin 2016; Kong et al. 2020). Adler and Kim (2018) examined how CT can be introduced through simulations and modelling within a science method class for preservice teachers using web-based simulation and Scratch. Results showed that after the intervention, teachers had a better understanding of the topic and realized how beneficial CT is in education, and wanted to integrate it into their future classrooms. Linde-koomen (2019) and Bean et al. (2015) ran an intervention module for pre-service teachers on how to use programming and CT as a teaching strategy within other subjects. The post-survey indicated that students intend to use Scratch in their future lessons. Falkner et al. (2018) and Bower et al. (2017) improved the CT skills of in-service teachers through workshops using programming as a teaching strategy and Hopscotch, Blockly, Scratch and Beebots as software tools. After the interventions the teachers had a detailed understanding of CT and its sub-components and have divergent ideas of various strategies they can use to teach CT in their classrooms. Marcelino et al. (2018) and Haduong and Brennan (2019) showed that Scratch can be taught and learnt through distance education effectively as teachers were able to develop quality teaching material that is useful for their classrooms. The literature shows that programming is mostly taught within a programming context, which does not help to dispel the perception that CT is about coding. Geldreich et al. (2018) and Zha et al. (2020) did a flipped module of coding activities to examine how it contributed to the pre-service teachers' knowledge, self-efficacy and attitudes towards CT. The teachers had an overall positive perception of the CT learning experience.

Project-Based Learning: Ozturk et al. (2018) has explored the use of Project-Based Learning (PBL) to integrate CT and results indicated that during collaborative project planning sessions, teachers worked closely to incorporate standards with grade-level team members. A meta-review by Hsu et al. (2018), shows that most CT training (in general), implement project-based learning, problem-based learning, cooperative learning, and game-based learning in the CT activities.

5 Discussions

The purpose of this review was to systematically examine what has been done to prepare primary school teacher to integrate CT into their lessons, explore what approaches and tools have been used and recognised research gaps.

The context of developing countries is insufficiently researched. Muñoz del Castillo et al. (2019) observed that while the teachers agree that CT skills needs to be incorporated into their training, there is a long way to go before well-organized training programs for teachers are put in place. Hence, the initial step towards preparing teachers to teach CT in developing countries is to have educational policies which promote professional development where CT takes center stage. Accordingly, teacher training curricular should be adapted to include CT skills teaching.

Fewer studies investigated the unplugged approach which can provide a way to CT skills in non-computing environments. Given Wing (2006)'s idea of CT being a fundamental skill that everyone should possess, this approach presents an opportunity to reach a large number of teachers and learners who don't have access to computing devices in developing countries. This is in agreement with Ouyang et al. (2018) and Rich et al. (2020) who indicated that using unplugged activities works to build confidence and fill teachers' knowledge gaps about CT. Unplugged activities not only introduce teachers to the CT concepts without overwhelming them with technologies and devices, but regardless of the teachers' socio-economic status, teachers can apply these concepts in different domains. This argument points in the same direction as the literature that although the subjects where CT can be integrated with the least effort are mathematics and science, some teachers saw the possibilities of integrating within content areas such as Arts and Social studies (Lamprou and Repenning 2018; Muñoz del Castillo et al. 2019; Gadanidis et al. 2017).

The literature has revealed that programming has been the most used approach to teach CT followed by robotics. However, the current dependence on programming as a CT teaching approach may deter teachers who don't have a computing background or from schools without access to computers and programming platforms to apply CT skills in other subject domains. Schools in developing countries are not only faced with lack of computers, but also the shortage of computing skills among the teachers, which makes it hard to use programming and robotics as teaching approaches.

While every study in this review described the research context, most interventions were provided to teachers who represented different contexts. One key strategy to ensure effective CT training in developing countries is to consider the context for each teacher when training them. Teachers should be able to reflect on the teaching approach and decide if it's suitable for their classroom. Accordingly, researchers should involve teachers in the design and development of the professional development programmes or interventions to incorporate different contexts.

Practice activities that teachers did through programming mostly used Scratch as a tool and teachers' perceptions toward such CT teaching strategies were positive and improved over time. However, teachers' perceptions in developing countries are likely to differ from what the literature has revealed as the tools and approaches may not be applicable to their context.

6 Conclusions

The absence of studies based in developing countries' context is clear and more research needs to be done in developing countries especially in an African context where the teaching strategies and tools should vary accordingly. Teachers' working contexts should be considered in the choice of teaching approach and tools used. Future research should involve teachers in designing and developing interventions. Teachers need to be well prepared to help learners develop and enhance their CT capabilities so they can create technology and not just consume it.

Acknowledgements. Funding for this research was made possible by the National Research Foundation, CSIR, South Africa.

References

Adler, R.F., Kim, H.: Enhancing future K-8 teachers' computational thinking skills through modeling and simulations. Educ. Inf. Technol. **23**(4), 1501–1514 (2018). https://doi.org/10.1007/s10639-017-9675-1

Bean, N., Weese, J., Feldhausen, R., Bell, R.S.: Starting from scratch: developing a pre-service teacher training program in computational thinking. In: Proceedings - Frontiers in Education Conference, FIE 2014, pp. 1–8 (2015). https://doi.org/10.1109/FIE.2015.7344237

Bower, M., Wood, L.N., Howe, C., Lister, R.: Improving the computational thinking teaching capabilities of school teachers. **42**(3) (2017)

Braun, V., Clarke, V.: Using thematic analysis in psychology. Qual. Res. Psychol. **3**(2), 77–101 (2006). https://doi.org/10.1191/1478088706qp063oa

Cetin, I.: Preservice teachers' introduction to computing: exploring utilization of scratch. J. Educ. Comput. Res. **54**(7), 997–1021 (2016). https://doi.org/10.1177/0735633116642774

Chalmers, C.: Robotics and computational thinking in primary school. Int. J. Child-Comput. Interact. **17**, 93–100 (2018). https://doi.org/10.1016/J.IJCCI.2018.06.005

Collins, S., et al.: A model for piloting pathways for computational thinking in a general education curriculum (2011). https://doi.org/10.1145/1953163.1953243

Crick, T.: Computing Education: An Overview of Research in the Field, April, pp. 1–38 (2017). https://royalsociety.org/-/media/policy/projects/computing-education/literature-review-overview-research-field.pdf

Curzon, P., Dorling, M., Ng, T., Selby, C., Woollard, J.: Developing computational thinking in the classroom: a framework (2014). https://eprints.soton.ac.uk/369594/

Desimone, L.M., Porter, A.C., Garet, M.S., Yoon, K.S., Birman, B.F.: Effects of professional development on teachers' instruction: results from a three-year longitudinal study. Educ. Eval. Policy Anal. **24**(2), 81–112 (2002). https://doi.org/10.3102/01623737024002081

Ertmer, P.A., Ottenbreit-Leftwich, A.T.: Teacher technology change. J. Res. Technol. Educ. **42** (3), 255–284 (2010). https://doi.org/10.1080/15391523.2010.10782551

Esteve-mon, F.M., Adell-segura, J., Ángeles, M., Nebot, L., Novella, G.V., Aparicio, J.P.: The development of computational thinking in student teachers through an intervention with educational robotics. J. Inf. Technol. Educ. Innov. Pract. **18**, 139–152 (2019). https://doi.org/10.28945/4442

Falkner, K., Vivian, R., Falkner, N.: Supporting computational thinking development in K-6. In: Proceedings - 2018 6th International Conference on Learning and Teaching in Computing and Engineering, LaTiCE 2018, pp. 126–133 (2018). https://doi.org/10.1109/LaTICE.2018.00031

Gadanidis, G., Cendros, R., Floyd, L., Namukasa, I.: Computational thinking in MATHS teacher education.pdf. Contemp. Issues Technol. Teach. Educ. 17(4) (2017). https://citejournal.org/volume-17/issue-4-17/mathematics/computational-thinking-in-mathematics-teacher-education/

Garneli, V., Giannakos, M.N., Chorianopoulos, K.: Computing education in K-12 schools: a review of the literature. In: 2015 IEEE Global Engineering Education Conference (EDUCON), pp. 543–551 (2015)

Geldreich, K., Talbot, M., Hubwieser, P.: Off to new shores: preparing primary school teachers for teaching algorithmics and programming. In: ACM International Conference Proceeding Series (2018). https://doi.org/10.1145/3265757.3265783

Gleasman, C., Kim, C.: Pre-service teacher's use of block-based programming and computational thinking to teach elementary mathematics. Digit. Exp. Math. Educ. **6**(1), 52–90 (2020). https://doi.org/10.1007/s40751-019-00056-1

Grover, S., Pea, R.: Computational thinking in K–12: a review of the state of the field. Educ. Res. **42**(1), 38–43 (2013). https://doi.org/10.3102/0013189X12463051

Guskey, T.R., Yoon, K.S.: What works in professional development? Phi Delta Kappan **90**(7), 495–500 (2009)

Haduong, P., Brennan, K.: Helping K–12 teachers get unstuck with scratch: the design of an online professional learning experience. In: Proceedings of the 50th ACM Technical Symposium on Computer Science Education, pp. 1095–1101. Association for Computing Machinery, New York (2019). https://doi.org/10.1145/3287324.3287479

Heintz, F., Mannila, L., Farnqvist, T.: A review of models for introducing computational thinking, computer science and computing in K-12 education. In: Proceedings - Frontiers in Education Conference, FIE, vol. 2016-November. Institute of Electrical and Electronics Engineers Inc. (2016). https://doi.org/10.1109/FIE.2016.7757410

Hew, K.F., Brush, T.: Integrating technology into K-12 teaching and learning: Current knowledge gaps and recommendations for future research. Educ. Tech. Res. Dev. **55**(3), 223–252 (2007). https://doi.org/10.1007/s11423-006-9022-5

Hsu, T.C., Chang, S.C., Hung, Y.T.: How to learn and how to teach computational thinking: suggestions based on a review of the literature. Comput. Educ. **126** (2018). https://doi.org/10.1016/j.compedu.2018.07.004

Ilic, U., Haseski, H.İ., Tugtekin, U.: Publication trends over 10 years of computational thinking research. Contemp. Educ. Technol. **9**(2), 131–153 (2018). https://doi.org/10.30935/cet.414798

ISTE© - International Society for Technology in Education. Computational Thinking Competencies (2014). https://www.iste.org/standards/computational-thinking. Accessed 18 June 2019

Jaipal-Jamani, K., Angeli, C.: Effect of robotics on elementary preservice teachers' self-efficacy, science learning, and computational thinking. J. Sci. Educ. Technol. **26**(2), 175–192 (2017). https://doi.org/10.1007/s10956-016-9663-z

Kallia, M.: Assessment in Computer Science Courses: A Literature Review (2017)

Kitchenham, B., Charters, S.: Guidelines for Performing Systematic Literature Reviews in Software Engineering (2007). https://doi.org/10.1145/1134285.1134500

Kong, S.C., Lai, M., Sun, D.: Teacher development in computational thinking: design and learning outcomes of programming concepts, practices and pedagogy. Comput. Educ. **151** (March), 103872 (2020). https://doi.org/10.1016/j.compedu.2020.103872

Lamprou, A., Repenning, A.: Teaching how to teach computational thinking. In: Annual Conference on Innovation and Technology in Computer Science Education, ITiCSE, pp. 69–74. Association for Computing Machinery (ACM) (2018). https://doi.org/10.1145/3197091.3197120

Lawless, K.A., Pellegrino, J.W.: Professional development in integrating technology into teaching and learning: knowns, unknowns, and ways to pursue better questions and answers. Rev. Educ. Res. **77**(4), 575–614 (2007)

Leonard, J., Barnes-Johnson, J., Mitchell, M., Unertl, A., Stubbe, C.R., Ingraham, L.: Developing teachers' computational thinking beliefs and engineering practices through game design and robotics. In: North American Chapter of the International Group for the Psychology of Mathematics Education (2017)

Leonard, J., et al.: Preparing teachers to engage rural students in computational thinking through robotics, game design, and culturally responsive teaching. J. Teach. Educ. **69**(4), 386–407 (2018). https://doi.org/10.1177/0022487117732317

Van Der Linde-koomen, D.: Computational thinking on primary education teacher education. In: EdMedia + Innovate Learning 2019 - Amsterdam, Netherlands, 24–28 June 2019 Computational, pp. 684–691 (2019)

Lockwood, J., Mooney, A.: Computational thinking in secondary education: where does it fit? A systematic literary review. Int. J. Comput. Sci. Educ. Sch. **2**(1) (2018). https://doi.org/10.21585/ijcses.v2i1.26

Lye, S.Y., Koh, J.H.L.: Review on teaching and learning of computational thinking through programming: what is next for K-12? Comput. Hum. Behav. **41**, 51–61 (2014). https://doi.org/10.1016/j.chb.2014.09.012

Mannila, L., et al.: Computational thinking in K-9 education. In: Proceedings of the Working Group Reports of the 2014 on Innovation & Technology in Computer Science Education Conference, pp. 1–29. Association for Computing Machinery, New York (2014). https://doi.org/10.1145/2713609.2713610

Marcelino, M.J., Pessoa, T., Vieira, C., Salvador, T., Mendes, A.J.: Learning computational thinking and scratch at distance. Comput. Hum. Behav. **80**, 470–477 (2018). https://doi.org/10.1016/j.chb.2017.09.025

Mason, S.L., Rich, P.J.: Preparing elementary school teachers to teach computing, coding, and computational thinking. Contemp. Issues Technol. Teach. Educ. (CITE J.) **19**, 790–824 (2019)

Muñoz del Castillo, A., Muñoz, M.A., Acosta Huertas, L.C., Herrera, E., Jimenez Toledo, J., Ramos, D.X.: Developing a teacher training curriculum including computational thinking skills: early advances on a study focused on Colombia. In: 2019 XIV Latin American Conference on Learning Technologies (LACLO), pp. 8–11 (2019). https://doi.org/10.1109/LACLO49268.2019.00012

Ng, W.S.: Coding education for kids: what to learn? How to prepare teachers? In: Proceedings of ICICTE, pp. 195–205 (2017)

Nickerson, H., Brand, C., Repenning, A.: Grounding computational thinking skill acquisition through contextualized instruction. In: Proceedings of the 2015 ACM Conference on International Computing Education Research, ICER 2015, pp. 207–216 (2015). https://doi.org/10.1145/2787622.2787720

Ouyang, Y., Hayden, K.L., Remold, J.: Introducing computational thinking through non-programming science activities. In: Proceedings of the 49th ACM Technical Symposium on Computer Science Education, pp. 308–313. Association for Computing Machinery, New York (2018). https://doi.org/10.1145/3159450.3159520

Ozturk, Z., Dooley, C.M.M., Welch, M.: Finding the hook: computer science education in elementary contexts. J. Res. Technol. Educ. **50**(2), 149–163 (2018). https://doi.org/10.1080/15391523.2018.1431573

Papert, S.: Mindstorms: Children, Computers, and Powerful Ideas. Basic Books (1980). https://dl.acm.org/citation.cfm?id=1095592

Rich, K.M., Yadav, A., Larimore, R.A.: Teacher implementation profiles for integrating computational thinking into elementary mathematics and science instruction. Educ. Inf. Technol. **25**(4), 3161–3188 (2020). https://doi.org/10.1007/s10639-020-10115-5

Rich, K., Strickland, C., Franklin, D.: A literature review through the lens of computer science learning goals theorized and explored in research. In: Proceedings of the Conference on Integrating Technology into Computer Science Education, ITiCSE, pp. 495–500. Association for Computing Machinery, New York (2017a). https://doi.org/10.1145/3017680.3017772

Rich, P.J., Jones, B.L., Belikov, O., Yoshikawa, E., Perkins, M.: Computing and engineering in elementary school: the effect of yearlong training on elementary teacher self-efficacy and beliefs about teaching computing and engineering. Int. J. Comput. Sci. Educ. Sch. **1**(1), n1 (2017b)

Selby, C.C.: How can the teaching of programming be used to enhance computational thinking skills? University of Southampton (2014). https://eprints.soton.ac.uk/366256/

Stanton, J., et al.: State of the States Landscape Report: State-Level Policies Supporting Equitable K–12 Computer Science Education | EDC (2017). https://www.edc.org/state-states-landscape-report-state-level-policies-supporting-equitable-k-12-computer-science

The Institute for the Advancement of Research in Education at AEL: Review of the Research: Nine Components of Effective Professional Development Prepared for Texas Instruments Educational and Productivity Solutions Division by The Institute for the Advancement of Research in Education at AEL (2004). www.ael.org

Waite, J.: Pedagogy in teaching Computer Science in Schools: A Literature Review (2017)

Wing, J.M.: Computational thinking. Commun. ACM **49**(3), 33 (2006). https://doi.org/10.1145/1118178.1118215

Yadav, A., Larimore, R., Rich, K., Schwarz, C.: Integrating computational thinking in elementary classrooms: introducing a toolkit to support teachers. In: Society for Information Technology & Teacher Education International Conference, no. 1, pp. 347–350 (2019)

Zha, S., Jin, Y., Moore, P., Gaston, J.: Hopscotch into coding: introducing pre-service teachers computational thinking. TechTrends **64**(1), 17–28 (2020). https://doi.org/10.1007/s11528-019-00423-0

A Classification Method of Inventory Spare Parts Based on Improved Super Efficient DEA-ABC Model

Na Xu[1(✉)] and Wei Xu[2]

[1] Hebei University of Science and Technology, Shijiazhuang 050000, China
xuna@163.com
[2] Shijiazhuang Rail Transit Co., Ltd., Shijiazhuang 050000, China
xuwei123@163.com

Abstract. Enterprises generally have the problem of high spare parts inventory costs at present. One of the main reasons for this problem is that the classification standard of spare parts inventory is single and the classification result is unreasonable. Based on the analysis of commonly used inventory spare parts classification methods, this paper proposes an improved ABC classification method based on super efficient DEA. It integrates the input-output efficiency of super efficient DEA into the supply chain of spare parts procurement and outbound use. In the process, the Delphi method is used to investigate the inventory management personnel, and the statistical results of the survey are added to super efficient DEA model as the weight restriction conditions, and than the improved super efficient DEA-ABC classification model was constructed. This model achieves a combination of subjective and objective, which increases the scientificity and practicality of the classification results. Finally, taking the inventory classification of subway spare parts in a certain city as an example, the effectiveness of the method is verified.

Keywords: ABC classification method · Super efficient DEA · Delphi method · Spare parts classification

1 Introduction

The standby materials stored in the warehouse that are temporarily idle to meet future needs are called inventory. In order to shorten equipment repair downtime or equipment maintenance and repair, the parts reserved for repair are called spare parts. The inventory of spare parts plays an important role in the continuous, safe and barrier-free operation of production and service processes. At present, many large enterprises generally have two major problems: high spare parts inventory costs and unreasonable inventory structure. One of the main reasons for these problems is the single standard

Supported by the National Natural Science Foundation of China (Grant No. 71771078) and the Social Science Research Project of Hebei Provincial Department of Education of China (Grand No. SD101020).

C. Pang et al. (Eds.): SETE 2020/ICWL 2020, LNCS 12511, pp. 214–224, 2021.
https://doi.org/10.1007/978-3-030-66906-5_20

of spare parts inventory classification. At present, domestic and foreign scholars have relatively mature theoretical results of inventory management research, but the research on the classification of special spare parts in inventory management cannot be closely integrated with the actual situation. In this paper, through the overall analysis of the spare parts inventory classification method, the super efficient Data Envelopment Analysis (DEA) is determined as the basic evaluation method for the importance of spare parts classification. Then use the expert experience value determined by Delphi method as the weight restriction condition and add it to the super efficient DEA model to construct an improved super efficient DEA-ABC classification method for inventory spare parts. This method combines subjective and objective classification, and supplements the shortcomings of traditional ABC spare parts inventory classification methods. The classification results can clearly reflect the characteristics of different types of spare parts inventory management, which is of practical significance to improve the efficiency of spare parts management and reduce the cost of spare parts inventory.

2 Literature Review

The research on inventory classification began with the ABC classification method created by Italian economist Pareto in 1906. Nowadays, Pareto's ABC classification is still the most widely used classification method for inventory spare parts in enterprises. The ABC classification method proposes to sort the spare parts according to the total annual consumption amount of spare parts. The value of category A accounts for the highest 20% of all spare parts, the value of category C accounts for the lowest 50% of all spare parts, and the rest are category B. Because it only considers the consumption amount of spare parts as a classification indicator, the classification results of spare parts are somewhat unreasonable.

2.1 Improvement of ABC Classification

On the basis of the ABC classification method, many scholars have improved and perfected the classification method of spare parts inventory. Gaipal.PP compared inventory classification methods comprehensively such as ABC, FSN, and VED, and concluded that ABC and FSN were not suitable for practical applications [1]. Flores improved the ABC classification by using material cost, procurement lead time, annual consumption and material importance as multiple indicators for material classification [2]. Chu proposed the ABC fuzzy classification method, which used the manager's experience, knowledge and judgment as indicators for inventory classification [3]. Jamshidi used exponential smoothing to calculate multiple indicators of inventory, and uses ABC classification to classify spare parts on this basis [4]. Partovi used neural network algorithms to classify spare parts [5]. Cui proposed a maintenance spare parts classification model that combines the Analytic Hierarchy Process (AHP) and ABC classification methods [6]. Fu combined the ABC classification with the VED classification, and proposed a KIO classification method for subway spare parts [7]; Taking Yangtze Power as an example, Jia combined the Analytic Hierarchy Process and the

ABC classification method to optimize the model, broke through the limitations of the traditional ABC classification method, and researched the material storage strategy of large hydropower companies [8].

Through analysis of the above-mentioned improved research on the ABC classification method of spare parts, there are few special studies on spare parts inventory derived from these studies, and the weights of classification indicators are still subject to subjective control. These methods rarely use objective indicator data as the basis for classification, which makes the classification results too dependent on the subjective opinions of people.

2.2 ABC Classification Based on DEA

Data Envelopment Analysis (DEA) is a comprehensive evaluation method of relative efficiency proposed by operations researchers A. Charn and W.W. Copper in the 1960s [9]. DEA is based on the concept of relative efficiency, using convex analysis and linear programming as tools to calculate the relative efficiency between Decision Making Units (DMU) and evaluate the evaluation objects. This method has the advantages of simple structure and convenient use, and the corresponding software development is also very mature. In recent years, people have begun to use this method as a new way of comprehensive inventory classification. Lu proposed a ABC classification of spare parts based on DEA, which uses input and output indicators to calculate the effectiveness of the classified objects and then classifies them [10]. Yu combined AHP with SE-DEA, sorted through grey correlation calculation, and applied it to the classification management of navigational lighting spare parts [11]. For spare parts management, Xu proposed AHP-DEA's ABC classification of spare parts [12]. Tao Yao used the DEA-ABC model as a classification method in the inventory management and classification of repairable spare parts for EMUs [13].

The classification method proposed in this paper uses the combination of Delphi method and super efficient DEA to determine the weight of the classification indicator. This method evaluates and analyzes the importance of stock spare parts from the perspective of super efficiency, which makes the classification results more scientific and more manipulable.

3 Improved Super Efficient DEA-ABC Classification Model

3.1 Proposing Model Building Ideas

The classification process of spare parts should consider both the upstream procurement process and the downstream use process of spare parts. The purchase of spare parts can be used as the input process, and the use of spare parts as the output process. In this way, the relative efficiency idea of DEA can just deal with the importance of inventory spare parts. Moreover, DEA has the advantages of automatic calculation of indicator weights, complete quantification, no need for a unified dimension, and special software support, which makes the classification results of spare parts more objective, real and efficient.

The evaluation results obtained by the traditional DEA model often show that multiple evaluation objects are effective simultaneously. The super efficient DEA model proposed by Andersen and Petersen solves this problem [14]. In the super efficient DEA model, the evaluation unit compares itself with the linear combination of other units so that all evaluation units can be fully sorted. Therefore, the super efficient DEA evaluation model is better applied in many fields [15]. Therefore, the super efficient DEA model is used as the basic model for the classification of spare parts.

Merja Halme believes that there is no restriction on the weight of the evaluation indicator in the DEA or super efficient DEA model, which is not only the advantage of the method, but also a disadvantage of it [16]. The weight of the indicator in DEA is the result of the solution in the linear programming model. The weight is automatically allocated for the relative optimal value of the evaluation object, and it depends on the indicator data completely. Therefore, the process of assigning weights in DEA is highly objective. However, in the calculation process of the model, the indicator weight assigned automatically is often equal to zero. That is, some input or output indicators are not considered at all, and only a few indicators have a weight value that is not zero, which is inconsistent with reality [17]. Therefore, some weight constraints need to be added to the DEA evaluation model. Since the procurement and use of spare parts in the company's inventory involves the company's operations and maintenance departments, the importance of spare parts will also change accordingly with the changes in operation and maintenance time. And take into account the differences in regional management, the indicator weights calculated automatically using DEA will be out of touch with the actual situation obviously. Therefore, the Delphi method will be used here to score the importance of the classification indicators systematically by the relevant personnel of the spare parts procurement department and the user department, and the statistical results will be added to the super efficient DEA model to establish an improved super efficient DEA model for spare parts classification. The model combines subjective and objective, which is a more effective and reasonable classification of inventory spare parts.

3.2 Building the Improved Super Efficient DEA-ABC Model

Stage 1: Super efficient DEA model of spare parts classification

When using DEA to evaluate the importance of spare parts, the evaluation indicators are first divided into two types: input indicators and output indicators. For the determination of the input and output indicators of spare parts classification, Wei Quanling gave the proof: the smaller the input value, the larger the output value, the greater the relative efficiency value at this time, which means that the more effective the decision-making unit and the more important the spare part [18]. However, it is also necessary to consider the procurement and use of spare parts in the actual spare parts inventory management process. Therefore, the principle of dividing the input and output indicators in this article is: if the indicator belongs to the procurement process of spare parts and the smaller the indicator value, the indicator will be classified as an input indicator; if the indicator belongs to the downstream use process and the indicator value is higher, the indicator should be classified as an output indicator.

The CCR model is the most original and most commonly used model in the DEA. Therefore, we discuss the super efficient DEA model improved based on the CCR model. For the j_0-th evaluation object, a relatively effective super efficient DEA model (1) can be established as follows.

$$\max \quad \theta = u^T Y_{j_0}$$
$$\begin{cases} v^T X_j - u^T Y_j \geq 0, \quad j = 1, 2, \cdots, m, j \neq j_0 \\ v^T X_{j_0} = 1 \\ u \geq 0, \ v \geq 0 \end{cases} \tag{1}$$

Here, $X_j = (x_{1j}, x_{2j}, \cdots, x_{pj})^T$ and $Y_j = (y_{1j}, y_{2j}, \cdots, y_{qj})^T$ are the input and output indicator data of the j-th decision-making unit. u and v are the indicator weight vectors.

Stage 2: Improved super efficient DEA model of spare parts classification

It can be seen that the input and output indicator weights in the model (1) are treated as unknown variables of the linear programming model, and there is no restriction on the indicator weight in its conditions. Here, we first use the Delphi method to score the importance of the input and output indicators for the relevant personnel in the spare parts procurement and use departments, and then calculate the results to obtain the weight constraints of the input and output indicators and add them to the model (1) to obtain the model (2) as follows.

$$\max \quad \theta = u^T Y_{j_0}$$
$$\begin{cases} v^T X_j - u^T Y_j \geq 0, \quad j = 1, 2, \cdots, m, j \neq j_0 \\ v^T X_{j_0} = 1, \\ u = \gamma \tau^T, \\ v = \delta \pi^T, \\ u \geq 0, \ v \geq 0, \gamma > 0, \delta > 0. \end{cases} \tag{2}$$

Here, $u = \gamma \tau^T$ and $v = \delta \pi^T$ are the limiting conditions for the weight of input and output indicators obtained through the Delphi method.

Stage 3: Classification of spare parts according to the ABC principle

It can be seen from the above that the importance of all spare parts can be fully sorted using the improved super efficient DEA model. Then we use the ABC classification principle to sort the relative importance of spare parts in descending order: the spare parts with the relative efficiency value in the top 20% are classified into the A core spare parts, the spare parts with the relative efficiency value in the bottom 50% are classified into C general spare parts, and the remaining 30% are classified into B important spare parts. This completes the classification of spare parts based on the improved super efficient DEA-ABC model.

4 Empirical Analysis

4.1 Application Instance

Rail transit has the advantages of high speed, punctual operation, safety and comfort. Efficient and rapid maintenance of rail transit operating equipment is the key to ensuring safe operation. In order to shorten the downtime of equipment maintenance and enable replacement or repair of equipment faulty and damaged parts timely, rail operating companies must reserve a certain amount of spare parts. The necessary inventory management of spare parts can guarantee the safety, continuity, and trouble-free operation of rail transit. This article takes some spare parts of a subway operating company in a Chinese city as a sample case, uses the improved super-efficient DEA-ABC classification model to comprehensively classify it, and compares the classification results with the results by traditional ABC classification and DEA classification to verify the effectiveness of the new method.

《Reliability, Availability, Maintainability and Safety Standard for Rail Transit》 (referred to RAMS standard) is a set of standards for rail transit evaluation and management established by the International Electrotechnical Commission based on the system life cycle theory [19]. RAMS standards include reliability, availability, maintainability, and safety. They are widely used in rail transit process design, equipment monitoring, data collection and other aspects. They are also applicable to the classification and management of subway spare parts.

Four dimensions are considered in RAMS, namely reliability, availability, maintainability, and safety. This article adds the economy indicator of ABC classification to construct an evaluation indicator system for the importance of subway spare parts from these five dimensions. Select the annual consumption of spare parts, purchase lead time, the importance of the equipment, repairability rate, and unit cost value as the evaluation indicators for the importance classification of spare parts, and select the data of 18 kinds of spare parts of the city subway in 2019 as the basic data of the spare parts classification indicators. The data description is shown in Table 1.

Table 1. Data of the spare parts classification indicators

Spare part model	Repairability rate (%)	Annual consumption (item)	Purchase lead time (day)	Unit cost (¥)	Importance for equipment (grade)
Height-adjusting gasket on rail bottom (50 track)	1	60	30	1.5	1
Insulated track gauge adjusting block (16#)	1	75	30	2.59	1
Type-I fastener of rail elastic bar (50 track)	1	200	30	1.2	1
2 mm height-adjusting gasket of Type-I fastener (50 track)	1	150	30	7.8	1
2 mm height-adjusting gasket of DJK5-1 fastener	1	100	30	10.8	1
Class A Ballast	1	10	60	106	7
Uninsulated coupler	1	20	30	18	1
P60 rail	50	1	300	7845.3	9
Motor of rolling shutter door	80	2	60	1170	5
P60 package plywood	1	6	60	104.5	5
Bolt cap for high-strength joint of 60 kg/m rail	1	18	30	6.6	3
Type-I Preservatively-treated sleeper	25	1	90	260	5
5 mm height-adjusting gasket of DJK5-1 fastener	1	20	30	19.5	1

(*continued*)

Table 1. (*continued*)

Spare part model	Repairability rate (%)	Annual consumption (item)	Purchase lead time (day)	Unit cost (¥)	Importance for equipment (grade)
Pull rod adjusting block of No. 9 switch rail	1	25	30	7.5	1
Guardrail adjusting sheet of No. 9 switch	1	40	30	3.8	1
High manganese steel integrated cast frog of 50 kg/m rail	65	3	300	25300	9
Straight switch rail of 50 kg/m rail	60	4	300	13975	9
Insulated gauge rod	40	2	90	230	5

4.2 Model's Performance

According to the selection principle of input and output indicators of DEA model, combined with RAMS standards, the input indicator is determined as the repairable rate; the output indicators are the importance degree of spare parts for equipment, purchase lead time, annual consumption and unit cost. According to the indicator data in Table 1, we use the ABC classification method, the DEA-ABC model and the improved super efficient DEA-ABC model constructed in this paper to classify the 18 kinds of spare parts using EMS software. The classification results are shown in Table 2.

Table 2. Classification results

Spare part model	ABC result	DEA value (%)	DEA result	Improved super efficient DEA value (%)	Improved super efficient DEA result
Height-adjusting gasket on rail bottom (50 track)	C	64.10	B	19.36	C
Insulated track gauge adjusting block (16#)	C	67.95	B	20.90	C
Type-I fastener of rail elastic bar (50 track)	C	100.00	A	31.04	B
2 mm height-adjusting gasket of Type-I fastener (50 track)	C	87.18	B	28.50	B
2 mm height-adjusting gasket of DJK5-1 fastener	C	74.36	B	25.06	B
Class A Ballast	A	100.00	A	140.00	A
Uninsulated coupler	C	53.85	C	20.16	C
P60 rail	A	45.44	C	42.13	B
Motor of rolling shutter door	B	4.47	C	4.37	C
P60 package plywood	B	100.00	A	81.55	A
Bolt cap for high-strength joint of 60 kg/m rail	C	53.33	C	42.86	B
Type-I Preservatively-treated sleeper	B	6.98	C	5.03	C
5 mm height-adjusting gasket of DJK5-1 fastener	C	53.85	C	20.54	C
Pull rod adjusting block of No. 9 switch rail	B	55.13	C	17.94	C
Guardrail adjusting sheet of No. 9 switch	C	58.97	C	18.26	C
High manganese steel integrated cast frog of 50 kg/m rail	A	100.00	A	167.05	A
Straight switch rail of 50 kg/m rail	A	62.61	B	60.82	A
Insulated gauge rod	B	4.18	C	2.95	C

4.3 Model's Effect Analysis

Through the comparative analysis of the classification results of the three methods in Table 2, the following conclusions are obtained.

(1) The classification results based on the improved super efficient DEA-ABC method are more scientific and more realistic than the traditional ABC classification results. For example, the P60 rail is classified into Class A in the traditional ABC classification due to its high cost, but the annual consumption of P60 rail is very low, and its repairability rate is high, so it is classified into Class B by the improved super efficient DEA-ABC method. Contrary to this, the P60 package plywood was classified into Class B due to the moderate amount occupied under the ABC classification method. However, its procurement lead time is long, and damage had a greater impact on the equipment, so it is classified into Class A by the improved super efficient DEA-ABC method, which is consistent with the actual situation.
(2) The classification result based on the improved super efficient DEA-ABC model realizes the sorting of all spare parts and is more in line with the actual situation than the classification result of the classic DEA model. For example, the Straight switch rail of 50 kg/m rail is classified into Class B using the classic DEA model, which is far from the actual situation and professional knowledge. The reason for this phenomenon is that the indicator weight of the classic DEA model is automatic calculated by DEA model. It can be seen from the calculation results of the EMS software that the weights of the two evaluation indicators of annual consumption and damage to the equipment are equal to zero, that is, these two indicators are not considered in the classification process, which is inconsistent with the actual situation. The improved super efficient DEA model adds the survey results of the professionals on the indicator weights as a restriction condition to the evaluation process. As a result, this spare part is classified into Class A, whose result combines professional qualitative cognition and quantitative analysis, makes the classification results more in line with the actual situation.

5 Conclusion

Based on combing the literature of the commonly classification methods for inventory spare parts, the traditional ABC classification method is improved in this paper. Firstly, combining RAMS standard and key factors in the procurement and use of spare parts, a classification indicator system is constructed. Then using the Delphi method to survey inventory management personnel as the weight restriction conditions and add it into the super efficient DEA model, the improved super efficient DEA-ABC classification model is built. This classification model can combine subjective and objective, which makes the classification results more scientific and practical. An empirical analysis of some spare parts of a subway operating company in a certain city in China is made. The 18 kinds of spare parts are classified by the traditional ABC model, the classic DEA-ABC model and the improved super efficient DEA-ABC model proposed in this paper, and the classification results are compared and analyzed. It is concluded that the classification method constructed in this paper is more scientific than the classification result obtained by the traditional ABC method and more in line with the actual situation. Moreover, this improved super efficient DEA-ABC classification method achieves a complete ranking of all spare parts, which is more practical than the results obtained by the classic DEA model, which verifies the advantages of the comprehensive classification method constructed in this paper.

References

1. Gajpal, P.P., Ganesh, L.S., Rajjendren, C.: Criticality analysis of spare parts using the analytic hierarchy process. Int. J. Prod. Econ. **35**(1–3), 293–297 (1994)
2. Flores, B.E., Olson, D.L., Dorai, V.K.: Management of multicriteria inventory classification. Math. Comput. Model. **16**(12), 71–82 (1992)
3. Chu, C.W., Liang, G.S., Liao, C.T.: Controlling inventory by combining ABC analysis and fuzzy classification. Comput. Ind. Eng. **55**(4), 841–851 (2008)
4. Jamshidi, H., Jain, A.: Multi—criteria ABC inventory classification: with exponential smoothing weights. J. Glob. Bus. Issues **41**(4), 389–404 (2002)
5. Patrovi, F.Y., Anandarajan, M.: Classifying inventory using an artificial neural network approach. Comput. Ind. Eng. **41**(4), 389–404 (2002)
6. Cui, N.F., Luo, X.: ABC classification model of maintenance spare parts based on AHP. Ind. Eng. Manag. **06**, 33–36 (2004)
7. Fu, Z.Y., Shi, C.H.: Research on classification of metro spare parts. Guangxi Light Ind. **05**, 118–120 (2008)
8. Jia, L.M., Yang, Y., Jiang, X.L.: Empirical research on material storage strategy of large hydropower enterprises based on AHP-ABC classification. Logist. Technol. **33**(09), 104–107 (2014)
9. Charnes, A., Cooper, W.W., Rhodes, E.: Measuring the efficiency of decision making units. Eur. J. Oper. Res. **32**, 429–444 (1978)
10. Cui, N.F., Lu, J.J.: ABC classification model of spare parts based on DEA. Logist. Technol. **03**, 55–58 (2007)
11. Yu, Z.J., Wang, K.P., Zheng, J.W., Tao, Y.K., Li, Y.L.: Management of spare parts for navigation aids based on improved ABC classification management method. Comput. Appl. Softw. **36**(07), 59–64 (2019)
12. Xu, X.Y., Li, N.L., Wang, J., Guo, J.J.: AHP-DEA's ABC classification of spare parts. Mech. Des. Manuf. **08**, 269–272 (2016)
13. Tao, Y.: Research on two-level inventory control scheme for repairable spare parts of EMU. Beijing Jiaotong University, Beijing (2018)
14. Andersen, P., Petersen, N.C.: A procedure for ranking efficient units in data envelopment analysis. Manag. Sci. **39**(10), 1261–1264 (1993)
15. Li, L., Liu, J.Y., Fu, C.Q.: MCDM model based on grey correlation analysis and super-efficiency DEA. Appl. Res. Comput. **27**, 526–528 (2010)
16. Halne, M.: A value efficiency approach to incorporating preference in data envelopment analysis. Manag. Sci. **45**(1), 103–115 (1999)
17. Xu, N., Lu, W.: Super efficiency DEA evaluation model with entropy weight restriction for expressway transport effectiveness. ICIC Express Lett. Part B: Appl. **8**, 2101–2106 (2015)
18. Han, S., Wei, Q.: DEA efficiency decomposition and scale return evaluation. Math. Pract. Knowl. **46**(24), 60–68 (2016)
19. Qu, J., Guo, Q.: Research on several issues in rail transit RAMS national standards. Railway Tech. Superv. **39**(12), 6–10 (2011)

MSNet: A Multi-scale Segmentation Network for Documents Layout Analysis

Bo Wang[1(✉)], Ju Zhou[1(✉)], and Bailing Zhang[1,2(✉)]

[1] Suzhou Aunbox Software Co., Ltd., Suzhou, China
wanghaixingbo@gmail.com, 1272735844@qq.com,
bai_ling_zhang@hotmail.com
[2] School of Computer and Data Engineering, Ningbo Tech University, Ningbo, China

Abstract. Layout analysis is often a crucial step in document image analysis and understanding. In this paper, we propose a deep learning-based layout analysis approach to identify and categorize the regions of interests in the scanned image of text document. Although semantic segmentation has been applied at pixel-level of document image for geometric layout analysis with much progress, many challenges remain with complex and heterogeneous documents which often have a sparse structure without closed boundaries and fine typologies with variable scales. We propose a multi-scale segmentation network, called MSNet, for high-resolution document image. The model is characterized by the enlarged receptive field size and multi-scale feature extraction. Experiments are conducted on a Chinese document dataset with satisfying performance.

Keywords: Layout analysis · Semantic segmentation · Multi-scale

1 Introduction

With the digital transformation of various industries, the production and storage of electronic documents have increased exponentially, which further increases the demand for the processing of documents. Due to the cost and inefficiency of manual document processing, it is significant to extract and analyze the massive data automatically by using artificial intelligence technology. Layout analysis is often a primary stage in the pipeline of document image analysis and understanding. By document layout analysis, a document image is separated into zones, which will be subsequently classified into one of the categories of text, tables, images or lines [1].

In recent years, deep learning technologies have been applied in computer vision, natural language processing and other fields. Object detection and semantic segmentation based on deep neural networks have also been successfully exploited to solve layout analysis problems. Most object detection methods use an anchor-box mechanism [2]. The positioning accuracy of general object detection methods is object-dependent, it is more appropriate for objects with closed boundary. In a document image, a text zone is a structure without closed boundary, and there may exist small

C. Pang et al. (Eds.): SETE 2020/ICWL 2020, LNCS 12511, pp. 225–235, 2021.
https://doi.org/10.1007/978-3-030-66906-5_21

objections in complex text. The diverse text structures and scales in different layouts would make it more difficult to select the right anchor-box.

Image semantic segmentation is pixel-level recognition which has been applied to layout analysis tasks. Most of the semantic segmentation networks are encoder-decoder architectures, such as FCN [3], SegNet [4], U-Net [5] and DeepLab [6–8]. The size of receptive field and multi-scale features are two important factors to improve the segmentation accuracy in semantic segmentation. Typically, in order to capture larger receptive field and reduce the computational complexity, the encoder reduces the feature maps by consecutive pooling operator or convolution with stride. However, the down-sampling operation loses a lot of detail information. A text zone usually contains small-scale elements such as words, numbers, symbols and others, and these factors are more likely to cause blurring, deformation and loss of detail information when the text image is scaled and down-sampled, and subsequently the low resolution of the encoder output is detrimental to the recovery of text details by decoder.

In order to solve the aforementioned problems, we propose a novel semantic segmentation network for layout analysis of high-resolution document images, with moderate cost of increasing the computation and memory consumption. Firstly, we adapt the U-type skip-connection structure of U-Net [5] network to recover more detailed information in the decoder; secondly, we improve the residual module of residual network [9] to increase the receptive field size and multi-scale features of the network; And finally, we use the attention mechanism [10–12] to capture long-range context information and enhance the representation ability of the network.

In summary, our main contributions include:

1. a novel light-weight semantic segmentation network MSNet is proposed for layout analysis of high-resolution document images.
2. the residual module is improved by combining methods such as multi-scale features fusion and channel-attention mechanism.
3. a novel layout analysis dataset is created for the experiment which is labeled by text lines.

2 Related Works

Semantic Segmentation. Semantic segmentation is one of the fundamental topics in computer vision, which assigns a semantic label from a set of classes to each pixel of the image [3]. Most modern methods regard it as a problem of dense prediction. Current state-of-the-art semantic segmentation approaches are mainly based on deep convolutional neural networks. A fully convolutional network (FCN) [3] uses a convolutional neural network such as VGG [13] and ResNet [9] to transform image pixels to pixel categories. To enlarge the receptive field without reducing the resolution of the feature map, the dilation convolution has been widely used in various semantic segmentation networks. Multi-scale feature pyramid fusion is another frequently discussed topic in semantic segmentation. For instance, the DeepLabv3 [8] devises an atrous spatial pyramid pooling that applies several parallel dilated convolutions with different dilation rates to capture multi-scale context information, while the PSPNet [14] adopts a pyramid pooling module that aggregates the feature maps into different sizes and concatenates them after up-sampling. Moreover, some encoder-decoder-based segmentation networks [15–17] adds extra top-down and lateral connections to recover the high-resolution feature maps in the decoder part. Recently, some methods [11, 12, 18–20] introduce the attention mechanisms which is helpful to focus on key information to capture long-range context information. In addition, there are also some segmentation networks that use multi-branches structure to maintain the high resolution, such as HRNet [21], BiSeNet [22, 23].

Documents Layout Analysis. As deep learning has revolutionized computer vision research, DL-based document image analysis methods have been extensively developed to tackle various challenging tasks. Recently, object detection technology has been applied for document layout analysis [24]. Most of these methods are based on the existing general object detection neural networks, such as Faster-RCNN [25], Mask-RCNN [26], and YOLO series [27]. For example, in [28] a preprocessed document image is first sent to the Region Proposal Network (RPN), and then a Faster-RCNN is used to detect whether the object is a table. In addition, semantic segmentation network has also been applied to document image layout analysis to achieve more refined prediction effect. PageNet [29] utilizes a Fully Convolution Network to identify the main page region of an image, so as to segment content from both textual and non-text border noise. Generally, semantic segmentation provides more accurate pixel-level positioning [30, 31].

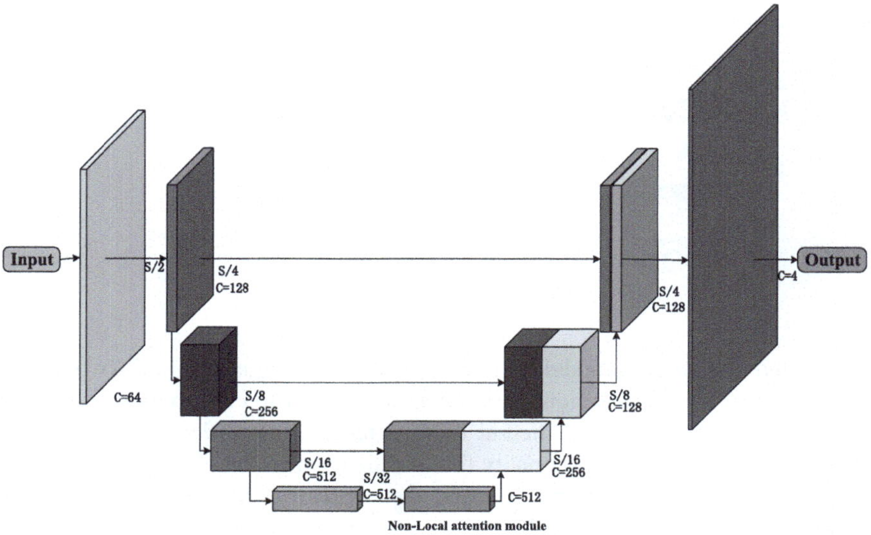

Fig. 1. MSNet architecture

3 Network Architecture

Our proposed Multi-Scale Segmentation Network (MSNet) for documents layout analysis, is shown in Fig. 1. The model employs an improved residual basic module to build the backbone network in encoder. The backbone network is divided into 4 stages, and each stage performs a 2 times down-sampling. The number of basic residual modules in each stage is 2, and the size of the feature map obtained by the final backbone network is 1/32 of original input image size. To avoid increasing too many parameters and calculations, we choose to add one non-local attention module after the last module of the backbone network to capture the long-range context information of spatial features. Then, in the up-sampling decoding stage, a U-shaped skip-connection structure similar to the U-Net network is used to recover more detailed information. Finally, the feature map is restored to the size of the input image by the up-sampling and classification module.

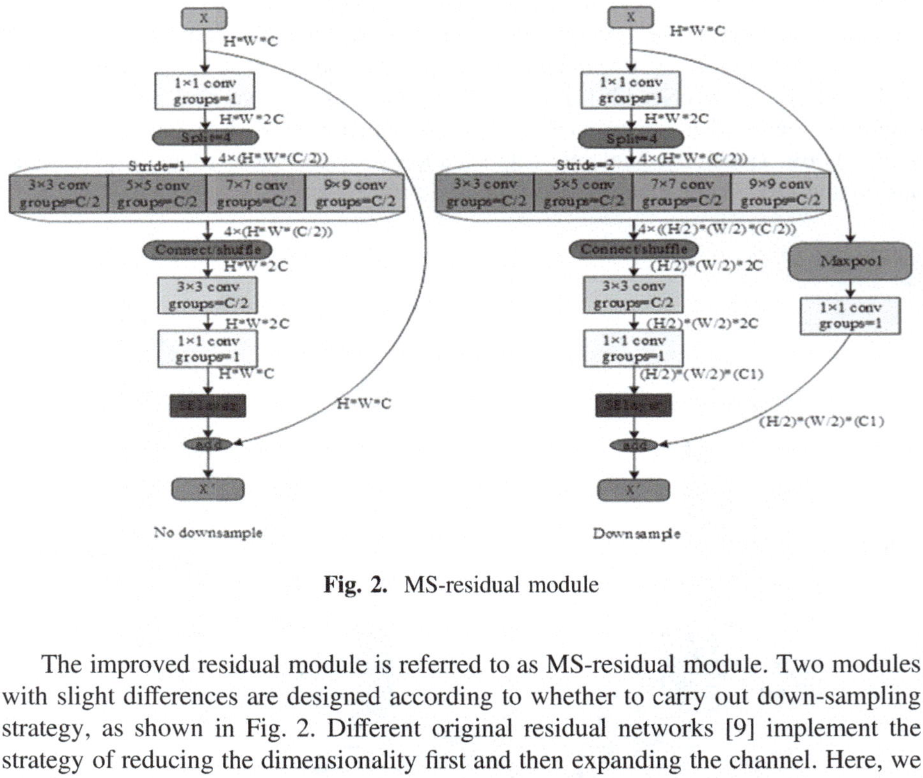

Fig. 2. MS-residual module

The improved residual module is referred to as MS-residual module. Two modules with slight differences are designed according to whether to carry out down-sampling strategy, as shown in Fig. 2. Different original residual networks [9] implement the strategy of reducing the dimensionality first and then expanding the channel. Here, we adopt the strategy of expanding first and then reducing the dimensionality of the channel. Specifically, we first use 1×1 convolution to increase the channel dimension of the input feature by 2 times, and then split the feature map into 4 small feature maps according to the channel dimension. Then we utilize the depth-wise separable convolution [32] with the convolution kernel size of $[3 \times 3, 5 \times 5, 7 \times 7, 9 \times 9]$ to perform the convolution operation on the 4 feature maps, and then connect the 4 feature maps into a new large feature map in the channel dimension. Next, we apply 3×3 group convolution [33] to fuse multi-scale features, and then use 1×1 convolution to reduce the channel dimension. Meanwhile, a channel attention module of SE-layer [12] is added after this step. Finally, the residual connection of input features is carried out by addition. It is worth noting that during down-sampling, the max-pooling and 1×1 convolution operations are performed on the input features before residual connection. At the same time, BN layer is added after each convolution layer. In addition to convolution layer which extracts multi-scale features, ReLU activation layer is added to each convolution layer.

4 Experimental Evaluation

In this section, we first provide implementation details and evaluation dataset. Then, we introduce the experimental results of several models on the evaluation dataset.

4.1 Implementation Details

We implement our network using PyTorch-1.5 framework. The models are trained on a TITAN-RTX GPU with 24 GB memory. To train the network, we adopt the mini-batch practice with Stochastic Gradient Descent (SGD) as the optimizer, where the batch size is set to 4, the momentum is set to 0.9 and the weight decay is set to 0.001. Inspired by [6], we employ the "poly" learning rate policy where the learning rate is multiplied by with power 0.9 and initial learning rate 0.06. In training, cross-entropy loss function is used. Due to the limited memory capacity on GPU cards, we reshape the image size to 960×1280. Finally, we randomly shuffle the training samples, and the total number of training is 60 epochs.

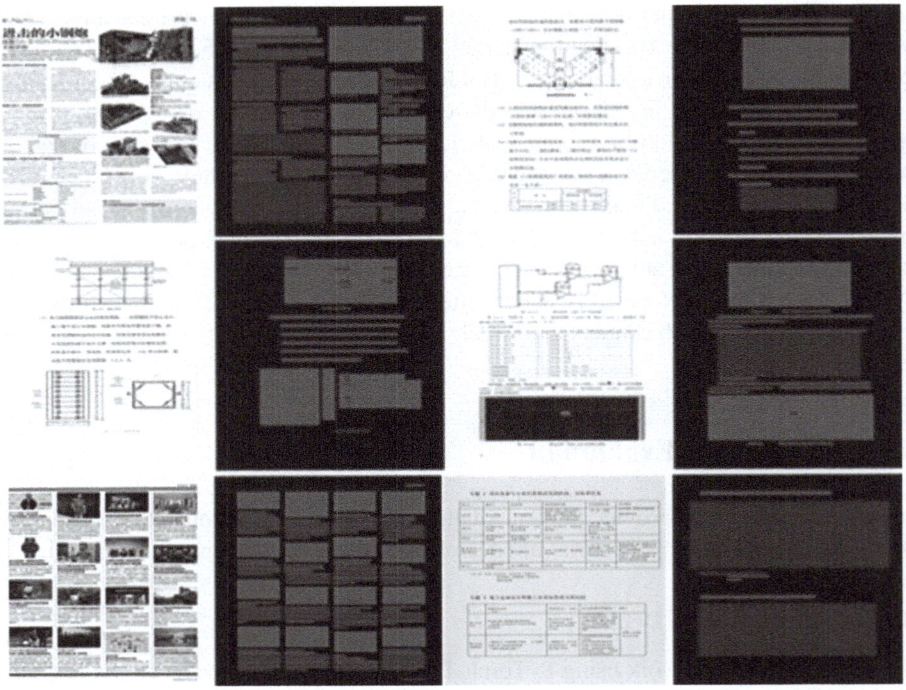

Fig. 3. Dataset training samples

4.2 Datasets and Evaluation Metric

Due to the lack of appropriate public datasets, we created a dataset for the subsequent layout analysis experiment, as shown in Fig. 3. Specifically, the dataset is divided into 750 document images for training, 177 document images for validation. There are totally 4 semantic categories: text, table, figure and background. It should be noted that the text label is mainly in the form of text lines, with a small number of irregular text blocks. As the dataset is relatively small, we apply more data augmentation methods such as random rotation, scaling, flipping, contrast and gray conversion to increase the diversity of training samples and reduce the risk of over-fitting.

For each document image, segmentation algorithms will produce a semantic segmentation mask, predicting the semantic category for each pixel in the image. In our experiment, we employ the average of the pixel-wise accuracy (Pixel Acc) and the mean of the classes-wise Intersection over Union averaged (Mean IoU) as the evaluation metric for semantic segmentation.

4.3 Experimental Results

The evaluation results are presented in Table 1. Compared with the DeepLab-v3+ [8], our method has achieved an obvious improvement of 1.38% (from 95.23% to 96.61%) and 3.32% (from 87.23% to 90.55%) on Pixal Acc and Mean IoU, respectively. The parameters of our model are only about 1/4 of the parameters of DeepLab-v3+, which is reduced to 9.7 M. Simultaneously, compared with DenseASPP [34], our method has achieved the improvement of 1.77% and 4.4% on Pixal Acc and Mean IoU, respectively. The parameters of our model are only about 1/3 of the parameters of DenseASPP. The backbone network of DeepLab-v3+ and DenseASPP is Resnet-50. It is worth noting that for the sake of fairness, we do not load the pretrained model parameters, but use the same random initialization strategy (after loading the resnet-50 pre-trained model, the performance of DeepLab-v3+ and DenseASPP models will have better performance). In addition, we also show the segmentation prediction results to make an intuitive comparison, as shown in Fig. 4.

Table 1. Segmentation results on the validation set of our own layout analysis dataset.

Methods	Parameters (M)	Time (s)	Pixel Acc (%)	Mean IoU (%)
DeepLab-v3+(ASPP)	39.14	0.0844	95.23	87.23
DenseASPP	27.94	0.0810	94.84	86.15
MSNet (our)	9.70	0.0933	96.61	90.55

From Table 1 and Fig. 4, we can see that our network model not only has fewer parameters, but also has better segmentation results. However, our inference time has slightly increased, probably due to the addition of more operations such as splitting, concatenation, shuffling and depth-wise separable convolution. It can be seen that DeepLab-v3+ and DenseASPP model are more likely to mistake the lines of text together, which also shows that the U-shaped network structure can better recover the detailed information. In general, the model shows that increasing the size of receptive field, multi-scale feature fusion and attention mechanism can improve the segmentation accuracy of network model.

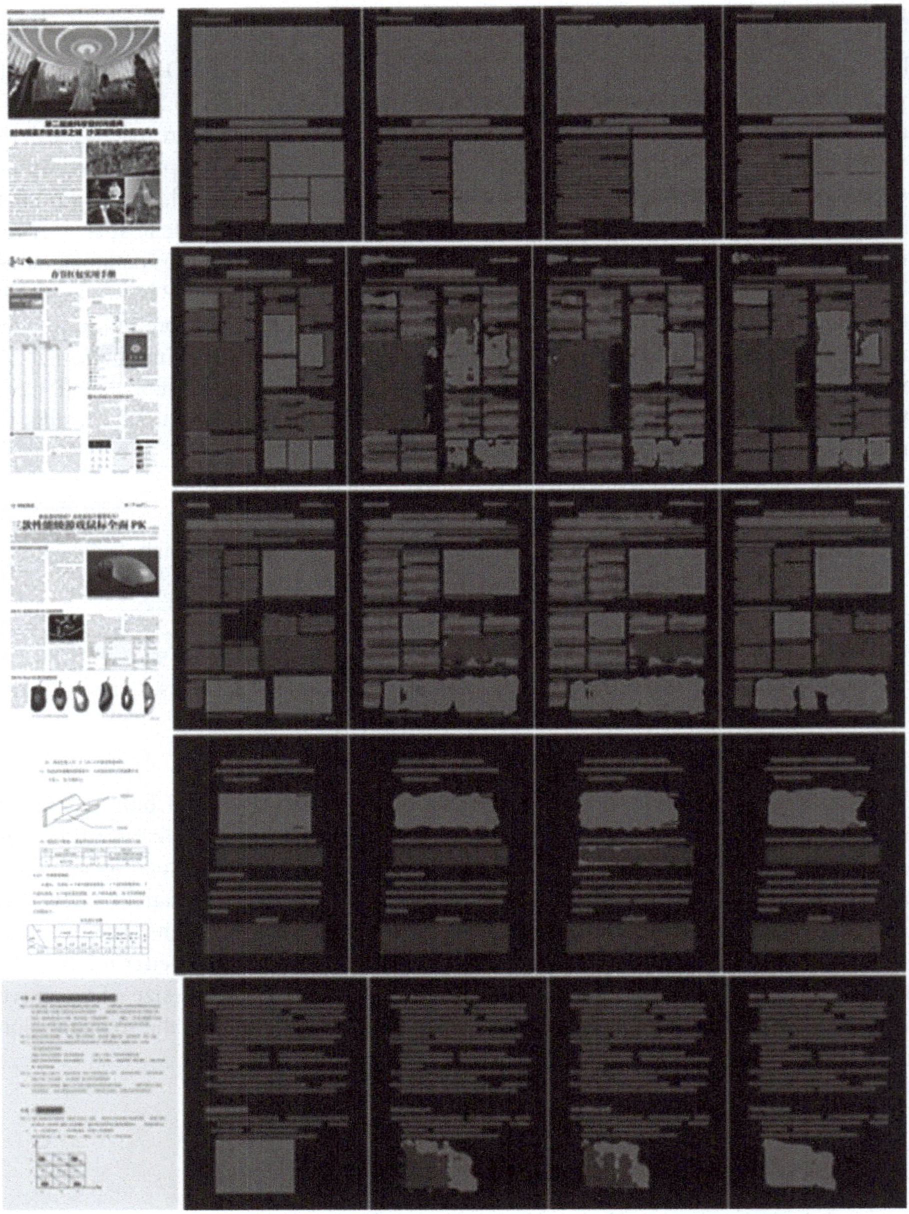

Fig. 4. Visual results. (a) Image. (b) GT. (c) DeepLabv3+. (d) DenseASPP. (e) MSNet (our).

5 Conclusion

In this paper, we propose a multi-scale segmentation network, called MSNet, which is suitable for high-resolution document images. Semantic segmentation network provides pixel-wise recognition and location, which is more suitable to handle the difficulty of complex document layout analysis. However, different layout elements are often nested in the document images, for example, table containing image, and image containing text, seal or watermark which overlap with other elements. There are still a number of issues to be solved by using semantic segmentation for layout analysis. Generally, semantic segmentation only assigns a category label to a pixel, while nesting and overlapping of different layout elements will cause difficulty in annotation and recognition, because existing annotation methods may destroy the complete semantic structure of some layout elements. We are working on new semantic segmentation methods to solve the nesting problem of layout elements.

References

1. Xu, Y., Yin F., Zhang Z.X., et al.: Page segmentation for historical handwritten documents using fully convolutional networks. In: 27th International Joint Conference on Artificial Intelligence, pp. 1057–1063. IEEE Computer Society, Kyoto (2017)
2. Ye, Q., Doermann, D.: Text detection and recognition in imagery: a survey. IEEE Trans. Pattern Anal. Mach. Intell. **37**(7), 1480–1500 (2015)
3. Long, J., Shelhamer, E., Darrell, T.: Fully convolutional networks for semantic segmentation. In: 2015 IEEE Conference on Computer Vision and Pattern Recognition, pp. 3431–3440. IEEE, Boston (2015)
4. Badrinarayanan, V., Kendal, L.A., Cipolla, R.: SegNet: a deep convolutional encoder-decoder architecture for image segmentation. IEEE Trans. Pattern Anal. Mach. Intell. **39**(12), 2481–2495 (2017). https://doi.org/10.1109/tpami.2016.2644615
5. Ronneberger, O., Fischer, P., Brox, T.: U-Net: convolutional networks for biomedical image segmentation. In: Navab, N., Hornegger, J., Wells, W., Frangi, A. (eds.) Medical Image Computing and Computer-Assisted Intervention, MICCAI 2015. MICCAI 2015. LNCS, vol 9351. pp. 234–241. Springer, Cham (2015). https://doi.org/10.1007/978-3-319-24574-4_28
6. Chen, L.C., Papandreou, G., Kokkinos, I., et al.: DeepLab: semantic image segmentation with deep convolutional nets, atrous convolution, and fully connected CRFs. IEEE Trans. Pattern Anal. Mach. Intell. **40**(4), 834–848 (2018). https://doi.org/10.1109/TPAMI.2017.2699184
7. Chen, L.C., Papandreou, G., Schroff, F., et al.: Rethinking atrous convolution for semantic image segmentation. arXiv (2017). https://arxiv.org/abs/1706.05587
8. Chen, L.-C., Zhu, Y., Papandreou, G., Schroff, F., Adam, H.: Encoder-decoder with atrous separable convolution for semantic image segmentation. In: Ferrari, V., Hebert, M., Sminchisescu, C., Weiss, Y. (eds.) ECCV 2018. LNCS, vol. 11211, pp. 833–851. Springer, Cham (2018). https://doi.org/10.1007/978-3-030-01234-2_49
9. He, K., Zhang, X., Ren, S., et al.: Deep residual learning for image recognition. In: 2016 IEEE Conference on Computer Vision and Pattern Recognition, pp. 770–778. IEEE, Las Vegas (2016)
10. Vaswani, A., Shazeer, N., Parmar, N., et al.: Attention is all you need. In: 30th Neural Information Processing Systems, pp. 6000–6010. Curran Associates, Long Beach (2017)

11. Fu, J., Liu, J., Tian, H., et al.: Dual attention network for scene segmentation. In: 2019 IEEE Conference on Computer Vision and Pattern Recognition, pp. 3146–3154. IEEE, Long Beach (2019)
12. Hu, J., Shen, L., Albanie, S., et al.: Squeeze-and-excitation networks. In: 2018 IEEE Conference on Computer Vision and Pattern Recognition, pp. 7132–7141. IEEE, Salt Lake City (2018)
13. Simonyan, k., Zisserman, A.: Very deep convolutional networks for large-scale image recognition. arXiv (2015), https://arxiv.org/abs/1409.1556
14. Zhao, H., Shi, J., Qi, X., et al.: Pyramid scene parsing network. In: 2017 IEEE Conference on Computer Vision and Pattern Recognition, pp. 6230–6239. IEEE, Honolulu (2017)
15. Lin, G., Milan, A., Shen, C., et al.: RefineNet: multi-path refinement networks for high-resolution semantic segmentation. In: 2017 IEEE Conference on Computer Vision and Pattern Recognition, pp. 1926–1934. IEEE, Honolulu (2017)
16. Zhang, H., Dana, K., Shi, J., et al.: Context encoding for semantic segmentation. In: 2018 IEEE Conference on Computer Vision and Pattern Recognition, pp. 7151–7160. IEEE, Salt Lake City (2018)
17. Yu, F., Koltun, V.: Multi-Scale Context Aggregation by Dilated Convolutions. arXiv (2016). https://arxiv.org/abs/1511.07122
18. Li, H., Xiong, P., An, J., et al.: Pyramid attention network for semantic segmentation. arXiv (2018). https://arxiv.org/abs/1805.10180
19. Yu, C., Wang, J., Peng, C., et al.: Learning a discriminative feature network for semantic segmentation. In: 2018 IEEE Conference on Computer Vision and Pattern Recognition, pp. 1857–1866. IEEE, Salt Lake City (2018)
20. Oktay, O., Schlemper, J., Folgoc, L.L., et al.: Attention U-Net: learning where to look for the pancreas. arXiv (2018). https://arxiv.org/abs/1804.03999
21. Sun, K., Xiao, B., Liu, D., et al.: Deep high-resolution representation learning for human pose estimation. In: 2019 IEEE Conference on Computer Vision and Pattern Recognition, pp. 5693–5703. IEEE, Long Beach (2019)
22. Yu, C., Wang, J., Peng, C., Gao, C., Yu, G., Sang, N.: BiSeNet: bilateral segmentation network for real-time semantic segmentation. In: Ferrari, V., Hebert, M., Sminchisescu, C., Weiss, Y. (eds.) ECCV 2018. LNCS, vol. 11217, pp. 334–349. Springer, Cham (2018). https://doi.org/10.1007/978-3-030-01261-8_20
23. Yu, C., Gao, C., Wang, J., et al.: BiSeNet v2: bilateral network with guided aggregation for real-time semantic segmentation. arXiv (2020). https://arxiv.org/abs/2004.02147
24. Zhong, X., Tang, J., Yepes, A.J.: PubLayNet: largest dataset ever for document layout analysis. arXiv (2019). https://arxiv.org/abs/1908.07836
25. Ren, S., He, K., Girshick, R., et al.: Faster R-CNN: towards real-time object detection with region proposal networks. In: 28th International Conference on Neural Information Processing Systems, pp. 91–99. Curran Associates, Montreal (2015)
26. He, K., Gkioxari, G., Dollár, P., et al.: Mask R-CNN. IEEE Trans. Pattern Anal. Mach. Intell. **42**(2), 386–397 (2018). https://doi.org/10.1109/TPAMI.2018.2844175
27. Redmon, J., Farhadi, A.: YOLOv3: an incremental improvement. arXiv (2018). https://arxiv.org/abs/1804.02767
28. Gilani, A., Qasim, S.R., Malik, I., et al.: Table detection using deep learning. In: 14th IAPR International Conference on Document Analysis and Recognition, pp. 771–776. IEEE, Kyoto (2017)
29. Tensmeyer, C., Davis, B., Wigington, C., et al.: PageNet: page boundary extraction in historical handwritten documents. arXiv (2017). https://arxiv.org/abs/1709.01618
30. Tuan, T.A., Oh, K., Na, I.S., et al.: A robust system for document layout analysis using multilevel homogeneity structure. Expert Syst. Appl. **85**(1), 99–113 (2017)

31. Oliveira, D.A.B., Viana, M.P.: Fast CNN-based document layout analysis. In: 2017 IEEE Conference on International Conference on Computer Vision, pp. 1173–1180. IEEE, Venice (2017)
32. Howard, A.G., Zhu, M., Chen, B., et al.: MobileNets: efficient convolutional neural networks for mobile vision applications. arXiv (2017). https://arxiv.org/abs/1704.04861
33. Xie, S., Girshick, R., Dollar, P., et al.: Aggregated residual transformations for deep neural networks. In: 2017 IEEE Conference on Computer Vision and Pattern Recognition, pp. 1492–1500. IEEE, Honolulu (2017)
34. Yang, M., Yu, K., Zhang, C., et al.: DenseASPP for semantic segmentation in street scenes. In: 2018 IEEE Conference on Computer Vision and Pattern Recognition, pp. 3684–3692. IEEE, Salt Lake City (2018)

Research on Intelligent Transportation Platform Based on Big Data Technology

Liu Bin[1,2(✉)], Si Liang[1,2], Liu Yu[3(✉)], Chen Li[4], Yuan Yijie[1,2], Liu Zengjie[1,2], Huang Ju[1,2], Liu Qing[1,2], and Yang Liwei[5]

[1] School of Economics and Management, Hebei University of Science and Technology, Shijiazhuang 050018, Hebei, China
tsingnlp@163.com

[2] Research Center of Big Data and Social Computing, Hebei University of Science and Technology, Shijiazhuang 050018, Hebei, China

[3] Library, Hebei Professional Collage of Political Science and Law, Shijiazhuang 050061, Hebei, China
446927742@qq.com

[4] Air Force Early Warning Academy, Wuhan 430019, Hubei, China

[5] Hebei Provincial Highway, Zhang-Cheng Chengde Management Office, Chengde, Hebei 0670001, China

Abstract. This paper focuses on the personnel education in intelligent transportation, aiming at the problems of high communication cost, poor coherence and unsatisfactory effect in the daily education and training of highway management departments. A mobile education platform providing data management and analysis is developed for 4 main fields as lean safety, rules and regulations, knowledge and experience and ideology. Since the platform was launched in September 2016, the practice has proved that the daily education of the management office has broken through the limitation of time and space, and has the characters of relevance, mutual assistance and sharing, timeliness and strong plastic effect, which has effectively improved the comprehensive quality of the whole staffs.

Keywords: Intelligent transportation · Big data · Mobile internet · Daily education

For the geographically distributed organization with the characteristics of scattered departments, long geographical distance and inconvenient face-to-face communication, there are many problems in their daily education, such as high communication cost (difficult to concentrate personnel), poor coherence (difficult to continue to participate in learning), poor effects [1–3]. With the rapid development of mobile Internet [4, 5] and social media [6, 7], various organizations have introduced various forms of mobile social media platforms in their daily communication. However, due to the essential characteristics of these platforms, such as pan entertainment, pan life and pan socialization, there are insurmountable difficulties in information security, communication security and secret related protection. This study develops an organization specific mobile education platform and takes Zhang-Cheng Chengde management office of Hebei provincial highway as the research and practice orientation, and realize the learning and education mode that

C. Pang et al. (Eds.): SETE 2020/ICWL 2020, LNCS 12511, pp. 236–243, 2021.
https://doi.org/10.1007/978-3-030-66906-5_22

breaks through the space-time constraints and meets the initiative, personalized and normalized needs.

1 Problems and Challenges

For Zhang-Cheng Chengde management office of Hebei provincial highway, due to the characteristics of long distance and scattered distribution of grass-roots departments such as toll stations, dispatching centers and management offices, it is difficult to carry out consistent education and training for improving the employees' skills and knowledge. In their daily work, various QQ or WeChat [8, 9] work groups have been established. However, due to the socialized nature of such groups, there exist many practical problems such as theme interpenetration, confusion of clues, and difficulty in refining communication themes [10]. Moreover, the relevant data are deposited in Tencent's database, which is not convenient for data control and analysis. Because of the objective differences of job requirements, the pertinence of education should be improved; in addition, the correlation between learning achievements and job matching, income, career development and so on is low, and learning or not, for employees, the results are the same, which leads to the lack of enthusiasm and initiative.

2 Requirement Analysis

The Chengde section of Zhang-Cheng Expressway runs from Chengde city to Zhangjiakou section, with a total length of about 203 km. Its basic units are distributed, which include 8 toll stations, 4 tunnel management offices, 4 maintenance work areas and 2 dispatch centers, etc., it is a typical geographically distributed organization. After investigation, its main daily education requirements management are listed as following [11]:

Deepen the Role of Online Training and Improve the Quality of Daily Learning. Develop the mobile education platform, make good use of internal and external resources, and conduct daily learning from 4 themes: lean safety, rules and regulations, knowledge and experience and ideology (see Fig. 1). The platform should support mobile learning and on-the-job training at any time. And each department can organize targeted learning independently; it should also support employees to use the fragmented time for learning, so as to alleviate the contradiction between work and study.

Support Data Analysis to Effectively Understand the Learning Needs of Employees. Data produced on the platform should belong to the management office, and the managers can count the data, view the learning progress of employees. Analyze the moments released by various departments and personnel, determine training needs, and realize the shift from traditional "spoon-feeding education" to "daily personalized learning". Promote the whole staff from passive learning to active learning, and improve the overall quality of the whole staff in the end.

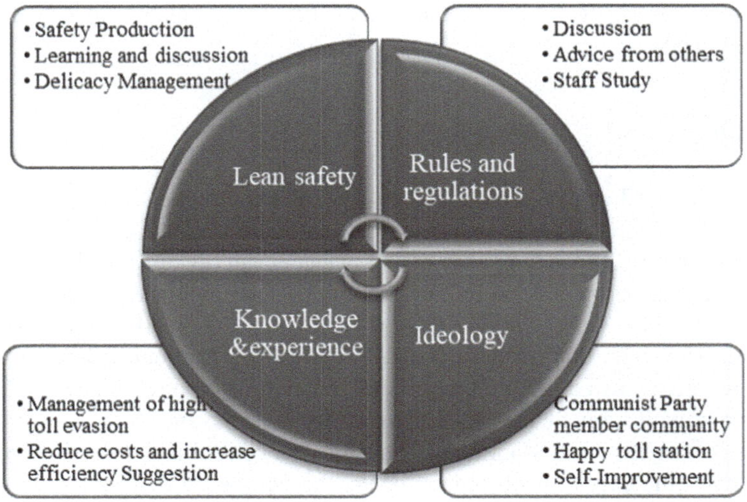

Fig. 1. Educational theme design of the platform – "i Zhang-Cheng"

Develop the Information Management System of the Whole Staff and Improve the Level of Information Management. The new platform should be convenient for the person in charge to manage the employee information, collect the employee basic information, ensure the authenticity and effectiveness of the information, and provide data support for analysis.

3 Platform Function Design

An online education platform "i Zhang-Cheng" is developed, which has a App as its front-end (responsible for publishing and collecting data) and a data management system (responsible for data analysis and maintenance) as its backstage (see Fig. 2). All of the platform data belongs to the office and can be analyzed. The main data analysis modules include exam management, sign-in statistics, moment analysis, panoramic data monitoring, activity analysis, ranking analysis, etc. Information management system has the user management, moment management, theme management, point management, circle management, etc.

Theme Release. For online learning, the tasks are released to the dedicated App, and centralized training is no longer required for staff if it is not necessary. With the dedicated App, everyone can use fragmented time to learn and post moments. The upper limit of each moment is 3000 Chinese characters and 9 graphs, which can provide rich information. The app can provide a variety of online learning forms and content, such as moment learning, rules and regulations learning, examination and test, discussion, etc., which is conducive to collaborative learning, discussion learning and autonomous learning among employees, reducing training costs and improving training efficiency.

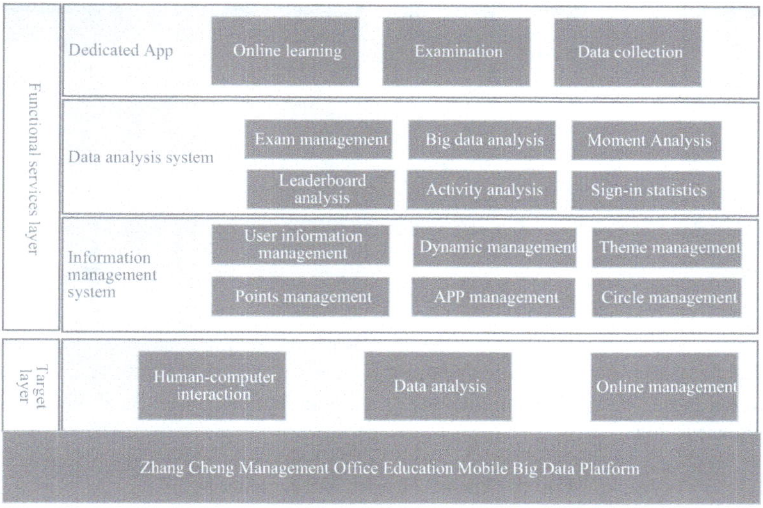

Fig. 2. Functional architecture of the platform – "i Zhang-Cheng"

Data Analysis System. The platform supports the analysis of all kinds of user data and visualizes the diversity, and intuitively represents the management department or department information of all levels and categories. At the same time, through the analysis of moments, hot discussion and other data, we can find various problems in the learning process, which helps to refine personalized learning needs.

Information Management System. Basic information management is helpful to ensure the integrity and accuracy of system information. For example, it is necessary to maintain the basic information of users, change the access rights of users, add, delete and modify the moments.

4 Platform Implementation

4.1 App-End Thematic Learning

In addition to the common learning themes, you can also publish other thematic moments. Everyone can share his idea, achievement and experience. Users can post daily work-related moments, opinions and ideas, and relevant personnel can express their opinions in the comments to form thematic discussions. Everyone who can see the topic can participate in the discussion. It is a good experience sharing process. Simultaneous sharing of information resources has invisibly improved the business capabilities of personnel. At the same time, it will increase communication and trust between personnel, and strengthen the team's spirit of collaboration and cohesion.

Rules and Regulations. The traditional rules and regulations learning is to use the propaganda board to display to the members or distribute the paper pamphlets for self-study, but these two methods are not convenient for employees to learn at any time, and

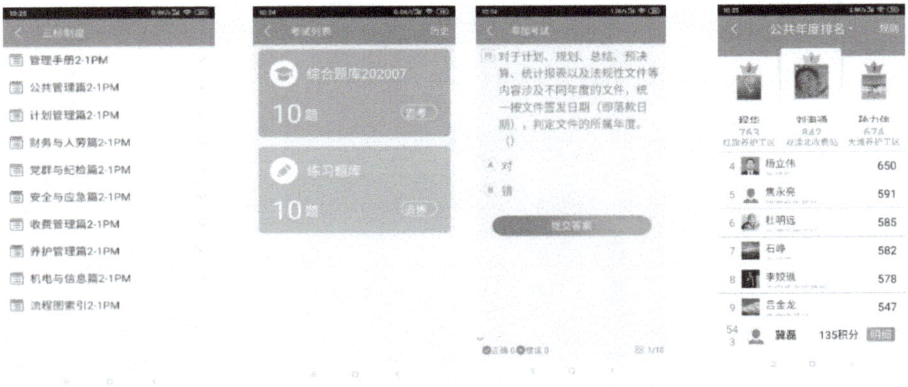

a. Rules and regulations learning b. Examinations c. Examination paper d. Points ranking

Fig. 3. Theme of rules and regulations

the effect is not ideal. As shown in Fig. 3, by publishing rules and regulations in the "i Zhang-Cheng" app, employees can effectively use the fragmented time to learn, and do not need to be centralized. Online examination can be used to test the learning effect and accurately collect the learning feedback of employees. Through the analysis of the feedback information, managers can find out the deficiencies in time and improve the quality of learning. At the same time, through the learning points system, the incentive rewards and punishments can be combined with the performance evaluation to stimulate the employees' initiative learning desire to the maximum extent.

Lean Safe Learning. In addition to daily popularization of safety knowledge, the management department should also truly change the safety awareness of employees and turn passive into active. Safety work cannot be done overnight, there is no shortcut to go, only in the daily work with a sense of safety, always pay attention to the string of safety, in order to avoid all kinds of accident risk (see Fig. 4).

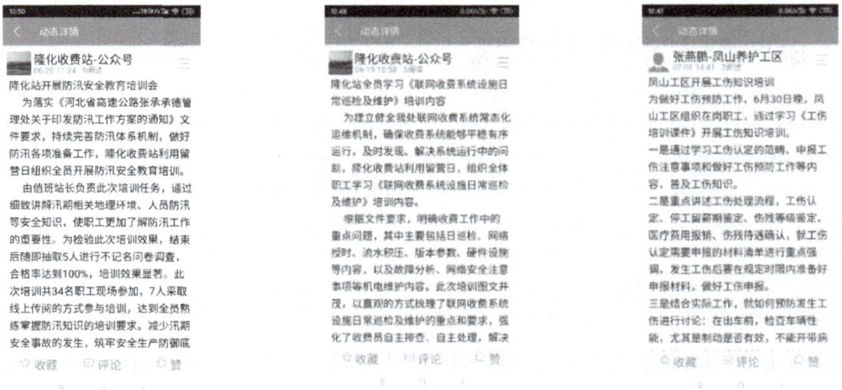

a. Flood control safety education b. Equipment maintenance c. Industrial injury knowledge training

Fig. 4. Theme of lean security

Knowledge Experience Learning. The problems, solutions and implementation process issued by any user are essentially knowledge sharing behaviors, and platform users can understand or learn these knowledges anytime and anywhere. Through this way, on the one hand, it realizes the dissemination and sharing of individual knowledge within the organization, forming the entire unit knowledge base; on the other hand, it constructs the internal cycle of knowledge extraction within the whole unit. As shown in Fig. 5, members share the problems and solutions encountered in their work.

a. Found issue b. Solve the problem c. Experience sharing

Fig. 5. Experience theme

Ideological Construction. The sense of belonging refers to the sense of identity, fairness, security, value, mission and achievement of the unit after a period of work, which is internalized into the sense of belonging of employees. It can mobilize one's internal driving force and form self-motivation. Through equal participation in the "happy station area, warm home", "moral online, reasoning" and other sections, each member has the right to speak, and can express his real feelings independently; at the same time, he can record and share the details of his work, thus generating team identification and strong sense of belonging in the discussion and interaction (see Fig. 6).

a. Party Building b. Warm as home c. Ethics online

Fig. 6. Ideological construction

4.2 Backstage Management System

Figure 7 shows the main functions of "i Zhang-Cheng" backstage system. The data analysis module is responsible for analyzing the user data collected from the app, such as the released moments and the main participants, the collected personal information (work experience, skill assessment, interests, etc.), post information (workflow, post skills, work methods, work authority). The basic information management module manages user information (name, position, title, mobile number, etc.), group information (member information, number of people, etc.), access rights, test paper creation and release, moment supervision, etc.

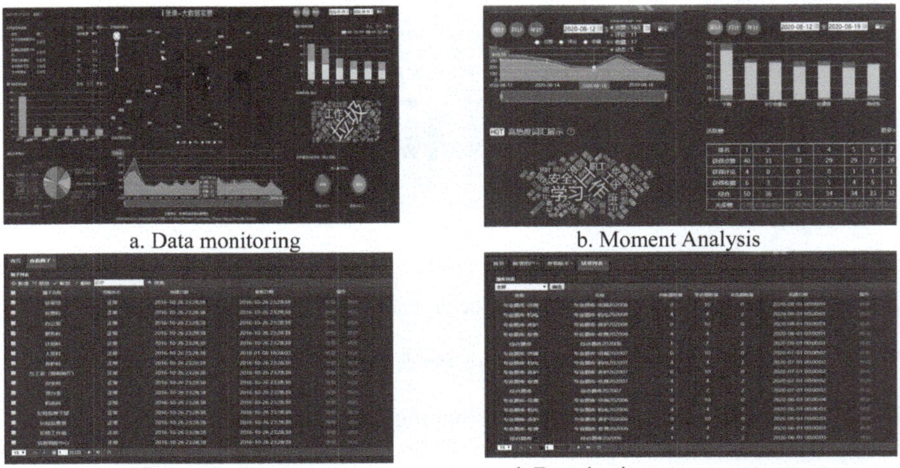

a. Data monitoring b. Moment Analysis

c. Group management d. Examination paper management

Fig. 7. Backstage management system

5 Conclusion

The daily learning education of the geographically distributed organizations has the problems of high communication cost, poor coherence and poor effect. Taking Chengde section of Zhang-Cheng Chengde expressway as the prototype, we designed and developed a dedicated daily work and education platform, which was launched in September 2016. Practice has proved that the learning efficiency and learning ability of the whole staff have been effectively improved through experience sharing, special examination, safety learning and other education sections.

Acknowledgement. This work is supported by S&T Program of Hebei (Grant No. 18210109D, 20310802D, 20551801K); High-level Talents Subsidy Project in Hebei Province (Grant No. A2016002015); Technological Innovation Fund Project of Technological Small and Medium-sized Enterprises of Shijiazhuang (Grant No. 19SCX01006); Science and technology research and development Program of Shijiazhuang (Grant No. 191130591A); 2014 Science and Technology Planning Project of Hebei Provincial Department of Transportation - Research on the Key Technologies of the Engineering Project Construction Management Platform Based on Mobile Internet (Grant No. Y2014025); Science and Technology Plan Project Of Chengde To Zhangjiakou Highway Chengde Period (Grant No. CZC-2014KY-1).

References

1. Muneera, B., Didar, Z., Natalie, S.: Empirical study of communication structures and barriers in geographically distributed teams. IET Softw. **10**(5), 147–153 (2016)
2. Salim, C., Christoph, M.: Prototyping a cognitive-centered model to improve knowledge creation in geographically distributed teams. In: 2016 IEEE International Conference on Knowledge Engineering and Applications (ICKEA), pp. 181–187. IEEE, Singapore (2016)
3. Liu, H.B., Hu, B.: Simulation of the coordinated operations mechanism for the geographically distributed organization: a perspective of interactive memory system and meta-networks. J. Syst. Manag. (2017)
4. Yang, J., Qiao, Y.Y., Zhang, X.Y., He, H.Y., Liu, F., Cheng, F.: Characterizing user behavior in mobile internet. IEEE Trans. Emerg. Top. Comput. **3**(1), 95–106 (2015)
5. Jiang, W.J., Wang, Y., Jiang, Y.R., Chen, J.H., Xu, Y.H., Tan, L.: Research on mobile Internet mobile agent system dynamic trust model for cloud computing. China Commun. **16** (7), 174–194 (2019)
6. Tarik, T., Adlen, K., Min, C., Riku, J.: Coping with emerging mobile social media applications through dynamic service function chaining. IEEE Trans. Wirel. Commun. **15** (4), 2859–2871 (2016)
7. Sharon, G.H., John, P., Ian, M.: Social media as a main source of customer feedback - alternative to customer satisfaction surveys. In: IEEE/ACM International Conference on Advances in Social Networks Analysis and Mining (ASONAM), pp. 829–832. IEEE, Vancouver (2019)
8. Ling, J.Y., Li, G.M.: A two-level stacking model for detecting abnormal users in Wechat activities. In: International Conference on Information Technology and Computer Application (ITCA), pp. 229–232. IEEE, Guangzhou (2019)
9. Xiong, X.L., Li, X.F., Wang, X.R., Wu, H.T., He, L.N.: Collective attention in WeChat public platform. In: 2018 5th International Conference on Behavioral, Economic, and Socio-Cultural Computing (BESC), pp. 261–264. IEEE, Kaohsiung (2018)
10. Liu, D.D., Wu, X.M.: Research on group social function and user differentiation – a case study of WeChat and QQ. In: IEEE 20th International Conference on High Performance Computing and Communications; IEEE 16th International Conference on Smart City; IEEE 4th International Conference on Data Science and Systems (HPCC/SmartCity/DSS), pp. 1147–1150. IEEE, Exeter, United Kingdom (2018)
11. Liu, B., Yuan, Y.J., Feng, T., et al.: A flat management system based on mobile internet. J. Hebei Univ. Sci. Technol. **40**(1), 86–96 (2019)

Data Synthesis for Document Layout Analysis

Lin Wan[1(✉)], Ju Zhou[1(✉)], and Bailing Zhang[1,2(✉)]

[1] Suzhou Aunbox Software Co., Ltd., Suzhou, China
654655902@qq.com, wanlin654655902@gmail.com,
bai_ling_zhang@hotmail.com

[2] School of Computer and Data Engineering, NingboTech University, Ningbo, China

Abstract. Layout analysis plays an important role in various document image processing tasks such as OCR and document understanding, and the methods based on deep learning have achieved significant achievements. In recent years, pre-training and transfer learning techniques have become a common practice in a variety of computer vision and natural language processing tasks. In this paper, we present an efficient approach of data synthesis for pretraining deep learning models in document layout analysis. The synthesized data is automatically annotated based on heuristic rules, and then applied to the PubLayNet pre-trained models. The models are fine-tuned with real document layout data. Three types of document elements are taken into account: text lines, tables, and figures/images. The experiments demonstrate that the pre-training model with synthesized data is very effective for transfer learning on different document domains.

Keywords: Layout analysis · Heuristic rules · Automatic annotations · Data synthesis · Fine-tuning

1 Introduction

Structural analysis and content extraction of electronic documents have become a research hotspot [1]. Electronic documents mainly include two categories: scanned or camera captured images of hardcopy documents, and computer-generated digital documents. With the development of optical character recognition (OCR) technology, automatic processing of documents has significantly progressed. An important step of an OCR system is the layout analysis of the document, which is a process of identifying and recognizing the logical and physical structure of digitized documents.

As a key step in document AI, traditional layout analysis is generally based on connected region analysis, which can be divided into two categories [2]: bottom-up methods [3, 4] and top-down methods [5], with different heuristic rules for labelling document images. Due to the lack of generally accepted customized rules and sufficient training data, these classical approaches are usually only applicable to some specific types of documents.

With the development of deep learning, convolutional neural networks have been widely used in document analysis and the advantage of pre-training and transfer learning has greatly promoted the progresses of document understanding. Earlier works

C. Pang et al. (Eds.): SETE 2020/ICWL 2020, LNCS 12511, pp. 244–252, 2021.
https://doi.org/10.1007/978-3-030-66906-5_23

often focused on detecting and analyzing some special parts of the document, such as the tabular area. Hao et al. [6] first proposed a table detection method using convolutional neural network (CNN). Later, more advanced models such as Faster R-CNN model [7] or Mask R-CNN model [8] were applied to document layout analysis, often with improved performance. Yang et al. [9] proposed an end-to-end multi-modal, fully convolutional network, which treated text semantic structure extraction as a semantic segmentation task. Kai Chen et al. [10] proposed a method for segmentation of handwritten historical document images, also based on convolutional neural net-work. Oliveira, D. A. B. [11] proposed a fast one-dimensional approach for automatic document layout analysis method based on Faster-RCNN, which takes text, figures and tables into account. D. He [2] proposed a method for the tasks of semantic page segmentation and element contour detection based on FCN and CRF and designed a novel way of synthesizing document.

Despite the achievement made in document layout analysis, the challenge remains: to accurately detect and classify document content into semantically meaningful categories. One of the biggest problems afflicting deep learning practitioners is that training deep neural networks models requires vast number of labeled datasets. Labelling the training samples is generally very time-consuming and costly. Some researchers have been conducted to alleviate the problem, for example, with the effort in areas of meta-learning, few-shot learning and self-supervised learning. Notwithstanding the progress, the availability of massive dataset is still the prerequisite of deep learning technology to be successfully applied.

In tasks where "off the shelf" datasets do not exist or accessing to real-world data is inconvenient or expensive, a possible solution is to use simulations to generate unlimited training data. Synthesis of data with some heuristic rules can be considered as a simplified way of simulating real-world dataset, which has been increasingly utilized recently [2, 12, 13]. Max Jaderberg et al. [14] use a synthetic text generation to produce highly realistic and sufficient synthetic data to replace real data, their experiments show that synthetic training data can be used for a real-world data problem. Inspired by this development, we proposed a data synthesis approach for document layout analysis.

Our method mainly includes three steps: 1) Synthesis of pseudo-document images with annotated information, with some heuristic rules; 2) Using the synthetic data to train the pre-trained deep neural network models provided in PubLayNet [15]; 3) Further fine-tuning the models with realistic document images with labelled layout information. With the proposed synthetic document images, the manual annotation costs can be considerably reduced. We experimented by first applying the synthetic images to train the pre-trained model from the PubLayNet, and then fine-tuning the models with real document images, achieving satisfying performances.

2 Methodology

Document layout analysis has always been a research hotspot. Early works on document physical layout analysis are based on heuristic rules [2–5]. These methods only rely on a small amount of manually annotated data, which have poor generalization ability and are only applicable to specific types of documents. At present, deep learning has been widely used in a different OCR or document image understanding tasks. Often, these deep neural network models rely on large number off real-world document images with appropriate annotation for different purposes. Realistic document images are not always available in real-world environments due to privacy or legal restrictions, and sometimes the annotations are not feasible (for example, too long or too expensive). And in many tasks, there are inherent data imbalances that most learning-based methods are difficult to deal with.

To avoid the problem of creating and labelling massive number of document images and thus overcome the development obstacle for deep learning models of document layout analysis, we dedicate to research of synthetic document images with simulated layout structures, i.e. the layouts not taken from the real document but artificially generated: the key idea is that with the comprehensive understanding and control of virtual document image layout structures, we can create "faked" document images with which to produce ground truth annotations. In another word, we take a surrogate approach of synthetic data generation to deal with massive image annotation problem in training deep neural network models.

In this work, we propose a novel method to simulate document layout structures and create an artificially annotated document image dataset. The synthetic data is subsequently applied to train an open source pre-trained models and transfer the text block labels to text line labels. Finally, a small real document layout ground-truth is used to fine-tune the model.

2.1 Data Synthesis

Synthesis of layout-realistic document image is not trivial as a number of requirements from real document layout structures have to be taken into account.

Collection of Document Images and Components. The first step is to collect document images with different layout structures and document components like figures and tables. We gathered 300 tables and 300 figures/images from part of real documents and different kinds of public tables and figures/images from the Internet. We select 2000 document images with Chinese text, which cover company documents, scientific papers, textbooks, newspaper and magazines, et al.

Data Preprocessing. The binarization method we use is local adaptive binarization [16], which does not need to set a fixed threshold. The threshold can be set adaptively based on the local characteristics of the image. Different brightness, contrast, and texture regions have different binarization thresholds. Firstly, the document image only with text is binarized, and the position information of the white area is extracted in the binarized image (white characters on black background), which corresponds to the text area in the original image. Then, the table images are binarized, and the position of the white area in the binary image is extracted (white text on black background), which corresponds to the text and table line area in the original table image.

Segmentation of Text Image with Rows and Columns. Firstly, the text binary image (white text on black background) is dilated once according to the scale (1, w/200) and eroded once, where w is the width of the document image, so that the text of the peers is combined. Then, the contour of the text line is detected to obtain the rectangle outside the contour, and the coordinates of the text line are obtained. Finally, the text line is projected to the vertical direction, the mean value of the projected histogram is calculated as doc_mean. We set the threshold thresh=doc_mean/20, and filter out the noise with smaller value in the histogram by the threshold. Finally, contour detection is then performed on the histogram image to get the coordinates of the text column and the corresponding column number.

Document Structural Components Replacement. For each column of text, we randomly select 0–2 structural components like image, figure, and table. The structural components replacement rules are as follows:

Algorithm 1 Document structural components replacement rules

Input: T_k: a subset of coordinate information of text lines of size m
W: the width of text field
H: the height of text field
T_k: a subset of coordinate information of text lines of size m
Output: A_k: a subset of coordinate information of element of size k
Initialize $p = 0$
for i=1:k **do**
  $\mathrm{H}_0 = W/E_i[0] * E_i[1]$
  //E_i represents the i-th element in set E_k
  $h_{num} = int(H_0/H)$
  if p+h_{num} <=M **then**
    q = randomint(p+1,M-h_{num})
    //randomint(a,b) Randomly select an integer between a and b
    p=q+h_{num}
    A_i=[T_p,T_q]
    //T_p,T_q are the coordinate information of p-th and q-th text lines respectively
  end if
end for

Algorithm 1. Document structural components replacement rules

The flowchart of the proposed data synthesis is further explained in Fig. 1. At first, a background image is randomly selected from the background dataset and resize to the size of the text image. After that, the position information of the text area is acquired by binarizing the text image, and the corresponding position pixels of the background image is then substituted with the text area pixels of the text image to obtain image A. And then, according to the structural components' replacement rules, if the illustration is an image or figure, the text block in the corresponding area in image A will be replaced with a picture/figure. If the illustration is a table, the text block in the corresponding area in image A will be replaced with table line and table text. We use this method to synthesize 10,000 layout analysis data with corresponding annotations.

Figure 2 shows some examples of our synthetic data.

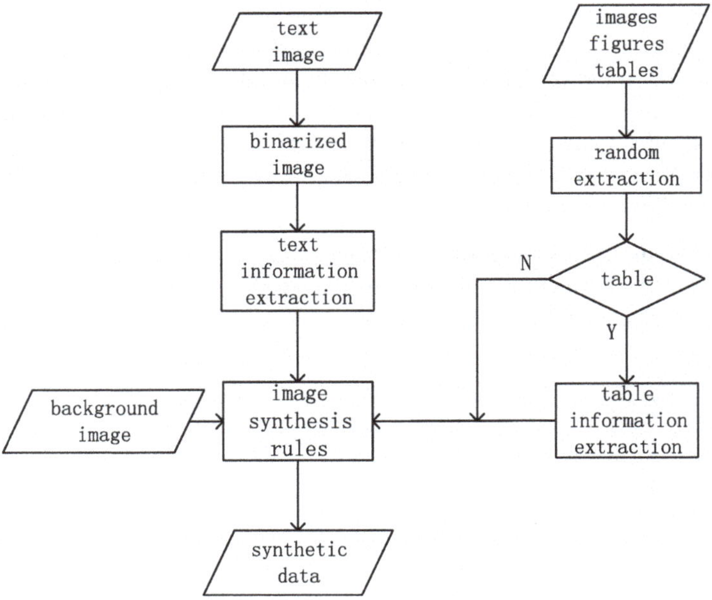

Fig. 1. Flowchart of data synthesis

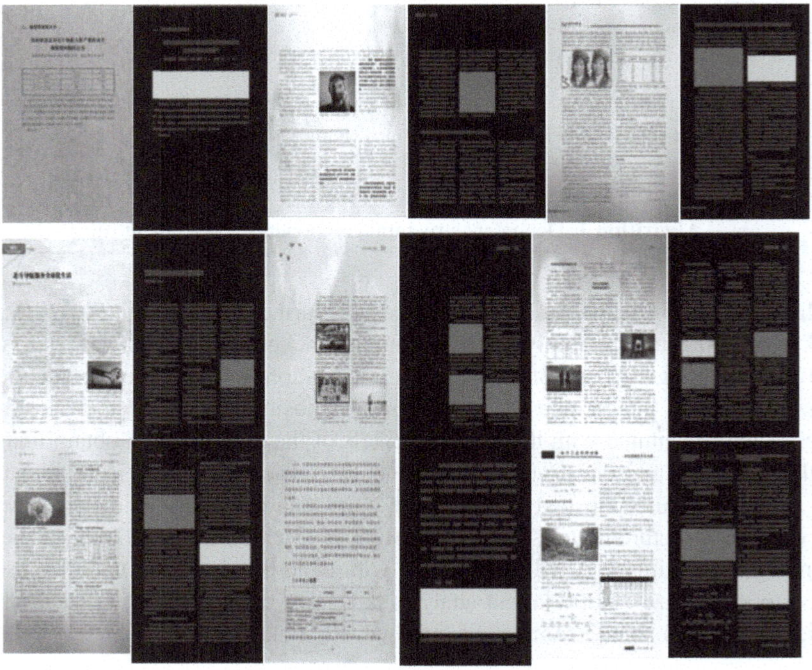

Fig. 2. Examples images from synthetic data

2.2 PubLayNet Pre-trained Model

PubLayNet is a high-quality document analysis dataset that contains more than 360,000 document images with annotations of typical document types, covering common document layout elements such as text blocks, titles, lists, and tables. The model pre-trained on the PublayNet data can be effectively transfer on other layout datasets [15].

2.3 Network Architecture

We use Faster-RCNN and Mask-RCNN as the baseline. Due to the success of ImageNet and COCO competition, Faster-RCNN and Mask-RCNN have been widely used in the field of computer vision. [17] In 2013, Girshick et al. first proposed RCNN to solve the problem of target detection. After that, Girshick proposed the Fast-RCNN [18] model, which improved the training speed and detection accuracy based on RCNN. Both of RCNN and Fast-RCNN use selective search to obtain candidate regions, which is very time-consuming. Therefore, Faster-RCNN [19] introduced a regional proposal network (RPN), which shares image convolution features with the detection network. In addition, the model shares the convolutional features of RPN and Fast-R-CNN, and merges them into a network to enable the network to be trained in an end-to-end manner. In 2017, Mask R-CNN [20] added a branch of prediction target binary mask parallel to regression of the existing target detection frame on the basis of Faster-RCNN.

3 Experiments

We used the Faster-RCNN and Mask-RCNN models trained on the PubLayNet dataset by Xu Zhong et al. and changed the model output from 6 categories to 4 categories. Both models use ResNeXt-101-64 * 4d model as the backbone. The models are trained on a TITAN-RTX GPU with 24 GB memory. Two different schemes are used to predict the real data. The evaluation metric is the mean average precision (MAP) @intersection over union (IOU) [0.50: 0.95] of the bounding boxes which is used in the COCO competition.

The synthetic document images are partitioned into 4800 training sets and 1200 testing sets according to 4: 1. The real document images are divided into 740 training sets and 187 testing sets. We used the following two schemes to train these models.

Scheme One. We used real data to train the models pre-trained on PubLayNet. Each model was trained for 20 k iterations with the initial learning rate of 0.001. at 15 k iteration, the learning rate was reduced by 10 times.

Scheme Two. We used synthetic data to train the PubLayNet pre-training models (Faster RCNN and Mask RCNN). Each model was trained for 50 k iterations with a base learning rate of 0.001. at 35 k iteration, the learning rate was reduced by 10 times. Then, we used the real data to fine-tune the Faster-RCNN and Mask-RCNN models for 10 k iterations with an initial learning rate of 0.001, and at 8 k iterations, the learning rate was reduced by 10 times.

The performance of the two schemes on the testing sets of real data is shown in Table 1. Figure 3 and Fig. 4 are some examples of Faster-RCNN and Mask-RCNN from the testing set of real data.

Table 1. MAP@IOU[0.5:0.95] of the F-RCNN and M-RCNN on our test sets

Category	Data	F-RCNN	M-RCNN
Text	Real data	53.9	55.1
	Synthesis and real data	77.1	77.8
Image	Real data	60.5	70.3
	Synthesis and real data	69.4	71.2
Table	Real data	89.5	97.0
	Synthesis and real data	96.8	97.8
Average	Real data	68.0	74.1
	Synthesis and real data	77.1	77.8

Fig. 3. The results achieved using our approach based on F-RCNN

Fig. 4. The results achieved using our approach based on M-RCNN

Table 1 summaries the performance comparison of fine-tuning different pre-trained models (F-RCNN and M-RCNN) on synthetic and real document data. The results show that synthetic training data can be used for document layout analysis and using synthetic and real document data to train these models is better than only using real document. Additionally, Mask-RCNN shows a small advantage over Faster-RCNN.

Figure 3 and Fig. 4 are the renderings of some real and complex layout testing sets. It can be seen that both models can generate accurate (MAP > 0.7) document layouts.

4 Conclusion

With the popularity of mobile devices and cloud services, the demand for intelligent document processing and understanding has kept increasing. The layout analysis of document image is often an important part of document AI system, e.g., OCR, document retrieval, and document generation. Deep learning has become the primary choice in the development of document layout analysis. To overcome the burden of creating high volumes of annotated document images for training deep neural networks, we propose a synthetic document image for layout analysis, the method synthesizes a large amount of layout data based on heuristic rules and automatically generate labels. We used the synthetic data to train the PubLayNet pre-trained model and changed the output of model from 6 categories to 4 categories, so that text block labels are transferred to text-line labels. Then a small amount of real document images is used for fine-tuning the model. Experiments have shown that our transfer learning is a promising approach. Synthetic data can help the model transfer different document domains.

References

1. Yiheng, X., Minghao, L., Lei, C., et al.: LayoutLM: Pre-training of text and layout for document image understanding. https://arxiv.org/abs/1912.13318, arXiv (2009)
2. Dafang, H., Scott, C., et al.: MultiScale multi-task FCN for semantic page segmentation and table detection. In: 2017 14th IAPR International Conference on Document Analysis and Recognition (ICDAR), IEEE Computer Society, Kyoto, Japan(2017)
3. O'Gorman, L.: The document spectrum for page layout analysis. Pattern Anal. Mach. Intell. IEEE Trans. **15**(11), 1162–1173 (1993)
4. Kise, K., Sato, A., et al.: Segmentation of page images using the area voronoi diagram. Comput. Vis. Image Underst. **70**(3), 370–382 (1998)
5. Nagy, G., Seth, S., Viswanathan, M.: A prototype document image analysis system for technical journals. IEEE Comput. **25**(7), 10–22 (1992)
6. Leipeng, H., Liangcai, G., et al.: A table detection method for pdf documents based on convolutional neural networks. In: 2016 12th IAPR Workshop on Document Analysis Systems (DAS), pp. 287–292. IEEE, Santorini, Greece (2016)
7. Shaoqing, R., Kaiming, H., et al.: Faster-RCNN: towards real-time object detection with region proposal networks. IEEE Trans. Pattern Anal. Mach. Intell. **39**(6), 1137–1149 (2017)
8. Kaiming, H.,Georgia, G., Piotr, D., et al.: Mask R-CNN. IEEE Trans. Pattern Anal. Mach. Intell. **42**(2), 386–397 (2017)
9. Yang, X., Yumer, E., et al.: Learning to extract semantic structure from documents using multimodal fully convolutional neural networks. In: 2017 IEEE Conference on Computer Vision and Pattern Recognition (CVPR), IEEE, Honolulu, HI, USA (2017)

10. Kai, C., Seuret, M.: Convolutional neural networks for page segmentation of historical document images. In: 2017 14th IAPR International Conference on Document Analysis and Recognition (ICDAR). IEEE, Kyoto, Japan (2017)
11. Viana, P.M., Oliveira, D.A.B.: Fast CNN-based document layout analysis. In: IEEE Conference on Computer Vision and Pattern Recognition, pp. 1173–1180. IEEE Computer Society, Venice, Italy (2017)
12. Nikolenko, S.I.: Synthetic Data for Deep Learning. https://arxiv.org/abs/1909.11512. arXiv (2019)
13. Kamilaris, A., van den Brink, C., Karatsiolis, S.: Training deep learning models via synthetic data: application in unmanned aerial vehicles. In: Vento, M., Percannella, G., Colantonio, S., Giorgi, D., Matuszewski, Bogdan J., Kerdegari, H., Razaak, M. (eds.) CAIP 2019. CCIS, vol. 1089, pp. 81–90. Springer, Cham (2019). https://doi.org/10.1007/978-3-030-29930-9_8
14. Max, J., Karen, S., et al.: Synthetic data and artificial neural networks for natural scene text recognition. In: Computer Vision and Pattern Recognition (2014)
15. Zhong, X., Tang, J., et al.: PubLayNet: largest dataset ever for document layout analysis. https://arxiv.org/pdf/1908.07836v1. arXiv (2019)
16. Romen, .S.T., Roy, S., Imocha, S.O., et al.: A new local adaptive thresholding technique in binarization. Int. J. Comput. **8**(6) (2011)
17. Girshick, R., Donahue, J., et al.: Rich feature hierarchies for accurate object detection and semantic segmentation. https://arxiv.org/abs/1311.2524. arXiv (2014)
18. Girshick, R.: Fast R-CNN. In: Proceedings of the IEEE International Conference on Computer Vision, pp. 1440–1448 (2015)
19. Ren, S., He, K., Girshick, R., et al.: Faster R-CNN: towards real-time object detection with region proposal networks. IEEE Trans. Pattern Anal. Mach. Intell. **39**(6), 1137–1149 (2017)
20. Kaiming, H., Georgia, G., Piotrk, D., et al.: Mask R-CNN. IEEE Trans. Pattern Anal. Mach. Intell. 2980–2988 (2017)

Education Technology and ICT for Education

An Experience of Teaching Advanced Control Engineering (ACE) for Postgraduate Students

Aberham Genetu Feleke and Luzheng Bi(✉)

Beijing Institute of Technology, Beijing 100081, China
bhxblz@bit.edu.cn

Abstract. Control engineering is a key engineering discipline to design systems with desired behaviors in control environments. Its application area is so vast that many fields of engineering students are taught the course. This paper presents the experience of teaching advanced control engineering for postgraduate students of mechanical engineering. It presents the methodologies employed by the instructors while they teach the course. It discusses the presented methodologies in comparison with the established principles of best teaching methodologies. This paper pinpoint student's lack of concentration or attention as the major problem associated with the existing teaching methods. Hence, a flexible teaching method such as online learning or hybrid learning was proposed to engage and motivate students' initiative.

Keywords: Advanced control engineering · Teaching methodology · Online learning · Hybrid learning

1 Introduction

Control engineering is a key engineering discipline that focuses on the design of controllers that will cause these systems to behave in the desired manner. Control system application ranges from simple such as home heating controller system to a complex industrial control system. It utilizes sensors to measure the output performance of the process being controlled; in most cases, these measurements are used to provide corrective feedback helping to achieve the desired performance.

According to [1] the development in control engineering methods is recognized by four stages. i.e., Conventional control strategies, Advanced Control I (classical), Advanced Control II, Advanced Control III (intelligent systems). In conventional control strategies, the majority of control systems rely on various forms of the ubiquitous proportional integral derivative (PID) controller which is simple and the most commonly used in the industry. However, there are cases where all control systems cannot be controlled with PID. Multivariable, nonlinear, and random processes require more advanced control techniques. Some of the methods at each stage of control engineering development are shown in Fig. 1.

The students who are taught of the subject need to know the mathematics behind the control concepts and develop the skill to implement the principles in real problems. The control engineering course should be delivered to students in the way to provide a synergetic integration of both theoretical and practical knowledge [2–5]. To solidify the

C. Pang et al. (Eds.): SETE 2020/ICWL 2020, LNCS 12511, pp. 255–264, 2021.
https://doi.org/10.1007/978-3-030-66906-5_24

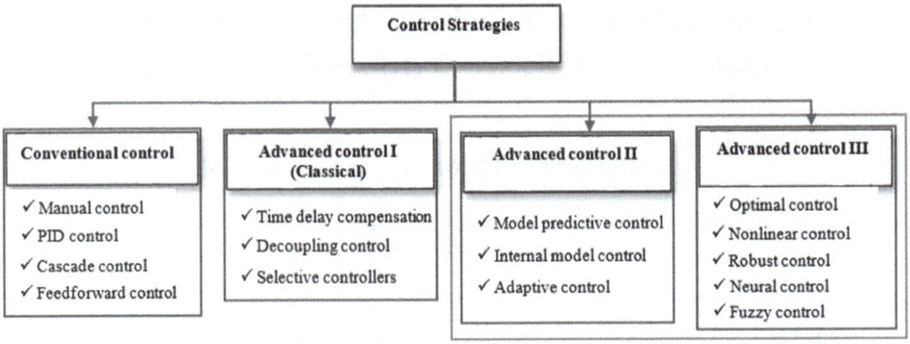

Fig. 1. Structure of control engineering course and some of its contents

theoretical concepts with practice, an experimental study is usually used in control education. The development of information and communication technology (ICT), creates new possibilities and expectations for teaching. In this regard, E-learning is a new trend. ICT in school learning can help improve the standards by improving the quality of teaching, management, and learning [6, 7]. E-learning can offer a suite of online learning opportunities for students. There is a growing acceptance of the potential of virtual and remote laboratories to enhance student engagement/learning and to supplement formal laboratory sessions [8–12].

Content Knowledge, Pedagogical Knowledge, and Technology Knowledge are the three core domains of a framework to understand and describe the kinds of knowledge needed by a teacher for effective pedagogical practice [6, 13]. Technological Pedagogical Content Knowledge (TPACK) which is the integration of the knowledge from the three domains, brings a deeper and richer understanding of the subject. It enables the students to understand the concepts of the subjects by bringing new technologies to deliver the contents of the subjects. The teacher will need to move away from over-reliance on the conventional didactic lecture to a more blended delivery style, making use of a variety of teaching techniques and resources [14–16].

The introduction of computers, the internet, software tools, and multimedia presentations have brought new teaching strategies for control engineering. Interactive simulation instruments such as Matlab/Simulink and LabVIEW are playing an important and increasing role in advanced control engineering education [17]. The use of software simulation tools can reduce the burden of expensive laboratory hardware and provide unique visualization and animation capabilities [18].

It is expected that the postgraduate students have some basic knowledge of control engineering in their undergraduate studies. Method of teaching, evaluation criteria, and feedback mechanism affects the effectiveness of teaching a particular course. The integration of research experience into the course structure enables students their problem-solving capacity. Moreover, the use of available technologies in teaching methodology enhances the quality of education. This paper is organized as follows. Section 2 describes the teaching methodology in the advanced control engineering course. Next, the discussion on the teaching experience of the advanced control engineering in comparison with the literature on the teaching principles is presented in Sect. 3. Section 4 presents the conclusion of the paper.

2 Teaching Methodology

The graduate students of mechanical engineering who taught the course are engaged in research activities. Therefore, the main aim of the course is to provide the students with the advanced control engineering principles so that they use it in their research works. In this regard, the teaching goals are the following; Provide the students with a brief review of the background information, provide the fundamental principles or theoretical concepts of the subject, introduce tools to implement the solutions and apply the topics in research problems. Postgraduate students require a deep understanding of engineering principles and problem-solving skills to be successful in their research activities. To achieve this, three key areas that need to be studied in the course are theory, implementation, and application. Figure 2 Shows how these three key areas are structured and integrated into the course.

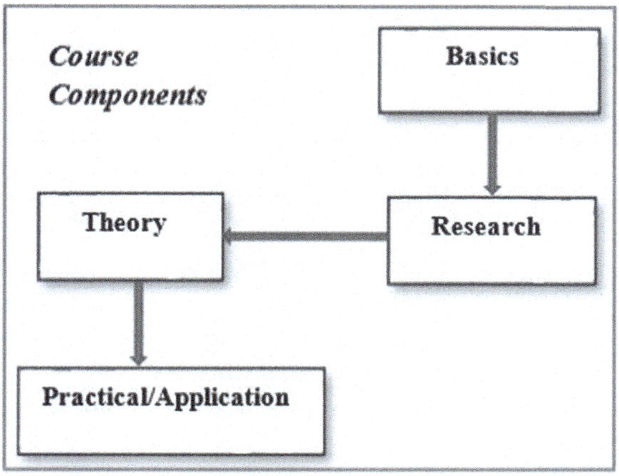

Fig. 2. Components of the course

In Fig. 2. the basics include the fundamental principle in control engineering. The research experience and findings of the teacher are integrated into the course. The finding from the research endeavor and the principles drawn enhance the scope of the contents. The application of the theory or findings in engineering is presented. Students also need to understand the significant theoretical concepts and be able to combine different concepts to find a solution for the given problem. Besides the theoretical concepts, another significant field of knowledge is how the theoretical concepts are implemented. Nowadays, implementation is usually done by computers which makes the management of some coding language essential. The last important field of knowledge is how to apply the learned concepts in real-life problems. Typically, real-life problems are complex and require a combination of different methods to find a suitable solution.

The courses focus on basic principles, research works, theory, and implementation. These components of the courses are covered in lectures, research article, oral presentations and case studies. The course schedule includes 32 h of guided teaching, consisting of lectures, demonstrations, exercises, and oral presentations.

2.1 Course Contents

To take this course, students should know the fundamentals of control Engineering or Automatic control theory as a prerequisite. After successful completion of this course, it is expected that students understand the characteristics of dynamic systems and feedback control; be familiar with classical methods for analyzing control system accuracy, stability, and dynamic performance; understand advanced control theory, including robust control, fuzzy logic control, and neural network control; be familiar with control system design methods. Table 1 shows the contents of the course.

Table 1. Course contents.

Contents	Topic
Introduction to Control Engineering	Fundamentals of control systems, Control system design
Mathematical Modeling of Systems	Differential equations of physical systems, Transfer function of linear systems, Block diagram models and Signal-flow graphs, State variable models and Sequential design
Time Domain and Frequency-Response analysis	Feedback control system characteristics, Dynamic performance, Steady-state error and concept of stability and Routh's stability criterion, Root locus method and Frequency response method
State-Space Methods for Control Systems Design	State-space methods, Control system design
Digital Control Systems	Shannon's sampling theorem, Z Transform, Digital control systems, Stability in the Z-plane, Digital compensator design
Sliding Mode Control	Fundamentals of sliding mode control, Mathematical description of sliding mode, Sliding mode control design
Model Predictive Control	The conceptual idea of model predictive control, optimal control synthesis, stability, and some application of model predictive control
Robust Control Systems	Robust control system design
Fuzzy Logic Control Systems	Introduction to fuzzy logic, design of fuzzy logic control systems, design of fuzzy logic control systems, and application of fuzzy logic control systems
Neural Network Control Systems	Introduction, Network architecture, Training neural network, and control system application

The course focuses on developing students' core skills and knowledge of control and systems. Students also put their engineering knowledge and skills into practice, applying them to a specific problem that interests them, as part of their research and dissertation.

2.2 Instructional Methods

The teaching room is equipped with adequate facilities for teaching such as computers, overhead projector, and writing board. The number of students is usually about 30. Therefore, the room size is appropriate for the specified number of students. Lecture, case studies, and journal reading are the major instructional method.

Lecture. Lecture is the main instructional method used in this course. The instructors use a PowerPoint presentation that is projected to the class by an overhead projector. Whenever necessary they use a whiteboard to explain and prove some mathematical algorithms. Moreover, the instructors present research articles published in reputable journals to provide students with research experience. The students ask, argue, and discuss each topic with the instructors and among themselves.

Case Studies. At the end of each chapter, instructors prepare case studies or exercise that students are expected to analyze and respond to the given scenarios or problems. In the next class, a specific number of volunteer students present their approach to the problem. The lectures initiate discussion on the presented solution for the given case studies.

Journal Reading. A selected topic from the course is assigned to each student. The students are required to review a journal article from top journals. Then, each student presents the result of the review to the class by using PowerPoint. There will be a question and answer session at the end of each presentation.

2.3 Evaluation

Assignments or case studies and oral presentations are used as the major evaluation criteria in this course. Sometimes the active participation of a student is rewarded and considered in the cumulative evaluation of the student. Table 2 shows the evaluation criteria along with their weight in percentage.

Table 2. Course contents.

Evaluation criteria	Weight in percentage
Assignments	45
Oral presentation	55
Total	100

Assignments. Usually one or two problems from a given topic are given to each student. Every assignment is associated with a deadline for submission, which varies from two to three weeks depending on the problem. Mostly, there is 5 such assignment for the course.

Oral Presentation. Each student is required to search for a journal article related to optimal control, model predictive control, robust control, neural control, or fuzzy control. Then, they conduct a review and prepare some slides to give an oral presentation on the paper at a designated time. Each student should submit his/her presentation and paper files via E-mail at least five days before his/her presentation. The presentation time allocated to each student is 15 min, which is followed by 5 min of question and answering session. Table 3 shows the oral presentation topic and the proportion of students assigned to a given topic.

Table 3. Presentation Topic.

Presentation topic	Proportion of Students assigned to the topic
Optimal control	15%
Robust control	15%
Fuzzy control	15%
Neural control	15%
Model predictive control	20%
Sliding mode control	20%

2.4 Teaching Resources

Textbooks and reference books are recommended to the students at the very beginning of the course. The textbooks and references are stored in a public e-mail. The materials are uploaded on the public e-mail (https://mail.qq.com) and students can download it from anywhere. Additional teaching resources delivered through this medium have text, PowerPoint presentations, assignments, case studies, and journals.

3 Discussion

This section discusses the experience of the authors while teaching the course in comparison with existing best practices of teaching methodologies. The comparison is mainly qualitative analysis, which is based on observation and literature review. The best experiences of the authors are presented along with some identified shortcomings.

3.1 Teaching Methods and Resources

Classroom. The layout of the classroom (i.e., the physical setup of chairs, tables, and presentations) can significantly influence learning. The arrangements of the seat determine the level of interaction that students have with the instructor and as well as among themselves. Moreover, students may prefer a particular seating arrangement. In postgraduate classes, communication between students and instructors is expected to be high. Therefore, a more flexible seating arrangement can be needed. The number of students in this course is usually about 30. The size of the room is not large and the seat arrangements are also in the way to enhance two-way communication.

Course Contents. Course contents might be taken from course catalogs of a specific program. However, it should be flexible to include current trends. Therefore, in this course, the contents are frequently updated to include current topics of engineering and technology. Especially much emphasis is given on intelligent controls, as the progress in the field is changing very fast.

Lecture. Research on teaching methodologies shows that lecture needs to build a bridge between students' knowledge base and subject matter of the lecture. Therefore, it is important to find out what students already know by collecting background information, review of the background information, and connect it with the lecture, use relevant examples related to students' experiences.

Moreover, it is beneficial for students if fewer points are presented for effective teaching. presenting too much material for a given class overloads students' information processing capacity. Students learn more effectively when they are actively engaged than when they passively receive information. In this regard, the discussion session, question, and answering session and oral presentation by students have greatly enhanced the effectiveness of the teaching.

Integration of Research Experience. Integrating the research experiences of the teacher into the contents of the course enables the students to grasp the reality of the principles of the subjects. Teachers can easily teach and cover the contents of the course from their vast research experience. It motivates the students to solve engineering problems with the acquired theoretical and practical foundation.

Therefore, the integration of the instructors' research works in this course is the best experience of teaching methods. At the end of each chapter, a published article of the instructors is presented to the students. The discussion initiated by the instructors can motivate the students to observe the problems of their surroundings in association with the theory that they have learned.

Demonstration. The courses can benefit from using virtual laboratories to support teaching activities. Such a system can be an effective tool to be used in practical classes and to enhance students' experimental skills. Multimedia activities included short videos that were very effective at illustrating concepts of the course and can be inserted at several points in the class. The videos can provide a simulation of a process that the students are unable to visit due to time and other constraints. In this regard, the authors use several demonstration videos in the class.

Evaluation Criteria. Evaluation criteria are used to evaluate the attainment of student outcomes. Effective assessment uses relevant direct, indirect, quantitative, and qualitative measures as appropriate to the outcome being measured.in this regard oral presentation and assignments are used in this course.

Oral Presentation. The students present a review of the journal article orally. The presentation is formal and can take 15 min of class time. The oral presentation enables the students not only to grasp the theoretical foundation of the course but also to gain critical reasoning. The discussion after presentation widens the understanding of the students from a multitude of directions. Moreover, it maintains the students' concentration on the course.

Case Studies or Assignments. Case studies or assignments can be selected from a book or journal article. Students are given adequate time to respond to case studies or to solve assignment problems. Since every student has their unique approach to solve a given problem, the presentation of the results initiates further discussion on the topic. There are many other possibilities as well. The goal is to enhance a critical review of the students that may help then in their research works.

Teaching Feedback. It is important to request feedback from students attending the courses on their learning outcomes and their satisfaction with various elements of the courses. Responding to students' feedback is effective in improving both student satisfaction and learning. In this regard, the instructors request students' feedback and respond accordingly.

3.2 Challenges and Possible Solutions

Challenges. Because of the increase in mobile technology, students might be engaged in tasks unrelated to class activities such as surfing, chat, access games, or be online during the lectures. The action of students involving in these technologies can cause disruptions inside the classroom that is considered as negative consequences.

A student's attention may shift in many ways while sitting in the classroom and taking lectures. Daydreaming, sleeping, looking, or gazing at windows are some of the examples of these activities. Students come to lectures with different levels of knowledge on the topic. Finding out what they already know to identify and correct student misconceptions are challenging.

Possible Solutions. To overcome some of the challenges that are associated with the teaching of advanced control engineering for postgraduate students the following solutions are suggested.

Online Teaching. Technology is the predominant factor for online teaching. Technologies that allow instructors to maintain audio and video control while giving students a chance to ask questions and engage in discussion using a live chat are common. There are several online teaching methods such as web conferencing tools, streaming video platforms, and live chats. First, the instructors should identify what they want to teach, and through what methods. Then, they can review the online learning technologies available and identify those that suit their course objectives.

Moreover, online teaching demands more independence, so students can learn at least some of the material by themselves. They can easily access downloadable lectures, Presentations, and assignments virtually from anywhere. One of the features that makes online teaching unique from the class-based lecture is that students do not necessarily have to report to class during scheduled sessions. Even though it is more flexible, the accessibility of technology from the students' side should be given due consideration.

However, not all online courses are structured the same way, where some allow a great deal of scheduling flexibility, others strive to recreate a more traditional campus-based course environment online. Professors must consider these factors when choosing instructional tools and methods.

Hybrid Learning. Hybrid learning combines various learning models, such as traditional learning and e-learning modeling. Traditional learning (face to face) in this study uses a direct learning model and task based learning. E-services are designed to maximize the results of traditional learning. The influence of technological development, then Hybrid learning also developed by combining traditional learning with learning that utilizes information and communication technology.

Make the class more interactive so that students are active. Therefore, the lecture should be designed in the way to enhance the communication between teacher and students. Moreover, the presentation should be made in a way to enhance visual attraction.

4 Conclusion

This paper discussed the experience of teaching advanced control engineering courses for postgraduate students. The teaching methodologies employed by the authors are presented. The presented methodologies are discussed in comparison to the best experience with in the field. Some of the challenges of the teaching process concerning students and delivery methods are discussed. Lack of concentration or attention from the students' side is identified as major challenges. To deal with such challenges online teaching or hybrid teaching is recommended as a possible solution. Generally, to improve the teaching quality of advanced control engineering course for postgraduate students, a flexible teaching methodology should be employed. The introduction of flexible teaching methods can help students to learn this course deeply and create a positive teaching atmosphere.

Acknowledgement. This work was supported in part by National Natural Science Foundation of China under Grant 51975052.

References

1. Kozák, S.: Advanced control engineering methods in modern technological applications. In: International Carpathian Control Conference (ICCC), pp. 392–397. IEEE (2012). https://doi.org/10.1109/CarpathianCC.2012.6228674
2. Precup, R.E., Preitl, S., et al.: Experiment-based teaching in advanced control engineering. IEEE Trans. Educ. **54**(3), 345–355 (2011). https://doi.org/10.1109/TE.2010.2058575

3. Zenger, K.: Challenges and new directions in control engineering education. In: Proceedings of the 9th EUROSIM Congress on Modelling and Simulation (EUROSIM 2016) and the 57th SIMS Conference on Simulation and Modelling (SIMS 2016), pp. 819–823. University Electronic Press, Linköping (2018). https://doi.org/10.3384/ecp17142819
4. Miller, A.S., Singhose, W., Glauser, U.: Integrating PLC theory and programming into advanced controls courses. In: Proceedings of the American Control Conference (ACC), pp. 7302–7307. IEEE (2016). https://doi.org/10.1109/acc.2016.7526825
5. Dittmar, R., Kahlcke, T.: A lab for undergraduate control engineering education equipped with industrial distributed control systems. Comput. Appl. Eng. Educ. **24**(2), 288–296 (2015). https://doi.org/10.1002/cae.21708
6. Jwaid, A.E., Clark, S., Ireson, G.: Understanding best practices in control engineering education using the concept of TPACK. In: 2014 IEEE Integrated STEM Education Conference. IEEE (2014). https://doi.org/10.1109/isecon.2014.6891027
7. Ravishankar, J., Epps, J., Ambikairajah, E.: A flipped mode teaching approach for large and advanced electrical engineering courses. Eur. J. Eng. Educ. **43**(3), 413–426 (2018). https://doi.org/10.1080/03043797.2017.1383974
8. Smutny, L., Farana, R.: The consortial approach to advanced control laboratory education. In: Proceedings of the 4th Iasme/Wseas International Conference on Engineering Education, pp. 211–213. WSEAS Press (2007)
9. Heradio, R., Torre, L., Dormido, S.: Virtual and remote labs in control education: a survey. Ann. Rev. Control **42**, 1–10 (2016). https://doi.org/10.1016/j.arcontrol.2016.08.001
10. Rampazzo, M., Cervato, A., Beghi, A.: Remote refrigeration system experiments for control engineering education. Comput. Appl. Eng. Educ. **25**(3), 430–440 (2017). https://doi.org/10.1002/cae.21810
11. Hu, W., et al.: Plug-in free web-based 3-D interactive laboratory for control engineering education. IEEE Trans. Indu. Electron. **64**(5), 3808–3818 (2017). https://doi.org/10.1109/TIE.2016.2645141
12. Matijevic, M.S., Jovic, N.D., Nedeljkovic, M.S., Cantrak, D.S.: Remote labs and problem oriented engineering education. In: 2017 IEEE Global Engineering Education Conference (EDUCON), pp. 1391–1396, IEEE (2017). https://doi.org/10.1109/EDUCON.2017.7943029
13. Koehler, M.J., Mishra, P., Cain, W.: What is technological pedagogical content knowledge (TPACK)? J. Educ. **193**(3), 13–19 (2013). https://doi.org/10.1177/002205741319300303
14. Hoernicke, M., Horch, A., Bauer, M.: Industry contribution to control engineering education: an experience of teaching of undergraduate and postgraduate courses. IFAC-PapersOnLine **50**(2), 133–138 (2017). https://doi.org/10.1016/j.ifacol.2017.12.025
15. Rossitera, J.A., et al.: A survey of good practice in control education. Eur. J. Eng. Educ. **43**(6), 801–823 (2018). https://doi.org/10.1080/03043797.2018.1428530
16. Juuso, E.K.: An advanced teaching scheme for integrating problem-based learning in control education. Open Eng. **8**(1), 41–49 (2018). https://doi.org/10.1515/eng-2018-0006
17. Benyo, I., Lipovszki, G., Kovacs, J.: Advanced control: simulation tools in LabVIEW environment. IFAC Proc. **36**(10), 237–241 (2003). https://doi.org/10.1016/S1474-6670(17)33686-8
18. Koku, A.B., Kaynak, O.: An internet-assisted experimental environment suitable for the reinforcement of undergraduate teaching of advanced control techniques. IEEE Trans. Educ. **44**(1), 24–28 (2001). https://doi.org/10.1109/13.912706

Modelling Change Towards Hybrid Learning

Johannes Petrus Rossouw(✉), Machdel Matthee, and Lizette Weilbach

Department of Informatics, University of Pretoria, Pretoria, South Africa
jp.rossouw@mweb.co.za, {machdel.matthee, lizette.weilbach}@up.ac.za

Abstract. Schools are often confronted with the complexities of change when implementing hybrid learning. Extant change theories provide a multitude of strategies to deal with change. As many of these strategies approach change in different ways, they often portray conflicting or opposing views. For schools to institutionalize hybrid learning, they require a unified view. This paper presents such a unified view of change and shows the development of a context-based change meta-model towards enabling schools to change to hybrid learning.

The change constructs making up extant change strategies were analysed to determine their scope and the way in which they are structured.

In combining a theoretical and real-life perspective, different modes of analysis and methodologies were applied to develop the change metamodel. This included a theoretical analysis of change constructs and their underlying structures as evident in extant literature; analyzing a real-life case; and integrating the theoretical view and the real-life view to mold the change metamodel.

Keywords: Change · Education · Metamodel · Hybrid learning

1 Introduction

Changes in the education domain has led to the concept of hybrid learning environments. Hybrid instructional models represent a mixture of face-to-face classrooms and concepts of e-learning [1]. Learners can normally adapt to such a hybrid approach where they still attend classes, while utilizing the courseware typical to e-learning.

E-learning represents a blanket term for different learning classifications. Anohina (2005) classifies e-learning as the collective of online learning, internet-based learning, web-based learning and computer-based learning. Irrespective of the method of e-learning, it presents unscaled change to schools [2]. Research has shown that educational institutions are not suitably prepared for new modes of education, as is also evident from the disruption caused by the COVID-19 pandemic [3]. This change must be managed to ensure the optimal outcome. When changing to hybrid learning, schools should adopt multiple change approaches based on models and theories [2].

C. Pang et al. (Eds.): SETE 2020/ICWL 2020, LNCS 12511, pp. 265–276, 2021.
https://doi.org/10.1007/978-3-030-66906-5_25

2 The Change Management Problem

This paper proposes a change metamodel to enable schools to analyze and structure change constructs towards developing a change model for their local school. In this section an overview of existing change models and change theories in both the education domain and the organisational domain is provided. Organizational change models were included since schools share some characteristics with business organisations. [4] These include power structures, organisational culture, managerial issues, formal and informal leadership, and managing processes to achieve organisational objectives [5–7].

A search of extant literature resulted in 24 change models/theories from the education domain and 12 models/theories from the organisational domain being considered for this paper. An analysis of the underlying strategies of each of these 36 models/theories showed that, although there are some similarities, a substantial variety of different change strategies became evident: 1) Change as a process – to transition or transform from a current situation (e.g. as-is or traditional practices) to an envisaged adopted situation [8–16]; 2) Change as a system – of components [17–19] inclusive of the different stakeholders [20–23], learning services and dimensions [24], organisational/structural arrangements [16, 25] and a holistic approach to change as a system [19]; 3) Change resulting from innovation and new technology – [26–34] and 4) Behavior based on context and types of change – [35, 36] inclusive of resistance to change [37], diffusion of change [38, 39] and adoption of change [12, 39].

These change strategies make the success rate for the implementation of change highly variable. In this regard, schools are faced with the challenge of deciding on an appropriate change strategy or model to address their local situation when contemplating change towards the implementation of technology related to establishing a hybrid learning ecosystem for their school.

Van der Waldt [40: 5] asserts that, although models imply a structure of relationships of variables, it is still characterised as incomplete "with different meanings in different contexts. …" and that "some dimensions or elements of a phenomenon can be over-emphasized, leaving the impression that it is more important/significant than other elements or dimensions".

Managing and understanding change in schools because of technology implementation appear to have all the elements of an ill-structured, wicked problem where the solution is not obvious. Researchers in the Information Systems field ascribe the messiness of such problems to the embeddedness of IS (e.g. e-learning systems) in social contexts. One particular school of thought considers Information Systems (IS) research (e.g. technology-induced change in schools) to be socio-technical by nature as opposed to the techno-centric nature of computer science [41]. The problem that should be addressed is therefore: the lack of a unified socio-technical view of the multitude of change concepts (as presented in the change models discussed above) and the need to encompass the different types of change while taking into account the migration (transition and transformation) approach towards hybrid learning. To address this problem, this paper proposes a change metamodel which schools can use and customise into a change model for their own situation. To reach this goal, it is firstly necessary to identify the constructs of a socio-technical change model towards hybrid learning, and secondly, to determine how these constructs should be structured.

3 Research Approach

With the Change Metamodel as the envisaged outcome, the step-by-step metamodel building process of Ribino et al. [42] was applied. These steps are:

- Step 1: Identify change constructs, comprising the identification of suitable change theories and change models towards composing a comprehensive knowledge base of change concepts.
- Step 2: Determine the basic change constructs (using basic concepts of an organisational change management model (PSIC)), qualitative inductive content analysis as a meta-analysis of categories and concepts to determine similarities amongst the different models/theories. Executing steps 1 and 2 produced the concept catalogue.
- Step 3: Identify relationships, comprising an explanation of the approach to consolidate the constructs.
- Step 4: Produce a theoretical abstraction of the change models based on the underlying structures that emerged from extant literature.
- Step 5: Construct a practical view of change models, comprising an analysis of the rich pictures and the theoretical view. Executing this step produced the two conceptual models.
- Step 6: Compose a change metamodel – towards a unified view. Executing step 6 comprised composing snippets that were reduced and repackaged as a metamodel.
- Step 7: Construct a metamodel description, comprising an explicit description of the Change Metamodel in the form of a narrative.

Each of these steps will be discussed in more detail in the sections to follow.

4 Development of the Change Metamodel

4.1 Step 1: Identification of Change Constructs

Inductive content analysis [43], as a qualitative data meta-analysis method, was used to extract key constructs from each of the 36 existing models and theories analysed. The analysis comprised open coding through formulating preliminary codes, coding the data, and developing the different categories.

This was followed by categorizing these constructs according to a set of basic concepts deduced from an organisational change model, the Punctuated Socio-technical Information System Change (PSIC) model of Lyytinen and Newman [17]. These basic concepts are explained in the next section.

4.2 Step 2: Determining the Basic Constructs Using PSIC

The PSIC model of Lyytinen and Newman [17] was opted for, firstly, because it "… affirms the socio-technical nature of existing educational research on ICT enabled change …" and secondly, because it deals with the "… episodic nature of change …" [44: 80].

The basic constructs or codes deduced from the PSIC model [17] comprised:

1. Environmental context: The external or outer context of an organisation, inclusive of the social context, economic context, political context and regulatory context, as well as the competitive context.
2. Organizational context: The inner context or environment of the organisation, inclusive of resources, authoritative structures, organisational culture, and internal political system within which change happens.
3. Antecedent conditions: Those conditions or relationships between the socio-technical components that are instrumental in or preceding the event that caused the change.
4. History: Patterns of events that manifested in the past in terms of successes and failures. It gives an indication of past outcomes of socio-technical systems and the trajectories that emerged in the past.
5. Technology: Represents tools, methods, infrastructure and other problem-solving inventions which form part of the development and implementation of systems.
6. Structure: The communication systems, authority systems and workflow as attributed by values, norms, behavior and roles. It includes the behavioral patterns of actors in the system, the work organisation, the decision-making structure and communication channels, and other structures such as frameworks and methodologies.
7. Task: The primary reason/purpose for the existence of the organisation i.e. the work being done. It normally has process-type features leading to a deliverable but related to socio-technical change in some or other way.
8. Actors: Members of an organization as well as the main stakeholders that execute or impact the work being done in the organisation. This includes customers, managers and technology workers.
9. Time: Systems follow dynamic trajectories as they move over time and resonate between stable states and unstable states. Systems follow different patterns of change, hence scalability over time becomes relative to the pliability of the system and the nature of the change.
10. Interactions: The interaction between the socio-technical components in a system and between the actors in the system. These interactions are reciprocal (give-and-take) and need to supplement each other. The relations must co-evolve as change events unfold.
11. Changes: Punctuated change results from the misalignment in systems. Two types of changes may manifest, i.e. incremental changes in the components of the system and re-configuration of system components through punctuations [45: 4]. Understanding how changes emerge over time, what the gaps are, and how the socio-technical components and structures are impacted by disequilibrium.

12. Outcome: Processes articulate sequential events that explain outcomes. It can be explained in terms of socio-technical configurations (in equilibrium, with a new deep structure) or a retained, misaligned deep structure. It relates to the type of change and change events. Change events and their associated outcomes are explained through analysis of the change impact on the current situation.

By using these basic constructs, a concept catalogue was formed of the identified change constructs.

4.3 Step 3: Relationship Identification Towards a Change Metamodel

The idea is to aggregate the constructs from the concept catalogue into a single model or unified view of change, articulated as a change metamodel. This aggregation started off with the theoretical modelling or structuring of the change constructs in the concept catalogue. The approach to modelling adopted the stance of [46: 2] when they said "[a] meta-model is an explicit model of the constructs and rules needed to build specific models within a domain of interest."

Models are linked to theory and should contain certain elementary concepts as building blocks. A metamodel provides relevant modelling information, contextual information, usage information and model parameters or restrictions [47]. The modelling layers (viewed as parameters) can be separated into three separate levels i.e. the strategic layer, the tactical layer and the support layer.

The Change Metamodel as a solution (a design proposition) to solve the wicked problem outlined in Sect. 1 is envisaged to have a heuristic character [48] as opposed to presenting the Change Metamodel as the only and absolute truth.

This aggregation started of with the theoretical modelling or structuring of the change constructs in the concept catalogue and proceeded with applying the theoretical view in a practical context. The change metamodel, therefore, represents a nexus (converging in a unified view) that combines constructs from both the theoretical approach and the real-life approach (explained in Sect. 5).

4.4 Step 4: Theoretical View of Change Models

The 36 models and theories were analyzed with the intension to determine the applicable categories and basic conceptual structure. This led to a theoretical structural view of the change models given in Fig. 1 on the next page.

4.5 Step 5: Practical View of the Change Models

The research context considered for establishing a real-life view entails a prominent private secondary school in South Africa (as a typical implementer and user and from here on referred to as The School) and its learning technology provider ITSI (a company providing support to some 250+ schools). The School is well advanced in the roll-out of hybrid learning and continuously endeavoring to change towards establishing a comprehensive hybrid learning ecosystem. The point of departure for the modelling of the change constructs, was to ascribe structure to the constructs by using tools from the Soft Systems Methodology (SSM).

The drafting of rich pictures was done to determine the links or relationships between the concepts in the concept catalogue. This provided structure that could not otherwise have been determined through the theoretical change construct analysis alone.

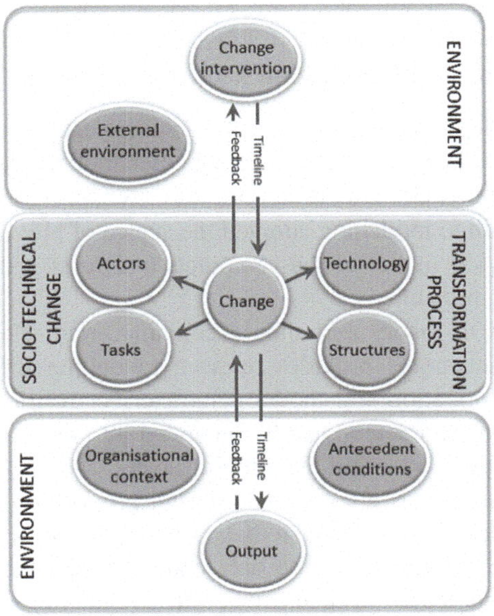

Fig. 1. Structural view of change models

Each category was assigned a graphical icon that gives a visual representation of that category. Towards drawing the rich pictures, separate facilitated focus group sessions were convened with participants from The School and ITSI. Figure 2 on the next page shows one of the rich pictures. In total, 19 of these rich pictures were drafted. The rich pictures culminated in two conceptual views: one for The School (see Fig. 3 on the next page) and one for ITSI (not included here).

4.6 Step 6: Composing a Change Metamodel – Towards a Unified View

The process to design the change metamodel yielded four meta-concepts, each with their own properties. The properties of each meta-concept are the culmination of snipping, refining and repackaging the data collected from literature in step and data collected from real-life practice as conceptual models. Simplistically, the change metamodel as depicted in Fig. 4 on the next page comprises a cycle that can be articulated as: **Contextualize** – **Intervene** – **Transform** – **Institutionalize**.

The change metamodel shows how the snippets (viewed as meta-concepts) were repackaged as a model. Each of these meta-concepts are discussed below.

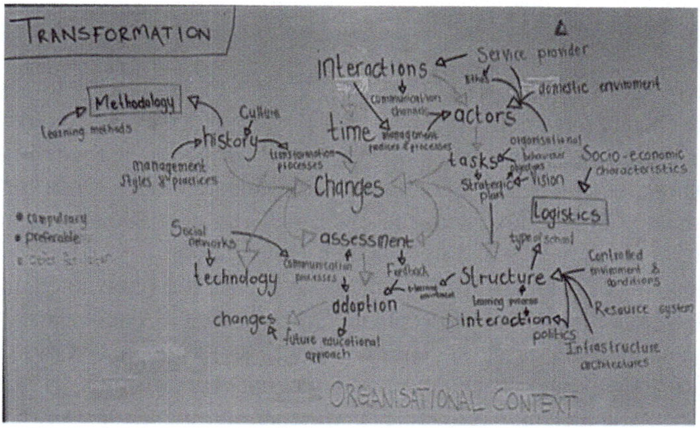

Fig. 2. Rich picture example

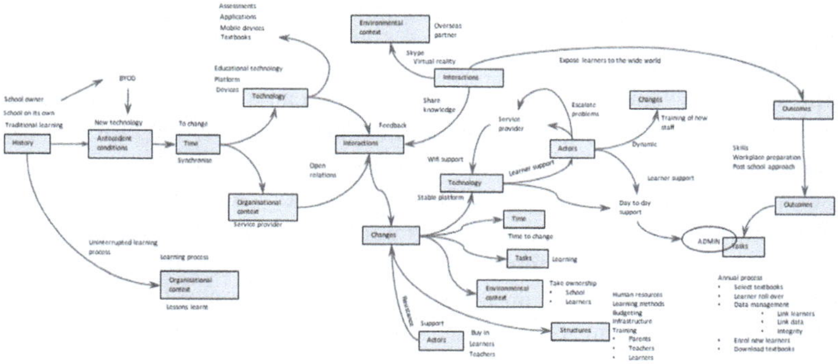

Fig. 3. Conceptual model for the school

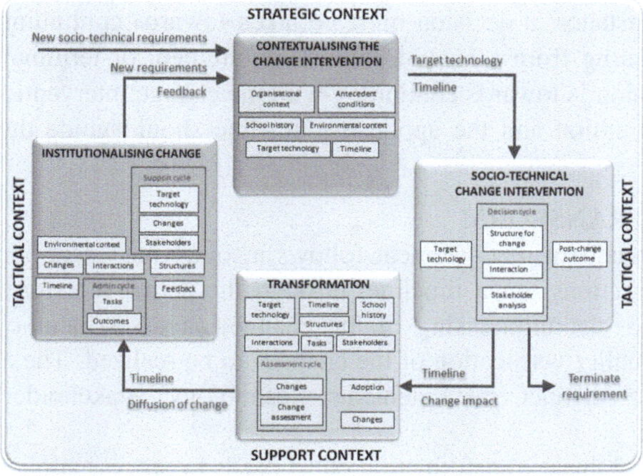

Fig. 4. Change metamodel

4.7 Step 7: Metamodel Description

Meta-concept: CONTEXTUALIZE

Although the change metamodel is shown as a closed change cycle, contextualizing the change interventions may be viewed as the primary meta-concept when first instantiating the change metamodel as a context-specific change model. Once the model has been de-abstracted and instantiated, the cycle will run continuously, generating and processing new requirements and new change interventions.

The meta-elements of **contextualize** are shown in Fig. 4. It is to be noted that the meta-elements align to the list of key concepts identified in step 2. These six meta-elements are: organisational context, antecedent conditions, the school's history, the environmental context, the target technology, and the relevant timeline. The descriptions assigned to the meta-elements provide a perspective on how the meta-elements should be interpreted or understood.

With the focus on socio-technical change, the link in the change cycle to initiate the change intervention (referred to as **intervene**) comprises of feedback from the contextual setting of the school and the target technology as an enabler of hybrid-learning.

Meta-concept: INTERVENE

The purpose of **intervene** as a meta-concept is to form an understanding of the relevant decisions and how to structure for change outcomes given the target technology and stakeholders. Deciding on the execution of socio-technical change interventions comprises a decision cycle to analyze the target technology and conclude the structures, interactions and stakeholders to proceed with transformation towards realizing the post-change outcome of the planned intervention. The decision regarding the intervention has both a time implication and an impact that must be factored into the **transform** meta-concept. Understanding the impact as envisaged by the decision to change (**intervene**), will assist in planning and preparing for the change.

The change metamodel makes provision for defining or articulating the context or parameters of the model (strategic, tactical and support), the context of the school (**contextualize**) and change decisions (**intervene**).

In the final instance, a decision must be taken towards continuing with the intervention as emanating from a socio-technical requirement or terminating the requirement. If the decision is towards continuing with the change intervention, the impact of the change intervention and the applicable timeline should guide the transformation effort.

Meta-concept: TRANSFORM

Although the **transform** meta-element follows a somewhat linear approach that runs from initial interactions, on a timeline, through the actual transformation to ensure adoption prior to institutionalising of the change, an assessment cycle should run iteratively as a quality verification of the changes to be realized. The assessment cycle feeds into three distinct meta-elements: technology, stakeholder adoption and structuring.

The end-to-end linear transformation effort needs to process applicable information regarding the target technology, the history of the school, relevant interactions between system components and stakeholders, the stakeholders involved, the applicable structures and the work or tasks being done at the school. Changes need to be affected towards an acceptable level of quality and adoption of the envisaged solution.

The timeline represents a continuous line of effecting the change from inception to institutionalization, but as informed by the integration of efforts from both the school and the learning technology provider, inclusive of project planning and synchronizing implementation to educational cycles.

Meta-concept: INSTITUTIONALIZE

Institutionalising the change or transformation represents the end of the timeline and needs to give the change some permanence and ensure equilibrium. **Institutionalising** change links to the environmental context and to the outcomes of learning. It is directly linked to the extent that the environmental context provides stable conditions for uninterrupted learning to continue towards the outputs of hybrid learning. The indirect link relates to the **institutionalising** of the transformation effected in the **transform** meta-concept. The changes and related interactions have an influence on the environment, the actors, the technology, the tasks executed and applicable timeframes. **Institutionalising** the change does not present itself as a linear set of meta-elements. The institutionalising of change includes three cycles that are executed (continuously). Firstly, the administration cycle runs on fixed time schedules. Secondly, the support cycle runs on an ad hoc basis and is only executed when changes impact stakeholders in relation to technology. The cycle may be executed multiple times until socio-technical stability is achieved. Thirdly, the feedback cycle is executed regarding corrective changes that result in the model being used as-is to effect such minor changes or second-order revolutionary changes that may warrant a total new change intervention based on new requirements leading to reviewing the total change model.

5 Conclusion

It is important to understand that the change metamodel presented in this paper is just that - a metamodel. It provides a unified view of change which user communities could use as a baseline to design their own respective context-based physical change models. In this regard, what happens during the execution of the activities required during each meta-concept is the product of a de-abstracted and refactored change model. The change metamodel needs to be refactored to show all the concepts relevant to each meta-element in the change model for each school. De-abstracting and refactoring the change metamodel towards designing context-based physical change models requires a formal, standardized approach which could assist schools when changing to hybrid learning.

References

1. Chandran, D., Kempegowda, S.: Hybrid E-learning platform based on cloud architecture model: a proposal. In: The International Conference on Signal and Image Processing, Chennai, India, pp. 534–537 (2010)
2. Parlakkılıç, A.: Change management in transition To E-learning system. Qual. Quant. Methods Libr. **3**, 637–651 (2014)
3. Chang, C., Fang, M.: E-Learning and online instructions of higher education during the 2019 novel coronavirus diseases (COVID-19) epidemic. In: Journal of Physics: Conference Series, vol. 1574 (2020). https://doi.org/10.1088/1742-6596/1574/1/012166
4. Turkkahraman, M.: Education, teaching and school as a social organization. Procedia – Soc. Behav. Sci. **186**, 381–387 (2015). https://doi.org/10.1016/j.sbspro.2015.04.044
5. Bidwell, C.E.: Analyzing schools as organizations: long-term permanence and short-term change. Sociol. Educ. **74**, 100–114 (2001). https://doi.org/10.2307/2673256
6. Imants, J.: Other approaches to schools as organizations. Other Educ.: J. Educ. Altern. **1**(1), 175–187 (2012)
7. Christie, P.: Landscapes of leadership in South African schools: mapping the changes. Educ. Manag. Adm. Leadersh. **38**(6), 694–711 (2010). https://doi.org/10.1177/1741143210379062
8. Menchaca, M., Bischoff, M., Dara-Abrams, B.: Model for systemic change management in education. Syst. Cybern. Inform. **2**(1), 1–6 (2001)
9. Ungureanu, A., Rascu-Pistol, S., Ungureanu, A.: Management and change in Romanian education. Contemp. Read. Law Soc. Justice **6**(1), 371–377 (2014)
10. Altunay, E., Arli, D., Yalcinkaya, M.: Qualitative study on change management in primary schools award winning and non-award winning schools case in study of TQM. Educ. Sci.: Theory Pract. **12**(2), 725–730 (2012)
11. Tearle, P.: A theoretical and instrumental framework for implementing change in ICT in education. Camb. J. Educ. **34**(3), 331–351 (2004). https://doi.org/10.1080/0305764042000289956
12. Sahin, I.: Detailed review of Rogers' diffusion of innovations theory and educational technology-related studies based on Rogers' theory. Turk. Online J. Educ. Technol. **6**(2), 14–23 (2006)
13. Weerakkody, V., Janssen, M., Dwivedi, Y.K.: Transformational change and business process reengineering (BPR): lessons from the British and Dutch public sector. Gov. Inf. Q. **28**(3), 320–328 (2011). https://doi.org/10.1016/j.giq.2010.07.010
14. Kotter, J.P.: Leading Change: Why Transformation Efforts Fail, pp. 1–10. Harvard Business Review, Brighton (2007)
15. Pryor, M.G., Taneja, R., Humphreys, J., Anderson, D., Singleton, S.: Challenges facing change management theories and research. Delhi Bus. Rev. **9**(1), 1–20 (2008)
16. Nadler, D.A., Tushman, M.L.: A model for diagnosing organizational behavior. Organ. Dyn. **9**(2), 35–51 (1980). https://doi.org/10.1016/0090-2616(80)90039-x
17. Lyytinen, K., Newman, M.: Explaining information systems change: a punctuated socio-technical change model. Eur. J. Inf. Syst. **17**(6), 589–613 (2008). https://doi.org/10.1057/ejis.2008.50
18. Ahmad, A., Lyytinen, K., Newman, M.: The evolution of process models in is research: from a punctuated social process model to a sociotechnical process model. In: European Conference on Information Systems (ECIS), Helsinki, Finland (2011)
19. Shin, N.S., Edington, B.H.: An integrative framework for the contextual factors affecting information technology implementation. J. Inf. Technol. Theory Appl. **8**(4), 21–38 (2007)

20. Speck, M.: The change process in a school learning community. Sch. Community J. **6**(1), 69–79 (1996)
21. Hord, S.M., Rutherford, W.L., Huling-Austin, L., Hall, G.E.: Taking Charge of Change. Southwest Educational Development Laboratory, Austin (1987)
22. Davis, N., Eickelmann, B., Zaka, P.: Restructuring of educational systems in the digital age from a co-evolutionary perspective. J. Comput. Assist. Learn. **29**(5), 438–450 (2013). https://doi.org/10.1111/jcal.12032
23. Ten Brummelhuis, A.C.A.: Models of Educational Change. Universiteit Twente, Delden (1995)
24. Aparicio, M., Bacao, F., Oliveira, T.: An e-learning theoretical framework. Educ. Technol. Soc. **19**(1), 292–307 (2016)
25. Appelbaum, S.H., St-Pierre, N., Glavas, W.: Strategic organizational change: the role of leadership, learning, motivation and productivity. Manag. Decis. **36**(5), 289–301 (1998)
26. Williamson, B.: Effective or affective schools? Technological and emotional discourses of educational change. Discourse: Stud. Cult. Polit. Educ. **33**(3), 425–441 (2012). https://doi.org/10.1080/01596306.2012.681901
27. Chan, T.: How East Asian classrooms may change over the next 20 years. J. Comput. Assist. Learn. **26**(1), 28–52 (2010). https://doi.org/10.1111/j.1365-2729.2009.00342.x
28. Song, J.S., Kim, S.J., Byun, G.S., Song, J.H., Lee, B.G.: Comparing wireless networks for applying digital textbook. Telecommun. Syst. **55**(1), 25–38 (2014). https://doi.org/10.1007/s11235-013-9748-4
29. Cardellino, P., Leiringer, R.: Facilitating change in primary education: the role of existing school facilities in ICT initiatives. Facilities **32**(13/14), 845–855 (2014). https://doi.org/10.1108/f-04-2013-0026
30. Hsu, S., Kuan, P.-Y.: The impact of multilevel factors on technology integration: the case of Taiwanese grade 1–9 teachers and schools. Educ. Tech. Res. Dev. **61**(2), 25–50 (2013). https://doi.org/10.1007/sl1423-012-9269-y
31. Maftuh, B.: Status of ICT integration in education in Southeast Asian countries. In: Tsukuba International Conference: Innovation of Classroom Teaching and Learning through Lesson Study. Tsukuba University, Japan (2011)
32. Aristovnik, A.: The impact of ICT on educational performance and its efficiency in selected EU and OECD countries: a non-parametric analysis. Turk. Online J. Educ. Technol. **11**(3), 144–152 (2012)
33. Cook, C.W., Sonnenberg, C.: Technology and online education: models for change. Contemp. Issues Educ. Res. **7**(3), 171–188 (2014)
34. Williams, M.D., Williams, J.: A change management approach to evaluating ICT investment initiatives. J. Enterp. Inf. Manag. **20**(1), 32–50 (2007). https://doi.org/10.1108/17410390710717129
35. Donnelley, P., Kirk, P.: Use the PDSA model for effective change management. Educ. Prim. Care **26**(4), 279–281 (2015)
36. Seijts, G.H., Roberts, M.: The impact of employee perceptions on change in a municipal government. Leadersh. Organ. Dev. J. **32**(2), 190–213 (2011). https://doi.org/10.1108/01437731111113006
37. Yılmaz, D., Kılıçoğlu, G.: Resistance to change and ways of reducing resistance in educational organizations. Eur. J. Res. Educ. **1**(1), 14–21 (2013)
38. Ng'ambi, D., Bozalek, V.: Leveraging informal leadership in higher education institutions: a case of diffusion of emerging technologies in a southern context. Br. J. Educ. Technol. **44**(6), 940–950 (2013). https://doi.org/10.1111/bjet.12108

39. Peslak, A., Ceccucci, W., Sendall, P.: An empirical study of social networking behavior using diffusion of innovation theory. In: Conference on Information Systems Applied Research, Nashville, Tennessee, USA (2010)
40. Van der Waldt, G.: Towards a typology of models in public administration and management as field of scientific inquiry. Afr. J. Public Aff. **6**(6), 38–56 (2013)
41. Ghaffarian, V.: The new stream of socio-technical approach and main stream information systems research. Procedia Comput. Sci. **3**, 1499–1511 (2011). https://doi.org/10.1016/j.procs.2011.01.039
42. Ribino, P., Seidita, V., Lodato, C., Lopes, S., Cossentino, M.: Common and domain-specific metamodel elements for problem description in simulation problems. In: Proceedings of the 2014 Federated Conference on Computer Science and Information Systems, Warsaw, Poland, pp. 1467–1476 (2014)
43. Cho, J.Y., Lee, E.-H.: Reducing confusion about grounded theory and qualitative content analysis: similarities and differences. Qual. Rep. **19**, 1–20 (2014)
44. Weilbach, L., Matthee, M.: Understanding change from a socio-technical perspective: the case of an e-textbook implementation. Int. J. Sci. Soc. **3**(1), 80–93 (2016). https://doi.org/10.4018/IJSS.2016010106
45. Lyytinen, K., Newman, M.: Punctuated equilibrium, process models and information system development and change: towards a socio-technical process analysis. Sprouts: Work. Pap. Inf. Environ. Syst. Organ. **6**(1), 1–48 (2006)
46. Fayoumi, A., Kavakli, E., Loucopoulos, P.: Towards a unified meta-model for goal oriented modelling. In: European, Mediterranean & Middle Eastern Conference on Information Systems 2015, Athens, Greece (2015)
47. Thalheim, B.: Towards a theory of conceptual modelling. J. Univ. Comput. Sci. **16**(20), 3102–3137 (2010). https://doi.org/10.1007/978-3-642-04947-7_7
48. Carlsson, S.A., Henningsson, S., Hrastinski, S., Keller, C.: Socio-technical IS design science research: developing design theory for IS integration management. Inf. Syst. E-Bus. Manag. **9**(1), 109–131 (2011). https://doi.org/10.1007/s10257-010-0140-6

How Instructors Initially Viewed Teaching Online in Higher Education in the UK During the COVID-19 Pandemic

Xin Zhao[1(✉)], Michael Kung[2], and Liang Cai[3]

[1] University of Sheffield, Sheffield, UK
xin.zhao@sheffield.ac.uk
[2] University of Florida, Gainesville, FL, USA
mkung@ufl.edu
[3] Ningbo Tech University, Ningbo, China
byroncai@sina.com

Abstract. Learning and teaching in higher education institutions around the world have been heavily affected by the outbreak of COVID-19 since the fall of 2019. Teachers were suddenly told to convert their classes online and to be prepared to teach virtually. An online focus group (n = 9) was conducted during the initial period of lockdown in the UK at the end of March 2020 to find out about their teaching experiences of transition into online education. A number of challenges were identified in both synchronous and asynchronous teaching processes, including unfamiliarity with the learning management system, privacy concerns, student engagement, preparation time and technological issues. A set of best practices was developed for instructors teaching online during the COVID-19 pandemic.

Keywords: Online teaching · COVID-19 pandemic · Educational technology

1 Introduction

1.1 COVID-19 and Higher Education

The COVID-19 pandemic has disrupted the personal and professional lives of virtually everyone on the planet. Massive disruptions have occurred everywhere from everyday businesses to educational institutions. Schools have been shut down and converted to virtual classes, forcing both the instructors and students to quickly adapt to a new digital norm [1, 2]. According to UNESCO, over 60% of the world's student population is estimated to be impacted by these nationwide closures [3]. According to Inside Higher Ed [4], the COVID-19 pandemic forced higher education institutions into four phases, namely, 1) rapidly transitioning to distance learning; 2) adding basics into emergency courses; 3) extending the transition; and 4) the emerging new normal. Challenges were most prominent for HEIs in phase one due to the unprecedented nature of the virus and urgency for taking immediate actions.

When the pandemic was first declared by the World Health Organization, Higher Educational Institutes (HEIs) from around the world immediately scrambled to find a

C. Pang et al. (Eds.): SETE 2020/ICWL 2020, LNCS 12511, pp. 277–286, 2021.
https://doi.org/10.1007/978-3-030-66906-5_26

way to guarantee the safety of their faculty, staff, and students. The rapid transition into remote teaching and learning proved to be problematic in many areas including learning and teaching, assessment, student services, university finance, and so on. In particular, HEIs were struggling with quickly moving learning and teaching from the traditional face-to-face model to an unfamiliar distance learning model. Research suggests that teachers and students were unprepared for the sudden transition into a new learning mode, and universities often lacked the technical capacity and infrastructure to handle the move [5, 6].

1.2 Distance Learning and Traditional Face-to-Face Teaching

Distance learning and traditional face-to-face teaching are two very different modes and require a different set of delivery methods, pedagogical considerations and educational technologies. Research suggests that although distance learning has advantages over traditional learning in terms of accessibility, convenience and flexibility, challenges remain for distance learning in the areas of quality of teaching, teacher-student communication, and ICT challenges [7]. Due to the rapid transition into a distance learning model, these issues become even more prominent for teachers and students. An expert face-to-face lecturer could easily have trouble teaching online; for instance, there is less direct feedback from students in an online course, there is difficulty sharing presentations, there can be technical issues and lag times, the conversation with students seems unnatural, and so on. When teaching in an synchronous session, the instructor may treat the class as if it were a traditional face-to-face lecture, although in actuality, they are quite different. If the faculty is unaccustomed to teaching online, they might just lecture by themselves for an hour without inserting appropriate breaks and waiting for student feedback.

Similarly, students also need to be prepared to take an online course, and previous students have shown that students with certain traits tend to perform better in an online course. To succeed in an online course, students are expected to be independent, responsible, self-motivated, adaptive, able to learn with minimal instructor guidance, and able to establish an online community group [8]. Students lacking these skills can struggle with an online course, and suddenly being forced to take an online course midway through a semester can add unnecessary stress to the student.

1.3 Synchronous and Asynchronous Teaching

Online classes can be synchronous (when everyone is logged in together at the same time) or asynchronous (when students can work at their own pace). Synchronous classes allow for direct interaction and immediate feedback from the instructor, while asynchronous classes allow students to access learning resources at their own time and speed [9].

Although research suggests that students welcome the synchronous sessions as they feel part of a learning community, Park and Bonk [10] pointed out the negative impact of students' internet connection and language barriers on their synchronous learning experiences, highlighting a lack of peer interactions. In the same vein, Romero-Hall and Rocha Vicentini [11] suggested that these challenges remain in the hybrid learning settings where face-to-face communication cannot be easily replaced by online interaction.

1.4 Classroom Interactions

Higher educational institutions around the world have long been developing and utilising interactive educational technology to promote active student engagement [12, 13]. Evidence from several cohort studies have already indicated the benefits of blended learning or flipped classrooms on students' academic performance [14]. Online discussion forums and social media apps have been used to promote student interaction outside traditional classrooms [15, 16]. However, several studies have shown that there is a significant decrease in student engagement in distance learning courses, ranging from students remaining silent during synchronising discussion activities to not logging into the virtual learning environments for months [17]. Interactivity, whether between the faculty and students or students to students are a key part to creating an online community. It is very easy for an online student to feel demotivated and fall behind in an online course, and having an interactive component gives the student a chance to engage with the online community and stay focused.

1.5 Educational Technology

With more and more universities including online classes, the use of educational technology becomes more and more relevant, and Higher Education Institutions have been a strong advocate for the use of interactive teaching methods to promote student learning. Traditional approaches such as teacher-fronted instructions and in-class group discussions were typically well-received by both teachers and learners [18, 19]. The emerging educational technology has brought new possibilities for addressing the different learning styles and needs of an increasingly diverse student body [20], although it is important to remember that pedagogy should always come before technology integration. Research suggests that learning technologies can enhance students' academic performance and engagement on STEM subjects [21]. For example, visual tools, such as mind mapping software, have been found to encourage student classroom participation [22]. Although scholars have frequently endorsed educational technology in empowering learning, much of the current literature has emphasised on how these technologies are deployed in face-to-face teaching contexts or in a blended learning context. With the increasing number of universities transitioning from traditional teaching to distance learning during the COVID-19 period, teachers and students have become more reliant on educational technology. However, lack of training for staff and/or insufficient IT resources remain problematic for HEIs.

2 Methodology

This paper adds to this emergent literature by using an exploratory study to investigate the university teacher' experiences of moving towards online teaching due to the COVID-19 outbreak. The researchers both work at large universities and experienced a swift move to distance learning. They employed a semi-structured focus group that allowed them to gain insight into the experiences of the participants. This method was best suited for the available study population since it allowed the researchers and participants to exchange viewpoints and confirm insights in order to achieve a robust discussion while addressing a complex topic and observing the group perception [23].

The focus group interview was held online in late March 2020 at the beginning of the lockdown in the UK in order to qualitatively explore the views of a group of university teachers (N = 9; 4F, 5M) on their transition into online teaching. The focus group consisted of a mixture of junior (n = 6) and senior teachers (n = 3) of social science disciplines. The interview was recorded, transcribed and analysed using the six phases of thematic analysis proposed by Braun and Clarke [24] including data familiarization, coding, searching of themes, reviewing of themes, defining themes, and producing the final report. The participants were anonymised using pseudonyms.

Specifically, there were three research questions that emerged from the preceding literature review:

1. What are the challenges in synchronous and asynchronous teaching sessions?
2. What are the impacts of these sessions on student engagement?
3. What recommendations can we provide for teachers in higher education during the outbreak of COVID-19?

3 Results

Although an asynchronous video recording can provide quality teaching content, students tend to be less engaged in those sessions, whereas synchronous live sessions can provide more opportunities for interaction [9]. Our participants offered both synchronous and asynchronous sessions for their students, and they witnessed a significant drop in student engagement in both sessions compared to the pre-lockdown campus-based face-to-face learning. They identified a number of challenges and proposed some potential solutions in the following three areas: synchronous sessions, asynchronous sessions and overall student engagement.

3.1 Challenges in Synchronous Sessions

According to our participants, they faced a number of issues with synchronous sessions at the beginning of the transition. These included unfamiliarity of the Blackboard Learning Management System (LMS), silence in student-centred activities, privacy concerns of the recording of live sessions, and so on. A number of technologies were proposed as solutions to help teachers engage the student in live sessions in Blackboard such as the polls feature, the interactive whiteboard feature and breakout rooms for group activities. However, challenges still remained with synchronous teaching. One participant described the challenges she faced while interacting with students in synchronous lectures:

'Students were reluctant to use their microphones during group discussions. We did ask students to mute their microphones during the lecture presentation part. That might have caused the issue. Some of them preferred to type in the comment sections, which again caused problems. We have to wait for students to type in their questions, it was awkward.' [T2, Female]

Low student participation is a common issue in online sessions, and the comment described by the teacher was not unusual. Some ways to help address this could be to provide the students with an instructional video about how to use the LMS system before the start of a session. The students need to be familiar with the tools and should participate in a dry run before attending classes. Additionally, some LMS systems have a raise hand feature that will allow the teacher to call on students in an orderly fashion. Teachers can encourage students to use this function more rather than relying on text comments to ask questions.

Privacy Concerns of Being Recorded. Another instructor reported that some students seemed to be concerned about being recorded while participating:

'I taught the first half of the session and used the second half for student-led discussions. But when I started recording, the student started to drop off of the course. There were fewer questions from students as a result.' [T1, Male]

After the first session was completed, the instructor informed the students that they would be recorded during the live discussion session. Students dropped from the session quickly after. While this could also be attributed to students not wanting to participate, it was also possible students did not want to be recorded. Students tend to be more shy when they know they are being recorded and may not feel comfortable truly speaking their minds if they know they will be on the record. As a result, some of the students chose to just leave when they knew their comments would be recorded.

One way to help prevent this is to inform the students earlier that they would be recorded and also provide a set of guidelines of what will be done with the recordings. By letting students know that the recording will only be seen by classmates and not publicly shared or that they will be deleted after a semester, they might feel more comfortable speaking. Another alternative is to stop recording during the live discussion. Many times, the lecture is what students really want to see recorded (so they can go back to it in the future) and not the live discussions. They could ease some of the students' fears about privacy as well. Lecturers can take notes about the common questions and provide a summary of answers to the frequently asked questions to students through email.

Facilitation of Live Sessions. A final concern that instructors had about synchronous teaching had to deal with the facilitation of the live sessions.

'I found that students were engaged with lecture presentations in the main delivery room. But as soon as we divided students into breakout rooms, their engagement dropped. It could be that there were no slides available to students in individual breakout rooms so they can't remember the tasks. It could also be that some students were not prepared to speak or didn't have a microphone. But once we moved between breakout rooms to clarify what they needed to do, they started to engage with tasks again. However, at this point, we've lost half of them.' [T2, Female]

In this comment, the instructor was concerned that students did not know how to properly engage in a breakout room. The instructor included possibilities as to why this happened. One possible solution is that the instructor can use teaching assistants or assign discussion leaders beforehand and let the discussion leader carry on the conversation in each breakout room.

3.2 Challenges in Asynchronous Sessions

Although according to our participants, while synchronous sessions offer more opportunities for student interaction, asynchronous teaching offers more flexibility to students, particularly in terms of video quality and accessibility when students are located in different time zones. Some teachers prefer asynchronous teaching as it can result in better video quality and allows teaching to be recorded all at one session.

'Having recorded all the sessions for asynchronous consumption, I found it easier. All I need to worry about is how to make them interactive.' [T4, Male]

'Recording quality is way higher than the live recording where there are lots of sound such as ums, ah, you know' [T6, Male]

However, instructors still reported some challenges when using asynchronous teaching. Traditional concerns with asynchronous teaching include lack of interactivity and instant feedback, as well as a higher demand for student responsibility. Our participants also specifically mentioned time concerns necessary for producing a quality asynchronous video.

Time Concerns. Some of our participants pointed out the extra preparation time needed for teachers compared to live sessions:

'But recording asynchronous sessions actually take up more time than synchronous sessions, and we are not given that much time for preparation.' [T5, Female]

While creating a quality online lecture does take more initial preparation, an instructor may find it worthwhile to invest the time at the start to create a good lecture. A quality lecture can be reused multiple times before needing to be updated and can provide a better learning experience for the student.

Recording Software. In order to create a lecture, some participants used PowerPoint to record lectures with audio narration. The advantage of this is that each slide has a separate audio recording, which can be re-recorded if necessary, without having to re-record the entire lecture. This allows for flexibility in terms of making changes per slide. Other presentation recording software such as Blackboard Kaltura, Google Meet, TechSmith Camtasia, and Blackboard Collaborate allow video to be recorded with a webcam, which provides a more personal touch and supports students with additional learning needs who may rely on lips reading and subtitles. Preparing scripts for recorded sessions is another way to improve the quality of teaching. Synchronous sessions tend to be more forgiving for umms and digressions than recorded lectures. Recorded sessions, however, require a smoother speech to create professional content, although instructors should also be careful not to be entirely flat or they risk sounding monotonous and losing the audience's attention.

3.3 Student Engagement

Participants noted that some students were not actively engaging with the school's LMS. The main concern was that it was difficult to get an overview of individual student engagement across modules to spot a struggling student who might need extra support. The overall student interactivity significantly dropped since the move to the distance learning model:

'Interactivity in a live session is quite reduced…but interaction in asynchronous in discussion boards is even worse' [T8, Female]

This could be potentially due to the length of the video, quality of the recording, features of the Learning Support System (or lack thereof) and lack of support for the student's transition into a distance learning model.

The Length of Video Lectures. One of our participants pointed out that long videos were not effective for maintaining student engagement:

'Student attention span is rather short. We might lose student engagement if the lecture lasts for an hour or two. It is more challenging for students to watch the recording that lasts that long' [T1, Male]

Although some research suggests that the optimal length of videos should be around 6 min [25], in order to maintain a student's attention, it is sometimes not practical to keep all videos within that length, particularly when delivering complex content. Strategies such as inserting interactive quizzes and labelling content by topic could potentially help to break up the monotony of longer content. The interactive quizzes could serve as short breaks, and labelling content could produce an index system, making it easier for students to navigate to a section of interest.

Lack of Support for Student Transition. Since the move to online teaching, more resources have been put in place to support teachers in terms of teaching design and delivery. Most students, however, were left with an overload of email announcements rather than practical training sessions supporting them to move to this new mode of learning. Our participants reported a lack of induction for students when the university transitioned into online teaching.

'We actually have no issues with our distance learners since locked down. I guess this is due to all the induction and support we offered for them at the beginning of their programmes. However, for our campus-based students, there was not much support for them during this transition. We kind of rely on teachers to instruct them within individual lectures' [T2, Female]

'Sometimes what I've done is a short welcome video of two or three videos about myself or the module. I may even consider doing that as a weekly introduction to the main lecture. To give students a bit of structure as well as a bit of personal touch' [T5, Female]

The confusion creates further challenges for students who have to navigate the learning system and lecture content themselves. A potential solution is to create short instructional videos to help students to navigate the unstructured learning resources that have been dramatically increased during the COVID-19 period.

4 Conclusion and Recommendations

The study provided further evidence of the challenges teachers faced when universities moved to online teaching at the beginning of the national lockdown in the UK due to COVID-19, ranging from technological barriers, privacy concerns, and teaching (facilitating) group activities in synchronous sessions to student interactivity, video length, extra preparation time for teachers in asynchronous sessions. A combination of these factors made it much harder for students to engage in online learning. Another barrier was that students were often neglected in the process of transitioning to online LMS, resulting in a further reduction of student engagement. Our recommendations include providing short instructional or walkthrough videos for students to navigate the LMS and often unstructured learning resources; assign group leaders or use teaching assistants to facilitate online group activities; choose appropriate recording software and record shorter videos; not to record student discussions for privacy concerns and student engagement rather providing a summary based on student discussions, and so on.

It is important to note that data for this research were collected at the start of the lockdown when universities were scrambling to get their courses online. Some of the concerns addressed in this research were significant for teachers during that period. Our sample size was limited due to difficulty to recruit participants during the lockdown period and is not generalizable for the general teaching population, particularly since our participants were all from social science disciplines. Future research should build on the results of this study by qualitatively and quantitatively exploring the teaching and learning experiences of various disciplines. To gain a better sense of the situation, a larger sample size is needed. Additionally, the perspective of university professional services teams should be investigated.

References

1. Viner, R.M., et al.: School closure and management practices during coronavirus outbreaks including COVID-19: a rapid systematic review. Lancet Child Adolesc. Health **4**(5), 397–404 (2020). https://doi.org/10.1016/S2352-4642(20)30095-X

2. Ahmed, S., Shehata, M., Hassanien, M.: Emerging faculty needs for enhancing student engagement on a virtual platform. MedEdPublish **9**, 1–5 (2020). https://doi.org/10.15694/mep.2020.000075.1
3. UNESCO Homepage. https://en.unesco.org/covid19/educationresponse. Accessed 01 Aug 2020
4. Inside Higher Education. https://www.insidehighered.com/news/2020/04/29/synchronous-instruction-hot-right-now-it-sustainable. Accessed 01 Aug 2020
5. Rapanta, C., Botturi, L., Goodyear, P., Guàrdia, L., Koole, M.: Online university teaching during and after the Covid-19 crisis: refocusing teacher presence and learning activity. Postdigital Sci. Educ. **2**(3), 923–945 (2020). https://doi.org/10.1007/s42438-020-00155-y
6. Institute of International Education: IIE releases survey on effects of COVID. https://www.iie.org/Why-IIE/Announcements/2020/03/IIE-Releases-Survey-on-Effects-of-COVID-19-on-International-Students-and-Study-Abroad. Accessed 01 Aug 2020
7. Hannay, M., Newvine, T.: Perceptions of distance learning: a comparison of online and traditional learning. J. Online Learn. Teach. **2**(1), 1–11 (2006)
8. Kung, M.: Methods and strategies for working with international students learning online in the U.S. TechTrends: Link. Res. Pract. Improv. Learn. **61**(5), 479–485 (2017)
9. Hrastinski, S.: Asynchronous & synchronous e-learning. Educ. Q. **31**, 51–55 (2008)
10. Park, Y.J., Bonk, C.J.: Synchronous learning experiences: distance and residential learners' perspectives in a blended graduate course. J. Interact. Online Learn. **6**(3), 245–264 (2007)
11. Romero-Hall, E., Rocha Vicentini, C.: Examining distance learners in hybrid synchronous instruction: successes and challenges. Online Learn. J. **21**(4), 141–157 (2017)
12. Markova, T., Glazkova, I., Zaborova, E.: Quality issues of online distance learning. Procedia – Soc. Behav. Sci. **237**, 685–691 (2017). https://doi.org/10.1016/j.sbspro.2017.02.043
13. Román, F., Delgado, R., Ubilluz, C., Bedón, C.: ICT integration in the teaching/learning process of natural sciences for seventh grade elementary students'. In: Rocha, Á., Ferrás, C., Paredes, M. (eds.) ICITS 2019. AISC, vol. 918, pp. 814–822. Springer, Cham (2019). https://doi.org/10.1007/978-3-030-11890-7_76
14. Fisher, R., Perényi, Á., Birdthistle, N.: The positive relationship between flipped and blended learning and student engagement, performance and satisfaction. Act. Learn. High. Educ. (2018). https://doi.org/10.1177/1469787418801702
15. Lam, J.: Collaborative learning using social media tools in a blended learning course. In: Cheung, S.K.S., Kwok, L.-f., Yang, H., Fong, J., Kwan, R. (eds.) ICHL 2015. LNCS, vol. 9167, pp. 187–198. Springer, Cham (2015). https://doi.org/10.1007/978-3-319-20621-9_15
16. Zhao, X., Reilly, P.: Breaking down barriers? ICTs, international students and intercultural communication within UK higher education institutions. In: Brassier-Rodrigues, C., Brassier, P. (eds.) A Collection of Pedagogical Approaches Aimed at Developing Student İntercultural Sensitivity by İnternationalization at Home. Peter Lang, Paris (2020)
17. Moorhouse, B.L.: Adaptations to a face-to-face initial teacher education course 'forced' online due to the COVID-19 pandemic. J. Educ. Teach. **46**, 1–3 (2020). https://doi.org/10.1080/02607476.2020.1755205
18. Armbruster, P., Patel, M., Johnson, E., Weiss, M.: Active learning and student-centered pedagogy improve student attitudes and performance in introductory biology. CBE—Life Sci. Educ. **8**(3), 203–213 (2009). https://doi.org/10.1187/cbe.09-03-0025
19. Zhao, X., Skidmore, D., Murakami, K.: Prosodic chopping: a pedagogical tool to signal shifts in the academic task structure. In: Skidmore, D., Murakami, K. (eds.) Dialogic Pedagogy: The İmportance of Dialogue in Teaching and Learning, pp. 203–216. Multilingual Matters, Bristol (2016)
20. Bates, A.W.: Teaching in a Digital Age: Guidelines for Designing Teaching and Learning. Tony Bates Associates Ltd., Vancouver (2015)

21. Biagi, F., Loi, M.: Measuring ICT use and learning outcomes: evidence from recent econometric studies. Eur. J. Educ. **48**(1), 28–42 (2013). https://doi.org/10.1111/ejed.12016
22. Wee, B., DePierre, A., Anthamatten, P., Barbour, J.: Visual methodology as a pedagogical research tool in geography education. J. Geogr. High. Educ. **37**(2), 164–173 (2013). https://doi.org/10.1080/03098265.2013.775568
23. Dilshad, R.M., Latif, M.I.: Focus group ınterview as a tool for qualitative research: an analysis. Pak. J. Soc. Sci. (PJSS) **33**, 191–198 (2013)
24. Braun, V., Clarke, V.: Using thematic analysis in psychology. Qual. Res. Psychol. **3**(2), 77–101 (2006)
25. Guo, P.J., Kim, J., Rubin, R.: How video production affects student engagement: an empirical study of MOOC videos. In: Proceedings of the First ACM Conference on Learning@ Scale Conference (L@S 2014), pp. 41–50. Association for Computing Machinery, New York (2014). https://doi.org/10.1145/2556325.2566239

Applying the Online Intelligent Grading System to College English Writing Teaching: An Empirical Study

Sun Juan(✉), Cui Li, Zhang Saiyun, and Zhou Xianwu

No. 1, Qianhu South Road, Yinzhou District, Ningbo, China

Abstract. Currently, much attention has been paid to the online intelligent grading system widely used in college English writing teaching. In an effort to make the best use of the system, the article conducted an empirical study of applying the system to college English writing teaching. The findings are: 1) the intelligent grading system does work efficiently in college English writing teaching, however, it needs to be optimized in assessing the students' writing texts and providing related feedback; 2) it is necessary to combine with teacher-assisted feedback and supervision on the basis of the intelligent grading system.

Keywords: Online intelligent grading system · College English writing · Empirical study

1 Introduction

Writing, as an essential part of the component of English proficiency, is of great significance in college English Teaching. No matter in the domestic English proficiency tests such as CET-4, CET-6, TEM-4 and TEM-8, etc. or the international English proficiency tests such as TOEFL, IELTS, and GRE, etc., the candidates' writing ability contributes much to the performance in the test. Besides, English writing ability is one of the most important communicative abilities that students should achieve in learning English. In the modern competitive society, proficiency in English writing is a must for the applied talents. Nevertheless, among the five basic skills of listening, speaking, reading, writing, and translating, writing ability is the most tricky to cultivate.

The traditional mode of teaching writing is still dominant in English teaching among Chinese universities. The teacher-centered mode can no longer satisfy the current demand of cultivating the applied talents. In the teacher-centered classrooms, students passively take writing tasks and mechanically practice with an aim to succeed in the test. This consequently leads to the loss of interest and a lack of creativity in writing. During the process, the focus is mainly on grammar, vocabulary and structure, etc. instead of the core factors, that is, the content and idea.

In the meantime, grading writing is a time-consuming task, which takes much energy of the college English teachers. Due to the reform of domestic universities, most of the college English teachers are now facing an expanding class. With limited time

C. Pang et al. (Eds.): SETE 2020/ICWL 2020, LNCS 12511, pp. 287–293, 2021.
https://doi.org/10.1007/978-3-030-66906-5_27

and energy, it is not practical for teachers to offer prompt and detailed feedback to every student after each writing practice. Thus the effect of writing is hard to achieve.

The trial version of *College English Curriculum Requirements* released in 2004 points out that colleges should take advantage of the opportunities brought by multimedia and web technology, employing new teaching mode to improve the teacher-centered monotonous mode in the classroom in order to make the learning process a much personalized one.

With an effort to lighten teachers' burden and promote students' performance in writing, online intelligent grading systems have gradually become popular in college English writing. Among them, pigai.org, a domestically developed automated writing evaluation (AWE) online system, has been increasingly applied to English writing scoring in China with over 5,000 universities as its users.

2 Previous Research on Online Intelligent Grading System

The study of online intelligent grading systems starts from 1960s. AWE (automated writing evaluation) is defined in Wikipedia as a network aided teaching system to evaluate and score compositions. It was first introduced to aid human raters to evaluate students' essays, which aimed to remove the heavy burden of grading essays and save the time of correcting (Warschauer and Ware 2006) [1]. Shermis and Barrera (2002) defined Automated Essay Scoring (AES) as using computer technology to evaluate and score the essays [2]. Original AWE or AES system, known as Project Essay Grade, was developed by Duke University in 1966 and designed to evaluate writing levels by analyzing linguistics features (Page 2003) [3]. Research of online intelligent grading system in China lags behind although it has enjoyed a history of over 50 years in western countries. Liang Maocheng is a pioneer in construction module researches in this field. His research team in developing EEE1.0 (EFL essay evaluator) has contributed valuable data to the actual design and development of China's own online intelligent grading tool. His new model fits Chinese students better (Liang 2011) [4]. Some Internet technology companies is developing AWE independently, like bingo (http://writing.bingoenglish.com) and pigai (http://www.pigai.org) in China now.

In recent years, many foreign scholars have shifted their focus to the application of online intelligent grading system in teaching and learning (Attali 2004; Warschauer and Grimes 2008; Grimes 2008) [5–7]. A drawback of essay writing assessments is that their evaluation requires a significant and time-consuming effort. These difficulties have led to a growing interest in the application of automated natural language processing techniques for the development of automated essay scoring (AES) as an alternative to human scoring of essays (Attali et al. 2010) [8]. The use of online intelligent grading system is not of a simple technological problem, instead, it is a reform concerning all aspects of teaching and learning (Grimes 2008) [7]. Some Chinese scholars (Liang Maocheng and Wen Qiufang 2007; Ge Shili and Chen Xiaoxiao 2007) [9, 10] pay attention to the analysis of the online intelligent grading system suitable for Chinese EFL learners. Profitable exploration has been attempted on the construction of online intelligent grading system suitable for Chinese learners (Liang 2011) [4]. The online intelligent grading tools, which employ natural-language processing, machine-learning or other computational methods in the analysis of the text, can provide both scores on writing quality as well as qualitative feedback on aspects of grammar, mechanics, style, discourse and organization. Burstein emphasizes that AES is mainly used because of reliability, validity, and cost issues in writing assessment. It is taking a great attraction from public schools, colleges, assessment companies, researchers and educators (as cited in Shermis 2006) [11]. Some researches indicate that the attitude of teachers and students is favorable toward online intelligent grading systems (Hoon 2006) [12]. Students actively use the system to improve the writing and the frequency of use is in sync with the improvement of the score (Schaber et al. 2008) [13].

Nevertheless, some other researches manifest that the application is not optimistic in the classroom because the validity and its operation in aiding teaching process are not clear yet. Yang et al. (2002) mention "the overreliance on surface features of responses, the insensitivity to the content of responses and to creativity, and the vulnerability to new types of cheating and test-taking strategies." [14] Moreover, critics have expressed concerns over the relevance of the features evaluated by AWE to the true qualities of writing, particularly the social and communicative dimensions (Ericsson 2006) [15].

3 Participants and Method

The data serving as support for the assumption were gathered from three freshmen classes. In this study, quantitative and qualitative data were collected and integrated during the analysis and interpretation phases.

Participants

The study was carried out with 97 participants (52 male-54% and 45 female-46%). They were freshmen, specializing in one of the following disciplines: International Trade (10%), Electrical Engineering and Automation (22%), Industrial Design (20%), Internet and New Media (10%), Computer Science (8%), Financial Management (15%) and Law (14%). The participants were students in the College English course in the fall of 2019 and in the spring of 2020 (Table 1).

Table 1. Participants' major and percentage

Major	No. participants	Percentage
Electrical engineering and automation	21	21.6%
Industrial design	19	19.6%
Financial management	15	15.5%
Law	14	14.4%
International trade	10	10.3%
Internet and new media	10	10.3%
Computer science	8	8.2%

4 Method

The study was based on the writing submitted to the online intelligent grading system-pigai.org by the above mentioned 97 students in the spring of 2020 semester. The scores graded by pigai.org and its feedback were analyzed. This study was conducted as part of a larger project investigating the validity of the online intelligent grading system-pigai.org.

The data included students' scores given by pigai.org, the online intelligent grading system, students' scores in the college entrance examination and final test of English for the 2019–2020 academic year. The essay students wrote depended on the teacher's design of the assignment. 97 students' essays were analyzed, among which some students offered several revisions.

5 Findings and Discussions

The case of pigai.org-based assessment was used to support the claim that online intelligent grading system is feasible in college English teaching.

Most students didn't revise their essay although feedback and suggestions were provided. Among the reasons for this tendency, the inaccuracy of the system is an obvious one. For example, some correctly written sentences were identified as run-on sentences or Chinglish. Nevertheless 35 of the students (36%) revised their essay based on the feedback, which supports the assumption that feedback offered by pigai.org is useful in making decisions on revisions even though improvement is still needed. Compared with male participants, female participants hold a more active attitude in revision, 40% female participants revised their essays and 26.7% female participants revised their essays more than three times (Table 2).

Table 2. Participants' revision

Item	Not	Once	Twice	Over 3 times
Male no. & percentage	35 (67.3%)	5 (9.6%)	5 (9.6%)	7 (13%)
Female no. & percentage	27 (60%)	3 (6.7%)	3 (6.7%)	12 (26.7%)
Total no.	62 (63.9)	8 (8.2%)	8 (8.2%)	19 (19.6%)

According to the statistic analysis, the correlation between the score graded by pigai.org and the students' average score of the two final examinations is 13.4%, the correlation between the score of pigai.org and the average score of the compositions in the two final examinations is 23.7%. For those who revised essays, the correlation is 34.2% and 30.9% respectively. Nevertheless, for those who didn't revise their essays, the correlation is 1.8% and 20.3% respectively. The correlation between the score of pigai.org and score of English in the College Entrance Examination is −8.6% for those who didn't revised their essays, 1.9% for all participants and 28% for those revised their essays.

Evidences suggest that the evaluation offered by pigai.org is not closely related to students' performance in the exams, which also raises the doubt about the validity of this online intelligent grading system. After the second thought, we understand that this is only one assignment on pigai.org and there is supposed to be some discrepancy between students' score and their English abilities. Also students' attitude is an important factor which definitely influences the correlation. We found from the study that the correl. between score of pigai.org of those revised their essays and their performance in examinations is much closer than those who didn't revise their essays. So we understand that attitude plays an important role here (Table 3).

Table 3. Correl. between score graded by pigai.org and exams based on revision or not

Item	Not revised	Revised	All participants
Correl. between score 1 and score 2	1.8%	34.2%	13.4%
Correl. between score 1 and score 3	20.3%	30.9%	23.7%
Correl. between score 1 and score 4	−8.6%	28%	1.9%

- Score 1: The score graged by pigai.org.
- Score 2: Participants' average score of College English in the final examination in the autumn of 2019 semester and the spring of 2020 semester.
- Score 3: Participants' average score of writing of College English in the final examination in the autumn of 2019 semester and the spring of 2020 semester.
- Score 4: Participants' score of English in College Entrance Examination.

In order to evaluate the validity of the online intelligent grading system, we first assessed the accuracy of the feedback on 10 error types commonly identified by the system in the students' writing. The 10 common error types are: subject-verb disagreement, run-on sentences, spelling mistakes, Chinglish, fragments, confused words, wrong punctuation, lack of a blank space, capital letter error and misuse of parts of speech. No matter what standard the grading system adopted, it generally did quite adequately. However, when we take a closer look at the errors, our findings are in agreement with those of Lavolette et al. (2015) [16], who also found low accuracy for run-on sentence errors and high accuracy for subject-verb agreement, and the confused word is another case in point. Pigai.org offers alternatives but it did not always fit the context.

Despite some of the low accuracy, there are also advantages of the system, for example, most of the online intelligent grading systems neglect the relevance to the subject while pigai.org does so. Besides, it also offers students the recommended expressions and tips to help them improve the writing. Students' good use of language is also pointed out by the system.

6 Conclusions

The online intelligent grading system actually offers prompt scores and feedbacks based on the cloud technology, relieve the burden of English writing teachers and arouse students' interest and enthusiasm in writing. Based on the survey of applications, there are some implications for the integration of online intelligent grading-system into college English classroom. First, online intelligent grading systems call for a further development in its intelligence of specific feedback. Run-on sentences and fragments may require additional instructions and practice to both diagnose and correctness. Second, it needs to combine with other grading methods to construct a better writing grading mode, for instance, the combined use of online-intelligent grading with teacher-assisted feedbacks is necessary. Third, more online help options are supposed to be developed to meet students' cognitive demands. However, owning to the limited sample size, a further research and evaluation is still needed.

Acknowledgement. This work was supported by the Preparatory Research of Special Topics: Anti-epidemic and resumption, 2020 Zhejiang Education Scientific Planning Program, and the 10th National Foreign Language Education Program (grant no. ZGWYJYJJ10A057).

References

1. Warschauer, M., Ware, P.: Automated writing evaluation: defining the class room research agenda. Lang. Teach. Res. **10**(2), 1–24 (2006)
2. Shermis, M.D., Koch, C.M., Page, E.B., Keith, T.Z., Harrington, S.: Trait ratings for automated essay grading. Educ. Psychol. Meas. **62**(1), 5–18 (2002)
3. Page, E.: Project essay grade: PEG. In: Shermis, M., Burstein, J. (eds.) Automated Essay Scoring: A Cross-Disciplinary Perspective, pp. 43–54. Erlbaum, Mahwah (2003)
4. Liang, M.: Constructing a Model for the Computer-assisted Scoring of Chinese EFL Learners' Argumentative Essays. Foreign Language Teaching and Research Press, Beijing (2011)
5. Attali, Y.: Exploring the feedback and revision features of criterion. J. Second Lang. Writ. **14**, 191–205 (2004)
6. Warschauer, M., Grimes, D.: Automated writing assessment in the classroom. Pedagog.: Int. J. **3**(1), 22–36 (2008)
7. Grimes, D.: Assessing automated assessment. ftp://ftp.ics.uci.edu/pub/grimesd/AutoAssessment.pdf. Accessed 20 June 2010
8. Attali, Y., Bridgeman, B., Trapani, C.: Performance of a generic approach in automated essay scoring. J. Technol. Learn. Assess. **10**(3), 4 (2010)
9. Liang, M., Wen, Q.: A critical review and implications of some automated essay scoring systems. Comput.-Assist. Foreign Lang. Educ. **5**, 18–24 (2007)
10. Ge, S., Chen, X.: Automated essay scoring for Chinese EFL learners. Foreign Lang. World **5**, 43–50 (2007)
11. Shermis, M.D.: Automated essay scoring. Turk. J. Distance Educ. **1**, 49–59 (2006)
12. Hoon, T.: Online automated essay assessment: potentials for writing development (2006). http://ausweb.scu.edu.aw06/papers/refereed/tan3/paper.html. Accessed 09 Aug 2010

13. Scharber, C., Dexter, S., Riedel, E.: Students experiences with an automated essay scorer (2008). http://escholarship.bc.edu/cgi/viewcontent.cgiarticle=1116&context=jtla. Accessed 10 Nov 2009
14. Yang, et al.: A review of strategies for validating computer-automated scoring. Appl. Meas. Educ. **15**(4), 391–412 (2002)
15. Ericsson, P.F.: The meaning of meaning: is a paragraph more than an equation? In: Ericsson, P.F., Haswell, R.H. (eds.) Machine Scoring of Student Essays: Truth and Consequences, pp. 28–37. Utah State University Press, Logan (2006)
16. Lavolette, E., Polio, C., Kahng, J.: The accuracy of computer-assisted feedback and students' responses to it. Lang. Learn. Technol. **19**, 50–68 (2015)

The Construction and Measurement of an Online Learning Evaluation System Based on ACSI Mode

Chunmei Han, Yuzhuo Zhang, Hongji Yang, Lu Yu, and Lei Zhang(✉)

People's Public Security University of China, Beijing, China
leizhang@ppsuc.edu.cn

Abstract. Affected by the COVID-19 outbreak, most of Chinese universities have made the transition to online learning. Based on the American Customer Satisfaction Index (ACSI) Model, theoretical model of factors influencing online learning is constructed from the perspective of learners, which contained learner quality, instructor quality, technical system and educational system quality, perceived interaction and online perceived performance. Main findings include: (i) Learner quality has great predictive effect on online learning performance; (ii) Technical system and educational system quality is an important factor in predicting Online Learning effectiveness; and (iii) The predictiveness of perceived interaction and instructor quality is not strong.

Keywords: E-learning · Learning evaluation · ACSI model · Perceived performance

1 Introduction

With the continuous development of education information network in China, more and more Chinese universities and teachers introduce information technology and big data technology into teaching. Since the outbreak of covid-19, the Ministry of Education proposed "suspended class, ongoing learning" during the epidemic period. It is an emergency response that we implement under the pressure of the epidemic, yet it is a concentrated display and inspection of "Internet plus Education" reformation previously implemented in universities.

Since the beginning of 21st century, the rapid development of information technology, especially the progress from the Internet to the mobile Internet, has created online learning which is not limited by time and space. Successful e-learning has been discussed and investigated in various studies, from different perspectives, and in different contexts and many researchers have attempted to identify e-learning success factors to stress the effectiveness of these systems. [1, 2] From the perspective of students, motivation is the most important driving force to explain online students' ability to pass exams. [3] The age of the students showed a positive impact. This means that the older the students are, the more confident they are in their ability to learn, which means they will perform better in the course. [4] Ismail et al. found that students'

C. Pang et al. (Eds.): SETE 2020/ICWL 2020, LNCS 12511, pp. 294–305, 2021.
https://doi.org/10.1007/978-3-030-66906-5_28

continuance intentions regarding e-learning were moderate and that this was due to a low level of interpersonal influence and the information quality offered by the e-learning services. [5] We now realize that adult students are task-oriented, working professionals who want feedback quickly, especially when in a mostly asynchronous online course where feedback is one of the primary forms of interaction with the instructors [6].

In terms of teacher teaching, studies reveal that instructor quality has a significant effect on the perceptions of satisfaction and usefulness of the system. This will aid instructors to gain an in-depth understanding and confidence using the e-learning system, in addition to increasing their awareness of the full features of the system. [7] The study revealed that the existence of communication and interactivity features, assessment and evaluation materials, and the diversity of learning styles positively influence utilization of the e-learning system, and aid students to be more engaged in their learning. A step forward is also needed in the analysis of the relationship between organizational structure in universities, ICT uses for administrative and educational purposes, and students' performance. [8] From the higher education perspective, we assume that if learners are to continue using e-learning, they need to continue to be satisfied with the e-learning services offered by higher education institutions. Evidently, there is a need for a comprehensive success model for multiple levels of success. [9] Demirkan discussed a reference model for sustainable e-learning systems. [10] Lin developed a model that examined users' e-learning continuance intention in relation to negative critical incidents. [11] This research aims at studying the relevant factors affecting the satisfaction of learners as a whole, and reveals the differences of the influencing factors of the degree of learners' satisfaction in different situations from an empirical perspective. We intend to construct a causal prediction model between environmental characteristics of online learning and learning effectiveness to systematically analyze and expose the main factors affecting the effectiveness of online learning.

2 Research Design and Model Construction

Learning effectiveness refers to the unique quality of education that learners who complete an online course. Online learning effect evaluation refers to the process of making value judgment on the whole learning process and learning effect of learners according to online learning objectives. The goal of online learning is to achieve at least as much as other transmission modes of the organization, especially the traditional face-to-face teaching. [12] Learning effectiveness should be a priority when measuring online education. Other important issues such as learning opportunities, teacher-student satisfaction, and cost-effectiveness will be meaningless if the network environment cannot achieve the same results as traditional face-to-face teaching.

In 1994, Fornell and et al. [13] built the American customer satisfaction model (ACSI) based on the SCSB model. The ACSI model comprehensively evaluated the level of customer satisfaction, which is an important indicator to measure the economic status of the United States. Taking customer satisfaction as an example, customer expectation, perceived quality and perceived value directly affect customer satisfaction.

At the same time, customer expectation and perceived quality have indirect influence on customer satisfaction with perceived value as the intermediate variable. The ACSI has a wider influence range, its measurement index and dimension design are more reasonable, and its operability is stronger. Therefore, this study will take the ACSI model as the basic model, referring to the core concepts of the ACSI model, and combining the elements of teaching effect in online learning to construct the theoretical model of this study (see Fig. 1).

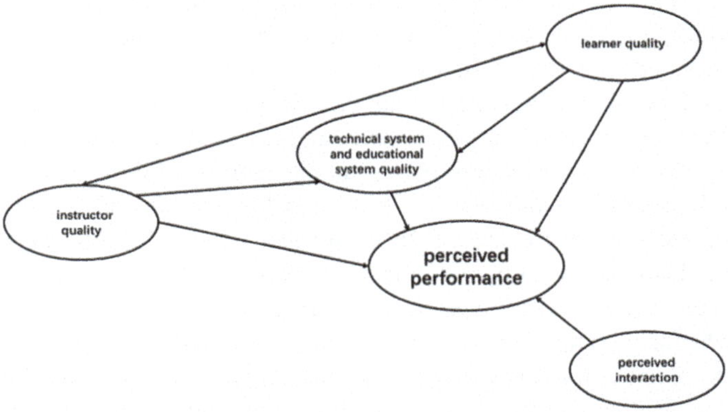

Fig. 1. Theoretical model of online learning evaluation

Based on the core concept of the American customer satisfaction model, this study divides the factors of online learning effect measurement into five dimensions, including learner quality, instructor quality, technical system and educational system quality, perceived interaction and online perceived performance, and builds the structural equation model which is used to verify the relationship between factors from the learner's perspective. Meanwhile, the relationship between these factors is systematically analyzed. The learners dimension selects self-efficacy, degree of self-regulation and engagement, teachers dimension selects online learning attitude, timely feedback and academic ability, technical system and educational system quality dimension variables including course quality and platform technology quality, perceived interaction dimension mainly focus on interactive, including the interaction between learners, between learners and teachers, and between learners and platforms, perceived effectiveness evaluation includes self-evaluation and external evaluation.

2.1 Learner Quality

Due to the considerable differences in online learners' learning motivation, learning needs, learning styles, learning cognition, etc., given the positive relations of the indicators associated with the variety of learner's characteristics, it is more likely that the quality of the learner will influence learning outcomes. [7] There are high requirements on students' autonomous learning ability. In addition, students may have little faith in their abilities to find information, or may lack the meta-cognitive awareness necessary to realistically assess their skills. [14] In this study, students' self-efficacy, self-regulation and engagement were selected as factors to measure. Self-efficacy refers to the individual's conjecture and judgment on whether one has the ability to complete a certain behavior. Bandura [15] defines self-efficacy as "the degree to which people feel confident that they can use their skills to complete a task." It will have an important influence on learning motivation, task choice, efforts, learning perseverance and learning outcome attribution. Learners with a high sense of self-efficacy are more confident in completing online learning tasks and help improve their online learning performance. Learners need self-regulation and commitment to ensure that their learning is regular, consistent and efficient. Therefore, Regulation and engagement are important factors to measure learners' learning effect. Accordingly, the following hypotheses are proposed:

H1a: Learners' sense of self-efficacy positively influences their learning outcomes.
H1b: Learner's self-regulation positively influences learning outcomes.
H1c: The degree of learners' involvement positively influences learning outcomes.

2.2 Instructor Quality

The influence of teachers on learners' learning outcomes is an indispensable dimension and factor. Students are experienced in online learning, and they can experience and feel teachers' teaching attitude, organizational ability and teachers' feedback. All these factors affect the effect and satisfaction of students' learning activities. Instructors will likely need training in this area, as it has not been given much emphasis in the past. [14] Teachers' attitude towards online learning has a significant impact on learners' learning outcomes. [16] For example, teachers with a negative attitude toward online learning should not expect high levels of satisfaction or motivation from students. Therefore, learners' learning outcomes are not only related to teachers' professional knowledge, but also teachers' awareness of online learning and related skills. In addition, previous studies have shown that teacher response has a significant impact on learners' learning outcomes. [17] Teacher response is defined as whether students think the teacher is responding in a timely manner to their problems. When learners encounter problems in online courses, teachers' timely help can encourage learners to continue learning, while teachers' failure to respond to students' problems in a timely manner has a negative impact on students' learning. Thus, the following hypotheses are proposed:

H2a: Teachers' online learning attitudes positively influence learning outcomes.
H2b: Teachers' timely feedback positively influences learning outcomes.
H2c: A teacher's knowledge reserve and professional ability positively influence learning outcomes.

2.3 Technical System and Educational System Quality

Since the online learning quality is based on the network, the measurement of the quality needs to include the course quality and the platform quality. Course quality includes course technology, course quality and platform quality. Course technology focuses on the course navigation, evaluation rules, course cases, and classroom quizzes provided by course design, which are very important for improving learners' learning effect. Course quality mainly refers to the attractiveness of course content, the concept of inheritance and innovation, and the cultivation of innovation ability, which can guide learners to think in multiple dimensions. The presentation of the course, including recording and broadcasting, has different effects on students' sense of social presence. Social interaction was employed as a key factor of success in computer supported collaborative learning (CSCL) and found to have a significant effect on student learning. [7] The stronger the sense of social presence, the higher the learner's satisfaction and the better the learning effect. The online learning platform interface runs smoothly and responds quickly. The interaction function is complete, the media interface operation procedure is concise to the learner study result has the influence. The online learning experience will also improve. Based on that, we propose the following hypotheses:

H3a: Course technology positively influences online learning outcomes.
H3b: The quality of the courses positively influences the learning outcomes.
H3c: Course live broadcast positively influences online learning outcomes.
H3d: The quality of online learning platforms positively influences online learning outcomes.

2.4 Perceived Interaction

Perceptual interaction refers to the learners' direct perception of the interaction quality in the learning process after online learning, including the interaction between learners and media, teacher-student interaction, and student-student interaction. The interaction between teachers and students in online learning helps to enhance the understanding between teachers and students, strengthen mutual trust, and create a warmer online learning environment. A lack of visible interaction between teachers and students can frustrate learners and make it difficult for them to concentrate on the material. According to Arbaugh [18], the more learners interact with others, the more satisfied they are with their learning. In the network learning environment, student interaction and the interaction between learners and media are both conducive to the solution of problems and the progress of learning. Many studies hold the views that participating in online class-based collaborative activities can improve students' learning effectiveness and learning satisfaction [20]. Therefore, the following hypotheses are proposed:

H4a: Students' perception of teacher-student interaction positively influences learning outcomes.
H4b: Students' perceived interactions with students positively influence their learning outcomes.
H4c: Students' interaction with the media positively influences their learning outcomes.

2.5 Perceived Performance

Learning effectiveness refers to the effect that learners who complete an online course can feel or reflect. Online learning effectiveness refers to the process in which learners make value judgments on the whole learning process and learning effect through self-evaluation and others' evaluation based on online learning objectives. The goal of online learning is to achieve at least as much as traditional face-to-face teaching. Learning effectiveness should be a priority when measuring online education. Other important issues such as learning opportunities and cost effectiveness are meaningless if the network environment cannot achieve the same learning results as traditional face-to-face teaching. Therefore, we make the following hypothesis:

H5a: A high rating of online learning services by learners positively influences learning outcomes.
H5b: High external ratings of online learning services positively influences learning outcomes.

3 Methodology

3.1 The Design of Survey

In the field of online education, researchers often use perception data of students to measure online learning effectiveness, and the study also proves that perceptual learning can reflect the actual online learning effectiveness. Thus, this research obtains students' subjective perception of online learning effectiveness by means of questionnaire measurement. The design of the scale firstly refers to the compilation of customer satisfaction or learner satisfaction questionnaires by domestic and foreign experts, and then explores the influencing factors of online learning effectiveness. Finally, based on the results of interviews with undergraduates of police college of China, the measurement scale is preliminarily formed. After completing the first draft of the questionnaire, experts in relevant fields or participants with rich online experience are invited to review and give suggestions. The questionnaire is revised and improved according to the feedback suggestions.

3.2 Scale Development and Testing

Combined with various references, this study divides the questionnaire into three parts. In order to enable learners to fill out the questionnaire truthfully, the main purpose of the survey is explained in the opening part of the questionnaire. The first part mainly includes the student's gender, age, major, daily online time and other information. The second part is mainly used to understand the latent variables corresponding to this research model, with a total of 33 questions. The measurement items are rated with the 5-point Likert scale with statements going from strongly disagree (1) to strongly agree (5). Most of the items involved in the research scale come from existing literature, including 8 items for learner quality, 7 items for Instructor quality, 6 items for Technical system and educational system quality, 8 items for perceived interaction and 4 items for perceived performance. The third part is the open-ended question, which mainly investigates what other comments and suggestions learners have about the current online learning.

3.3 Data Collection

This questionnaire is distributed and recalled through Internet. Ensuring the randomness of the sample area, the samples are distributed in all provinces of the country, and each sample is an independent personality, ensuring the independence of the sample. A total of 347 questionnaires have been recalled, and the invalid questionnaires have been removed. 346 valid questionnaires are analyzed.

3.4 Descriptive Statistics for Each Item

Descriptive statistics for each item are shown in Fig. 2. It can be found from the table that the mean value of all variables is around 4, ranging from 3.84 to 4.4, indicating that the overall response of learners to the measurement variables in this study is relatively positive. The standard deviation of each item is between 0.739 and 1.02, indicating that different learners have some differences in their perception of different items and have their own views on the problem.

Learner quality

Question	Mean value	Standard deviation
Q1	3.84	0.943
Q2	3.98	0.938
Q3	4.01	0.873
Q4	4.23	0.799
Q5	3.87	0.935
Q6	3.89	0.93
Q7	3.87	0.961
Q8	3.92	0.939

Perceived interaction

Question	Mean value	Standard deviation
Q1	4.14	0.829
Q2	4.03	0.912
Q3	4.23	0.835
Q4	4.17	0.876
Q5	4.35	0.739
Q6	4.04	0.914
Q7	4.08	0.915
Q8	3.97	0.973

Instructor quality

Question	Mean value	Standard deviation
Q1	4.38	0.76
Q2	4.38	0.756
Q3	4.4	0.747
Q4	4.38	0.768
Q5	4.32	0.801
Q6	4.44	0.74
Q7	4.32	0.801

Technical system and educational system quality

Question	Mean value	Standard deviation
Q1	4.38	0.76
Q2	4.38	0.756
Q3	4.4	0.747
Q4	4.38	0.768
Q5	4.32	0.801
Q6	4.44	0.74
Q7	4.32	0.801

Perceived performance

Question	Mean value	Standard deviation
Q1	4.05	0.885
Q2	4.1	0.85
Q3	3.97	0.922
Q4	4.01	0.893

Fig. 2. Descriptive statistics of variables

3.5 Reliability Analysis

This study uses SPSS to carry on the reliability and components analysis. We carry on the Cronbach α coefficients tests through questionnaire from five different dimensions, which includes learner quality, instructor quality, technical system and educational system quality, perceived interaction and perceived performance. As for the table, the data show that the structural variables of the five different dimensions of this questionnaire, Cronbach α coefficients are above 0.9, which indicates that the questionnaire has good internal reliability, and the composite reliability of the questionnaire reached 0.986, indicating that the questionnaire has a high reliability.

3.6 Validity Analysis

Discriminant Validity. Figure 3 is the correlation coefficient matrix of variables in this study, in which the diagonal value 1 is replaced by the AVE of potential variables. The criterion for discriminant validity between variables is that if the value of the cell corresponding to the diagonal position is higher than the value of the row and column corresponding to it, the discriminant validity between different variables can be considered to be higher. As can be seen from Table 5, each variable in this study has a good discriminant validity at the concept hierarchy.

		Learner quality	Instructor quality	Technical system and educational system quality	Perceived interaction	Perceived performance
Learner quality	Relationship	1	.615	.598	.636	.712
	P	.000	.000	.000	.000	.000
	N	346	346	346	346	346
Instructor quality	Relationship	.615	1	.582	.639	.603
	P	.000	.000	.000	.000	.000
	N	346	346	346	346	346
Technical system and educational system quality	Relationship	.598	.582	1	.661	.706
	P	.000	.000	.000	.000	.000
	N	346	346	346	346	346
Perceived interaction	Relationship	.636	.639	.661	1	.603
	P	.000	.000	.000	.000	.000
	N	346	346	346	346	346
Perceived performance	Relationship	.712	.603	.706	.603	1
	P	.000	.000	.000	.000	.000
	N	346	346	346	346	346

Fig. 3. Correlation coefficient matrix of variables in study

Convergent Validity. Bartlett's Test of Sphericity statistic is 16817.160, and the p is close to 0, indicating that there is a strong relationship between the 33 variables, while the KMO statistic is 0.969, greater than 0.9, which indicates applicability for factor analysis.

The common measurement of variables indicates that the common measurement of all variables is above 80%. Therefore, the extracted common factor should have a strong explanatory ability to the original variables. It shows that each item has high convergent validity.

4 Analysis and Results

4.1 Predictors of College Students' Online Learning Effectiveness

This study takes the students' experience, instructor quality, technical system and educational system quality and perceived interaction that influence college students' online learning effectiveness as the independent variables of the multi-linear regression model and takes the perceived performance as the dependent variable to carry on multi-linear regression analysis.

We can find from the table that four predictive variables have significant prediction on influencing factors of college students' online learning. The multivariate correlation coefficient of the four predictive variables and the dependent variable is 0.703, the determination coefficient(R^2) is 0.611, and the F value of the overall significance test of regression model is 78.062($p = 0.000 < 0.05$). Therefore, the four predictors can effectively explain the variation of 61.1% of the "influencing factors of college students' online learning" (Table 1).

Table 1. Stepwise multiple regressions analysis of influencing factors of college students' online learning effectiveness

Predictive variables	Multivariate correlation coefficient	Determination coefficient (R^2)	Increment	F	ΔF	B	β
Learner quality	1.298	.600	.582	48.126***	62.758	.121	.229
Technical system and educational system quality	.971	.638	.416	68.642***	97.933	.233	.253
Instructor quality	.736	.451	.035	29.288***	39.002	.302	.285
Perceived interaction	.703	.611	.088	78.062***	102.617	.221	.237

Note: ***$p < .001$

From the predictive power of each variable, the biggest predictor is learner quality, whose amount of variation is 58.2%; secondly is technical system and educational system quality, whose amount of variation is 41.6%; and the other two predictive power of independent variables is 3.5% and 8.8%.

4.2 Standardized Regression Equation of College Students' Online Learning Effectiveness

From the standardized regression coefficients, the β values of the four predictors in the regression model are 0.229, 0.253, 0.285, 0.237, all positive, which indicates that they are positive for "the influencing factors of college students' e-learning". Thus we can get the standardized equation as follow, the influencing factors of college students' e-learning = 0.229 × learner quality + 0.253 × technical system and educational system quality + 0.285 × instructor quality + 0.237 × perceived interaction.

Besides, we can get the standardized residual scatter plot. We find the residual error ranging from −4 to +2, which can explain mostly predictive value, and indicate the regression model is effective.

5 Discussion

5.1 Correlation Analysis of Factors Influencing College Students' Online Learning Effectiveness

The data showed that the correlation between learner quality and perceptual effectiveness was the largest, with a value of 0.712 and a significant positive correlation. The next highest correlation between technical system and educational system quality and perceived effectiveness was 0.706, which was also significant. The correlation between instructor quality and perceived interaction and perceived effectiveness is the same, both of which are 0.603. From the results, we can infer some behavior of students' online learning time. Only when students voluntarily experience and perceive online learning can they have a high degree of satisfaction with online learning education. Moreover, the quality of online learning is positively correlated with the effectiveness. At the same time, instructor quality and perceived interaction are also significantly correlated with the perceived effects of college students.

5.2 Predictive Power Analysis of the Four Factors

Learner quality, technical system and educational system quality, instructor quality and perceived interaction have predictive power.

Learner Quality has Great Predictive Power. This study found that the learner quality of online learning was significantly correlated with learning outcomes, which was consistent with the existing research findings. As such, going online is not seen as a big change for many universities in the world. However, students' readiness and access to technological gadgets needs to be considered at all times.

Learner quality is mainly measured from internal variables such as students' self-efficacy, self-regulation and engagement. The results clearly show that if adults have a strong sense of self-efficacy and control, they can easily overcome difficulties and setbacks in participating in online learning and learn effectively. In addition, it also shows that the stronger the students' desire to participate in online learning, the more helpful it is to generate the behavioral desire to learn. Desire is the internal driving force for people to carry out a certain activity. Learner autonomy can be seen as an essential component of an individual's ability to take greater responsibility for his or her own learning and, thus, engage in lifelong learning.

Technical System and Educational System Quality are Important Factors. Consistent with previous studies, the quality of courses and the quality of online platforms are important factors affecting learning outcomes. The quality of course design, course structure, course content decide the quality of course. In order to enable students to have higher learning satisfaction and achieve better learning results, teachers should carefully arrange the course schedule, teaching methods and select course materials. Design and development of online learning programs requires many decisions, such as pedagogy and learning environment. At present, with the rapid development of network technology, the technology used in online learning environment has been quite mature, but students have high requirements for network bandwidth. Technology is still an important factor in the online learning environment. Universities need to be aware of the critical impact of IT infrastructure services and consider how investment in these services could improve system and information quality.

The Perception of Instructor Quality and Perceived Interaction is Not Strong. The difference between this study and other studies is that the predictive power of instructor quality and perceived interaction is not strong. Most of the respondents in this study are students from police college of China. Compared with other colleges and universities, students pay more attention to politics, self-discipline and self-awareness. Therefore, students' feedback to teachers and interaction with peers are not important or key factors affecting students' learning. On the contrary, they are more likely to accept the current teaching situation, regulate and adjust themselves from internal factors, so as to improve their learning efficiency. In the learner-focused online classroom, course evaluations should refrain from focusing on whether or not the student liked the instructor, but should instead focus on whether the course provided an opportunity for learning.

References

1. Fathema, N., Shannon, D., Ross, M.: Expanding The Technology Acceptance Model (TAM) to Examine Faculty Use of Learning Management Systems (LMSs) In: Higher Education Institutions (2015)
2. Islam, A.K.M.N.: Investigating e-learning system usage outcomes in the university context. Comput Educ. **69**(nov), 387–399 (2013). https://doi.org/10.1016/j.compedu.2013.07.037
3. Chua, Y.P., Don, Z.M.: Effects of computer-based educational achievement test on test performance and test takers' motivation. Comput. Hum. Behav. **29**(5), 1889–1895 (2013). https://doi.org/10.1016/j.chb.2013.03.008
4. Castillo-Merino, D., Serradell-López, E.: An analysis of the determinants of students' performance in e-learning. Comput. Hum. Behav. **30**, 476–484 (2014). https://doi.org/10.1016/j.chb.2013.06.020
5. Ismail, N.Z., Razak, M.R., Zakariah, Z., Alias, N., Aziz, M.N.A.: E-Learning continuance intention among higher learning institution students' in Malaysia. Proc. Soc. Behav. Sci. **67**, 409–415 (2012). https://doi.org/10.1016/j.sbspro.2012.11.345
6. Hodge, E., Chenelle, S.: The challenge of providing high-quality feedback online: building a culture of continuous improvement in an online course for adult learners. Transform.: J. Incl. Scholarsh. Pedagogy **28** (2018). https://doi.org/10.1353/tnf.2018.0013

7. Al-Fraihat, D., Joy, M., Masa'Deh, R., Sinclair, J.: Evaluating E-learning systems success: an empirical study. Comput. Hum. Behav. **102**, 67–86 (2020). https://doi.org/10.1016/j.chb.2019.08.004
8. González-González, I., Serradell-López, E., Castillo-Merino, D.: The Implementation of Process Management: A System to Increase Business Efficiency - Empirical Study of Spanish Companies (2012). https://doi.org/10.4018/IJKSR.2012010102
9. Eom, S.B., Ashill, N.J.: A system's view of e-learning success model. Decis. Sci. J. Innov. Educ. (2018). https://doi.org/10.1111/dsji.12144
10. Demirkan, H., Goul, M., Gros, M.: A reference model for sustainable e-learning service systems: experiences with the Joint University/Teradata Consortium. Decis. Sci. J. Innov. Educ. **8**(1), 151–189 (2010). https://doi.org/10.1111/j.1540-4609.2009.00250.x
11. Lin, K.M., Chen, N.S., Fang, K.: Understanding e-learning continuance intention: a negative critical incidents perspective. Behav. Inform. Technol. (2011). https://doi.org/10.1080/01449291003752948
12. Swan, K.: Learning effectiveness: what the research tells us. Sloan Center Online Educ. **4**(3), 106–116 (2003). https://doi.org/10.1145/1117417.1117423
13. Fornell, C., Johnson, M.D., Anderson, E.W., Cha, J., Bryant, B.E.: The American customer satisfaction index: nature, purpose, and findings. J Mark. **60**(4), 7–18 (1996). https://doi.org/10.2307/1251898
14. Rimland, E.: Assessing affective learning using a student response system. Portal Libr. Acad. **13**(4), 385–401 (2013). https://doi.org/10.1353/pla.2013.0037
15. Bandura, A., Freeman, W.H., Lightsey, R.: Self-efficacy: the exercise of control. J. Cogn. Psychother. (1999). https://doi.org/10.1016/S0091-7435(02)00048-8
16. Sun, P., Tsai, R.J., Finger, G., Chen, Y., Yeh, D.: What drives a successful e-Learning? An empirical investigation of the critical factors influencing learner satisfaction. Comput. Educ. **50**(4), 1183–1202 (2008). https://doi.org/10.1016/j.compedu.2006.11.007
17. Arbaugh, J.B.: Managing the on-line classroom: a study of technological and behavioral characteristics of web-based MBA courses. J. High Technol. Manag. Res. **13**(2), 203–223 (2002). https://doi.org/10.1016/S1047-8310(02)00049-4
18. Arbaugh, J.B.: Virtual classroom characteristics and student satisfaction with internet-based MBA courses. J. Manag. Educ. **24**(1), 32–54 (2000)

Effective Cultivation of Cross-Border E-Commerce Talents via Virtual Communities of Practice with Multi-media, Multi-layered WeChat Groups

Yulu Wang[1], Ying Lu[1], Ping Liu[2], Hailun Wang[1,3], Xiaoli Shen[1], Kewen Shi[1], Yuhui Pan[1], Lin Su[1], and Wen Zhou[1](✉)

[1] NingboTech University, Ningbo, China
1457109691@qq.com, luying@nit.zju.edu.cn, helenw0703@163.com, 810402745@qq.com, 837005905@qq.com, 365975194@qq.com, 3080650691@qq.com, allwin_idc@sina.com

[2] Zhejiang Wanli University, Ningbo, China
lpxc2001@sina.com

[3] Zhejiang University, Hangzhou, China

Abstract. Cross-Border eCommerce (CBeC) is now one of the most dynamic and important phenomena in international business, and the cultivation of talents in this area has become a pressing issue. This paper discusses our experiments in which a Virtual Community of Practice (VCoP) with multi-media, multi-layered WeChat groups was established. Results show that this VCoP has been proved both suitable and effective: it solved the problem of multi-location of participants in this programme by setting up a convenient and almost free-of-charge VCoP; the structure of 3-layered WeChat groups has been both efficient and effective, with students concentrating on the specific product and function in which they specialize; and the use of multi-media functions available on the WeChat platform for lecture delivery and discussion has improved the teaching effects and motivated students in learning and practice.

Keywords: Cross-Border eCommerce · Virtual Community of Practice · WeChat groups · Multi-media & multi-layered

1 Introduction

With the development of the Internet, mobile technology and various business platforms, and the increase in the number of Netizens (4.39 billion in 2019), Cross-Border eCommerce (CBeC), as a new type of international business, has been developing very quickly, and has become one of the most dynamic and important economic phenomena in world business. While B2B is still the major mode, B2C is accelerating its momentum, with the estimated CBeC B2C total value reaching one trillion US dollars for 2020 [1]. In China, CBeC transaction volume exceeded RMB 186 billion (almost 27 billion US dollars) in 2019, an increase of 5 folds than that of 2015 [2].

C. Pang et al. (Eds.): SETE 2020/ICWL 2020, LNCS 12511, pp. 306–317, 2021.
https://doi.org/10.1007/978-3-030-66906-5_29

A typical CBeC transaction involves market participants (buyer, seller and business facilitators) from different countries/economies, with the purchase order being placed via the Internet, and an individual parcel being delivered to the buyer after being shipped from an overseas warehouse. There are often problems such as high logistics cost, long delivery time, language barriers, and different legal and tax conditions [3]. Therefore, talents with CBeC competences need to be cultivated.

In China, since the establishment of the first CBeC Comprehensive Pilot Area in Hangzhou in March 2015, 12 more were approved by the State Council of China in January 2016 [4], and 22 more approved in July 2018 [5]. This rapid development has led to an even more serious shortage of talents with CBeC competences. As recent statistics show, 80% of CBeC companies in China are in need of such talents, and the gap between the supply and the need is as high as 4.5 million [6].

The cultivation of CBeC talents, however, requires a learning model that is different from traditional ways of instruction. Except for the basic language skills of listening, speaking, reading and writing, CBeC talents need to acquire basic business knowledge and hands-on experience in transaction, in addition to modern technology (esp. mobile technology and social networks like WeChat).

2 Literature Review

2.1 Mobile Technology in Learning

With mobile learning's five principal characteristics: portability, accessibility, personalization, social connectivity and increased learning motivation, there is possibility for the teacher to transform the traditional teacher-centered instruction into one focused on the learner [7]. In Zheng's learning model, language learning is languaging in place, a 3D holographic experience. Experiencing events together creates community, one that is emergent, dynamic, place-making and ecological in nature [8]. It is shown that three factors should be attended to in order to achieve the desired effects of the mobile network learning mode: the change of the teacher's role in the teaching process, the acquisition of mobile network applications in the new learning mode, and the support of exterior service technologies [9]. Learning models utilizing mobile technologies, however, are not without demerits. While educational application of mobility may experience a "productive shift" in the teacher-learner relationships, there might be objections and even resistance from both teachers and students who are confronted with new mobile educational practices [10].

2.2 Learning Models Using WeChat

With the popularity of social networks, educators are exploring the possibility of using social network platforms in some learning models. Some analyzes learning interactions between members of Facebook communities [11]. Some explores the design of WeChat-based blended learning [12]. Some even proposes a new cellular learning automata-based algorithm for community detection in complex social networks [13]. In China, the availability and popularity of the social network WeChat has expedited experiments in new learning models. WeChat's micro-innovation strategy makes it possible that one can have such tools as text messaging, hold-to-talk voice messaging, broadcast (one-to-many) messaging, photo and video sharing, and others in one platform [14], and with virtually no cost. The five advantages of using WeChat in learning, i.e., multifunctionality, individuality, accessibility, interactivity, and affordability, enable better immersion and enhanced motivation in language learning [15]. Also under examination is the effect of WeChat-assisted problem-based learning (PBL) model on the critical thinking. Courses other than EFL are also experimented with positive results, for example, Folk Literature by Shang [16]. All these learning models are illuminating, but the learning model for would-be CBeC talents, with its goals of both the acquisition of foreign language skills and the cultivation of CBeC competences, need something more comprehensive. In this case, the concept of "Community of Practice" comes to assistance.

2.3 Community of Practice (CoP) in Learning

Jean Lave and Etienne Wenger first coined the term "community of practice" in 1991. Seven years later, they redefined it as "groups of people who share a concern or a passion for something they do and learn how to do it better as they interact regularly" [17]. In 2015, Wenger & Trayner, further explained that a community of practice implies a commitment to the domain and a shared competence that distinguishes members from other people. Members engage in joint activities and discussions, help each other, and share information [18].

Over the years, more and more educators have become interested in this concept of CoP. The new generation of learning theory has emphasized "learn-to-be" more than "learn about" [19]. Sleeter [20] thinks it a central idea that learning is socially and culturally constituted. Gee [21] explores into the possibility of students' learning through interacting with others and with technology. A study of women and minorities in STEM at MIT shows that learning communities do benefit learning [22].

When CoP is expanded online, new practices emerge. McAllister proposes a curriculum innovation in nursing with the help of an online learning community to cultivate clinical skills and confidence in practice. It is potentially relevant and useful to other groups that seek to develop communal competence, shared identity and aggregated knowledge [23]. Some educators also study project-based flipped learning in out-of-class activities that are based on communities in real and virtual spaces [24]. The factors that affect intention to use an online learning community include: Online Course Design, Perceived Ease of Use, Perceived Usefulness, Previous Online Learning Experience, etc. [25].

3 VCoP with Multi-media, Multi-layered WeChat Groups

In the third semester of 2017, we implemented a Virtual CoP (VCoP) with multi-media, multi-layered WeChat groups for Business English majors for the effective cultivation of their CBeC competences. Altogether 60 students from our university and another vocational college participated in this programme, in which 2 university teachers were acting as directors, 2 CBeC professionals from local companies in Ningbo, 1 from company headquarters in Shenzhen were appointed as tutors for hands-on CBeC training, and 1 teaching assistant based in Paris was invited to facilitate communication between students and overseas warehouses in UK, France, Germany and the U. S. (Fig. 1).

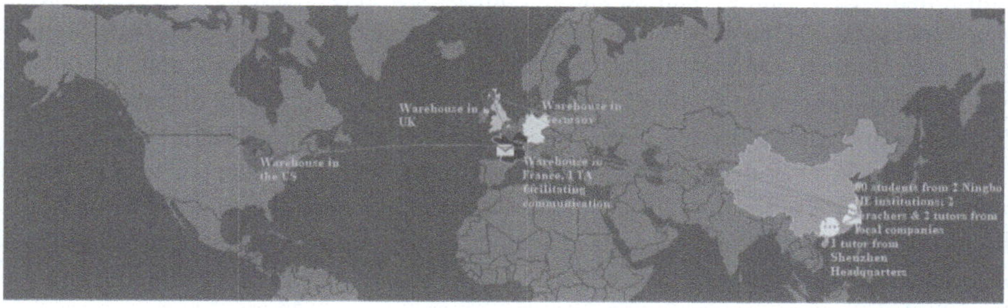

Fig. 1. Location of students, supervisors and overseas warehouses

3.1 3-Layered WeChat Grouping

The first layer is a single big chat group (hereinafter referred to as the Big Group) with all students, teachers and tutors in it. It was kept for instruction of general knowledge, demonstration of basic skills and issue of notices for all students. We then sub-divided the Big Group into 6 product groups as the second layer, and 5 function groups as the third layer (Fig. 2).

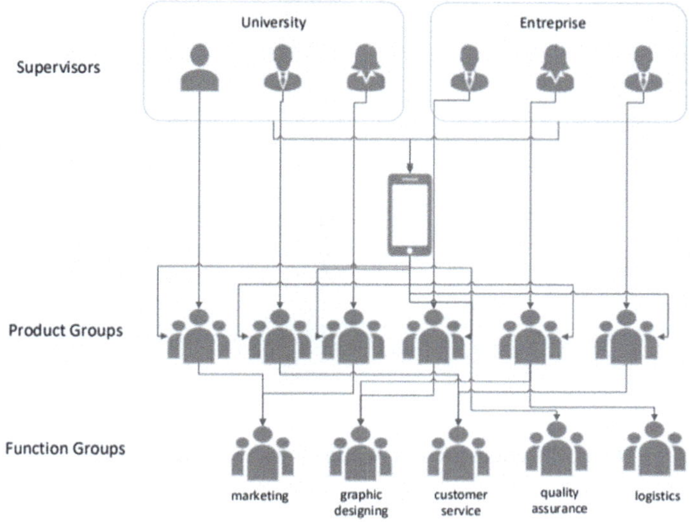

Fig. 2. 3-layered WeChat grouping

Each of the 6 product groups at the second layer specializes in one of the most needed commodities via Chinese CBeC platforms in 2017: 3C electronic products, clothing, home & garden, outdoor products, health & cosmetics, footwear & bags, etc. [2] In each product group, every two members have specialization in different functions: marketing, graphic designing, quality assurance, customer service, and logistics. Thus, for each product group, a sort of intra-group competition could be established with 5 members forming Team A and the other 5 forming Team B, each team having their "specialists". One supervisor was assigned to each product group. Just as study shows, an optimal composition of multidisciplinary learning groups, that is, quantifying various indicators, such as background diversity and similarity of interest among the participants can positively impact the e-learning communities [26].

The third layer is made up of 5 function groups, with 12 students and one or two supervisors in each. In these groups, students specializing in the same function could discuss what they had learned, raise questions to the supervisor/tutor, and share their experience in the course. As has been discovered already, cooperative relationships are the foundation of a model of online learning communities [27].

3.2 VCoP with WeChat's Multimedia Functions

WeChat, initially designed to be a social networking platform, has shown more potential in creating free interactive space of learning due to its communicative and platform-forming functions. In our course, all multimedia functions of WeChat were utilized as follows (Fig. 3).

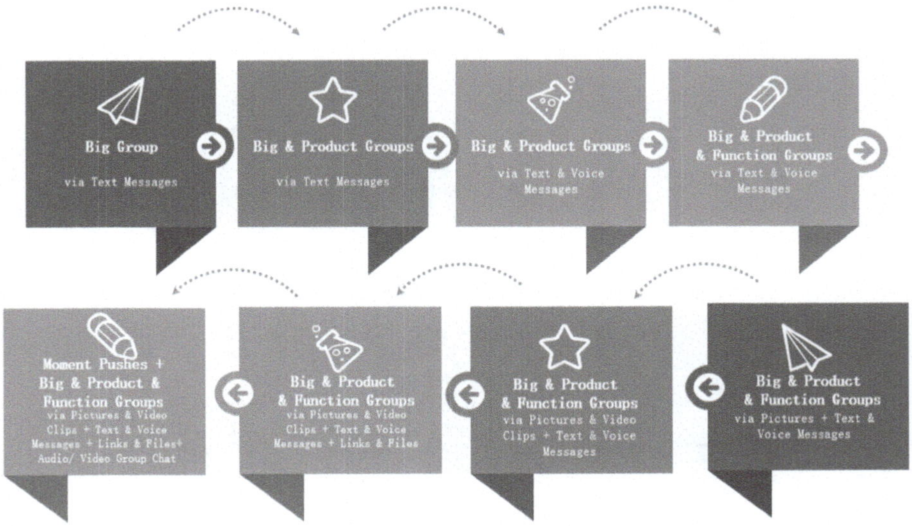

Fig. 3. Various uses of WeChat's multimedia functions

Group Chat. In each chat group, lectures were delivered in the form of voice messages and short text messages, with the aid of pictures (prepared screen captured PPT slides). When it was necessary, ready-made video clips were posted. For example, a clip showing how the spare parts were assembled can help the customer service group better know their products. Besides, students were encouraged to raise questions in any formats that WeChat platform accepts. Supervisors or even students' answers were shared among all group members. Besides, related files (in PDF, Word or Excel formats) were also shared, helping to build up an active VCoP.

Audio/Video Group Chat. As an important feature in the WeChat platform, audio/video group chat was utilized as audio/video conferencing in our VCoP. While this was seldom used in the first period of lecture delivery, it became more and more frequent when students were entering the stage of practical exercises, especially the challenging stage of running their own eBay stores. For each participant of a certain conference (group discussion), the initiator's call was like a phone call, loud enough to arouse his/her attention. And the almost free use (with Wi-Fi connection) of all the functions on the WeChat platform, especially audio/video group chatting, enabled our students to communicate freely with VCoP members, who were scattered in different cities of the country and even overseas, without any fear of financial burden incurred by long distance and even international phone calls. Audio/video group chat could also be simplified as a one-to-one dialogue, when some student wanted to seek help from the supervisor, or the supervisor wanted to provide tutorial to the individual student. Of course, there were limitations for the function of audio/video group chat: 1) the maximum number of audio/video conferencing participants was 9, which meant not all the group members (either the 10 product group members or the 12 function group members) could participate in an audio/video conference; 2) for VCoPs with members in different countries, the time difference posed a sort of limitation for summoning an audio/video group chat, and sometimes our students needed to wait for their overseas supervisor to wake up from her sleep and then seek help from her.

Moment Pushes on the WeChat Public Platform. In our practice, this function was not widely used, because most learning materials could be shared in the 3-tiered WeChat groups. We did use this function, however, when we released some news on this programme, reporting the students' progresses, especially their achievements in their CBeC sales. This actually served as a kind of publicity for enhancing team morale, encouraging students to aim higher and achieve more while the low achievers were encountering some difficulties and even thinking of quitting. And it seems we may make full use of this function in our future practice and research, as teachers can apply for a pubic account specially for the class and push lecture notes and learning materials onto this public service account, which students can follow and receive all the pushes automatically and instantly.

4 Effectiveness Analysis of VCoP

In order to evaluate the effectiveness of our practice, we carried out a questionnaire survey among the students who participated in this programme, asking them 9 questions, 8 being multiple choice statements and questions, and the last an open question. We also carried out interviews with the five students who have successfully run their online CBeC stores after they participated in the programme, and their achievements will be reported.

4.1 Survey Results

The survey was done 6 months after the end of the programme, hoping that students could have a more objective view of both the teaching mode via WeChat and the traditional mode. All the 60 students who participated in the programme were surveyed and they all submitted their answers to the 8 statements and questions on an online survey platform. As to the last open question, 19 students provided some comments and suggestions, most of which being brief comments like "very good", "none", etc.

The survey results show that as many as 81.66% of the students agree completely or basically that the teaching mode via WeChat has motivated them (Statement 1). This is reinforced by some students' comments in Question 9: "We could have more time to learn & practice in our own pace, without the restriction of class hours, thus enhancing our initiative in learning" and "[We] Need our self-discipline to ensure the completion of the learning process and tasks".

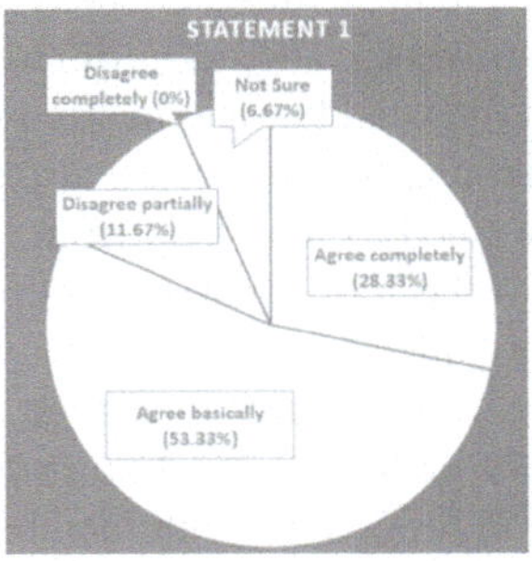

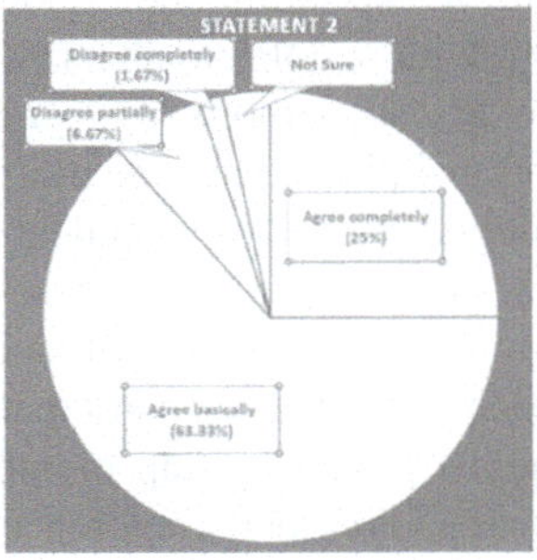

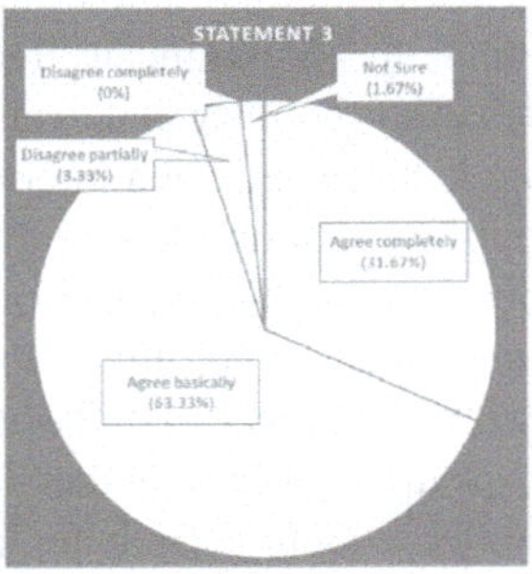

The survey results also show that, as many as 95% of the students agree completely or basically that in the course of the programme, the teachers have made full use of the multi-media functions available on the WeChat platform (Statement 3). Therefore, they enjoyed the convenience and instantaneity of this teaching mode, with 98.33% and 90% of them agree completely or basically with Statements 4 & 5 respectively. Related to these statements, 88.33% of the students say the programme has met their expectations of learning and practice in cultivating CBeC competences (Statement 2).

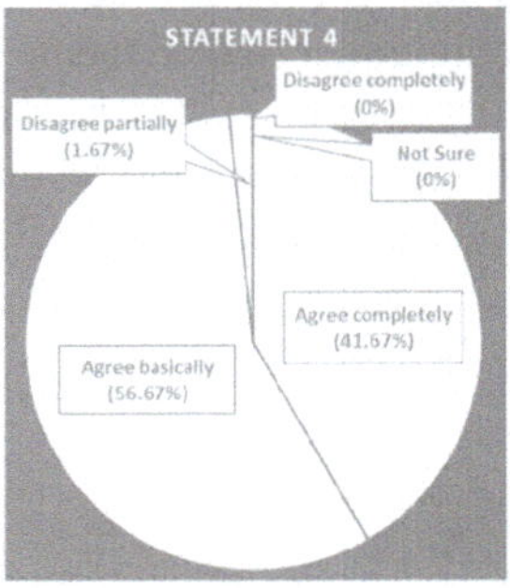

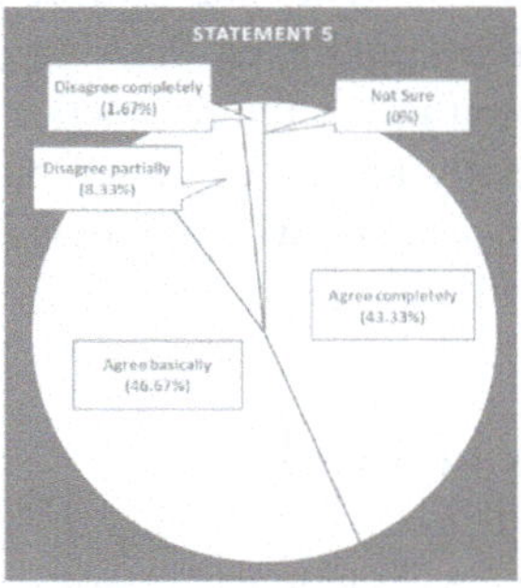

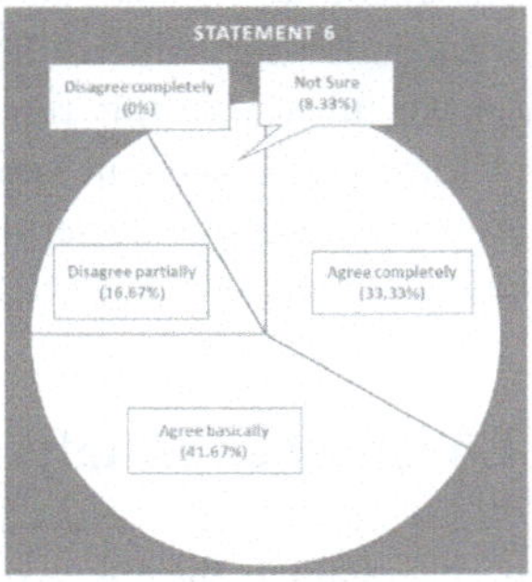

And in terms of effectiveness, 75% of the students say the teaching mode via WeChat is better than the traditional mode in this programme (Question 6). This figure is comparatively low if we consider the previous figures, and this implies that our first experiment on a teaching mode via WeChat could have done better, perhaps just as one of the comments in Question 9 says, "A combination of the teaching mode via WeChat with the traditional mode might achieve better effects." A blended mode is what we are considering in our future programme.

Interestingly enough, when the students were asked to make a choice between the teaching mode via WeChat and the traditional one in the programme of cultivating CBeC competences, 93.33% of them preferred the former (Question 7). But when they were asked whether they would like their teachers to adopt the same teaching mode via WeChat in other courses, only 75% of them gave the positive answer. The discrepancy between the two could be explained that the students might be fully aware of the limitations of the teaching mode via WeChat, which might not be suitable for other courses (Table 1).

Table 1. Questionnaire survey data [No. of students (%)]

Statements & questions	Agree completely	Agree basically	Disagree partially	Disagree completely	Not Sure
1. The teaching mode via WeChat has motivated us in participating in the programme of cultivating CBeC competences	17(28.33)	32(53.33)	7(11.67)	0(0)	4(6.67)
2. This programme has met my expectations of learning and practice in cultivating CBeC competences	15(25)	38(63.33)	4(6.67)	1(1.67)	2(3.33)
3. In the course of the programme, the teachers have made full use of the multi-media functions (text & voice messages, pictures & video clips, etc.) available on the WeChat platform	19(31.67)	38(63.33)	2(3.33)	0(0)	1(1.67)
4. In the course of the programme, the teaching mode via WeChat has fully displayed its convenience	25(41.67)	34(56.67)	1(1.67)	0(0)	0(0)
5. In the course of the programme, the teaching mode via WeChat has fully displayed its instantaneity	26(43.33)	28(46.67)	5(8.33)	1(1.67)	0(0)
6. In terms of effectiveness, the teaching mode via WeChat is better than the traditional mode in this programme of cultivating CBeC competences	20(33.33)	25(41.67)	10(16.67)	0(0)	5(8.33)

(continued)

Table 1. (*continued*)

Statements & questions	Agree completely	Agree basically	Disagree partially	Disagree completely	Not Sure
7. Which teaching mode do you prefer, the one via WeChat or the traditional one, in the programme of cultivating CBeC competences?	WeChat mode: 42(70)	Traditional mode: 14(23.33)		Not sure: 4(6.67)	
8. Would you like your teachers to adopt the same teaching mode via WeChat in other courses?	Yes: 40(66.67)	No: 5(8.33)		Not sure: 15(25)	
9. What suggestions or comments do you have for your teachers who used the teaching mode via WeChat in the programme of cultivating CBeC competences?	Only meaningful comments are shown here: 1) Need our self-discipline to ensure the completion of the learning process and tasks 2) We could have more time to learn & practice at our own pace, without the restriction of class hours, thus enhancing our initiative in learning 3) A combination of the teaching mode via WeChat with the traditional mode might achieve better effects 4) Could have provided more prepared learning materials 5) Should ensure Wi-Fi connection in advance				

4.2 Successful Cases

Among the students who participated in the programme, five of them have continued with their internship after the training programme (both theoretical and practical stages) and successfully run their own CBeC stores, with a total sales value of $300,000 for just 6 months. This was quite an achievement for university students, who had little knowledge of international business or mastery of CBeC competences, and had never thought of running a CBeC store before the start of this programme. Through the systematic and effective training in cultivating CBeC competences via WeChat platform, they acquired substantial and sufficient knowledge and competences, gained hands-on experience, and finally became quite experienced in the CBeC field.

The following data are from their UK store (online store open to UK customers), which displayed the store status and listings as of April 2018. On the one hand, Table 2 shows the bright side of their store operation: beginning from September 2017, after only a little more than 6 months, their store reached the level of “Top-rated Seller”, with very low “Transaction defect rate” (0.28%), very low percentage of “Cases closed without seller resolution” (0.28%), fairly low “Late delivery rate” (4.17%) for a total of 2,246 transactions completed and a sales value of £135,361.20 realised. On the other hand, Table 3 shows the back-end diligence and hardship behind the front-end achievement and glory: within less than 200 days, they uploaded more than 400 product listings, an average of 2 listings per day. Among these listings, only 47.36% (197) were active while 52.64% (219) were unsold, which means they had to adjust their listings constantly to ensure their product listings’ attractiveness to the buyers. That’s why they would rate hard-working and perseverance as one of the most important qualities in CBeC practices when being interviewed about their success.

Table 2. UK store status as of 10 April, 2018

Seller level (Region: UK)	>
Current seller level	Top-rated Seller
If we evaluated you today	Above Standard
Transaction defect rate	0.28%
Late delivery rate	4.17%
Cases closed without seller resolution	0.28%
Transactions (last 12 months)	2,246
Sales (last 12 months)	£135,361.20
Next evaluation 20 Apr	

Table 3. Product listings in UK Store as of April 2018

Listings	>
Create listing	
Drafts	
Active listings	197
with questions	101
with open offers from buyers	4
all auctions	5
with reserve met	0
ending today	10
running out of stock	1
underperforming	3
Scheduled listings	0
Unsold ended listings	219
Show more	

5 Conclusion

For a training programme cultivating talents with CBeC competences, a VCoP with multi-media, multi-layered WeChat groups has been proved both suitable and effective. First of all, it solved the problem of multi-location of the supervisors and students. Where it could have been very inconvenient and costly to organize a traditional training programme of such scale and variety in participants, a VCoP with multi-media, multi-layered WeChat groups has been acclaimed by most students to be convenient and almost free of charge. Neither supervisors nor students had to travel to and convene in some place, find suitable classrooms, rent and fix necessary facilities for international conferencing. Secondly, the structure of multi-layered WeChat groups has been both efficient and effective, with students concentrating on the specific product and function in which they specialize. Thirdly, the use of multi-media functions available on the WeChat platform for lecture delivery and discussion has improved the teaching effects and motivated students in learning and practice.

However, it should be pointed out that a VCoP with multi-media, multi-layered WeChat groups has its own limitations, and may not be as efficient and effective when it is applied to other courses whose focal point is, for instance, face-to-face interaction (such as debating). Also, tools and functions available on the WeChat platform could be more fully used, as is the case of using moment pushes on WeChat. These are rooms for improvement in our future research and practice.

References

1. Institute of China International Electronic Commerce Center: Global E-Commerce 2017 Annual Report (2017). http://www.ec.com.cn/article/zhlt/swblh/201804/27259_1.html?from=singlemessage. Accessed 18th June 2018
2. E-commerce in China 2019. http://www.ec.com.cn/article/yjfx/hybg/202007/273_1.html
3. Kawa, A., Zdrenka, W.: Conception of integrator in cross-border e-commerce. Logforum **12** (1), 63–73 (2016). https://doi.org/10.17270/J.LOG.2016.1.6
4. Lu, B., Wang, H.: Research on the competitive strategy of cross-border E-Commerce comprehensive pilot area based on the spatial competition. Sci. Program. 1–9 (2016). http://dx.doi.org/10.1155/2016/6216052
5. Han, H. (ed.): Two decisions were made by the PRC State Council Executive Meeting. http://www.gov.cn/xinwen/2018-07/13/content_5306333.htm
6. Anonymous. Successful abutment between CBeC companies and talents, favoring the interdisciplinary talents. http://www.nbzsq.gov.cn/news/news!detail.do?id=IF7f62fa1219c1451b9cb35e0bd96032a6
7. Sung, Y., Chang, K., Yang, J.: How effective are mobile devices for language learning? A meta-analysis. Educ. Res. Rev. **16**, 68–84 (2015). https://doi.org/10.1016/j.edurev.2015.09.001
8. Zheng, D., et al.: An ecological community becoming: language learning as first-order experiencing with place and mobile technologies. Linguist. Educ. **44**, 45–57 (2018). https://doi.org/10.1016/j.linged.2017.10.004
9. Lin, W.: Exploring the convergence of the mobile learning mode in network environment and the traditional classroom teaching mode. Int. J. Emerg. Technol. Learn. **12**(7), 170–181 (2017). https://doi.org/10.3991/ijet.v12i07.7248
10. Rankin, W.: Mobility and learning: the challenge of differentiating between being “new” and being a “novelty”. Int. J. Mob. Blended Learn. **10**(2), 1–7 (2018)
11. Giglio, C., Palmieri, R.: Analyzing informal learning patterns in Facebook communities of international conferences. Proc.: Soc. Behav. Sci. **237**, 223–229 (2017). https://doi.org/10.1016/j.sbspro.2017.02.067
12. Zhang, L., Yi, W., et al.: Analysis on WeChat-based blended learning in network marketing course. Int. J. Emerg. Technol. Learn. **17**(14), 85–96 (2019). https://doi.org/10.3991/ijet.v14i17.11277
13. Khomami, M.M.D., et al.: A new cellular learning automata-based algorithm for community detection in complex social networks. J. Comput. Sci. **24**, 413–426 (2018). https://doi.org/10.1016/j.jocs.2017.10.009

14. Yang, X., Sun, S.L., Lee, R.P.: Micro-innovation strategy: the case of WeChat. Asian Case Res. J. **20**(2), 401–427 (2016). https://doi.org/10.1142/S0218927516500152
15. Shi, Z., Luo, G., He, L.: Mobile-assisted language learning using WeChat instant messaging. Int. J. Emerg. Technol. Learn. **12**(2), 16–26 (2017). https://doi.org/10.3991/ijet.v12i02.6681
16. Shang, W.: Construction and application of WeChat learning platform in "Folk Literature" teaching. Int. J. Emerg. Technol. Learn. **11**(5), 10–15 (2016). https://doi.org/10.3991/ijet.v11i05.5688
17. Lave, J., Wenger, E.: Communities of Practice: Learning, Meaning, and Identity. Cambridge UP, Cambridge (2011)
18. Wenger-Trayner, E., Wenger-Trayner, B.: Communities of Practice: A Brief Introduction (2015). http://wengertrayner.com/wp-content/uploads/2015/04/07-Brief-introduction-to-communities-of-practice.pdf
19. Syed Mustapha, S.M.F.D.: Building learning system for content knowledge and social knowledge. Int. J. Emerg. Technol. Learn. **13**(1), 4–22 (2018). https://doi.org/10.3991/ijet.v13i01.6912
20. Sleeter, C.: "Foreword." In: Pease-Alvarez, L., Schecter, S.R. (eds.) Learning, Teaching, and Community: Contributions of Situated and Participatory Approaches to Educational Innovation, pp xi–xiii. Lawrence Erlbaum Associates (2005)
21. Gee, J.P.: Situated Language and Learning: A Critique of Traditional Schooling. Routledge, New York (2004)
22. Russell, L.: Can learning communities boost success of women and minorities in STEM? Evidence from the Massachusetts Institute of Technology. Econ. Educ. Rev. **61**, 98–111 (2017). https://doi.org/10.1016/j.econedurev.2017.10.008
23. McAllister, M., Moyle, W.: An online learning community for clinical educators. Nurse Educ. Pract. **6**(2), 106–111 (2006). https://doi.org/10.1016/j.nepr.2005.10.002
24. Yamashita, K., Yasueda, H.: Project-based learning in out-of-class activities: flipped learning based on communities created in real and virtual spaces. Proc. Comput. Sci. **112**, 1044–1053 (2017). https://doi.org/10.1016/j.procs.2017.08.108
25. Liu, I.-F., et al.: Extending the TAM model to explore the factors that affect intention to use an online learning community. Comput. Educ. **54**(2), 600–610 (2010). https://doi.org/10.1016/j.compedu.2009.09.009
26. Dascalu, M.-I., et al.: Improving e-learning communities through optimal composition of multidisciplinary learning groups. Comput. Hum. Behav. **30**, 362–371 (2014). https://doi.org/10.1016/j.chb.2013.01.022
27. Bassani, P.B.S.: Interpersonal exchanges in discussion forums: a study of learning communities in distance learning settings. Comput. Educ. **56**(4), 931–938 (2011). https://doi.org/10.1016/j.compedu.2010.11.009

A Study on the Autonomous Learning Model of English Pronunciation for Business English Major Students

Chen Ou[1(✉)], Kaijing Zhao[2], Xiang Li[1], and Li Jia[1]

[1] NingboTech University, Ningbo, Zhejiang, China
ocean79ou@163.com, lilixiangx@foxmail.com, jl27li@163.com
[2] Zhejiang Normal University, Jinhua, Zhejiang, China
ramona_zkj@126.com

Abstract. Students' autonomous learning of English pronunciation plays an important role in students' acquisition of English pronunciation. This paper intends to demonstrate, through an empirical investigation, how the establishment of a pronunciation center can facilitate business English major students' autonomous learning. The result shows that the implementation of the pronunciation center not only increases students' interest and confidence in learning English pronunciation, but also promotes students' self-reflection and enhances the interaction, negotiation and cooperation between teachers and students, which surely help the students achieve the sustainability of their autonomous learning of English pronunciation.

Keywords: English pronunciation · Autonomous learning · Pronunciation center

1 Introduction

With the development of economic globalization, international business communication is becoming increasingly important. More and more foreign trade companies are in great need of staff who can carry out international business communication smoothly. Thus it is essential that more and more business English major students, the potential international business conductors, be paid great attention to in college education. These students ought to be trained to have both language skills and business knowledge.

Pronunciation is the foundation for language learning and it plays an essential role in facilitating communication. For business English major students, their pronunciation largely reflects whether they have laid a solid foundation for English learning and how well they can communicate in international business transactions. However, as a result of test-oriented education in middle school and high schools, many business English major students have problems with their English pronunciation before they are enrolled in colleges or universities. To make the matter worse, some students are not given enough time practicing English pronunciation in class after they are enrolled. In order to achieve sustainable improvement of pronunciation proficiency, these students need to spend more time conducting autonomous learning of English pronunciation.

C. Pang et al. (Eds.): SETE 2020/ICWL 2020, LNCS 12511, pp. 318–326, 2021.
https://doi.org/10.1007/978-3-030-66906-5_30

However, according to a survey conducted by the author, many business English freshmen's current situation of English pronunciation autonomous learning is not satisfying. Although, they are all fully aware of the importance of English pronunciation learning and are eager to learn pronunciation autonomously, they do not find their own ways effective or efficient. Hence, it is necessary to think about how, in the process of English pronunciation teaching, teachers can help these students improve their autonomous learning of English pronunciation. This study calls for the application of autonomous learning to English pronunciation acquisition, and proposes the establishment of a pronunciation center to facilitate business English majors' autonomous learning, the effectiveness of which is demonstrated through an empirical investigation.

2 Theoretical Foundation

The discussion on autonomous learning originated in western countries in 1970s. Holec [1] introduces the theory of autonomous learning into the field of foreign language teaching. He defines autonomous learning as "the ability to take charge of one's own learning". Later, Boud [2] identifies that the main characteristic of autonomous learning is "that students take some significant responsibility for their own learning over and above responding to instruction." As autonomous learners, they ought to be able to formulate their own learning objectives, select and make use of appropriate learning strategies, monitor their use of these strategies and assess their own learning [3].

An important development in the definition of autonomous learning in recent years has been a growing emphasis on the role of the teachers in the development of learner autonomy. Ganza [4] argues "it is not sufficient to define learner autonomy as the learner's taking control, or taking responsibility, or knowing how to exercise learning strategies, or being self-directed: the extent to which the learner can realize these achievements depends upon his or her relationship with the teacher." Benson and Voller [5] put forward that teachers remain an integral part of students' autonomous learning process. They state that teachers should interact with learners and provide assistance to them with the aim of helping learners direct their own learning. Galloway and Labarca [6] recommend that teachers provide "scaffolding" for their learners, gradually withdrawing support as learners gain greater task autonomy. Pang Jixian [7] proposes that in autonomous learning, students expect the teachers to be their helpers, promoters, instructors, consultants, coordinators, participants and communicators.

Meanwhile, in autonomous learning process, learners often need to cooperate, interact and negotiate with their peers to promote their learning [8]. When describing a self-access center, Murray [9] mentions that learner autonomy will be promoted significantly if learners are provided with a stress-free environment. Richards [10] points out learners will not be handicapped by their limited English proficiency or fear of making mistakes in front of their peers. Therefore, cooperative learning is also an important approach to facilitate and enhance autonomous language learning.

Since autonomous learning has been regarded as a vital aspect of learning process [11], a number of issues concerning autonomous learning in language learning have come to the fore [12]. Substantial researches relevant to autonomous learning have been carried out from different perspectives [13]. Researchers identify several factors which may influence autonomous learning. They argue that attitude, belief, motivation, reflection, interaction, cooperation, and cultural factor will greatly affect autonomous learning. They also propose that some methods can be applied to promote students' autonomous learning, including setting up syllabus and courses, integrating learner autonomy into classroom teaching, carrying out strategy training, establishing self-access centers, and utilizing multimedia to assist teaching.

3 Implementation of the Pronunciation Center

Gardner and Miller [14] propose to establish self-access centers to facilitate students' self-access learning. According to them, a self-access center has two major functions. First, it can provide self-study language learning materials (grammar, listening, etc.) which learners can use to satisfy their own needs. Second, it can cultivate learner autonomy by encouraging learners to develop individualized strategies, to reflect on learning, and to take responsibility. A self-access center can provide an environment where learners can immerse themselves in their target language. In recent years, self-access centers both at home and abroad have played a positive role in promoting learners' autonomous learning ability [15]. Among all kind of self-access centers, English writing centers get a lot of attention from the public.

In 1970s, English writing centers began to develop rapidly in the United States, which were dedicated to provide a variety of writing assistance to college students and improve students' writing skills [16]. These writing centers are always composed of writing workshops, face-to-face tutoring, online resources and writing groups. Based on the needs and common problems of college students, writing workshops are held regularly by the teachers of English writing centers. The workshops can offer students the opportunities to learn from feedback and discussion obtained from mutual communication between teachers and students. Face-to-face tutoring, which provides students with the opportunity to get face-to-face writing consultation from a teacher, can meet students' individual needs. Online resources offer students rich learning materials, and writing groups facilitate peer evaluation of students' writing assignments. From what has been mentioned above, the English writing centers not only provide more opportunities for students to contact and communicate with teachers, but also help teachers to monitor students' writing processes, enabling students to keep writing effectively and continuously. In particular, individual difference and personal needs are given full attention to in the writing center, thus personalized education can be easily achieved. Practices show that the establishment of these English writing centers can stimulate students' autonomy in writing [17].

With reference to the model of the English writing center, the author establishes a pronunciation center for autonomous learning of English pronunciation. An empirical investigation, lasting for one semester, has been conducted among 56 business English major freshmen of NingboTech University. The following shows how the pronunciation center is carried out:

(1) Setting up English pronunciation workshops. The workshops are regularly held by the teachers of the pronunciation center to meet students' needs and to help students overcome common difficulties. In the workshops, students can discuss the pronunciation acquisition process with their teachers and obtain feedback from their teachers.
(2) Having face-to-face tutoring. Teachers will give guidance to the students who ask for help with their pronunciation learning. Each student, through making an appointment in advance, can have a private session with a teacher. The teacher will conduct a diagnostic assessment on student's pronunciation, establish an English pronunciation portfolio for the student, and give personalized guidance, suggestions and evaluations to students.
(3) Building an online English pronunciation resource bank. In order to facilitate students' autonomous learning, the resource bank provides the related courseware, audios and videos as well as the corresponding self-study plans (learning hours per week, learning content, etc.) for reference. Online resources encourage autonomous learning by allowing students to choose personalized learning materials and adjust their learning progress based on their needs and interests [10].
(4) Establishing English pronunciation learning groups. The groups provide a platform for students to communicate and learn from each other. In the learning group, students can have peer assessment, help each other diagnose pronunciation errors and formulate plans to improve English pronunciation proficiency. Senior students who have enthusiasm, strong sense of responsibility and good pronunciation can be selected as the group leaders. After a short-term pronunciation teaching skills training, those senior students will be responsible for organizing pronunciation learning activities and correcting the pronunciation errors of the group members.

4 Findings and Discussion

In the empirical investigation of the pronunciation center, the author designs questionnaires to evaluate the influence of the pronunciation center on students' autonomous learning. The questionnaire is made up of three parts. The first part is designed to get the basic information of respondents. The respondents are asked to write down their name, age, gender, and other important personal information. The second part is designed to collect respondents' view on the influence of the pronunciation center on autonomous learning. It consists of eight statements. Each statement is given three options for the respondents to choose from: A=Always or almost always true of me; B=Sometimes or somewhat true of me; C=Never or almost never true of me. The eight statements are listed as follows:

1. I am very interested in learning English pronunciation.
2. I find it fun to learn English pronunciation.
3. I am confident in learning English pronunciation.
4. I often reflect on my pronunciation learning.
5. I will pay attention to the progress of my pronunciation acquisition.
6. I can communicate with teachers frequently.
7. I can get the attention from teachers and fellow students.
8. I can get the encouragement from teachers and fellow students.

The third part is designed to collect respondents' self-evaluation of the implementation of the pronunciation center. It consists of three statements. Each statement is given YES/NO options for the respondents to choose from. The three statements are listed as follows:

- I have benefited a lot from the implementation of the pronunciation center.
- I think that my pronunciation proficiency has been greatly improved.
- I think the implementation of the pronunciation center is very helpful to promote the autonomous learning of English pronunciation.

The questionnaires are given to 56 business English freshmen. The first patch of questionnaires is completed by students at the beginning of the semester (before the implementation of the pronunciation center). The second patch of questionnaires is completed by students at the end of the semester (after the implementation of the pronunciation center has been conducted for one semester). Both patches of questionnaires are collected for analysis.

4.1 The Influence of the Pronunciation Center on Autonomous Learning

Cohen [18] identifies that the success of second language acquisition depends largely on language learners themselves, their internal factors and their ability to make full use of various learning opportunities. Therefore, it is significant for learners to hold positive attitudes toward autonomous learning. In the pronunciation center, learner autonomy is fully developed. Students establish learning objectives, make plans, and choose one or several methods to practice their pronunciation according to their own requirements. They can adjust time to join in the pronunciation workshops, face-to-face tutoring and learning groups in order to obtain guidance and assistance. Meanwhile, online resource bank is available for students to choose learning materials they need (English textbooks, books, magazines, multimedia software and online learning resources, etc.). With the help of these online materials, students' learning interest can be stimulated and their learning process can be adjusted based on their own learning ability. Therefore, it can be seen that the implementation of the pronunciation center can improve students' confidence and stimulate students' enthusiasm in pronunciation learning, the result of which is verified by the data presented in Table 1 and Table 2. As is shown in Table 2, after the implementation of pronunciation center, all the respondents show interest in English pronunciation learning. Among them, 41 respondents (73.21% of the total respondents) show great interest in English pronunciation learning. Compared with the data collected before the implementation of pronunciation center, this figure reflects a significant increase in the number of respondents (35.71% of the total) who have acknowledged their keen interest in English pronunciation learning. The percentage of the respondents who find it fun to learn English pronunciation reaches to 69.64%, which is also a sharp rise compared with the data before the implementation (30.36%). At the meantime, the number of the respondents who do not think that learning English pronunciation is fun drops from 26 to 2. Additionally, the percentage of respondents who are always confident in English pronunciation learning has increased from 26.79% (before the implementation of pronunciation center) to 48.21% (after the implementation of pronunciation center).

Changes of the above data reflect that after the implementation of pronunciation center, respondents become more positive towards pronunciation learning and are more willing to take part in pronunciation learning activities. Since it is acknowledged that students' learning interest will affect their emotions as well as learning effects [19], that beliefs and attitudes learners hold will have a profound influence on their learning behavior [20], and that learners cannot develop autonomous approaches to learning without their confidence in their ability to learn successfully [21], it is reasonable to believe that the increase of respondents' interest, confidence and other emotional factors will contribute to their autonomous learning of English pronunciation.

Table 1. Students' feedback on English pronunciation learning before the implementation of pronunciation center

Students' view	Option A		Option B		Option C	
	Number	Percentage	Number	Percentage	Number	Percentage
Statement 1	20	35.71%	17	30.36%	19	33.93%
Statement 2	17	30.36%	13	23.21%	26	46.43%
Statement 3	15	26.79%	18	32.14%	23	41.07%
Statement 4	8	14.29%	19	33.93%	29	51.79%
Statement 5	25	44.64%	20	35.71%	11	19.64%
Statement 6	15	26.79%	11	19.64%	30	53.57%
Statement 7	10	17.86%	13	23.21%	33	58.93%
Statement 8	7	12.5%	10	17.86%	39	69.64%

Table 2. Students' feedback on English pronunciation learning after the implementation of pronunciation center

Students' view	Option A		Option B		Option C	
	Number	Percentage	Number	Percentage	Number	Percentage
Statement 1	41	73.21%	15	26.79%	0	0%
Statement 2	39	69.64%	15	26.79%	2	3.57%
Statement 3	27	48.21%	16	28.57%	13	23.21%
Statement 4	30	53.57%	26	46.43%	0	0%
Statement 5	49	87.5%	7	12.5%	0	0%
Statement 6	33	58.93%	13	23.21%	10	17.86%
Statement 7	26	46.43%	16	28.57%	14	25%
Statement 8	22	39.29%	15	26.79%	19	33.93%

One of the characteristics of "good language learners", according to Stern [22], is that they "consciously monitor their performances". In the process of foreign language acquisition, learners need to constantly reflect and evaluate themselves, including their learning motivation, learning strategies and learning effects [8]. According to Table 1 and Table 2, after the implementation of the pronunciation center, all the respondents learn to keep an eye on their pronunciation learning process. Among them, 53.57% of the respondents will often reflect on their own learning, and 87.5% of the respondents will pay attention to their pronunciation acquisition progress. By contrast, before the implementation of the pronunciation center, only 14.29% of the respondents will often reflect on their own learning, and only 44.64% of the respondents will pay attention to the progress of their pronunciation acquisition progress. This proves that the implementation of the pronunciation center can foster the respondents to reflect on and to pay more attention to their own pronunciation acquisition progress, which will further facilitate them in adjusting their learning strategies and methods to promote their autonomous learning of English pronunciation.

In order to promote autonomous learning, learners often need the guidance, feedback, encouragement and assistance from teachers. In the English pronunciation center, teachers' assistance has been fully offered. In English pronunciation workshops and face-to-face tutoring, teachers communicate with students and help students to know why they should learn, what they should learn and how they should learn. Through online English pronunciation resource bank, teachers provide students with abundant learning resources. Besides, in English pronunciation learning groups, teachers give guidance to students in various learning activities. Likewise, many scholars have stressed the vital importance of the interaction and collaboration between the students in autonomous learning. In the pronunciation center, the establishment of English pronunciation learning groups enables students to enhance autonomous learning of English pronunciation through cooperative learning.

The figure in Table 2 shows that after the implementation of the pronunciation center, 58.93% of the respondents believe they really benefit from the mutual communication between the student and the teacher; 46.43% of the respondents feel that their teachers and their fellow classmates really care about them; and 39.29% of the respondents believe that they actually get the encouragement from their teachers and their fellow classmates. However, the percentages of these three items are just 26.79%, 17.86% and 12.5% respectively before the implementation of the pronunciation center (see Table 1). Hence, it can be proved that the implementation of the pronunciation center is conducive to the interaction and communication between teachers and students as well as between students and their classmates. With the attention and the encouragement from their teachers and peers, students' autonomous learning of English pronunciation can be more effective.

4.2 Students' Self-evaluation of the Implementation of the Pronunciation Center

According to the questionnaire, most respondents' self-evaluation of their pronunciation is relatively low before the implementation of the pronunciation center. After the implementation of the pronunciation center, all the respondents say that they have benefited from this model because their pronunciation proficiency has been improved. They also feel that the implementation of the pronunciation center model is conducive to the autonomous learning of English pronunciation. Among all the respondents, 45 of them (80.36%) think that they have benefited from the pronunciation center, 25 of them

(44.64%) think that their pronunciation proficiency have been greatly improved, and 41 of them (73.21%) think that the implementation of the pronunciation center is very helpful to promote the autonomous learning of English pronunciation.

The above statistics show that the implementation of the English pronunciation center plays an influential role in promoting business English students' English pronunciation acquisition, and that it has been welcomed by the students.

5 Conclusion

Pronunciation learning plays an important role in business English majors' English acquisition. This paper calls for the establishment of a pronunciation center to facilitate business English majors' English pronunciation autonomous learning. Practice proves that the implementation of the pronunciation center not only increases students' interest and confidence in learning English pronunciation, but also promotes their self-reflection, enhance the interaction, negotiation and cooperation between teachers and students and between students and their peers. To conclude, the pronunciation center can effectively facilitate business English majors' autonomous learning of English pronunciation. Its construction may shed light on the effectiveness of English pronunciation teaching and learning.

Acknowledgments. This research was financially supported by the Zhejiang Provincial Planning Office of Philosophy and Social Science (Grant No. 17NDJC033Z) and by the Ministry of Education of the People's Republic of China (Grant No. 17YJC740072).

References

1. Holec, H.: Autonomy and Foreign Language Learning. Pergamon Press, Oxford (1981)
2. Boud, D. (ed.): Developing Student Autonomy in Learning. Kogan Page, New York (1988)
3. Dickinson, L.: Talking shop: aspects of autonomous learning. J. ELT J. **47**(4), 330–336 (1993). https://doi.org/10.1093/elt/47.4.330
4. Ganza, W.L.: Learner autonomy—Teacher autonomy: interrelating and the will to empower. In: Lamb, T., Reinders, H. (eds.) Learner and Teacher Autonomy: Concepts, Realities, and Responses, pp. 63–79. John Benjamin, Amsterdam (2008)
5. Benson, P., Voller, P.: Autonomy and Independence in Language Learning. Longman, London (1997)
6. Galloway, V., Labarca, A.: From student to learner: style, process and strategy. In: Birckbichler, D.W. (ed.) New Perspectives and New Directions in Foreign Language Education, pp. 111–158. National Textbook Company, Illinois (1990)
7. Pang, J.X., Ye, N., Zhang, Y.L.: Learner autonomy: identity and self. J. Foreign Lang. Teach. **2004**(6), 22–25 (2004). https://doi.org/10.13458/j.cnki.flatt.002900. (in Chinese)
8. Deng, L.M.: Developing learner autonomy in foreign language acquisition. J. Sichuan Int. Stud. Univ. **2004**(2), 156–159 (2004). (in Chinese)
9. Murray, G.: Older language learners, social learning spaces and community. In: Benson, P., Reinders, H. (eds.) Beyond the Language Classroom, London, pp. 132–145. Palgrave Macmillan, Houndmills (2011)

10. Richards, J.: The changing face of language learning: learning beyond the classroom. J. RELC J. **46**(1), 5–22 (2015). https://doi.org/10.1177/0033688214561621
11. Orawiwatnakul, W., Wichadee. S.: An investigation of undergraduate students' beliefs about autonomous language learning. J. Int. J. Instr. **10**(1), 117–132 (2017)
12. Pemberton, R.: Introduction. In: Pemberton, R., et al. (eds.) Taking Control: Autonomy in Language Learning, pp. 1–8. Hong Kong University Press, Hong Kong (1996)
13. Xu, J.F., Zhan, X.H.: Research on learner autonomy both at home and abroad. J. Foreign Lang. World **2004**(04), 2–9 (2004). (in Chinese)
14. Gardner, D., Miller, L.: Establishing Self-Access: From Theory to Practice. Cambridge University Press, Cambridge (1999)
15. Hua, W.F.: Investigation report on the establishment a self-access language learning center. J. Foreign Lang. World **2003**(06), 43–48 (2003). (in Chinese)
16. Elizabeth, H.B.: "Our little secret": a history of writing centers, pre- to post-open admissions. J. Natl. Council Teach. Engl. **50**(3), 463–482 (1999). https://doi.org/10.2307/358861
17. Gu, J.X., Ding, Y.: A new means of teaching English writing—Online writing lab. J. Media Foreign Lang. Instr. **2002**(5), 37–40 (2002). (in Chinese)
18. Cohen, A.D.: Language Learning: Insight for Learners, Teachers and Researchers. Newbury House/Harper Row, New York (1990). https://doi.org/10.2307/3586956
19. Shu, D.F., Zhuang, Z.X.: Modern Foreign Language Teaching-Theory, Practice and Method. Shanghai Foreign Language Education Press, Shanghai (1996). (in Chinese)
20. Cotterall, S.: Readiness for autonomy: investigating learner beliefs. J. Syst. **23**(2), 195–205 (1995). https://doi.org/10.1016/0346-251X(95)00008-8
21. Wenden, A.: Learner Strategies for Learner Autonomy. Prentice Hall, New York (1991)
22. Stern, H.H.: What can we learn from the good language learner? J. Can. Mod. Lang. Rev. **31**(3), 304–318 (1975)

Online Language Education

Distance Education and MOOCs for Language Learning in China

Weihua Du[1] and Shili Ge[2](✉)

[1] Faculty of European Languages and Cultures, Guangdong University of Foreign Studies, Guangzhou 510420, China
[2] Laboratory of Language and Artificial Intelligence, Guangdong University of Foreign Studies, Guangzhou 510420, China
geshili@gdufs.edu.cn

Abstract. Massive open online course (MOOC) is an online learning tool, especially for distance learning. It has attracted a great deal of attention by higher education institution around the globe. It gave rise to academic discussion on MOOC impact, design and research. However, researches on MOOC's impact on language learning are still lacking. Based on the analysis of the external and internal defects of MOOCs, this paper attempts to analyze the methods to solve these defects and put forward some methods, such as paying equal attention to business and knowledge goal, arranging teaching plan reasonably, arranging teacher-student ratio, and adopting multidimensional formative evaluation, to promote the development of Chinese foreign language MOOCs.

Keywords: MOOCs · Distance education · Foreign language learning

1 Introduction

China online courses have mobile platforms, the first one is the online conference tools such as Tencent conference, Zoom, BiliBili, and the second one means all kinds of small video platforms, such as Tik Tok, Kwai, and the third is mobile phone Apps, such as daily English listening, daily German listening, Podcast, Coco English Himalaya. The most important is MOOC platform.

Since the first MOOC in 2013, the construction of foreign language online open courses in China has made rapid progress. More than 12 500 courses have been put into use online, and more than 200 million people have chosen courses for learning. In 2018 and early 2019, the Ministry of Education selected and launched 1291 online open courses as MOOCs with national best quality (referred to as NSOOC, National Superior Online Open Courses) [1].

C. Pang et al. (Eds.): SETE 2020/ICWL 2020, LNCS 12511, pp. 329–333, 2021.
https://doi.org/10.1007/978-3-030-66906-5_31

There are as many as 20 platforms for NSOOC courses, most of which come from three platforms: icourses[1] (MOOC of China University), zhihuishu[2] and xuetang[3]. In particular, icourses carry the most courses, 916 of 1291 courses come from icourses platform, accounting for 71%. According to the results of the Chinese Ministry of Education's recognition in early 2018 and early 2019, 44 foreign language online open courses were selected to the NSOOC, accounting for 3.4% of the total number selected best MOOCs.

Among 44 foreign language courses, 28 are language skills courses, accounting for 63.6%. Most of these courses are for the training of language listening, speaking, reading, writing and translation. Secondly, there are 8 literature and culture courses, 4 special purpose courses, 2 business foreign language courses and 2 other language courses. And there is not a linguistics course that has been recognized as the national best MOOCs.

The Course names for MOOCs with national best quality as following:

A. Language skill (28 courses)

They are English listening and speaking, New Scientist English: speech and writing, Translation with "Tao", Direct English speaking, Expanding English vocabulary, College English speaking, Higher Vocational Public English, English reading and writing, English science thesis writing and academic report, College English self-study course, General English (I), College English process writing, College English academic reading, English-Chinese mutual translation Methods and skills, Academic communication English, College English writing, English speech, English public speech, College English, English grammar and writing, New standard college English, English effective expression: Language rhetoric and logic, Workplace communication English, College English comprehensive course, College English writing, Oral English talk, Higher vocational English, American drama, Learning oral language.

B. Literature and culture (8 courses)

These courses are Impression of Britain and the United States, History of foreign literature, Selected readings of foreign literature classics and realistic considerations, Selected readings of famous British and American poems, Appreciation of English works of Nobel Prize writers, British and American poems, General Lesson of Chinese culture (in English), and English speaking about China.

C. Special purpose (4 courses)

International Communication English, IT Workplace English, Police English, Workplace English.

D. Business foreign languages (2 courses)

Business English, Business English.

[1] http://www.icourses.cn/home/.

[2] www.zhihuishu.com.

[3] https://next.xuetangx.com/.

E. Other languages (2 courses)

Introduction to Japanese, Japanese culture and Korean.

Some of the online open courses have not been effectively used in the teaching after the completion of the construction. They are few responses to the students' problems, the problems in the discussion area have not been updated in time, and they ignore the learning effect of the students.

2 The Internal and External Characteristics of MOOCs

2.1 The External Characteristics of MOOCs

The contradiction between commonweal and Commerce is different. The development of MOOCs has brought impact to the offline traditional courses with profitability. The development of courses is time-consuming and laborious, and there is also a need to break in between teachers and MOOCs platforms [2].

If there is no sustained commercialization and only the support of the national or provincial education departments, the continuity of MOOCs is not guaranteed. If the business goal is too strong, learning becomes business, this kind of MOOC is actually a closed online course, without the characteristics of MOOCs.

2.2 The Internal Characteristics of MOOCs

Talent Training Program. It is convenient to promote the development of MOOC in our Society, but MOOCs cannot replace the learning of the whole course content, because at present, even in a quite long time, the offline application practice of foreign language is necessary and irreplaceable [3].

There are more difficulties in the promotion of MOOCs developed by the universities and colleges in other universities. How to get the recognition and support of other universities? How to integrate the talent training plans of different universities? How to distinguish it from the existing teaching contents in Colleges and universities? All these questions should be answered.

Students' Ability of Self-control. Low completion rate of students, high proportion of teachers and students lead to obstacles in online interaction. When students use mobile phones and computers to study, there are some difficulties in their concentration. Many students are often interrupted by mobile games or information when using online courses.

In terms of students' learning habits, they often rely on external supervision. If the teacher's supervision is not timely enough, some students' learning status is not very good.

3 External and Internal Characteristics of MOOCs

3.1 Dealing with the Lack of Externalization

Conduct more SPOCs Education. It is an online course for small-scale and small group students. For large-scale learners, the course content of MOOCs is usually a course with high repetition rate, strong general knowledge and meeting the needs of the public. On the contrary, SPOC is aimed at the courses with low repetition rate and strong specialty.

Carry out Degree education. Qualified colleges and universities can cooperate with relevant technical service institutions to carry out MOOC teaching for some degree courses. If the degree awarding standards are met, appropriate degree will be awarded to attract more people to participate in MOOC learning.

3.2 Dealing with the Lack of Internalization

Hybrid Teaching in MOOCs. For example, online teaching focuses on innovative basic theory explanation, while offline teaching focuses on application of the theory. The teacher-student ratio of MOOC is much higher than that of offline courses, which can make up for this defect [4].

Protect the Diversity of MOOC. We should design different contents, forms and methods of offline teaching and online teaching. Do a lot of online practices and after class contact counseling in foreign language courses. Foreign language teaching attaches great importance to the cultivation of basic language ability, especially for the students who start from scratch. They should master vocabulary and grammar skillfully, and make a lot of connections to consolidate and test their learning level [4].

Different Levels for Foreign Language MOOCs. These MOOCs should emphasize the different levels, and divide the level and difficulty of the courses according to the Chinese foreign language ability scale or the European Union language ability level. Step by step, foreign language learning can have better results.

Reduce the Ratio of Teachers to Students. This mode is also called "1+N" teaching mode, that is, 1 foreign language MOOC teacher teaches online, and N foreign language teachers of various colleges and universities conduct flipped teaching in N classes offline. Through the cycle and mutual promotion of learning before, during and after class, the online and offline communication and interaction between teachers and students are integrated to form a deep learning field of MOOCs and flipped classroom integration [5].

Reduce the Single Teaching Time. General offline courses are usually 40–50 min. MOOCs can't. The Students mainly study MOOCs on mobile terminals (especially mobile phones), and their contents should conform to the habit of fragmented learning. Curriculum design should be micro curricular, and the duration of each section should be controlled within 10 min as much as possible, with 5 min as the best.

Dynamic Formative Evaluation. The evaluation method needs to be changed, focusing on formative evaluation and improving the evaluation system. For foreign language teaching in the technology era, especially when using online open courses for foreign language teaching, foreign language teachers should actively abandon the traditional final evaluation method with one test, adopt innovative teaching evaluation method, use learning portfolio, interview, classroom display and other ways to conduct dynamic formative evaluation of students' performance. At the same time, teachers should adopt the ways of interaction between teachers and students, peer evaluation, online testing, etc., track and record students' learning in real time, focus on their usual performance, cultivate students' independent learning ability, and form a more comprehensive, objective and multi-dimensional formative evaluation method.

4 Conclusion

MOOCs are the product of the deep integration of information technology and higher education. It leads the reform of teaching and promotes education equity. Good MOOCs have the characteristics of high-level, innovative and challenging. At present, the quality of foreign language MOOC courses in China is uneven, the number of courses in Colleges and universities in various provinces and cities is few, the influence needs to be strengthened, and it is still in the initial stage.

There are many internal and external characteristics by MOOCs. The arrival of 5G era brings new development opportunities for the construction of online open courses. Colleges and universities should actively participate in the construction of MOOCs, lead foreign language teaching with information technology, improve the quality of foreign language teaching, promote the improvement of undergraduate and graduate teaching quality, and create a new paradigm of foreign language teaching and Reform in the technology era.

References

1. Xu, S.: Reflection on the teaching of broadcasting theory based on MOOC. J. News Res. **4**, 201–202 (2020)
2. Qin, J.: Foreign language MOOCs in China: current situation, problems and countermeasures. TEFLE **6**, 55–61 (2019)
3. Zhang, F.: Exploration of the mixed teaching mode based on MOOCs and traditional teaching mode. Educ. Teach. Forum **22**, 316–317 (2020)
4. Guerrero, M., Heaton, S., Urbano, D.: Building universities' intrapreneurial capabilities in the digital era: the role and impacts of Massive Open Online Courses (MOOCs). Technovation **1**, 1–19 (2020)
5. Poquet, O., Jovanovic, J., Dawson, S.: Differences in forum communication of residents and visitors in MOOCs. Comput. Educ. **156**, 1–17 (2020)

Evaluating Online Courses: How Learners Perceive Language MOOCs

Zixuan Ye(✉) and Rong Luo(✉)

Hangzhou Normal University, Hangzhou, China
hshalyy@163.com, Longrong2684@gmail.com

Abstract. Course evaluation is a mirror of the quality of courses. In recent years, Language MOOCs has become an emerging field of online language education. How learners perceive language MOOCs (LMOOCs) and what constitutes their evaluation of these online courses is vital to the development of online foreign language education. The present study investigates the evaluation of China's LMOOCs from learners' perspective and reveals existing problems of current LMOOCs. Based on 600 evaluation comments from learners on China's biggest MOOC platform "icourse", our study analyzes distinctive features of language MOOCs that impact on learners' course evaluation and identify key factors in their evaluation. The research findings show that most learners give positive evaluation of LMOOCs in the study. Meanwhile technology and online management are two dimensions that have brought negative evaluation. The results shed light on the status quo of current LMOOCs and expand the scope of course evaluation to the cyberspace.

Keywords: Course evaluation · LMOOCs · Learners · Grounded theory

1 Introduction

The rapid development of Internet and digital technologies have brought a new era of E-learning in higher education. Online courses have played an important role in the spread of E-learning all around the world. Among all the courses, Massive Open Online Courses (MOOCs) provide unlimited learning opportunities for learners and have received the most attention in the past decade.

As an emerging field in online foreign language education, Language MOOCs (LMOOCs) are "dedicated Web-based online courses for second languages with unrestricted access and potentially unlimited participation." [8]. An ideal Language MOOC is a good practice of language teaching and learning that engages students in the use of the target language in meaningful and authentic ways. Language MOOCs provide innovative learning experiences for learners, especially during the outbreak of COVID-19 around the world. Up until 2020, there are more than 500 LMOOCs in China provided by more than 20 MOOC platforms, ranging from big

This paper is supported by The National Social Science Foundation of China (19CYY049) and Special Project of Social Science Planning & Talent Development of Hangzhou (2018RC2X14).

C. Pang et al. (Eds.): SETE 2020/ICWL 2020, LNCS 12511, pp. 334–343, 2021.
https://doi.org/10.1007/978-3-030-66906-5_32

national platforms to smaller local platforms. The current LMOOCs include all kinds of foreign language courses, represented by ESL speaking, writing, listening, reading and integrative skills, as well as ESP and cultural studies courses. Now LMOOCs are playing a more important role in foreign language education than any time in human history.

Compared with traditional face-to-face courses, research on online courses did not receive much attention until recent years. Most of the studies focus on the features of distance or online courses and few of them have touched upon the evaluation of online courses. Online course evaluation is an indispensable component of educational ecosystem to guarantee the quality of online teaching and learning. There are multiple ways to effectiveness of online courses, such as questionnaires, interviews, log files [2]. However, how to evaluate online open courses is a new topic that has just aroused people's interest and is now under heated discussion.

2 Online Course Evaluation: Related Studies

The study of online course evaluation started in the beginning of the 21st century. From that time, scholars noticed the difference between distance learning and face-to-face learning [1–3, 5, 6]. These studies made a comparison between students' evaluation towards online and face-to-face learning environments. They concluded that online course evaluation involves much more complex factors. Most of these previous studies on online course evaluation focused on traditional online courses like distance education instead of open online courses.

Research on the evaluation of MOOCs did not appear until recent years. Grover and his colleagues [4] proposed an evaluation framework grounded in CSCL along with a discussion of the unique aspects of MOOCs. Zhao and Dong [9] designed a teaching quality assessment model for the MOOC platform based on comprehensive fuzzy evaluation. Lin and Cantoni [7] assigned twelve indicators to the model's four levels of evaluation and designed pre-, in- and post-course surveys and post-course interviews to investigate the effectiveness of the course. The above course evaluation system includes the evaluation of the dimensions like MOOC materials, the learning process and the results.

Most of the above studies were based on personal experience and these MOOC quality evaluation frameworks are mostly based on experts' perspective. Right now the empirical research into the evaluation of MOOCs is rare. While experts and important stakeholders of MOOCs, learners' evaluation are always neglected. As for LMOOCs, which are often the most popular MOOCs on MOOC platforms, the evaluation of these courses are valuable to guarantee the good quality of online foreign language education.

The current study probes into the evaluation of LMOOCs in China from learners' perspective. The present study probes into the three questions step by step and reveals the features and problems of current language MOOCs. What is learners' overall evaluation of China's LMOOCs and how are the evaluations of three types of LMOOCs different from each other? Which factors have influenced learners' evaluation of LMOOCs? And what are the existing problems of current LMOOCs and what can be done to solve these problems?

3 Evaluation of LMOOCs from Learners' Perspective

3.1 Theoretical Framework

The present study investigated learners' evaluation of LMOOCs based on grounded theory. Grounded theory is a systematic and inductive research method for qualitative research and was first proposed by Barney Glaser and Anselm Strauss. Grounded theory aims to build a theory based on empirical data. In order to build up the substantive theory from the bottom up, the theory abstracts the core concepts reflecting the essence of phenomena based on systematic data collection, and then construct the relevant social theory through the connections between these concepts. The procedures involved in the process of grounded theory generally include: 1) generate concepts from the data, and register the data step by step; 2) continuously compare data and concepts, and systematically ask generative theoretical questions related to concepts; 3) develop theoretical concepts and establish connections between concepts; 4) theoretically sample then systematically encode data; 5) construct the theory that strive to obtain the density, variability and high integration of theoretical concepts.

3.2 Data Collection and Analysis

The current study collected 600 evaluation comments from LMOOCs learners and analyzed the comments using grounded theory. Among all MOOC platforms, "icourse" provides more than 350 LMOOCs, which is now the biggest provider of language MOOCs. This study chose 6 representative LMOOCs involving 3 different themes on this platform[1], which include 2 ESL speaking courses 2 ESL writing courses, and 2 ESL reading courses. Those courses are the most popular ones among all the courses with same theme like ESL reading courses.

On the platform of "icourse", there is an evaluation page to show learner's evaluation of the course. Each course in our study has more than 300 comments from anonymous learners. To some degree, those comments represent learners' overall evaluation and attitude of China's current LMOOCs (Fig. 1).

[1] The six LMOOCs chosen in the present study are the most popular LMOOCs on the "icourse" platform with the largest number of learners on the platform. Some of the courses have more than 1 million learners up until 2020.

Fig. 1. The webpage with learners' evaluation of LMOOCs on "icourse"

The present study includes three steps of analysis. The study makes use of Nvivo 12 in the analysis of learners' comments. The first one is data collection. To collect 600 evaluation comment from ten courses, 100 comments were chosen randomly in each course. Most of the comments were written in Chinese and the study coded the comments according to their meaning.

The second step is to judge whether the comment is positive or negative. Simple comments like "This course is great" or "This course is too boring" clearly express the learners' positive or negative attitude toward the course. While complex comments like "I like teacher's pronunciation, but the peer review is unreasonable and unfair." would be divided into two categories, the first part being a positive praise of the teacher and the second part being a negative complaint about the peer review.

The third step is to code learners' comments into five dimensionalities based on their specific content. Overall and irrelevant evaluation like "I like this course" is not taken into consideration in this step. Comments like "This course is interesting, and I have learned a lot from its resources" is coded as "positive+teaching content". "The course itself is very useful, but this platform is really hard to use" is coded as "positive +teaching content & negative+technology". The coding procedure via Nvivo is shown in Fig. 2.

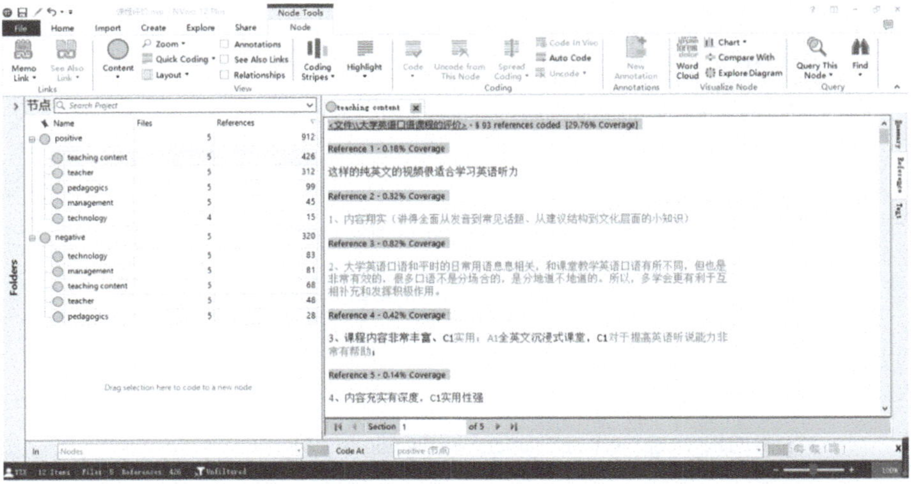

Fig. 2. The coding procedure via Nvivo

3.3 Results

The Overall Evaluation of LMOOCs from Learner's Perspective. After coding all 600 comments into the nature of positive or negative, the result shows that most of the evaluation comments of six LMOOCs are positive[2]. The positive evaluation of online courses is common in China partly because Chinese people attach great importance to "face" or dignity. Learners generally give positive evaluation or five stars to show politeness but their specific comments contain many detailed suggestions and complaint.]. To be more specific, the ratios of positive/negative comments of 3 types of courses are shown in Table 1. With 72.4% positive comments on average, the two ESL Speaking LMOOCs has the highest ratio of positive evaluation among the six LMOOCs in our study. Although ESL reading LMOOCs obtain a lower ratio of positive evaluation than other types of courses, the percentage of positive comments of the two courses is still nearly 70%.

[2] The positive evaluation of online courses is common in China partly because Chinese people attach great importance to "face" or dignity. Learners generally give positive evaluation or five stars to show politeness but their specific comments contain many detailed suggestions and complaint.

Table 1. The ratio of positive/negative comments of each type of courses

	The ratio of positive comments	Examples of positive comments	The ratio of negative comments	Examples of negative comments
ESL Speaking LMOOCs	72.4%	This teacher's oral English is proficient and I have learned many English communicative skills in this course	27.6%	The teacher spoke so fast that I couldn't follow him
ESL Writing LMOOCs	70.5%	The course told us how to write a proper English essay and I really like this course	29.5%	The workload of this course is too heavy, I can hardly finish homework in time
ESL Reading LMOOCs	69.9%	The course is entertaining. Those reading skills taught in this course are very useful	30.1%	If the teacher can give more examples when he explains knowledge points, that would be better

The Overall Evaluation of LMOOCs from Learner's Perspective. Table 2 show the 5 dimensions that learners' evaluation comments mostly describe and focus on.

Table 2. Five dimensions that learners' evaluation focus on

Key dimensions in learners' evaluation	Statements	Specific examples
Teacher	Teachers have good teaching ability and language proficiency	Teacher's oral English proficiency; Teacher's teaching style; Teacher's teaching ability
Teaching content	The course provides rich and useful teaching content, which helps learners to realize effective learning	Authentic educational resources; Use of multimedia/tech; Variety of activities that promote all basic language skills and support cultural awareness
Pedagogy	The course provides effective and innovative teaching methods	Instructional design; Collaborative learning; task-based teaching; Scaffolding teaching; Reflected learning; Connected community
Technology	The course and the platform provides helpful and accessible functions to assist online language learning	Quality of videos; Subtitles in the videos; The availability of supporting materials; The convenience of the platform; Functions of the platform
Management	The course organizers and teachers manage and implement the teaching online appropriately and in a careful manner	Communication (peer-peer, student-teacher, open class community); The assignment of exercises and teacher's feedback; Peer review; Providing assessment criteria and appropriate implementation

Bearing in mind the five key dimensions or factors in learners' evaluation, the present study continued with an in-depth analysis of positive/negative comments in the five dimensions of three types of LMOOCs.

Firstly, for ESL speaking LMOOCs, teaching content and teacher are two dimensions that gain most positive comments. And "management" and "technology" are two dimensions that receive more negative comments than others. The ratios of negative comments of these two dimensions are both over 70%, which means that learners of ESL speaking LMOOCs have come across many difficulties concerning course management and technology. Learners' evaluation of the five dimensions in ESL speaking MOOCs is shown in Fig. 3.

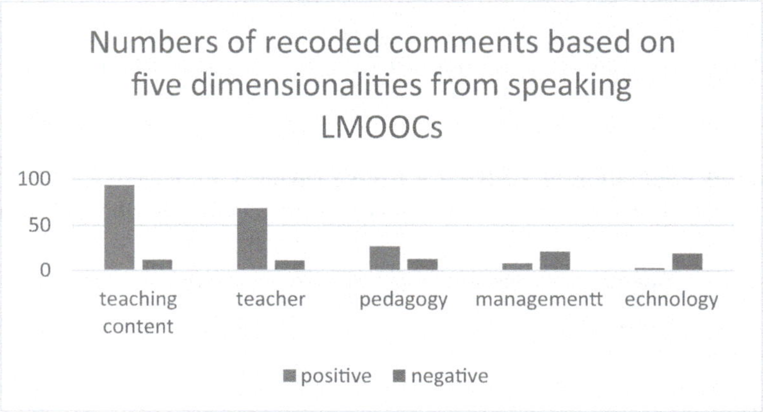

Fig. 3. Numbers of positive/negative comments in five dimensions in ESL speaking LMOOCs

For ESL writing MOOCs, there are still more negative comments concerning "management" and "technology" in writing LMOOCs as Fig. 4 shows. However, one thing which is worthy of attention is the number of negative comments of "teaching content" is larger than that of "management" and "technology", which means many learners find teaching content unsatisfying and the teaching content is not good enough to meet the needs of learners.

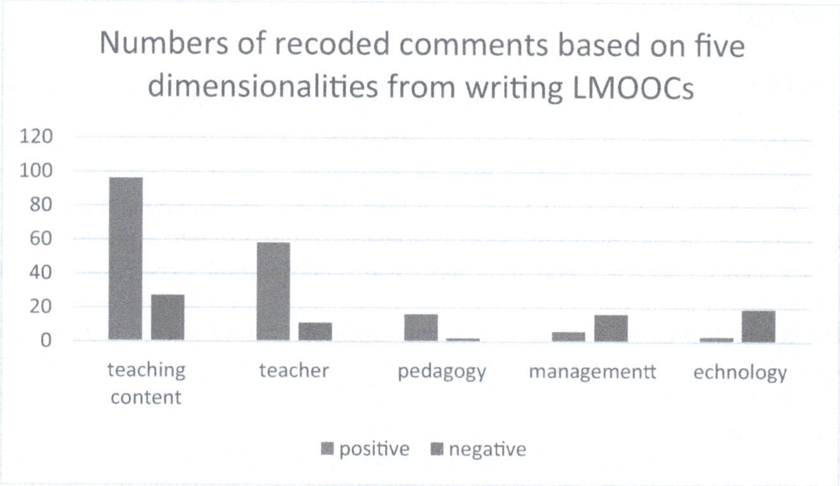

Fig. 4. Numbers of positive/negative comments in five dimensions in ESL writing LMOOCs

For ESL reading LMOOCs, the distribution of positive/negative comments in five dimensions are unbalanced as Fig. 5 shows. Dimensions of "teaching content" and "teacher" obtain many positive comments while none of comments related to technology is coded as positive. It is suggested that these two LMOOCs may have serious problems with technology issue. Many learners hope to have English subtitles to assist their learning while the two courses did not provide it. Besides, negative comments about management should not be ignored since the number is over 20.

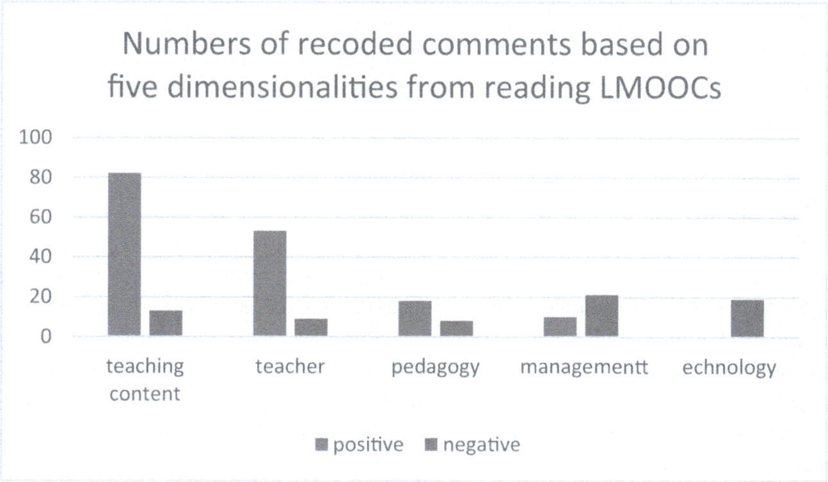

Fig. 5. Numbers of positive/negative comments in five dimensions in ESL reading LMOOCs

4 Discussion

The present study investigates the overall evaluation towards China's LMOOCs from learners' perspective and find out existing problems of China's LMOOCs. The results show that the overall evaluation of current LMOOCs is generally positive since the ratio of positive comments is higher than negative comments. To some degree, the six courses can represent China's LMOOCs as they are the most popular LMOOCs on China's biggest MOOC platform. Most learners in this study hold a positive attitude towards China's LMOOCs and they have had a meaningful experience of online foreign language learning.

The research findings show that different types of LMOOCs have different problems. For ESL speaking LMOOCs, many negative comments of ESL LMOOCs lie in online course management and technology issues. For example, complaints about peer review and load of exercise appear frequently. Many learners feel discontented with the online peer review system. Other learners complain that the load of homework is too heavy to finish before deadline. As for negative comments of technology issues, learners usually complain about the absence of video subtitles and the difficulties of handing in the self-made recorded homework.

As for ESL writing LMOOCs, learners complain most about teaching content. The present study finds teaching content taught in English put much pressure on learners and the content itself is beyond learners' capability so learners could not comprehend the knowledge. The reason why students in writing LMOOCs care most about teaching content lies in the objective of the course to enable students employ more effective and useful writing skills, while students find it difficult to grasp it within limited online learning time. Thus, the assignment of writing tasks and teacher's feedback seem to be a problem that ESL writing MOOCs should take into consideration.

Among negative comments of ESL reading LMOOCs, management and technology problems are put forward frequently. The present study finds that what affect learner's experience of reading courses are the unreasonable setting of peer review, the heavy load of homework, the absence of subtitles of videos and the instability of platform. Most of them are similar with the problems that would happen in other types of LMOOCs.

5 Conclusion

The present study analyzed 600 learners' evaluation comments of six LMOOCs on China's biggest MOOC platform based on grounded theory. To sum up, the overall evaluation of China's LMOOCs is positive. Some problems are universal and technology and management are the two dimensions that contain most negative comments across different types of LMOOCs. The problems include the absence of subtitles of videos, unreasonable setting of peer review, and heavy load of exercises. However, because of the special features of different types of LMOOCs, specific problems arise in each type of LMOOCs. All of these problems may hinder the sustainable development of LMOOCs. To guarantee the quality of LMOOCs, course designers and organizers should take them into careful consideration and make more efforts to improve the pedagogy, management, implementation, as well as technology issues of LMOOCs in the future.

References

1. Benigno, V., Trentin, G.: The evaluation of online courses. J. Comput. Assist. Learn. **16**, 259–270 (2000)
2. Chan, A.Y.K., Chow, K.O., Jia, W.: A framework for evaluation of learning effectiveness in online courses. In: Zhou, W., Nicholson, P., Corbitt, B., Fong, J. (eds.) ICWL 2003. LNCS, vol. 2783, pp. 383–395. Springer, Heidelberg (2003). https://doi.org/10.1007/978-3-540-45200-3_36
3. Dziuban, C., Moskal, P.: A course is a course is a course: factor invariance in student evaluation of online, blended and face-to-face learning environments. Internet High. Educ. **14** (4), 236–241 (2011)
4. Grover, S., Franz, P., Schneider, E., Pea, R.: The MOOC as distributed intelligence: dimensions of a framework evaluation of MOOCs (2013)
5. Jones, S.J.: Reading between the lines of online course evaluation: identifiable actions that improve student perceptions of teaching effectiveness and course value. J. Asynchronous Learn. Netw. **16**(1), 49–58 (2012)
6. Kavadella, A., Tsiklakis, K., Vougiouklakis, G., Lionarakis, A.: Evaluation of a blended learning course for teaching oral radiology to undergraduate dental students. Eur. J. Dent. Educ. **16**(1), e88–e95 (2012)
7. Lin, J., Cantoni, L.: Assessing the performance of a tourism MOOC using the Kirkpatrick model: a supplier's point of view. In: Schegg, R., Stangl, B. (eds.) Information and Communication Technologies in Tourism 2017, pp. 129–142. Springer, Cham (2017). https://doi.org/10.1007/978-3-319-51168-9_10
8. Martín-Monje, E., Bárcena, E. (eds.): Language MOOCs: Providing Learning, Transcending Boundaries. Walter de Gruyter GmbH, Berlin (2015)
9. Zhao, C., Dong, X.: An applicable way of teaching quality evaluation based on MOOC platform. Int. J. Emerg. Technol. Learn. (iJET) **12**(03), 57–67 (2017)

Exploring the Effectiveness of Fully Online Translation Learning During COVID-19

Wenchao Su[1], Defeng Li[2], Xiaoxing Zhao[3], and Ruilin Li[1](✉)

[1] Guangdong University of Foreign Studies, Guangzhou 510420, China
201410046@oamail.gdufs.edu.cn
[2] University of Macau, Macau, China
[3] Sun Yat-sen University, Guangzhou 510275, China

Abstract. Universities worldwide have started the fully online teaching mode in response to the outbreak of COVID-19. This paper reported on an empirical study, using the Kirkpatrick model, on the effectiveness of fully online translation learning during COVID-19, and how online teaching influenced effectiveness in translation learning. Results showed that the online translation course was effective in that students achieved most of the stated goals for the translation course and applied what was learnt to translation practices. Also revealed are the perceived difficulties of online translation learning, such as staying concentrated and interacting with peers and the teacher, and its advantages in encouraging students to use online tools in their translations. Pedagogical implications for translation courses are also discussed.

Keywords: Effectiveness · Online learning · Translation learning · Kirkpatrick model · COVID-19

1 Introduction

Since the outbreak of COVID-19, schools and universities worldwide have experienced shutdowns and most initiated the mode of full online education. This special period has given us researchers a unique opportunity to probe into the performance and effectiveness of learning activities that are based fully on E-learning tools and platforms. The investigation of learning effectiveness has been a central topic in online language education, but research that assesses and evaluates effectiveness in online, especially entirely online translation training is still scarce. Using the Kirkpatrick model of training assessment (Kirkpatrick and Kirkpatrick 2006), this study attempts to explore students' perspectives of the effectiveness of full online translation training during COVID-19. The study is hoped to help instructors improve similar online translation training programs in the future.

C. Pang et al. (Eds.): SETE 2020/ICWL 2020, LNCS 12511, pp. 344–356, 2021.
https://doi.org/10.1007/978-3-030-66906-5_33

2 Literature Review

2.1 Theoretical Framework of Effectiveness Evaluation

Learning effectiveness can be defined as "the degree to which intended outcomes were attained" (Huang et al. 2019: 95). One way to assess the extent to which the intended outcomes are achieved in the learning process is to investigate learners' reflections on their learning outcomes, and Kirkpatrick's four levels model is one of the most useful theoretical frameworks for understanding learning effectiveness in education research (Lin and Cantoni 2017; Blum et al. 2020). In the Kirkpatrick model, learning effectiveness is operationalized into the levels of ***reaction, learning, behavior and results***. In the context of translation training, the first level of reaction measures how learners react and respond to the translation course, i.e., whether they consider the course satisfactory, useful and relevant to their careers. The second level of learning examines how learning goals and objectives are accomplished, i.e., what translation knowledge, skills and theories they have or have not learnt. The third level of behavior investigates the extent to which learners can apply the acquired knowledge, skills and theories to their translation practices and future jobs. The fourth level of results measures to what degree translation quality, productivity, efficiency and other aspects of translation performances are improved after training.

Since the present study is intended to gauge the learning effect of online training courses as opposed to traditional classroom teaching, an additional level of learners' online learning experience should be added to assess how the method of online learning may influence the students' perceived effectiveness of translation learning. Building on Kirkpatrick's four levels model, we propose an extended framework to evaluate translation training effectiveness in the mode of online learning (see Fig. 1). This additional level aims at investigating the particular advantages, strengths, difficulties, problems and challenges that are brought about by the full online learning mode in translation training.

Fig. 1. Effectiveness evaluation in online translation learning (extended from Kirkpatrick and Kirkpatrick 2006)

2.2 Learning Effectiveness of Online Translation Learning

Online learning may be defined as learning that "uses electronic technologies via the Internet to engage learners and facilitate their learning" (Sunal and Wright 2012: 2499). Online learning usually implies that learners and their course instructors are in different locations and the instruction and therefore the learning takes place over the online teaching platforms such as Tencent, Zoom, DingTalk etc. In light of this, online translation learning describes online learning that involves the teaching, practices and discussions of translation knowledge, skills and/or theories, delivered synchronously (i.e., the instructor teaches the translation course in real time) and/or asynchronously (i.e., there is a time gap between translation teaching and learners' reception and feedbacks).

Previous studies of online translation and interpreting learning mainly focus on the mode of blended learning or flipped learning (Chan 2014; Kim 2017), where face-to-face teaching in the classroom and online teaching aided by educational technologies combine and support each other (Ifenthaler 2012). Only a few have attempted to investigate predominantly or fully online learning in the field of translation and interpreting education (see, for instance, Skaaden 2017). Lee and Huh (2018) investigated the perspectives of trainers and trainees towards a business interpreting and translation training program which might be considered offered mostly online. The online program was offered asynchronously, with audio lectures made by trainers in advance and then sent to the trainees. By means of surveys and interviews, both the trainees and trainers were invited to provide their feedback on the advantages, drawbacks and effectiveness of the online learning experience. The findings showed that overall, the trainees considered that the online translation and interpreting learning had more advantages than disadvantages, and more than half of them viewed online translation learning and face-to-face translation learning equally effective. It will be interesting to see whether such positive experience will also hold true in a synchronous mode of full online translation teaching and learning.

The present study is concerned with the learning effectiveness of fully online translation teaching and learning, taking as a case of example the teaching of translation courses during the outbreak of COVID-19 in 2020. In the teaching of this course, there was no face-to-face trainee-trainee or trainee-trainer interaction or communication. The following two research questions were used to guide the exploration of the effectiveness of fully online translation learning:

(1) How effective is the learning of the fully online course on translation skills?
(2) How in particular, does the fully online teaching mode affect the effectiveness of translation learning?

3 The Present Study

3.1 Participants

The participants were forty-seven students in Year 3 who attended a fully online translation course offered by a reputable university in the Chinese mainland during the outbreak of COVID-19. They were English majors with comparable Chinese and English proficiency. Before the online translation learning, they had developed basic understanding of translation skills and strategies. None of them had taken any fully online translation courses before.

3.2 The Fully Online Translation Course

The 14-week mandatory course of Chinese-English translation was offered between April and July 2020. The students attended the course once a week, and each session lasted for 90 min. This online course was delivered synchronously via the Ten-cent Meeting online platform. The teacher uploaded course materials such as PPT slides and files of study materials on to the platform and shared the screen when necessary for demonstration of translated texts. When students had questions or wanted to voice their opinions, they could activate their microphone to speak, or they could just type their questions or opinions in the chat box area. They could share their screen or display their translation in the chat box area so that all could see it and make comments and ask questions. The course is skill-based, which aims to help students develop translation skills and strategies, especially those specific to various text types; it also aims to help them improve their ability to solve problems in translation with translation tools so that they could apply them in their future translation projects or practices.

3.3 Research Instrument and Procedure to Gather Students' Perceptions

To gather information, we conducted a questionnaire survey consisting of two parts with 17 questions in total (see Appendix). The first part comprised 2 questions. The first question asked the students to indicate their perceptions towards the overall effectiveness of the online translation course, using a five-point Likert scale (1 = very ineffective, 3 = neutral, 5 = very effective); the second is a multiple choice question that asked the students to compare the effectiveness of online translation learning with that of face-to-face learning. The second part consisted of 15 open-ended questions asking the students to report on five slightly different aspects of learning effectiveness based on their experience of learning the online translation course. The open-ended questions were developed by one of the researchers based on the extended Kirkpatrick's model of learning effectiveness (Fig. 1) as discussed earlier.

All the students were invited to complete the survey questionnaire at the end of the online translation course on a voluntary basis with consent forms in Helsinki declaration format signed and they were encouraged to honestly share their experience and reflections in detail. 46 students completed and returned the questionnaire, a return rate of 97.9%.

3.4 Data Analysis

We took a two-step process to analyze the students' perceptions. First, we referred to the questionnaire data to identify possible themes. Two of the researchers read the written comments in the questionnaires, and independently identified themes with regard to students' perceived effectiveness of the online translation course. The researchers then held a discussion and drew up the possible themes regarding the learning effectiveness. Second, we calculated how many times each of the identified themes were referred to in the data. For that, the two researchers read a quarter of the written comments on the questionnaires, compared their coding and resolved their slight differences before they reread and coded the entire written comments in the questionnaires and summarized the total mentions of each of the themes identified earlier.

4 Findings of the Research

Overall, the students considered the fully online translation learning effective (mean = 3.65, SD = 0.71). The original data showed that 30 out of the 46 students (65.2%) viewed the online translation course effective or very effective, with only 3 students (6.5%) seeing the course ineffective. The results from the open-ended questions are reported as per the five aspects of learning effectiveness in the earlier model.

4.1 Reaction

The Reaction factor examined the trainees' perceptions on the relevance and usefulness of the online translation course. When asked what they learnt from the online translation course, the responses revealed 9 major themes or areas (see Table 1). Many reported that they learnt skills and strategies for general translation tasks and news translation, as well as how to avoid using Chinglish in their Chinese to English translations. Only a few said that they learnt how to appreciate and criticize translated texts and translation theories.

When asked whether the course was useful, all responded positively. Most (35, 76.1%) viewed the course relevant to their studies and/or future jobs, some (8, 17.4%) were not sure whether the translation course was relevant to their future careers, and a few (6, 13%) didn't think the course was related to their future professions. When asked which of the course contents was not relevant or useful, three students mentioned the Office 365 dictation tool, two students mentioned the experience sharing sessions by more advanced learners, and one student mentioned translation theories. The rest (40, 87%) did not refer to anything in the online translation course as not relevant or useful.

Table 1. Students' reaction on relevance of the online translation course

What they learnt	No.	%	What they learnt	No.	%
News translation skills	39	84.8	Translation tools & search skills	13	28.3
Strategies to avoid Chinglish	39	84.8	Tourism translation skills	11	23.9
General translation skills	35	76.1	Translation criticism	6	13
Language register for translators	18	39.1	Translation theories	3	6.5
Translation market	16	34.8			

4.2 Learning

After we evaluated the trainees' reaction to the online translation course, we assessed students' goals and actual gains as the result of the translation training provided in the course. When asked what the learning goals of the online translation course were, the students reported a variety of goals (see Table 2), with most students replying that their goals were to properly use translation skills and to avoid using Chinglish in their Chinese to English translation. When asked whether they were able to achieve the stated goals, 37 of 46 students (80.4%) reported that they achieved most of their learning goals, 8 (17.4%) said that they did not achieve the stated goals, and one student offered no response to this question. Additionally, the students expressed that they still needed much more practice in order to facilitate their use of the learned translation skills and strategies, to improve their translation quality, Chinese and English proficiency and so on (see Table 2).

Table 2. Students' learning

Learning	No.	%	Learning	No.	%
The stated goals			*The goals not achieved*		
To properly use translation skills	36	78.3	To master translation skills	22	47.8
To avoid using Chinglish	25	54.3	To improve translation quality	19	41.3
To improve translation performance	17	37	To improve translation efficiency	11	23.9
To study news translation skills	14	30.4	To improve language proficiency	8	17.4
To study tourism translation skills	14	30.4	To learn to use translation tools	4	8.7
To learn to use translation tools	13	28.3			
To improve language proficiency	11	23.9			
To know about translation market	9	19.6			
To study different text types	6	13			

4.3 Behavior

We further evaluated how the course might have changed their behaviors of translation practice. When prompted, 4 students (8.7%) responded that they could apply all that they learnt to translation practices, whereas 42 (91.3%) said that they could apply most of what they learnt in the online translation course to their practices. When asked what exactly they could apply, 23 students (50%) mentioned that they could apply some general translation skills, such as literal translation and paraphrasing to their translation exercises and practices; 22 (47.8%) stated that they started to pay attention to the heads, dates and leads when translating news; 18 (39.1%) reported they were now more sensitive to Chinglish and knew how to avoid it, such as redundancy of nouns and verbs in their English translation;17 (37%) could apply some translation tools and search skills when doing translation tasks (see Table 3). When asked what they still could not apply well in their practice, 15 students (32.6%) said they still were not able to apply some translation skills such as those for translating long sentences and idiomatic expressions, and 10 students (21.7%) said some specific skills for news translation, such as how to maintain proper text structure, remained a challenge for them (see Table 3).

Table 3. Changes in behaviors

Behavior	No.	%	Behavior	No.	%
What they could apply			*What they could not apply*		
General translation skills	23	50.0	General translation skills	15	32.6
News translation skills	22	47.8	News translation skills	10	21.7
Avoidance of Chinglish	18	39.1	Knowledge of registers	5	10.9
Translation tools & search skills	17	37.0	Translation tools & search skills	4	8.7
Knowledge of registers	9	19.6	Application of theories to practices	4	8.7
Chinese/English differences	7	15.2	Knowledge of translation market	3	6.5
Tourism translation skills	3	6.5	Chinese/English differences	2	4.3

4.4 Results

We asked the students to evaluate how their translation quality and efficiency improved as a result of taking this online translation course. 26 students (56.5%) replied that they could translate better because of the online training. They noted for example that the quality of their translation exercises improved as indicated in their higher grades as the course progressed and that they produced less Chinglish in their translations as reflected in their teacher's comments. However, 10 students (21.7%) said their translation could be much better, mostly because they believed that their language proficiency was still not good enough. 5 students (10.9%) said they could not translate better and another 5 students (10.9%) were not sure. 5 out of these 10 students claimed that they were not confident in their translation ability.

As for translation speed, 25 students (54.3%) thought they translated faster after taking the online training, and 15 students said they could tell from their improved speed of finishing the translation exercises while 8 students felt they were more efficient in using translation tools to solve problems in translation tasks. However, 15 students (32.6%) did not feel the course helped them translate faster, with 9 students saying that they spent too much time on searching relevant information for the translation assignments. 3 students (6.5%) said their speed could be much faster and another 3 (6.5%) replied their translation speed depended on different text types.

4.5 Effectiveness of Online Learning

Finally, we addressed how fully online teaching influenced the effectiveness of translation learning. Although most students considered the online translation course effective or very effective, they did not view the online learning experience as effective as the face-to-face learning experience. 25 students (54.4%) found the online translation course was less effective than the face-to-face course, while 11 students (23.9%) found they were equally effective and 6 (13%) considered the online translation course was more effective. 4 students (8.7%) were not sure which mode of learning was more or less effective.

When asked what difficulties the students encountered during the fully online translation training, 5 potential learning barriers emerged. The greatest difficulties identified by the students were to stay focused during the online translation course (23, 50%) and to communicate, interact and collaborate with peers and the teacher (23, 50%), followed by problems caused by networks and platform instability (15, 32.6%) as well as learning environment being interrupted as the result of taking online lectures at home (10, 21.7%). 4 Students (8.7%) also experienced physical and mental fatigue from the online translation learning.

Distraction

Most students commented that one of the biggest challenges in online translation learning was to maintain their attention. They reported that they were more likely to mind-wander when attending online translation classes than face-to-face classroom teaching. The following excerpts by the students attested to this distraction.

(1) There are a lot of distractions at home, and without the supervision of the teacher and classmates, I am easily distracted during the online classes and unable to resist the temptation of playing the mobile phone. (Q14: Steve) (All the names that appeared in the quotations are pseudonyms.)
(2) I am not concentrated enough when I am sitting in front of the computer alone. (Q14: Amy)

Some students reported that because of the distractions, they needed to catch up with the teacher from time to time but might still miss some knowledge points and had to turn to their classmates for help. Interestingly, a few students reported that they were more likely to focus when they were engaged in class activities during the online translation course.

(3) Honestly, I am easily distracted during the online classes, but when the teacher asked us to finish translation exercises in real time, I become focused right away. (Q14: Leo).

Difficulties of Communication, Interaction and Collaboration
Most students pointed out that the communication became slow, the interaction seemed unreal, and the collaboration was inefficient in online translation learning.

(4) Fully online communication feels unreal because I cannot see the reactions from my classmates and teacher. Also, it takes a long time to get a reply during group discussions. (Q14: James)

Some students felt it was a pity that they could not discuss about the translations with their deskmates in online classes the way they did in a classroom setting.

(5) When I am doing translation exercises, I'd like to see how my classmates deal with the source text. I used to discuss with my deskmate in offline classes, but now I do my translation exercises all on my own. (Q14: Tom).

Several students also mentioned that collaboration in doing translation projects became difficult in the online translation course.

(6) Without face-to-face meetings and communications, it is hard to ensure group members' participation when doing translation projects fully online. (Q14: Mary)

Consequently, due to such communication and collaboration difficulties in fully online translation learning, the students did not see much improvement in their teamwork, communication and collaborative skills.

Technical Problems and Learning Environment
It may not be surprising that the students encountered technical problems such as network instability, malfunctioning of the online course platform, and equipment problems during the online translation training, because many studies on online learning reported similar technical challenges. However, it was worth noting that nearly one fourth of the trainees mentioned their undesirable learning environment during online translation learning. Many reported on disruptions of all sorts during their online lessons. Below was a typical comment by the students.

(7) During the online classes at home, I have been distracted by all kinds of interference, such as the noisy flat decoration project nearby, door knocking by neighbors and my family guests in the house. Sometimes I even had to deal with them in the middle of the course. I feel very upset by the disruption. (Q14: Tina).

Fatigue
A few students felt exhaustion and tiredness during the fully online translation learning. Staring at the computer screen for a long time caused physical and mental fatigue for some learners.

(8) I get very tired from looking at the screen all the time during the online classes. I even feel resistant to online learning just because I don't want to look at the monitor. (Q14: Jim).

Advantages of Online Translation Learning
When asked what aspect(s) of this fully online translation course they found most helpful to their translation learning, 7 themes emerged (see Table 4). Most students (25, 54.3%) said that it was easy and convenient to take notes for review in online translation learning by capturing the screens or recording the lectures. Although some expressed that it was hard to get immediate feedback in online translation learning, 15 students (32.6%) felt less nervous or embarrassed to speak up their ideas, raise questions or share their translated texts with peers, because studying online gave them a sense of invisibility and they felt less psychological stress when voicing their ideas online.

Another advantage referred to by the students was the convenience of searching information online. It was helpful because the students could clear their doubts immediately by searching relevant information online, and the process of checking, correcting and revising was particularly useful for enhancing their translation competence. Several students mentioned that the online learning experience motivated them to use online tools when doing translation practices. Some also reported that learning translation online was good for sharing translations, because students could type their translated texts in the chat box of the platform, and everyone could see different translation versions of the source text clearly and immediately.

Table 4. Advantages of online translation learning

Helpful aspects identified by the students	No.	%
Easy to take notes	25	54.3
Less nervous	15	32.6
Easy to search information online	14	30.4
Can ask questions anytime and anywhere	8	17.4
Easy to share the translated texts	7	15.2
Ready to try the online translation tools	6	13
Can see PPTs clearly	5	10.9

5 Summary and Conclusion

The present study set out to evaluate the effectiveness of fully online translation learning during COVID-19. The first research question aimed to investigate the effectiveness of the online translation course by implementing Kirkpatrick's four levels model. It was found that overall, the fully online translation learning was considered effective. Students' reaction to the online course was favorable. Most considered it useful and relevant to their studies and future careers, and what they learnt from the training covered the major contents of the course. The learning goals were clearly understood and mostly achieved, as reported by most students. They could also apply most of the learned translation skills, strategies and techniques to their practices. Translation quality was improved, and productivity was increased because of the online learning.

It is important to note that some students felt less positive. Some students felt that translation practices were not sufficient during the online learning, and their Chinese and English proficiency were still not good enough for them to produce translations efficiently with high quality. These concerns reflected that students attached great importance to practices and language proficiency in translation training, which was also reported by Li's (2002) research on trainees' attitudes towards translation training. We suggested that more real-time translation practices and exercises be delivered in the synchronous classes and the content concerning translation knowledge and theories be largely offered asynchronously.

The second question attempted to examine the influence of online teaching on the effectiveness of translation learning. Contrary to Lee and Huh's (2018) findings, the online translation course was less effective than the face-to-face course, as reported by 54.3% of the students. Distraction and absent-mindedness occurred frequently during the online translation course. The difficulty to concentrate and focus has been widely reported in online education research (see Szpunar et al. 2013 for systematic review). Another major perceived difficulty was the delayed, unreal and inefficient communication, interaction and collaboration in the online translation learning, which is caused by the lack of regular physical contact during online learning and is considered as "the inherent problem in fully online learning" (Sun 2014: 31). It seemed that the difficulties encountered by the students in online translation learning are similar to those in general online education (see also Roddy et al. 2017).

It is delightful to see that students were motivated to use online tools to help with their translations after the online translation course. Some students also pointed out that the online teaching platform provided functions of chat box and screen sharing for sharing translated texts. Translation instructors could make use of these functions for comparing different translated texts either in online or face-to-face classes.

To conclude, online teaching of translation shares similar problems with online teaching of many other subjects. But it also has distinct advantages. It encourages students to share their translations and offer their comments for their peer's translations, which appears to be much easier for translation students than in a traditional classroom setting. It is also conducive to their adoption of translation tools to their translation assignments, as they are attending lectures online and having the computer readily available. For the same reason, they were more ready to look online for background knowledge or other information to help them with their translation assignments, especially when dealing with texts of different subject matter. Therefore, we can conclude that online translation teaching has positive effects on the students and their learning despite some disadvantages as experienced in all online teaching. The crux for enhancing this new mode of teaching is to carry out further research to combat the disadvantages while some problems such as instability of internet access will be solved as technology advances in the future.

Acknowledgements. This work was supported by Guangdong Planning Office of Philosophy and Social Sciences (grant number GD20YWY01) [广东省哲学社会科学 "十三五" 规划 2020 年度青年项目 "同声传译视觉信息加工过程的眼动研究", 项目编号: GD20YWY01] and The National Social Science Fund of China (grant number 19BYY126).

Appendix

PART ONE: 1. How do you consider the effectiveness of the fully online translation course (1) Very ineffective (2) Ineffective (3) Neutral (4) Effective (5) Very effective 2. How do you compare the effectiveness of the online translation course (O) to the face-to-face course (F) (1) O is more effective than F. (2) O is as effective as F. (3) O is less effective than F. (4) Not sure.

PART TWO: **Reaction**: Q1: What did you learn in the translation course?/Q2: Do you think the course contents are relevant to you and your future career?/Q3: Do you think what is taught in the online translation course useful?/Q4: Is there anything not relevant or useful? Why? **Learning**: Q5: What were the goals for this translation course?/Q6: Do you think you were able to achieve the stated goals for the translation course?/Q7: Is there any goal you are not able to achieve? Why? **Behavior**: Q8: Do you think you can apply what you have learned in the course to translation practice?/Q9: What can you apply?/Q10: What can you not apply?/Q11: Why and why not? Results: Q12: Do you think you can translate better? How do you know?/Q13: Do you think you can translate faster? How do you know?/**Online Learning**: Q14: Did you encounter any difficulties during the online translation learning? What are they? How did it affect the effectiveness of your translation learning?/Q15: What aspect(s) of this fully online translation course did you find most helpful to your translation learning and why?

References

Blum, E.R., Stenfors, T., Palmgren, P.J.: Benefits of massive open online course participation: deductive thematic analysis. J. Med. Internet Res. **22**(7), e17318 (2020)

Chan, C.H.: Building an online library for interpretation training: explorations into an effective blended-learning mode. Comput. Assist. Lang. Learn. **27**(5), 454–479 (2014)

Huang, R., Spector, J.M., Yang, J.: Educational Technology: A Primer for the 21st Century. Springer, Singapore (2019)

Ifenthaler, D.: Blended learning. In: Seel, N.M. (ed.) Encyclopedia of the Sciences of Learning, pp. 463–465. Springer, New York (2012)

Kim, D.: Flipped interpreting classroom: flipping approaches, student perceptions and design considerations. Interpret. Transl. Train. **11**(1), 38–55 (2017)

Kirkpatrick, D., Kirkpatrick, J.: Evaluating Training Programs: The Four Levels, 3rd edn. Berrett-Koehler Publishers, San Francisco (2006)

Lee, J., Huh, J.: Why not go online? A case study of blended mode business interpreting and translation certificate program. Interpret. Transl. Train. **12**(4), 444–466 (2018)

Li, D.: Translator training: what translation students have to say. Meta **47**(4), 513–531 (2002)

Lin, J., Cantoni, L.: Assessing the performance of a tourism MOOC using the kirkpatrick model: a supplier's point of view. In: Schegg, R., Stangl, B. (eds.) Information and Communication Technologies in Tourism 2017, pp. 129–142. Springer, Cham (2017). https://doi.org/10.1007/978-3-319-51168-9_10

Roddy, C., Amiet, D.L., Chung, J., Holt, C., Shaw, L., McKenzie, S., Mundy, M.E.: Applying best practice online learning, teaching, and support to intensive online environments: an integrative review. Front. Educ. **2**, 59 (2017)

Skaaden, H.: "That we all behave like professionals": an experiential–dialogic approach to interpreter education and online learning. In: Cirillo, L., Niemants, N. (eds.) Teaching Dialogue Interpreting: Research-Based Proposals for Higher Education, pp. 323–340. John Benjamins, Amsterdam (2017)

Sun, S.Y.: Learner perspectives on fully online language learning. Distance Educ. **35**(1), 18–42 (2014)

Sunal, C.S., Wright, V.H.: Online learning. In: Seel, N.M. (ed.) Encyclopedia of the Sciences of Learning, pp. 2499–2502. Springer, New York (2012)

Szpunar, K.K., Moulton, S.T., Schacter, D.L.: Mind wandering and education: from the classroom to online learning. Front. Psychol. **4**, 495 (2013)

From Offline to Online: The Reduced Embodiment in Teacher-Initiated Turn-Taking in GSL Instructions

Enhua Guo[1] and Fuxin Zhang[2(✉)]

[1] Ocean University of China, Songling Road 238, Qingdao, China
[2] Tianjin Foreign Studies University, Machang Road 117, Tianjin 300204, China
fxzhang@tjfsu.edu.cn

Abstract. This paper looks into the systematic organization of teacher-initiated turn-taking in online GSL (German as second language) sessions conducted via one of the video-conferencing platforms (Tencent Meeting), and cross-examines the virtual instructional environment created by it with face-to-face GSL instructions. Drawing on the framework of Multimodal Conversation Analysis (MCA) as primary research method, we analyze and compare both screen-recorded online GSL sessions collected during Covid-19 pandemic, and in-person GSL classroom interactions collected prior to Covid-19 outbreak. Our findings demonstrate that technological affordances provided by video-conferencing platforms compare unfavorably to the face-to-face learning environment. Online technology typically undermines the availability of embodied resources needed to establish co-present participation frameworks (Goffman 1983) which are experienced by social actors in face-to-face classroom interactions. Changes in the embodiment of participation frameworks, in turn, lead to systematic differences in how the speakership and recipiency are managed in teacher-initiated turn-taking.

Keywords: Teacher-initiated turn-taking · Embodiment · Online/offline instruction · GSL

1 Introduction

The pandemic of Covid-19 has fundamentally reshaped the educational landscape of the year 2020: school closures have necessitated online delivery of courses previously offered in a face-to-face learning environment. This necessity has fueled the appearance of many video-conferencing solutions—e.g. Tencent Meeting (also known as Voov outside China), and Dingtalk, etc.—to offer technological support for online education. While online education was not a novel thing and had long existed before Covid-19—i.e. MOOC or Zoom courses, it only goes viral due to the outbreak of Covid-19. However, the widespread online delivery of courses has posed many challenges to both the teachers and learners. This paper is an attempt to understand the systematic differences between the two paradigms of instruction by paying close attention to the moment-by-moment interactional details therein.

C. Pang et al. (Eds.): SETE 2020/ICWL 2020, LNCS 12511, pp. 357–369, 2021.
https://doi.org/10.1007/978-3-030-66906-5_34

2 Face-to-Face Interactions vs Mediated Interactions

Technological advances have provided numerous cutting-edge media solutions (both voice and video) supporting virtual communication across space and time. However, human beings always have a natural inclination for intercorporeal, interbodily co-presence between *alter* and *ego* [1–3]; face-to-face encounters constitute the primordial site of human sociality and the major form of human existence. As Goffman [3, p. 2] puts it:

> *It is a fact of our human condition that, for most of us, our daily life is spent in the immediate presence of others; in other words, that whatever they are, our doings are likely to be, in the narrow sense, socially situated.*

According to Goffman [4, 5], co-present social encounters, or face engagements, are not merely immediate physical co-existence of human bodies, but are, more importantly, themselves interactional achievements. They are characterized by an "eye-to-eye ecological huddle" [4, p. 95] where interactants are jointly committed to "doing co-presence", by mobilizing a stock of multimodal, embodied practices including body movements, eye gaze, gestures, and facial expressions, etc. [6]. Normally, being physically co-present is the precondition for the co-interactants to establish embodied participation frames—achievements of mutual orientation and coordination—which in turn are essential to incipient focused interactions.

Goffman might not have predicted all of the technological advances of the 21st century that allow people to do "being together" online. However, technology, per se, is disembodied, nonhuman and alienating. In comparison with face-to-face interactions, all technology-mediated interactions impose constraints, albeit to varying degrees, on our embodied capability of marshalling a full range of multimodal, intercorporeal resources (including artefacts in the immediate physical environment) which are indispensable to establishing participation frames in face-to-face interactions. Any technological progress in affordances [7] of virtual communication media has been an outcome of persistent endeavors to restore the resources mainly available in (hence not easily extrapolatable from) co-present interactions. From less interactive Emails, to monomodal instant messaging, and further to more powerful, multimodal communication applications such as WhatsApp, Wechat, QQ, Skype, etc., the technological trajectory represents our inclination to burst the confinement imposed by virtual environment on interactions, as well as our desire to maximize embodied participation frames as experienced in face-to-face interactions. Therefore, in order to gain a deeper understanding of how people engage in technology-mediated interactions, it's necessary to cast it against the default situation of unmediated, face-to-face interactions, where our primordial form of human sociality exists. This constitutes the *raison d'etre* of our paper to examine teacher-initiated turn-taking—a very significant organizational resource—in online courses against the background of face-to-face sessions.

Although it seems impossible to achieve telecommunication solutions which can deliver 100% emulation and modelling of co-present encounters, applications can be designed to be infinitely close to them. One example is the video-chatting feature which allows many application users to monitor their interactants' gaze and facial expressions, and to coordinate their interaction on the basis of this monitoring.

Notwithstanding that, it should be noted that video-chatting environment is still far from real, co-present interactional context. Given the fact that modern smart phone technology has substantially minimized the distance between the front camera and the screen, different degrees of "being gazed at" can still be easily perceived by recipients when their interactants shift their direct gaze between front camera and the screen. More importantly, co-present encounters allow far more tactile, visuospatial and embodied resources relevant to interactions than in video-chatting environment. As will be shown in the following sections, this is particularly true in multi-party online instructed learning settings.

3 Embodiment and Turn-Taking Organization

There has never been a lack of interest among interaction researchers in the role of nonverbal, embodied behaviors—such as eye gaze [8–10], gestures [6, 11], etc.—in turn-taking and turn-allocation practices. For example, Auer [10] shows that gaze is one of the most fundamental resources for turn-allocation, hence the most common next-speaker-selection technique. Mondada [6] demonstrates that pointing gesture can function as a technique for projecting self-selected speakership in turn-taking.

However, studies on the import of embodied resources in classroom interactions—including the 2nd language (L2) in-person classroom interactions—are relatively scarce. Even fewer studies have examined embodied resources in online classroom environment. Previously, researchers such as Mehan [12] observed that gaze, head nods or pointing gestures are frequently used in classroom for next speaker turn-allocations. Such nonverbal, embodied resources are also mobilized by teachers as floor-holding methods for the already-nominated speaker. Recently, the role of embodied interactional resources in teacher-fronted L2 classroom interactions has been also studied within the framework of multimodal conversation-analytic method. Kääntä [13] suggests that gaze is the most prevalent embodied resource used by teachers in allocating turns to students. The teacher always directs gaze towards the selected student as next speaker at some point while allocating turns, and if the teacher's gaze is directed towards the whole class, it serves as a pre-step for locating a potential respondent. Studies also exemplify how students turn to gaze and use it as a resource for communicating with the teacher. Mortensen [14, 15] showed that in Danish L2 classrooms gaze can be systematically deployed by students to show willingness to be the selected as next speaker. Following this line of work, researchers [16–18] have turned their attention to how students use gaze withdrawal or mutual gaze avoidance to display their unwillingness to be selected as the next speaker in L2 classrooms. Apart from gaze behaviors, the students in L2 classroom often use gestures to solicit teacher's gaze before bidding for the speakership [19].

These findings offer a convenient starting point for the analysis of online L2 instructional environment where a large proportion of such embodied resources become either limited or simply unavailable to both teachers and students. They also provide a backdrop against which we can apprehend better the consequences of compromised embodied resources in online L2 whole-class interactions. Furthermore, moment-by-moment adaptations to the deteriorated embodiment in online L2 context come into focus. In the next section, I will describe the data and method of the current study.

4 Data and Method

Our data set consists of two parts: (1) Approximately 30 h of video-recorded naturally occurring classroom interactions over the course of Fall semester from a university undergraduate GSL program in 2019, and (2) approximately 30 h of screen-recorded naturally occurring online GSL sessions during the Spring semester of 2020 from the same program of the same university. Both courses were opened to the first-grade German majors. The classroom interaction data were gathered with two cameras. One camera was set up in the back of the class, and the other was facing the classroom in order to capture both students' and teachers' embodiment. The transcription and analysis of the data were informed by a multimodal conversation-analytical approach [6, 20]. Such approach was chosen as it allows to perform moment-by-moment microanalysis of interactional sequences where the teacher and students orient to and employ multimodal, interactional resources in order to coordinate their participation frame and accomplish pedagogical goals. Data transcription was conducted with reference to GAT2 [21]. All the video-recorded participants have signed informed consent form giving permission to use recordings for academic purposes.

5 Analysis and Findings

In this section, we will present our findings in two parts: (1) analysis of fully embodied multimodal interactions in GSL classroom, and (2) analysis of reduced embodiment in online GSL sessions. Due to the space limitation, this paper only focuses on the interrelationship between embodiment and teacher-initiated turning-taking.

5.1 Embodied Co-participation in GSL Classroom Turn-Taking and Turn Allocation

In the 30-h classroom interaction data, 733 teacher-initiated turn-taking instances have been identified and categorized into two major types: Type 1 (N = 482) is teacher's addressing the whole class where at least one student volunteers for an answer[1], and Type 2 (N = 251) is teacher's next speaker turn-allocation where teacher nominates a next speaker. Both types of turn-taking involve a rich variety of multimodal resources. Excerpt 1 below illustrates Type 1 classroom interactions, while Excerpt 2, which happens immediately after Excerpt 1, exemplifies Type 2. Both excerpts occur during the same pedagogical activity—brainstorming, in which the teacher is trying to solicit more words from the class about different sports names.

[1] Note that if no response is delivered from the students, it is not treated as teacher-initiated turn-taking by definition.

Excerpt 1: J (student) responds to T (teacher)'s whole-class elicitation.

```
48    T:   {weitere  sportARten  oder s:(0.2){porttraining?
             other     sport types  or           sport    training
           'Other sport types or sport-training?'
49         {((T gazes at the class on her left side))
50                                                {((T shifts gaze to her right
                                                       side))
51         {(1.0)
52         {((T sustains gaze to the class on her right side))
53→J:      {<<p>SCHWIMmen>;
                swim
           'Swimming.'
54         {Fig #1 ((J gazes at T who sustains gaze to the class on her
           right side))
55    T:   {SCHWIMmen? {(1.9)
            swim
           'Swimming.'
56         {Fig #2 ((T shifts gaze to J))
57                       {((T walks to the blackboard))
```

Fig #1

Fig #2

In Excerpt 1, after the teacher has obtained two brainstorming responses from the class and has written them on the blackboard, she resumes the brainstorming in Line 48. This invitation for more answers is multimodally achieved by both a verbal question (Line 48) and a scanning gaze over the whole classroom from the left side to the right (Line 49–50). After one-second silence, student J begins to gaze at the teacher (Fig #1) and makes an attempted answer *schwimmen* (swimming) in a low voice (Line 53). J's bid for speakership immediately attracts the teacher's gaze (Fig #2) which was previously fixed upon the right side of the classroom (Fig #1). So, a mutual gaze has now been established where the teacher and J are mutually oriented to each other in Line 55 (also shown in Fig #2). This mutual gaze is essential in that it has changed the teacher's participation frame with whole class into a focused, dyadic interaction with J. As is evidenced in Line 55, the teacher does not need to use address term (by calling out J's name) to confirm her answer. In Line 57, the teacher immediately walks to the blackboard to write the answer down, which functions as a positive ratification of J's answer. In Excerpt 1, the teacher's [verbal question + scanning gaze] has successfully secured a response from J, hence a teacher's initiated turn-taking. However, the teacher's inviting questions sometimes receive no response. When this happens, the teacher can either deliver the answer by him/herself or nominate a student as next speaker (namely, allocates the turn to a student). Excerpt 2 is a typical example.

Excerpt 2: T (teacher)nominates Leo (student) after whole-class elicitation fails.

```
69   T:    =oka:y ANdere sportarten? {      (0.3)
            okay  other   sport types    sport types
           'Okay, other types of sport? '
70                                     {((T shifts gaze from her right
                                          to left))}
71         {sportarten?  (2.7){(.)
           sport types
72         {((T slowly shifts gaze from her left to right))
73                               {Fig #3((T's gaze falls on Leo who is gazing
                                  at blackboard))
74   T:    {LEO,
75         {((T points her right hand at Leo and quickly withdraws))
76→L:      {<<p>laufen>;
            walk/run
           'Walk/Run.'
77         {Fig #4((Leo shifts gaze from blackboard to teacher))
```

Fig #3

Fig #4

In Excerpt 2, the teacher continues to solicit brainstorming answers from the whole class, again using the multimodal format of [verbal question (Line 69) + scanning gaze from her right side of the classroom to her left (Line 70)]. After a 0.3-s pause, she makes an effort to pursue response by repeating the core element of the question *sportarten* (sports type) in Line 71, and simultaneously launching a longer scanning gaze (2.7 s) from her left side of the classroom to her right (Line 72) before it falls upon the student named Leo (Line 73). This pursuit is still multimodally performed and embodied. However, still no response is volunteered from any student. Consequently, when the teacher's gaze finally falls on Leo (Fig #3), she conveniently nominates him as next speaker by calling out his name (Line 74) and delivering a quick pointing gesture to him. This multimodal nomination, formatted as [verbal address term (Leo's name) + pointing gesture], has drawn Leo's attention from the blackboard (Fig #3) to meet the teacher's gaze (Fig #4), hence achieving a mutual gaze. Similar to Excerpt 1, the mutual gaze here also shifts the student's participation frame with the blackboard—an artefact in the classroom environment—to a dyadic participation frame with the teacher. Once the participation frame is visually established by mutual gaze, a further focused interaction can proceed smoothly. In Line 76, Leo immediately delivers the answer, albeit not in a confident manner (indicated by the low voice).

The above analysis shows that Type 1(Excerpt 1) and Type 2 (Excerpt 2) in classroom interactions share many organizational features of teacher-initiated turn-taking: (1) All involve deployment of multimodal, embodied resources, such as verbal question, gaze, gesture, etc. (2) Shifts in participation frames systematically occur as a preparatory step for the teacher-initiated turn allocations: either the teacher shifts gaze from the whole class to an incipient student, or the students often change their attention from the artefacts in the classroom (i.e. textbook or blackboard) to the teacher. Such shifts in participation frame are unanimously accomplished by multimodal, embodied practices, most typically a mutual gaze[2]. They enable the establishment of a dyadic interactional formation, which in turn guarantees smooth teacher-initiated turn-taking and turn-allocational practices. Among all the 733 teacher-initiated turn-takings in GSL classroom, there is not a single case of interactional trouble occurring with regard to the transition of turns. The picture is significantly different in online GSL sessions, as will be shown in the next section.

5.2 From Face-to-Face GSL Classroom to Virtual GSL Sessions: Reduced Embodiment

In the 30-h screen-recorded online GSL sessions, 417 cases of teacher-initiated turn-taking have been singled out, which are also categorizable into two major types: Type 1 (N = 171) is teacher's addressing the whole class where at least one student volunteers for an answer, and Type 2 (N = 246) is teacher's next-speaker turn allocation where teacher nominates a next speaker. It is interesting to compare this statistic with that of classroom interactions. Even though it is the same GSL course opened for the same-grade students from the same program of the same university, the total number of teacher-initiated turn-taking (N = 417) is significantly lesser than that of classroom interactions (N = 733). Also, the proportion of teacher-initiated turn-allocations vis-à-vis teacher-initiated addressing of the whole class in online GSL settings (246/171) is larger than that of in-person GSL sessions (251/482). It seems that online education has reduced the opportunities or willingness for instructors to initiate students' contributions to the sessions and that online instructors tend to favor student nomination over whole class addressing. This phenomenon, albeit deserving further investigations, will not be addressed in this paper. As mentioned above, online teaching-learning environment has inevitably excluded some embodied interactional resources and shut down many multimodal channels. This has been fundamentally reshaping the way in which the instructed learning is conducted. One typical dimension of this impact is reified in the unsmooth, sometimes fractured organization of teacher-initiated turn-taking, as is shown in Excerpt 3.

[2] It should be noted, however, that among the 733 instances of teacher-initiated turn-taking there are a very small number of exceptions (N = 27) where the nominated student assumes the speakership but never returns gaze to the teacher. Detailed discussion of these cases goes beyond the scope of this paper.

Excerpt 3: T checks C (a student named Christina)'s availability to answer.

```
32   T:   okay 这    是    呃(.)这个
          ok   this  is    eh   particle
          'OK, this one is done'
33        我 再     问   一   个              同学      啊
          I  again  ask  one  classifier     student   particle
          'OK,I will ask another student.'
34        (1.7)Christina,
          'Christina?'
35        (6.0)
36        Christina?
          'Christina?'
37→C:     hh.老师    能      听见    [吗?]
             teacher can     hear    particle
          'Teacher, can you hear me?'
38   T:   [啊       听得到
          particle can hear
          'Yes, I can hear you.'
```

In Excerpt 3, all students are in a camera-off status, leaving only vocal channel available for online interaction. In fact, this is a common situation in our online GSL data. Although Tencent Meeting provides both visual and audio channels for online interactions, students often have to shut off the camera to save the internet bandwidth. This issue can only be resolved with faster internet connection technologies. The unavailability of the visual channel has two consequences: (1) when most students' visuals are not available, their competing activities off camera cannot be monitored by the teacher during online sessions. (2) More importantly, the teacher cannot use embodied resources, such as scanning gaze, to check the students' embodied willingness or availability to contribute answers, as is frequently seen in L2 classroom data [14, 16, 22, 23]. The only source remaining within the teacher's command is a vocal callout of a targeted student's name (Line 34). Because no embodied participation frame is established, teacher-initiated turn-taking often becomes fragile and sometimes unsmooth. This is evidenced by the 6-s gap (Line 35) after the teacher's callout of Christina's name. It is not until the teacher re-launches the vocal callout of her name (Line 36) that Christina delivers the response (Line 37). Interestingly, Christina's first response is not to express her willingness or availability to answer the teacher's question, but to check whether she can be heard by the teacher (Line 37). Only after the teacher confirms this does Christina begin to proceed with further interaction. Such instances in which the nominated student checks his/her speakership while taking the teacher-miniated turn are by no means rare in our online GSL instruction data, as is shown in Excerpt 2.

Excerpt 4: T(teacher) instructs N (a student named Nele) to read the paragraph on the shared screen.

```
29   T:   {接下来         接下来     这     一      段
           next           next        this  one    paragraph
          'The next paragraph.'
30        {Fig #5 ((T gazes into the screen while all students' camera sta-
          tus is OFF))
31        我们  先   找个   同学     来     读    一下 (.) 这个 文章
          we first look for student  come  read  once   this  text
          'Frist, let's find someone read the text.'
32        le- nele,
          'Nele.'
33        (5.0)
34→N:     em (0.4){als   wir  wieder  in unserem zimmer
          em          when  we  again   in  our    room
          "Wenn we were back in our room again."
35                 {((Fig #6 ((N's name is first shown in the
                     chat box))
36→       <<all>老师         是      我        吧>
               teacher     is     me     particle
          'Am I speaking teacher?'
```

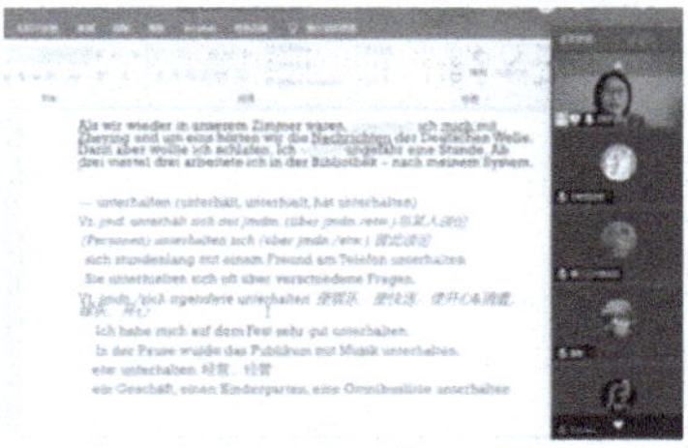

Fig #5

Fig #6

```
37   T:   是     你    nele;
          is    you   nele
          'Yes, Nele, it is you speaking.'
38   N:   哦 (是)       in unserem zimmer waren;
            oh (it is)  in  our     room    are
            'OK, then. We were in our room.'
```

In Excerpt 4, after the teacher explicitly expresses her intention to find a student to read the text on the shared screen, she nominates Nele as next speaker by calling out her name. After 5-s gap, Nele takes the turn and begins reading the text. However, at the beginning of her turn, her speakership is not immediately shown in the chat box (see Fig #6). This might constitute an additional reason, apart from unavailability of embodied participation resources such as mutual gaze with the teacher, why she suddenly suspends her reading and inserts a verbal check of her speakership (Line 36). After the teacher confirms her speakership (Line 37), Nele resumes her reading of the text. Different from Excerpt 3 and 4 which illustrate Type 2 teacher-initiated turn-taking, Excerpt 5 is an example of Type 2, in which the teacher's whole class addressing is met with a volunteering response from a student named Felix.

Excerpt 5: T(teacher) instructs the students to make sentences using the grammatical structure listed on the shared screen.

09 {**他 允许** Monika **其实 是就 他 让** Monika;
he allow Monika actually that is he let Monika
'Actually, "he allows Monika" is "he lets Monika". '
10 {**Fig #7** ((T facilitates the sentence-making task by giving a Chinese sentence))

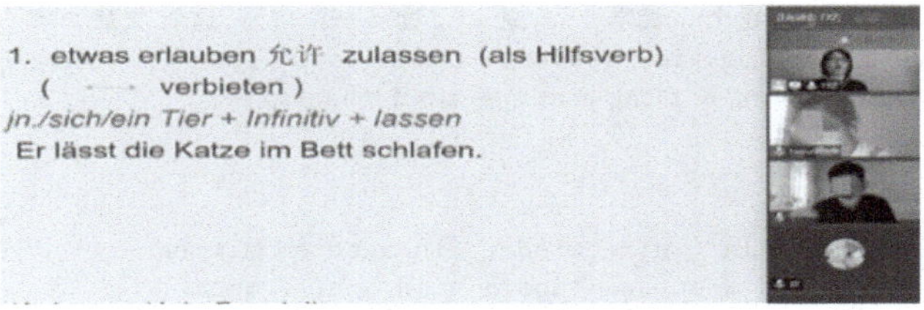

Fig #7

11 **其实 就 用** lassen;
actually exactly use let
'Actually we use the word "let".'
12 Monika **读 这 封 信**
Monika read this classifier letter
'Monika read this letter.'
13 {(10.0)
14 wer möchte proBIEren?
who want try
'Who wants to try?'
15 {(11.0)
16 T: **他 允许**(.) **者或 他 让** Monika **读 这封 信**
He allow or he let Monika read this letter
'He allows or he lets Monika read this letter.'
17 (3.0)
18→F: **啊 我 说 吧 老师**
ah I say particle teacher
'I will answer it teacher. '
19 T: **来** felix **是 吧**
come felix is particle
'Come on. It's Felix, right?'
20 {(0.6)
21 {**Fig #8** ((Felix's name first appears on the talking box indicating he is currently responding to T)).
22→F: [**嗯**]
Yeah
'Yeah.'
23 T: [**来**] bitte;
come *please*
'All right. Please.'
24 F: **嗯** er LÄSST monika den brief lesen;
en he let monika the letter read
'He lets Monika read the letter. '

Fig #8

In Excerpt 5, the teacher assigns the whole class a pedagogic task – that is, to make sentences using the Garman grammatical structure displayed on the shared screen (Fig #7). The teacher facilitates students' sentence-making by uttering a sentence which contains a Chinese equivalent of the German grammatical structure (Line 9–12). However, this facilitation, which functions as an implicit invitation for students' contribution, meets with a long silence (10 s) from the students (Line 13). In Line 14, the teacher explicitly verbalizes her invitation for students' response. Again, another long gap occurs (Line 15). In Line 16, the teacher repeats the Chinese sentence, expecting a student to make an attempt. After three seconds, a student named Felix takes the turn by expressing his willingness to make a sentence (Line 18). Interestingly, the teacher does not respond to Felix with a go-ahead token as is often seen in classroom interactions. Rather, she delivers a confirming question to check Felix's speakership. From the recorded screen on the teacher's side, it shows clearly that Felix's name does not show up in the chat box (Fig #8) until her confirming question in Line 11 is finished. This lapse could be caused by the limit of internet connection. Whatever the reason is, this technological contingency triggers the teacher's checking of Felix's speakership. This interpretation is supported by the following overlap between the teacher's go-ahead token in Line 23 and Felix's confirmation in Line 22. That is, the timely appearance of Felix's name in the chat box in Line 21 serves as a solid indicator to the teacher of Felix's current speakership, thus rendering Felix's verbal confirmation redundant. In Line 24, Felix delivers the sentence, and successfully finalizes the task.

The above analysis shows that Type 1(Excerpt 3 and 4) and Type 2 (Excerpt 5) in online GSL sessions share many organizational features of teacher-initiated turn-taking: (1) The availability of multimodal and embodied resources is substantially reduced in virtual environment. (2) The reduced embodiment has ruled out the possibility of establishing embodied participation frames which are systematically used in classroom interactions. Consequently, both the teacher and the students have to use more verbal checks of each other's speakership in order to establish focused, dyadic participation frames, which in turn often leads to unsmooth, fragile organization of teacher-initiated turn-taking.

6 Conclusion

This paper has explored the systematic differences in the organization of teacher-initiated turn-taking between offline, in-person GSL instructions and online, mediated GSL sessions. We argue that findings from face-to-face classroom interactions are both heuristic and consequential to our understanding of online education: they offer a good vintage point to examine how technology leaves its footprints on teacher-initiated turn-taking mechanism in online, whole-class instructions. Technology-mediated education, i.e. via the platform of Tencent Meeting, cannot avoid sacrificing the range of multi-modal channels and embodied resources which are normally present in the in-person classroom interactions and aid in establishing and adapting effective participation frames. Such embodied participation frames, as have been discussed above, play a vital role in guaranteeing smooth teacher-initiated turn-taking and turn-allocations. When most of the students are visually unavailable due to technical difficulties, i.e. internet connection, they are practically no different from being positioned in a "dark box" from the teacher's perspective. According to Gibson [7], the affordance of an object refers to the potential actions that the object enables its user to perform. Technology is fundamentally inhuman and disembodied, and technological progress follows a path of fighting against this fundamental feature of disembodiment. In fact, Tencent Meeting platform has already provided many solutions to make up for such disembodiment. For example, the student' microphone-status icon can serve an indicator of his/her willingness to take the turn. However, the spreading display of whole class' microphone-status icons over a limited screen has made it nearly impossible for the teacher to monitor each one of them on a moment-by-moment basis. This difficulty in monitoring students' status is enhanced by competing use of the screen such as screen sharing. Also, technological contingencies, such as lapse in registering the speaker's name in the chat box, make the monitoring even harder.

References

1. Merleau-Ponty, M.: Phenomenology of Perception. Routledge, London (1962)
2. Mead, G.H.: Mind, Self and Society. University of Chicago Press, Chicago (1934)
3. Goffman, E.: The interaction order. Am. Sociol. Rev. **48**(1), 1–17 (1983)
4. Goffman, E.: Behavior in Public Places. Notes on the Social Organization of Gatherings. The Free Press, Glencoe (1963)
5. Goffman, E.: Interaction Ritual. Essays on Face-to-Face Behavior. Doubleday Anchor, New York (1967)
6. Mondada, L.: Multimodal resources for turn-taking: Pointing and the emergence of possible next speakers. Discourse Stud. **9**(2), 194–225 (2007)
7. Gibson, J.J.: The Ecological Approach to Perception. Houghton Mifflin, London (1979)
8. Goodwin, C.: Restarts, pauses, and the achievement of a state of mutual gaze at turn-beginning. Sociol. Inquiry **50**(3–4), 272–302 (1980)
9. Rossano, F.: Gaze in conversation. In: Sidnell, J., Stivers, T. (eds.) The Handbook of Conversation Analysis, pp. 308–329. Wiley-Blackwell, Chichester (2013)
10. Auer, P.: Turn-allocation and gaze: a multimodal revision of the "current-speaker-selects-next" rule of the turn-taking system of conversation analysis. Discourse Stud. (2020, forthcoming)
11. Streeck, J.: Gesturecraft: The Manufacture of Meaning. John Benjamins, Amsterdam (2009)
12. Mehan, H.: Learning Lessons. Social Organization in the Classroom. Harvard University Press, Cambridge (1979)
13. Kääntä, L.: Teacher turn-allocation and repair practices in classroom interaction: a multisemiotic perspective. Jyväskylä Studies in Humanities (137) (2010)
14. Mortensen, K.: Selecting next speaker in the second language classroom: how to find a willing next speaker in planned activities. J. Appl. Linguist. **5**(1), 55–79 (2008)
15. Mortensen, K.: Establishing recipiency in pre-beginning position in the second language classroom. Discourse Process. **46**(5), 491–515 (2009)
16. Sert, O.: Social Interaction and L2 Classroom Discourse. Edinburgh University Press, Edinburgh (2015)
17. Lauzon, V.F., Berger, E.: The multimodal organization of speaker selection in classroom interaction. Linguist. Educ. **31**, 14–29 (2015)

18. Evnitskaya, N., Berger, E.: Learner' multimodal displays of willingness to participate in classroom interaction in the L2 and CLIL contexts. Classroom Discourse **8**(1), 71–94 (2017)
19. Lee, Y.A.: Negotiating knowledge bases in pedagogical discourse: relevance of identities to language classroom interactions. Text Talk **35**(5), 621–642 (2015)
20. Sacks, H., Schegloff, E.A., Jefferson, G.: A simplest systematics for the organization of turn-taking for conversation. Language **50**(4), 696–735 (1974)
21. Selting, M., et al.: Gesprächsanalytisches Transkriptionssystem 2 (GAT 2). In: Gesprächforschung – Online-Zeitschrift zur verbalen Interaktion **10**, 353–402 (2009). http://www.gespraechsforschung-online.de/fileadmin/dateien/heft2009/px-gat2.pdf
22. Sert, O.: 'Epistemic status check' as an interactional phenomenon in instructed learning settings. J. Pragmat. **45**(1), 13–28 (2013)
23. Sert, O.: Mutual gaze, embodied go-aheads and their interactional consequences in second language classrooms. In: Hall, J.K., Looney, S.D. (eds.) The Embodied Work of Teaching. Multilingual Matters, pp. 142–159. Blue Ridge Summit, Bristol (2019)

An Online English Learning Community for College Students Based on Community of Inquiry Framework

Cuiying Liao and Lu Tian(✉)

School of Interpreting and Translation Studies, Guangdong University of Foreign Studies, Guangzhou, China
ivytianlu@gdufs.edu.cn

Abstract. Establishing an online English learning community based on Community of Inquiry (CoI) framework provides an effective way to cope with problems existing in collegiate English teaching and learning in China. By participating in designed activities and discussing learning issues with other learners, participants can not only enhance their overall competences, in particular, productive competence, but also receive timely and effective feedback from their peers. This paper discusses the feasibility and practicality of building such an online English learning community by using WeChat group and Xiaodaka mini-program as the platform and following the elements of social presence, cognitive presence and teaching presence of CoI as the theoretical guidelines. Our pilot study shows that the established community received generally positive feedback from the participants who appreciated its effectiveness in improving their integrated English skills and acknowledged that the community helped with their motivation and self-regulation in English learning.

Keywords: Online education · Learner community · Collaborative learning · Community of inquiry · Integrated English skills

1 Introduction

With emerging mobile technologies and the use of social media, online English education has witnessed innovation in both technological and pedagogical dimensions. Various online teaching and learning modes such as MOOC and SPOC are gaining increasing popularity. Blended English learning, which combines online and offline learning, stimulates the growth of English education. However, despite such reforms and innovations in English education, problems remain in collegiate English teaching and learning in China.

Although communicative teaching is advocated in college education, teachers still play the dominant role in many English classes in China, elaborating on language points, explaining meaning of texts, etc. Due to the limit of class time, there is usually rather limited time for students' output training and practicing (speaking and writing). Such insufficient engagement in active learning leads to inefficient English learning, especially in terms of productive competence. It is found that when training on input

C. Pang et al. (Eds.): SETE 2020/ICWL 2020, LNCS 12511, pp. 370–379, 2021.
https://doi.org/10.1007/978-3-030-66906-5_35

and output fail to integrate, students show unbalanced development in listening, speaking and writing [1].

Another problem is that even if there are some output practices, students can hardly receive adequate feedback from teachers due to its time-consuming nature. It is reported that for most teachers it takes them at least 20 to 40 min to comment on an individual student paper, which adds up to an enormous amount of time [2]. This problem is particularly prominent in MOOC. With massive enrollment, teachers hardly have time to give feedback to individual students and monitor individual learning. In fact, MOOC is seen as an autonomous learning approach, requiring learners' high level of self-regulation. Thus, it is not surprising that the dropout rate is usually high [3].

To cope with these problems, we propose an online English learning community for college students. Relying on WeChat, a popular social media network in China, the community is mainly constructed and operated by students. A student instruction team is responsible for the design of activities and supervision of the learning process. The learning community features its tailored instruction, balanced input and output, and most important of all, peer interaction. Through peer assessment, participants will either feel pressured or gain encouragement from their peers, both of which have proved helpful in regulating their everyday English learning. Instead of relying solely on teachers, the community satisfies participants' need for sufficient output practices and timely feedback. Based on the framework of Community of inquiry, this paper is a report on the rationale and a trial operation of such an online English learning community, discussing its resign and operation as well as sharing our experience in its management.

2 Community of Inquiry

2.1 Definition and Application

Community of inquiry (CoI) is defined as "a group of individuals who collaboratively engage in purposeful critical discourse and reflection to construct personal meaning and confirm mutual understanding" [4]. The CoI framework was first proposed by Garrison, Anderson and Archer in 2000. The philosophical premise of the framework is a collaborative constructivist approach to teaching and learning. Garrison [5] believes that "socially situated" and "thinking collaboratively" are innately human characteristics, and technology advancement makes possible the connection and knowledge sharing across time and space.

CoI has been a widely adopted framework in research and pedagogy of online learning to enrich students' learning experiences. Despite its wide application in scientific fields, such as biology and medicine, the adoption of CoI framework in English learning is still insufficient. Among the few studies, Goda and Yamada [6] give some suggestions on how to use CoI to design computer-supported collaborative learning (CSCL) for English as a foreign language (EFL). Their suggestions include to support open communication and to help students establish social presence first then shift to academic focus. Different from Goda and Yamada's study, which is based on blended learning and focuses on Japanese English learners, this study pays attention to the application of CoI to full online English learning and emphasizes on students' dynamic roles in constructing the community based on the reality of English higher education in China.

2.2 Three Elements of CoI

The CoI provides a theoretical framework to study and understand the influence of a purposeful learning environment on the cognitive processes of individuals. This is accomplished through the concurrent consideration of three interdependent elements—social presence, cognitive presence, and teaching presence—operating in technological means of communication.

Social presence is defined as "the ability of participants to identify with the group or course of study, communicate purposefully in a trusting environment, and develop personal and affective relationships progressively by way of projecting their individual personalities" [4]. Cognitive presence, the core thinking and learning element, is the extent to which learners are able to construct and confirm meaning through sustained reflection and discourse [7]. Teaching presence provides the essential leadership dimension that keeps a learning community functioning effectively and efficiently [8].

The categories and indicators of each element are summarized and listed by Garrison as below (Fig. 1).

ELEMENTS	CATEGORIES	INDICATORS (Examples only)
Social Presence	Open communication Group cohesion Personal/affective	Risk-free expression Group identity/collaboration Socio-emotional expression
Cognitive Presence	Triggering event Exploration Integration Resolution	Sense of puzzlement Information exchange Connecting ideas Applying new ideas
Teaching Presence	Design & organization Facilitating discourse Direct instruction	Setting curriculum, methods Shaping exchange Resolving issues

Fig. 1. Community of inquiry categories (Source: [5])

It is worth noting that the three elements are not separated but coordinated to sustain an efficient and collaborative community. Figure 2 is a clear demonstration of their connections and interrelationships.

Based on the three elements of CoI and their operating mechanisms, this study explores the construction of an online English learning community for college students to enhance their English competences (teaching presence), exchange ideas, and share

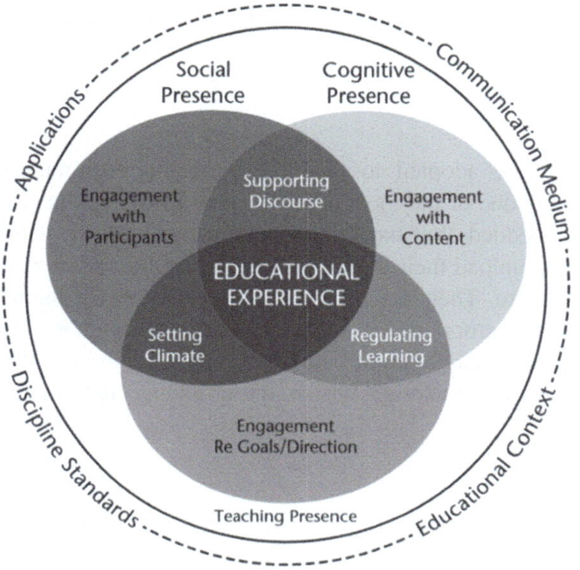

Fig. 2. Community of inquiry framework (Source: [5])

experience in English learning (social presence). By joining the community, participants are expected to cultivate better self-regulation in English learning and obtain the ability to find solutions to problems by themselves (cognitive presence).

Apart from the three elements of CoI which mainly involve thinking in a community, the shared metacognition focuses on "thinking on the thinking". By providing students with a new approach to English learning, the learning community is expected to raise students' awareness of English output and help them get into the habit of routine English learning.

In the following sections, under the guidance of CoI framework we will elaborate on the design of such an online English-learning and student-centered community for Chinese college students that aims to motivate participants to learn English more actively through fulfilling well-designed collaborative tasks.

3 An Online English Learning Community

3.1 Participants

This online community is created for Chinese college students to practice all-round English skills. In our pilot study, 24 students joined the community and participated in activities for two months. All the participants are translation majors with Chinese as the native language. They share the unified purpose of improving English. Among the 24 participants, 10 are sophomores and 14 freshmen. Tested by the Common European Framework of Reference for Languages (CEFR), their English levels range from B2 to C1. Sophomores generally have better command of English (listening, speaking, writing, and reading) than freshmen. The study was conducted during February and March in 2020.

3.2 Instruments

Two network tools are adopted to facilitate the online community—WeChat and Xiaodaka. WeChat is one of the most popular social media applications in China, and Xiaodaka is an embedded WeChat mini-program for check-in management. Participants are required to upload their daily assignments (audio and/or word documents) to Xiaodaka mini-program. Their assignments are open to all the members of the community. Both the instructors and participants can leave comments and/or show their support by clicking the "like" button below the post. Meanwhile, WeChat groups are set up for sharing learning materials, discussing problems, and communicating among participants (Fig. 3).

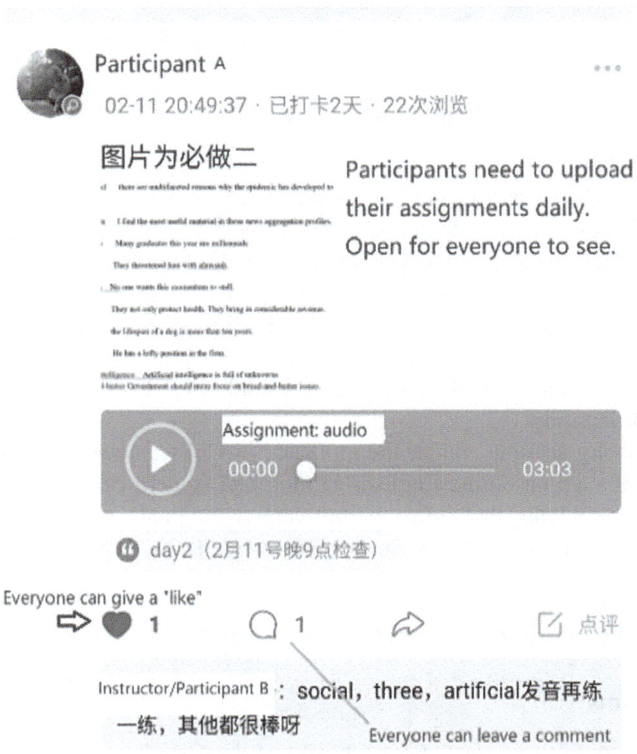

Fig. 3. Display of Xiaodaka Mini-program

3.3 Teaching Presence in the English Learning Community

In this learning community, an instruction team comprising advanced learners is responsible for the operation and management of the community. Some preliminary tasks include setting up a WeChat group and a Xiaodaka circle, issuing rules and guidelines, etc. Participants are required to upload their daily assignments on Xiaodaka before stipulated deadline. If one fails to hand in daily assignment for more than two days in a week, a warning will firstly be given. If the situation continues, they might be removed from the community. Participants are asked to be responsible for their words and behavior. Irrelevant information is forbidden in the community.

A major task for the instruction team is to organize learning activities which are generally categorized into input and output practices. Input refers to listening and reading, and output refers to speaking and writing. Scholars believe that integration of input and output practices could help improve efficiency in English learning [9]. Following such a principle, our English learning community is designed to improve participants' integrated English abilities through integrated input and output training, viz. speaking after listening and writing after reading. The instruction team will distribute learning materials, specify training requirements, and make timely adjustments as the project moves on. The instruction team will also give feedback to participants, although peer assessment is more encouraged.

3.4 Social Presence in the English Learning Community

Social presence refers to participants' engagement and communication in the community, including open communication, group cohesion and affective expression [5]. First, open communication is realized through WeChat group discussion. In this community, there are no teachers, so the participants are less pressured and will not feel embarrassed or intimidated to express their views. By sharing their experience as well as methods and materials in English learning, participants can not only learn from each other but also gain comfort and confidence. By raising questions and giving answers to others' inquiries, participants' active thinking will be greatly stimulated and their teamwork spirit will be cultivated.

Second, group cohesion is what motivates participants to study harder in the community. According to Garrison, motivation and psychology are important for promoting cognitive process. A sense of belonging and emotional satisfaction are critical sources of motivation. To sustain motivational and psychological satisfaction, the purpose of the learning community must be identified and the climate that participants are valued in the community must be created [5]. As participants are from the same major and have the same purpose of enhancing their English competence, they have shared emotions and identity. Stronger group cohesion helps in raising participants' awareness of keeping pace with other participants in the community. When they see others doing well, they are also be motivated to keep on studying. As anyone who breaches the rules or fails to finish the assignment on time runs the risk of being removed from the community, participants are motivated to study hard and behave well in the community for fear of punishment. This helps in fostering co-regulation among participants and establishing an active and efficient English learning ambience in the community.

Third, participants can express their emotions during the course of learning. On Xiaodaka mini-program, students can give "like" to show their appreciation of others' performance or effort. Participants who receive this "like" will be motivated, feeling approved and encouraged. As the learning materials cover a wide range of topics, such as climate change and consumerism, participants can also express their opinions about the topics and show their concerns for various social issues.

Fourth, peer assessment together with automated essay scoring are employed to give participants efficient assessment. Participants can receive timely feedback and advice from their peers as well as the instruction team. Such feedback may include grammatical error correction, sentence structure improvement, textual organization suggestion, etc. As students may not always be able to receive individual feedback from their teachers, this would function as a supplement to classroom teaching and satisfy participants' need for feedback after training.

3.5 Cognitive Presence in the English Learning Community

Cognitive presence in the English learning community starts from posing questions. Then through discussion with other participants and their own introspection, participants learn to solve problems with the help of others as well as with their own effort. In the process, their English will be improved and their self-regulation be enhanced.

For example, when participants have doubt about a new phrase in an article and find it difficult to understand, they can raise their question in the chat group. Anyone can respond and share their understanding. If there are divergent answers to this question, participants can discuss, exchange information, and work collaboratively to solve the problem. Therefore, if they encounter the same phrase next time, they could apply what they have learned to the understanding of the text. Ultimately, participants will have a better command of English through such an inquiry-exploration-resolution process.

In addition, metacognition refers to individuals' reflection on their own thinking [5]. It helps participants have positive attitudes towards learning and adjust their learning methods as needed. In this community, participants have access to various learning methods shared by other participants, and they are trained in a new approach to English learning—the integration of input and output. Therefore, participants are encouraged to compare different learning approaches and methods in order to choose the ones that suit them the best.

3.6 Coordination Between Elements

As Fig. 4 shows, social presence, cognitive presence and teaching presence are coordinated in a balanced manner to sustain an efficient English learning community.

Teaching presence consists of the design and organization of the community. To create an online English learning community, the instruction team is responsible for recruiting participants, setting up WeChat group and Xiaodaka circle, and issuing guidelines and rules of the community. The learning activities run under the design of

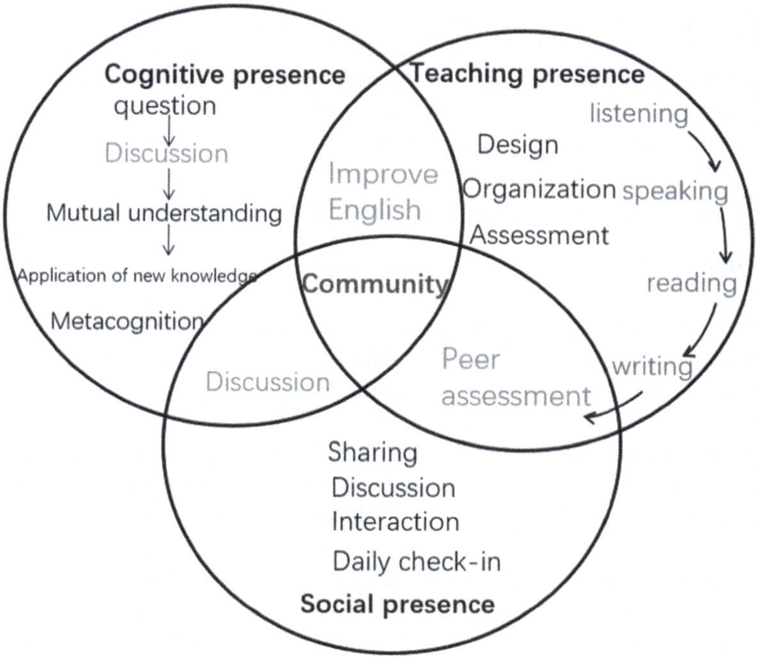

Fig. 4. Coordination between elements in English learning community

integrating input and output, and participants follow the routine tasks of listening, speaking, reading, and writing. Peer assessment is required for output tasks, i.e. speaking and writing assignments. Participants have to assess their partners' assignments according to a set of standards, such as vocabulary, grammar, collocation usage, and text structure. They are encouraged to leave comments under each other's assignment on Xiaodaka. Peer assessment connects social presence and teaching presence, involving interaction between participants and the design of the learning routine in the community.

Social presence includes daily check-in and sharing, discussion, and interaction among the participants. On Xiaodaka mini-program, participants are required to submit their daily assignments which are open to everyone in the community. Giving "like" to each other and leaving comments contribute to participants' motivation for working harder in the group as well as mutually beneficial interactions among participants. In the chat group, participants share their English learning methods and experience, which creates a climate that encourages participants to either bravely raise their queries or actively participant in the discussions. Discussion connects social presence with cognitive presence as discussions facilitate participants' deeper thinking into their questions.

Cognitive presence starts from participants' questions or puzzlement in English learning. Through discussions with others, participants can inquire deeper into their questions, construct meaning, and confirm mutual understanding. The process will help them build up a new understanding of the original problem. Therefore, when encountering a similar problem next time, they will be able to apply their acquired knowledge to its solution. Moreover, in the community participants have access to new learning approaches and methods, so they can adjust their learning strategies by comparing the efficiency and effectiveness of their ways of learning with those of others.

4 Feedback

We have conducted a survey on participants' opinions of the community. The feedback is generally positive, showing that our English learning community based on the CoI framework is helpful and effective in their English learning.

In terms of teaching presence, participants report that "the design of this learning plan helps me improve my English skills especially the output", "the instructor creates a community where learning materials and experience can be exchanged", and "I can know my mistakes from the instructor's feedback".

In terms of social presence, participants report that "I feel that I can learn from others' learning experience", "the feedback from other participants and the instructor helps me correct my errors", "I get a lot of useful learning materials and resources", "we can exchange different ideas in this community", "I have a sense of belonging", "I will keep on studying with my companion because it is hard to do so when studying alone", "the learning ambience is great for fostering a good learning habit", "I'm motivated when I see others doing well in English", and "I'm motivated when seeing other participants working hard and having self-regulation".

In terms of cognitive presence, participants report that "I have learned to make full use of the materials and assignment tasks to improve my English", "I feel proud that I can persist in submitting my assignment every day", "I can explore questions on English with my companions", "I can relate the newly learned expressions with the ones I already know and make comparisons between them", and "I can apply the newly acquired English expressions and knowledge to my speaking and writing".

5 Conclusion

The online English learning community under discussion is based on the Community of Inquiry framework, coordinating social presence, cognitive presence, and teaching presence. Using the two Internet instruments, WeChat chat group and Xiaodaka mini-program, we created such a learner-dominated community and conducted a pilot study. Both organizers and participants of the community are students. Teaching presence includes design of the learning routine and organization of the community. Advanced learners build up an instruction team responsible for the operation and management of the community. Social presence involves peer assessment, information sharing, discussion and interaction. This creates a sense of belonging which motivates students to work harder. Cognitive presence refers to participants' inquiry into questions they encounter in learning English and find solutions through discussions with others in the community.

The coordination of the three presences sustains an effective online English learning community, which has been proved to satisfy students' need for output practice and effective assessment. Feedback from the participants shows that an online English learning community can stimulate students' enthusiasm in English learning and enhance their integrated English skills and overall competence.

Acknowledgements. The work reported in this paper was substantially supported by The Undergraduate Innovation and Entrepreneurship Training Program of GDUFS.

References

1. Wang, N., Chen, J., Zhang, D.: SPOC-based flipped classroom of college English: construction of an efficient learning model. Technol. Enhanc. Foreign Lang. Educ. **16**(3), 52–57 (2016)
2. Sommers, N.: Responding to student writing. Coll. Compos. Commun. **33**(2), 148–156 (1982)
3. Xu, T.: Literature review on the dispute about MOOC. China Univ. Teach. **15**(7), 22–27+21 (2015)
4. Garrison, D.R.: E-Learning in the 21st Century: A Framework for Research and Practice, 2nd edn. Routledge, New York and London (2011)
5. Garrison, D.R.: Thinking Collaboratively: Learning in a Community of Inquiry. Routledge, New York and London (2015)
6. Goda, Y., Yamada, M.: Application of CoI to design CSCL for EFL online asynchronous discussion. In: Akyol, Z., Garrison, D.R. (eds.) Educational Communities of Inquiry: Theoretical Framework, Research, and Practice, pp. 295–316. IGI Global, Hershey (2013)
7. Anderson, T., Rourke, L., Garrison, D.R., Archer, W.: Assessing teaching presence in a computer conferencing context. J. Asynchronous Learn. Netw. **5**(2) (2001)
8. Tsai, C.W.: Applying web-based co-regulated learning to develop students' learning and involvement in a blended computing course. Interact. Learn. Environ. **23**(3), 344–355 (2015)
9. Wang, C.: An analysis of three L2 teaching preferences in light of an efficient route to L2 learning. Foreign Lang. Teach. Res. **43**(4), 540–549+640 (2011)

Language Learning and NLP

Predicting Potential Difficulties in Second Language Lexical Tone Learning with Support Vector Machine Models

Juqiang Chen[1(✉)], Catherine T. Best[1,2], and Mark Antoniou[1]

[1] The MARCS Institute for Brain Behaviour and Development, Western Sydney University, Penrith, NSW 2751, Australia
{J.Chen2,C.Best,M.Antoniou}@westernsydney.edu.au
[2] Haskins Laboratories, New Haven, CT, USA

Abstract. Second language speech learning is affected by learners' native language backgrounds. Teachers can facilitate learning by tailoring their pedagogy to cater for unique difficulties induced by native language interference. The present study employed Support Vector Machine (SVM) models to simulate how naïve listeners of diverse tone languages will assimilate non-native lexical tone categories into their native categories. Based on these simulated assimilation patterns and extrapolating basic principles from the Perceptual Assimilation Model (Best 1995), we predicted potential learning difficulties for each group. The results offer teachers guidance concerning which tone(s) to emphasize when instructing students from particular language backgrounds.

Keywords: Lexical tones · Perceptual assimilation · Second language learning · Support Vector Machine models

1 Introduction

In a traditional second language classroom, teachers present fixed contents to students and engage them in exercises and activities to facilitate learning. However, this may be problematic when there are students coming from different language backgrounds, each with their unique difficulties. Thus, it may be helpful to provide teachers with some information about potential learning difficulties derived from native language interference so that teachers can tailor their teaching to address these potential issues beforehand and can better understand students' mistakes and help them overcome those difficulties at later stages. Ideally, when teaching pronunciation, language teachers would examine phonetics of their students' native languages and conduct cross-language categorisation tests to predict native language influences. But that would be time-consuming and impractical, especially in large classrooms with students of varying language backgrounds. This study aims to explore the possibilities of employing machine learning algorithms to simulate how naïve listeners of differing language backgrounds initially perceive non-native phonetic categories, allowing predictions about their learning difficulties in light of second language speech learning theories.

C. Pang et al. (Eds.): SETE 2020/ICWL 2020, LNCS 12511, pp. 383–392, 2021.
https://doi.org/10.1007/978-3-030-66906-5_36

Second language (L2) speech perception and production is affected by learners' native language backgrounds [1]. Consequently, many theoretical models, such as the Perceptual Assimilation Model (PAM [2], PAM-L2 [3]), Speech Learning Model (SLM) [4], and Native Language Magnet model (NLM) [5], have been developed to account for the native language influences. However, PAM provides the most comprehensive theoretical considerations of how native phonological and phonetic factors modulate L2 learners' perception and production of non-native consonants, vowels and tones. SLM has instead focused on phonetic properties and on production and perception of individual phones whereas NLM has focused on perception of within category phonetic variations. Thus, we employed PAM to interpret simulated assimilation patterns and predict potential learning difficulties.

PAM proposes that if a non-native phone is perceptually assimilated to a single native phonological category, i.e., Categorised, and as a good exemplar of that native category, then no further perceptual learning will happen for that non-native phone. If the non-native tone is perceived as a deviant example of that native category, however, some learning will be possible for the non-native phone. If a non-native phone is not assimilated to any single native phonological category, i.e., is Uncategorised, it will be easier to learn.

In addition, PAM maintains that if two non-native phones are assimilated into two native categories (TC assimilation), the discrimination of this contrast should be better than if two non-native phones are assimilated into a single native category as equally good exemplars (SC assimilation). But neither TC nor SC contrasts are expected to show much change over time in L2 learning. If two non-native phones are assimilated into a single native category but one as a good exemplar and the other as a deviant exemplar (Category-Goodness or CG assimilation), the discrimination of this contrast is better than SC but worse than TC. CG contrasts will be the easiest type to acquire as an L2 contrast.

Despite years of research on native language effects in L2 perception and production, there has been little application of these findings to L2 teaching, which this paper aims to address. We argue that classroom teaching will benefit if students' native language backgrounds are taken into consideration and pedagogy is tailored to specific learning difficulties induced by differences among native languages. Ideally teachers could design psycholinguistic studies [6, 7] to discover perceived similarities between non-native phones and students' native languages, and use those to predict potential learning difficulties. However, this is impractical as it would add too much workload to teachers and students. Thus, it would be helpful to simulate non-native perception with machine learning algorithms to uncover potential difficulties by L2 learners.

Classification algorithms, such as Linear Discriminant Analysis (LDA) [8, 9], or Support Vector Machine (SVM) models [10, 11], allow acoustic similarities to be modelled across languages as an estimate of the extent to which non-native listeners may use specific acoustic features to categorise non-native items into their native categories. For example, Strange has used LDA models in cross-language vowel perception, in which the model was trained with American English (AmE) vowels and tested with Parisian French and North German vowels [12]. The AmE-trained model acted as a "virtual listener" to determine the acoustic distances between vowels in AmE and the other two language systems. In addition, Styler used SVM models to classify

English and French nasal vowels and determine the most important acoustic correlates for nasality [10]. Chen and He [11] used SVM model to classify and examine non-native production of English information foci based on temporal and pitch contour characteristics.

However, very few studies have used machine learning models to simulate cross-language categorisation of lexical tones, although more than 60% of the languages use tones, in which pitch variations change the lexical meanings of words [13]. In this paper, we present simulations of perceptual assimilation of Thai tones to native Mandarin tones (Experiment 1) and of Mandarin tones to native Thai tones (Experiment 2).

Thai (five tones) and Mandarin (four tones) differ phonologically in terms of the number and types of tones they contrast. Phonologically, Thai has three level tones contrasting in height (high, mid and low) and two dynamic contour tones (rising and falling) that do not contrast in height [14]. Mandarin has one level and three dynamic contour tones in citation form: a level, a rising, a falling-rising and a falling tone (see Fig. 1).

Chao notation, in which F0 height at tone onset, offset and midway turning point if there is one, indicates general phonetic properties of tones using the numbers 1–5 (low to high). The three Thai (T) level tones are phonetically characterised as T45 for the high-level tone, T33 for the mid-level, T21 for the low-level, and the Thai contour tones as T315 for the rising tone and T241 for the falling tone [15]. The four Mandarin (M) tones are notated as level M55, rising M35, falling-rising M214 and falling M51 [16].

Figure 1 shows the Lobanov-normalised [17] and time-normalised mean F0 contours of the Thai (left panel) and Mandarin (right panel) tones, with their names and Chao numbers (figure adapted from [18]).

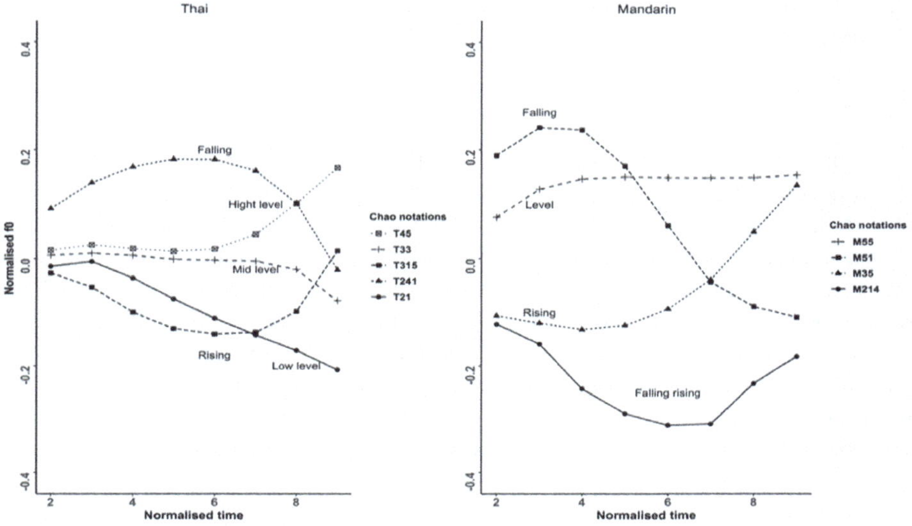

Fig. 1. Time- and Lobanov-normalised [17] mean F0 contours of Thai (left), Mandarin (right).

In this study, we employed native-data-trained SVM models to classify non-native tones based on the acoustic measures to simulate non-native perception of lexical tones. Although SVM models will not reflect all the underpinning processes of human tone perception, they model how specific acoustic information from non-native lexical tones contribute to acoustics-based classifications of the tones into a language's tone categories.

2 Experiment 1

In this experiment, we trained an SVM model with Mandarin tone acoustic features and tested it with Thai acoustic data. This simulates Thai-naïve Mandarin listeners' assimilation of Thai tones to Mandarin tone categories. We also compared our model outputs with real Mandarin listener data [18].

2.1 Data Collection

We recorded productions of two syllables (/ma/and /mi/) by native speakers of each language with each of their native tones in testing booths at the MARCS Institute for Brain, Behaviour and Development, Western Sydney University. All target syllables were meaningful morphemes in both languages. Mandarin items were elicited via Pinyin, and Thai items via Thai orthography, in random order.

Native Thai speakers (n = 9, 4 females) were recruited for a separate study [19] to produce the Thai syllables (/ma/and /mi/ $\times$ 5 tones $\times$ 4–6 repetitions, 430 Thai tokens in total) in citation form, at the sampling rating of 48 kHz 16 bit for a separate study [19], and were used here with permission.

Native Mandarin speakers (n = 8, 4 females) were recorded at a sampling rate of 44.1 kHz 16-bit, each producing 64 tokens (/ma/and/mi/ $\times$ 4 tones $\times$ 8 repetitions, 511 tokens in total[1]).

2.2 Feature Selection

All of the recorded syllable boundaries were first automatically marked with a Praat [20] function and manually checked. The Praat script *ProsodyPro* [21] provided syllable duration and 10 equidistant points of F0 values (in Hz). Raw F0 (in Hz) was normalised using the Lobanov method [17], which reflects how much an F0 value for a tone varies from the F0 mean of the speaker. The most stable part of the normalised tone (from 10% to 90% of the syllable) was used to calculate all F0-related measures.

We calculated syllable duration, F0mean, F0onset, F0offset, and F0excursion (the range of F0 variation of the tone), which were used to characterise level tone contrasts in a previous study [22], and we added one more measure, F0max_location (i.e., F0 peak location as a proportion of the tone's duration) to distinguish differently-timed

[1] One token was mispronounced and thus was deleted.

peaks in convex and concave contours (e.g. T241 and T315). Data were scaled so that all the features contributed equally to the results.

2.3 Model Specifications

Due to a relatively small dataset (511 tokens), we used the "10-fold cross-validation" [10], in which the analysis was run 10 times on the data, each time using a different 9/10ths of the data as "training" and the remaining 1/10th as "test". The *e1701* package [23] in *R* was used for building the SVM classifiers in this study. SVM works by finding a line that best separates the different classes in the data. When the data is not linearly separable, a kernel can be used to handle the non-linear relations. In this study, a radial basis function kernel was used, following [10].

2.4 Results

The overall accuracy of the Mandarin-trained SVM model is 96.67%, comparable (or even better) to previous studies with vowel productions [24]. There were variations in machine classification accuracy among different Mandarin tones. M35 was classified perfectly (100%), whereas M214 (accuracy 96.1%) was occasionally misidentified as M35 ($\sim$3.9%). M55 and M51 were identified with high accuracy 99.2% and 98.4% respectively.

Our Mandarin-trained SVM model (see Fig. 2, panel A) assimilated T45 as M35 (64%) and M55 (27.9%), which is similar to results in human perception using the same data as stimuli [25]. In that study, Mandarin listeners Categorised T45 as M35 (88.4%, see Fig. 2 panel B). T33 was split among M55, M214, M51 by the SVM model whereas Mandarin listeners Categorised T33 as M55 with high percent choice (92.5%). T21 was assimilated as M214 (75.6%) and M51 (23.3%) by the SVM model similar to Mandarin listeners with slight difference in percent choice, M214 (59.1%) and M51 (27.9%). T315 was assimilated mainly as M35 (70.9%) but M214 was also selected by the SVM model. This is similar to what Mandarin listeners did as they split the choice between M35 (79.6%) and M214 (20.4%). T241 was assimilated with high percent choice as M51 (91.9%) by the SVM model, whereas Mandarin listeners in the previous study split their responses between M55 (48.1%) and M51 (50.8%).

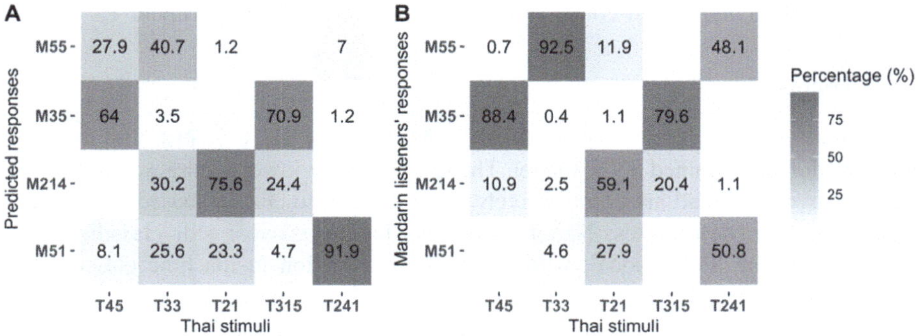

Fig. 2. SVM predicted results (panel A) and Mandarin listeners' responses (panel B [18]).

Although our SVM model results were not absolutely identical to the native Thai listener results, the SVM model covered all the potential native response categories and with largely comparable percent choices.

In order to identify Categorised assimilation as in PAM, two approaches have been used in previous studies. One approach is to test the percent choice via a set of statistical tests as in [21]: first, a given native tone must be selected significantly more than chance level; second, that single native tone category must be chosen significantly more often than any other native categories. However, our SVM model results do not show the group variations among individual listener responses (it was essentially a simulation of a single artificial listener) that are required for such statistical tests. Therefore we had to use the second approach, which defines Categorised assimilation as performance above an arbitrary pre-set threshold [1], 50% or 70%. We used the stricter threshold of 70% to determine Categorised assimilation.

In addition, we also considered overlapping of response categories that were above chance level (25% for the Mandarin model) because recent studies have shown this affects discrimination of non-native contrast [7, 26], thus inducing potential learning difficulties.

According to our criteria of Categorised assimilation, T21 is Categorised as M214, T315 is Categorised as M35, and T241 is Categorised as M51. T45 and T33 overlapped on M55; T33 and T21 overlapped on M214; and T315 and T45 overlapped on M35.

3 Experiment 2

This experiment reversed the training/test languages, that is, training the model with Thai tone acoustic features and tested the model with Mandarin acoustic data. This will reveal how native language inputs shape the SVM model and simulate how Thai listeners with no Mandarin experience would assimilate Mandarin tones to Thai categories based on the acoustic measures. Symmetry or asymmetry between the classification of Mandarin tones by the Thai-trained SVM model and the classification of Thai tones by the Mandarin-trained SVM model will show whether this method reflects simple acoustic correspondences between two datasets or the influence of the trained language tone system on classification of tones in a different language. Data collection, feature selection and model construction were the same as in Experiment 1.

3.1 Results

The overall accuracy of the Thai-trained SVM model is 91.16%, just slightly lower than the Mandarin-trained SVM model. The reason for the lower accuracy could be that Thai has five tones and are comparatively more difficult to classify based on the same number of acoustic features. Nonetheless, all Thai tones were still classified quite accurately by the SVM model, with some minor variation in machine classification accuracy of different Thai tones. T241 was classified almost perfectly ($\sim$98%) whereas T21 (93%) and T33 (93%) were occasionally misidentified as each other ($\sim$5%). T45 (94.2%) and T315 (95.3%) were confused even less often ($\sim$3%). Having achieved a relatively high accuracy in classifying native production data, the same Thai-trained SVM model was then used to classify Mandarin tone data as a simulation of Thai listeners' perception of Mandarin tones.

In the model results, Mandarin level tone M55 was split across Thai T45 (43%), T33 (32%), and T241 (25%) (see Fig. 3 panel A). In a previous study [27] (see Fig. 3 panel B) with actual Thai listeners who had no experience with Mandarin, they assimilated M55 as split between T33 (59%) and T45 (27.8%). It should be noted that the Mandarin stimuli used in [27] were different from this study. The SVM model's classification split Mandarin rising tone M35 between T45 (38.3%) and T315 (57%). This was consistent with the native listener findings, in which M35 was split between T45 (61.3%) and T315 (22.3%). The falling-rising tone M214 was classified by the SVM model more as T21 (64.1%) and less as T315 (27.3%) whereas the native Thai listeners had shown the opposite bias, choosing T315 (50.5%) more than T21 (26.8%). The SVM model assimilated M51 primarily as Thai falling T241 (71.7%), while native Thai listeners split the choices among T241 (54.5%), T45 (20.8%) and T21 (18.8%).

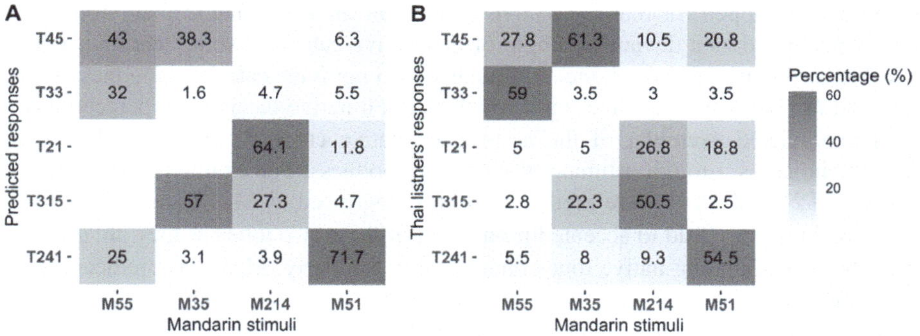

Fig. 3. SVM predicted results (panel A) and Thai listeners' responses (panel B, based on [27]).

According to our criteria of Categorised assimilation (a pre-defined threshold 70%), we found the only Categorised assimilation by the SVM model was M51 as T241. Other Mandarin tones were all Uncategorised by the SVM model. It is important to note that M55 and M35 overlapped in the SVM Thai response category T45 (the chance level for the Thai SVM model is 20%); M214 and M35 overlapped in the response category T315; and M55 and M51 overlapped in the response category T241.

4 Discussion

The present study bridges basic research and language pedagogy by employing machine learning techniques together with PAM [2] theoretical principles to predict learning difficulties in discrimination and accented production of non-native tones. Recent studies have indicated that overlapping in native responses could lead to problems in discrimination of non-native contrasts. In light of this, those non-native contrasts with overlapping native response categories should receive teachers' attention. For example, the Mandarin-trained SVM model's categorization of T45 and T33 overlapped in Mandarin category M55; T33 and T21 overlapped in M214; and T315 and T45 overlapped in M35. Thus, when teaching Mandarin students, Thai teachers may find that they are likely to confuse these contrasts T45-T33, T33-T21, T45-T315 and should tailor their teaching to emphasize these L2 contrasts. On the other hand, the Thai-trained SVM model's categorisation of Mandarin M55 and M35 overlapped in Thai category T45; M214 and M35 overlapped in T315; M55 and M51 over-lapped in T241. Consequently, Mandarin instructors should focus primarily on these contrasts when teaching students whose native language is Thai. The differences (or asymmetry, compare panel A in Fig. 2 and panel A in Fig. 3) between the two SVM models reflect both the native tone systems (in each training dataset) and the tone systems of the target languages, suggesting that the results were not based on simple acoustic correspondences between two datasets.

In addition, according to PAM-L2 [3], if a non-native phone is perceptually assimilated to a single native phonological category, i.e., Categorised, and if it is perceived as a good exemplar of that native category, then no further perceptual learning will happen for that non-native phone. On the other hand, if the non-native tone is perceived as a deviant example of that native category, some learning will be possible for the non-native phone. Although we do not have category-goodness rating data that are often used to determine whether a non-native category is perceptually a good or deviated exemplar of the assimilated native category, we can rely on the percent choice as a rough estimate of category goodness. For example, for Mandarin listeners, T21 and T241 were predicted to be Categorised as M214 and M51 respectively, and this may lead to accented production of these two tones as they are different from the corresponding native tones phonetically. Similarly, M51 was predicted to be Categorised as T241 with very high percent choice. Thus, it is likely that Thai-native leaners of Mandarin will use their native T241 to replace M51. This will lead to accented production due to phonetic deviation of T241 from M51, as can be seen in Fig. 1.

SVM models will not reflect all the underpinning processes of human tone perception, but they model how specific acoustic information from non-native lexical tones may contribute to listeners' classifications of the tones into their native tone categories. The dataset of the present study is admittedly small; however, it is comparable to what teachers could obtain in a foreign language classroom. The present study shows that the general predictions of the SVM models are comparable to native listener results, except for some variations in percent choice. In practice, it appears that language teachers can rely on SVM predictions to focus their training of students from different language backgrounds.

5 Conclusion

By combining machine learning algorithms and PAM principles, the present study provides an illustrative example as how to bridge basic research and teaching practices in language teaching classrooms. SVM models trained with the lexical tones of specific languages can be used to simulate how listeners of those languages will initially perceive non-native tones. The SVM predictions can be interpreted with reference to principles from PAM to offer further guidance on how to tailor teaching materials and designs to resolve unique problems of students from different language backgrounds. This method could also be applied in consonant and vowel learning and teaching.

Acknowledgements. This research was supported by a China Scholarship Council and Western Sydney University Joint scholarship awarded to the first author, Juqiang Chen.

References

1. Tyler, M.D., Best, C.T., Faber, A., Levitt, A.G.: Perceptual assimilation and discrimination of non-native vowel contrasts. Phonetica **71**, 4–21 (2014)
2. Best, C.T.: A direct realist view of cross-language speech perception. In: Strange, W. (ed.) Speech Perception and Linguistic Experience: Issues in Cross-Language Research, pp. 171–204. Timonium, York Press (1995)
3. Best, C.T., Tyler, M.D.: Nonnative and second-language speech perception: commonalities and complementarities. In: Bohn, O.-S., Munro, M.J. (eds.) Language Experience in Second Language Speech Learning, pp. 13–34. John Benjamins Publishing Company, Amsterdam (2007)
4. Flege, J.E.: Second-language speech learning: theory, findings, and problems. In: Strange, W. (ed.) Speech Perception and Linguistic Experience: Issues in Cross-Language Research, pp. 229–273. York Press, Timonium (1995)
5. Kuhl, P.K., Iverson, P.: Linguistic experience and the "Perceptual Magnet Effect." In: Speech Perception and Linguistic Experience: Issues in Cross-Language Research, pp. 121–154. York Press (1995)
6. Chen, J., Best, C.T., Antoniou, M., Kasisopa, B.: Cognitive factors in perception of Thai tones by naïve Mandarin listeners. In: Calhoun, S., Escudero, P., Tabain, M., Warren, P. (eds.) Proceedings of the 19th ICPHS, Melbourne, pp. 1684–1688 (2019)
7. Faris, M.M., Best, C.T., Tyler, M.D.: Discrimination of uncategorised non-native vowel contrasts is modulated by perceived overlap with native phonological categories. J. Phon. **70**, 1–19 (2018)
8. Strange, W., Bohn, O.-S., Trent, S.A., Nishi, K.: Acoustic and perceptual similarity of North German and American English vowels. J. Acoust. Soc. Am. **115**, 1791–1807 (2004)
9. Escudero, P., Simon, E., Mitterer, H.: The perception of English front vowels by North Holland and Flemish listeners: acoustic similarity predicts and explains cross-linguistic and L2 perception. J. Phon. **40**, 280–288 (2012)
10. Styler, W.: On the Acoustical and Perceptual Features of Vowel Nasality (2015)
11. Chen, J., He, X.: Computational perception of information foci produced by Chinese English learners and American English speakers. In: Proceedings of APSIPA Annual Summit and Conference 2019, Lanzhou, pp. 1780–1785 (2019)
12. Strange, W.: cross-language phonetic similarity of vowels: theoretical and methodological issues. In: Bohn, O.-S., Munro, M.J. (eds.) Language Experience in Second Language Speech Learning, pp. 35–55. John Benjamins Publishing, Amsterdam (2007)
13. Yip, M.: Tone. Cambridge University Press, Cambridge (2002)
14. Gandour, J.T.: The perception of tone. In: Tone: A Linguistic Survey, pp. 41–76. Academic Press (1978)

15. Reid, A., et al.: Perceptual assimilation of lexical tone: The roles of language experience and visual information. Atten. Percept. Psychophys. **77**(2), 571–591 (2014). https://doi.org/10.3758/s13414-014-0791-3
16. Chao, Y.R.: A system of tone-letters. Maitre Phon. **45**, 24–27 (1930)
17. Lobanov, B.M.: Classification of Russian vowels spoken by different speakers. J. Acoust. Soc. Am. **49**, 606–608 (1971). https://doi.org/10.1121/1.1912396
18. Chen, J., Best, C.T., Antoniou, M.: Native phonological and phonetic influences in perceptual assimilation of monosyllabic Thai tones by Mandarin and Vietnamese listeners. J. Phon. **83**, 101013 (2020)
19. Burnham, D., Kuratate, T., McBride-Chang, C., Mattock, K.: Making speech three-dimensional: adding tone to consonant- and vowel-based speech perception and language acquisition research, quantification and theory (2009). http://purl.org/au-research/grants/arc/DP0988201
20. Boersma, P.: Praat, a system for doing phonetics by computer. Glot Int. **5**, 341–345 (2001)
21. Xu, Y.: ProsodyPro—a tool for large-scale systematic prosody analysis (2013)
22. Kuang, J.: The tonal space of contrastive five level tones. Phonetica **70**, 1–23 (2013)
23. Dimitriadou, E., Hornik, K., Leisch, F., Meyer, D., Weingessel, A.: Misc functions of the Department of Statistics (e1071), TU Wien. R Package, 1, 5–24 (2008)
24. Escudero, P., Vasiliev, P.: Cross-language acoustic similarity predicts perceptual assimilation of Canadian English and Canadian French vowels. J. Acoust. Soc. Am. **130**, EL277–EL283 (2011). https://doi.org/10.1121/1.3632043
25. Chen, J., Best, C.T., Antoniou, M., Kasisopa, B.: Cross-language categorisation of monosyllabic Thai tones by Mandarin and Vietnamese speakers: L1 phonological and phonetic influences. In: Epps, J., Wolfe, J., Smith, J., Jones, C. (eds.) Proceedings of the Seventeenth Australasian International Conference on Speech Science and Technology, Sydney, Australia, pp. 168–172 (2018)
26. Best, C.T., Avesani, C., Tyler, M.D., Vayra, M.: PAM revisits the articulatory organ hypothesis: italians' perception of english anterior and Nuu-Chah-Nulth posterior voiceless fricatives. In: Nyvad, A.M., Hejná, M., Højen, A., Jespersen, A.B., and Sørensen, M.H. (eds.) A Sound Approach to Language Matters: In Honor of Ocke-Schwen Bohn, pp. 13–40 (2019)
27. Wu, X., Munro, M.J., Wang, Y.: Tone assimilation by Mandarin and Thai listeners with and without L2 experience. J. Phon. **46**, 86–100 (2014)

Corpus-Based Sociolinguistic Study of Corporate Emails and the Implication for Business English Teaching

Xiaoxiao Chen[1] and Shili Ge[2](✉)

[1] School of English for International Business, Guangdong University of Foreign Studies, Guangzhou 510420, China
[2] Laboratory of Language and Artificial Intelligence, Guangdong University of Foreign Studies, Guangzhou 510420, China
geshili@gdufs.edu.cn

Abstract. Corporate emails, as a register, have specific linguistic features, which are important for business English learners and can be comprehensively studied under sociolinguistic framework. In order to better teach business English, especially business email writing, a corpus of corporate email corpus was compiled and linguistic features of intra-company email messages extracted and investigated. Statistical data of lexical, grammatical and conventional textual features show that corporate email messages have characteristics of both personal email communication and professional information exchange. The research result also shows that business English includes various sub-registers, which should be widely studied for teaching advanced business English.

Keywords: Email · Corpus-based study · Sociolinguistic study · Business English teaching

1 Introduction

Emails are one of the most common methods of communication in business, especially for internal office correspondence. "Most people in companies use emails for a wide range of purposes: to confirm appointments and meetings, request help or action, provide information, etc." [1]. Emails, just like letters, are a register, i.e., "varieties according to use" [2] of international business. The relatively new register of writing is often confusing for business English learners in China, who are more adept at formal writing. Therefore, a comprehensive corpus-base study of corporate emails in comparison with other registers under sociolinguistic theoretical framework shall be helpful for Business English teaching.

C. Pang et al. (Eds.): SETE 2020/ICWL 2020, LNCS 12511, pp. 393–400, 2021.
https://doi.org/10.1007/978-3-030-66906-5_37

2 Theoretical Backgrounds

2.1 Emails in Business English Teaching

Business English, as a practice-driven program or major in colleges and universities in China, has experienced a long time debate on its disciplinary orientation, i.e. business-oriented, language oriented, or both? [3]. In fact, no matter what the orientation is, "regardless of where you live, you will need to speak and write Business English with proficiency if you want to have a successful career" [4]. This is especially true for students who are learning business English and want to pursue their career on the global business market.

Business English writing is an essential way of professional communication. The four types of writing usually taught in business English programs are internal company routine writing, such as notices, invitations, memos, and minutes; external company publicity writing, such as company profiles, name cards, product advertising copies, and product descriptions; foreign trade communication, such as sales letters, sales contracts, complaint letters, and adjustment letters; and personal job hunting materials, such as resume and job application letters; etc. Among them, "the most common and demanding writing task today relates to electronic messaging, or email" [4]. Email has its advantages in quick information exchange and communication among busy professionals, which facilitates problem solving effectively. Yet, all these advantages can be achieved only with the condition of using email effectively, which poses challenges for professionals, as well as Business English learners. "When composing e-mail, writers must make effective decisions quickly. Decisions must also be objective: a business relationship can end with the push of a button" [4]. Therefore, Business English learners have to be fully prepared before they step into global business market.

2.2 Sociolinguistic Study of Register and Emails

For teaching and learning of email writing, the major differences between email and other registers are worth exploring. The differences mainly lie in the field of sociolinguistic studies. Business English is one type of English for Specific Purposes (ESP), but it "differs from other varieties of ESP in that it is often a mix of specific content (relating to a particular job area or industry), and general content (relating to general ability to communicate more effectively, albeit in business situations)" [5].

The term of Business English makes it clear immediately that it is the language of the workplace, i.e., a language variety according to use. "A distinction is commonly made between linguistic variation according to the user and variation according to use. The latter has been … referred to as register" [6]. "A register is a publicly recognized cluster of linguistic features (e.g. pronunciation, specific words, syntactic constructions, morphology, intonation patterns, sometimes also gestures) associated with particular cultural practices and types of people who engage in them" [7]. Various studies have shown that proper and strategic use of registers can promote inter-group dialogue [8].

Email, as a specific type of Business English register, has its unique linguistic features. Without mastering of these features, BE learners can hardly achieve the goal of high proficiency in Business English communication. As for the analysis of linguistic features, corpus is a valuable, even indispensible, tool.

2.3 Corpus-Based Sociolinguistic Study of Register and Emails

Corpus-Based Study and Sociolinguistics. The combination of corpora and sociolinguistics dates back to 1980s. For instance, Altenberg [9] examines the differences in the ordering of cause-result constructions between spoken language and written language, and Kjellmer [10] examines masculine bias in American and British English based on Brown and LOB corpora. While McEnery and Wilson [11] explores the possibility of sociolinguistic study on corpora, Hunston [12] discusses how corpora can be applied in the description of sociolinguistic, diachronic and register variation.

Baker [13] lists five reasons that corpus linguistics, as a research method and resources, can be used in sociolinguistics study. First, In the process of linguistic analysis, the best practices of corpus linguistics and sociolinguistics share some basic principles: Both approaches involve the collection and analysis of naturally occurring language data and both approaches emphasize language use and social context. Secondly, both sociolinguistics and corpus linguistics make use of quantitative methodologies, supported with statistics, to study the differences and similarities between different populations. Thirdly, both approaches often adopt sampling techniques and inferential procedures to test hypotheses about populations. Fourthly, both approaches study language variation and change, and both focus on a lot of linguistic features (phonetics, morphology, lexis, grammar, discourse and pragmatics). And finally, sociolinguists and corpus linguists both try to explain their findings from their research if possible. In a word, corpus linguistics and sociolinguistics share a number of similarities in their epistemology, focus and scope and, therefore, can and should be combined in the register study of Business English emails.

Corpus-Based Study of Register. The term of register is widely used in systemic functional grammar and sociolinguistic studies. Halliday and Hasan [14] define a register from the perspective of textual functions as "a concept of the kind of variation in language that goes with variation in the context of situation". Spolsky [15] defines a register from the perspective of sociolinguistics as "the special variety especially marked by a special set of vocabulary (technical terminology) associated with a profession or occupation or other defined social group and forming part of its jargon or in-group variety". Biber and Conrad [16] combine both perspectives to define a register as "a variety associated with a particular situation of use (including particular communicative purposes)", but they also emphasize the objective description of linguistic features for registers, including typical lexical and grammatical characteristics. Context plays a large part in the sort of language people use and people are likely to adapt their language use to these different context [13]. In other words, people usually adopt different register under different situations. Therefore, language learners should proactively learn about proper use of different registers and their linguistic features. This is also true for emails in corporate communication.

3 Methodology

3.1 Research Questions

Though email is a rather general register like face-to-face conversation with various purposes, business email, especially intra-corporate email messages also have distinct linguistic features as a special register, such as the interactiveness vs. directness, and spoken vs. written, etc. [16] Teachers and learners of Business English should pay attention to the possible differences representative for the special register of business emails. Therefore, the following questions were investigated:

1) What are the lexical characteristics of corporate emails?
2) What are the grammatical characteristics of corporate emails?
3) What are the textual conventions for corporate emails?

3.2 Corpus

To investigate the linguistic characteristics of email messages, a large set of email messages from a certain corporation were collected, cleaned, annotated, and a corpus was compiled. The corpus contains 10,019 messages belonging to 32 users with an average of about 334 messages per user.

3.3 Procedures

As discussed in register study, there are a full range of linguistic characteristics could be investigated. Among them, the most relevant are lexical and grammatical features related to the major situational characteristics, that is, "the interpersonal and interactive nature of email (which is similar to conversation but not as directly and immediately interactive), and the production circumstances and lack of shared physical context (which makes email more like prototypical written registers)" [16]. Certainly, for business emails, textual conventions used for opening and closing a message are also important because it is part of etiquette for digital communication. Therefore, the investigation was carried out in 2 steps:

The Compilation of the Email Corpus. Since the focus of this study is on the writing of the main body of emails, the large amount of irrelevant sections, such as the heading information like date, address, subject, etc. were removed from documents in the compilation of email corpus. Some irrelevant information in the body text, such as the hyperlinks was replaced with tags as [WEBLINK]. And some other irrelevant information, such as the tag of [IMAGE] and Original Message were removed. Certainly, emails without body information or less than 5 words were not included in the corpus at all. The data of the self-compiled corporate email corpus is listed in Table 1.

Table 1. Brief description of the email corpus.

Parameter	No. of words	No. of mails	Ave. mail length	Standard deviation
Value	990,300	10,019	98.84	60.88

The Comparison of Corporate Email with Other Registers. Biber and Conrad [16] carry out a study on 76 personal email messages, with a total of 15,840 words, with conversation and academic prose. The lexical features include mail length, the grammatical features include major word classes, and the textual convention features include the forms of email opening and closing. The email corpus in Biber and Conrad [16] is quite small, and the corporate email corpus in this study is much larger to represent business email register. These same features were identified and extracted from the corporate email corpus with self-coded Python programs after the corpus was part-of-speech tagged and some pattern rules were manually written. These features are compared with their counterparts in Biber and Conrad [16] to find out linguistic characteristics of corporate emails.

4 Results and Discussion

4.1 Lexical Features of Corporate Emails

All space separated words in emails are counted. Except for the main parameters listed in Table 1, we also know that the shortest emails have 6 words due to the selection of emails and the longest email has 523 words, which is a news report about the company for the receiver's reference. The length distribution of all emails can be seen from Fig. 1.

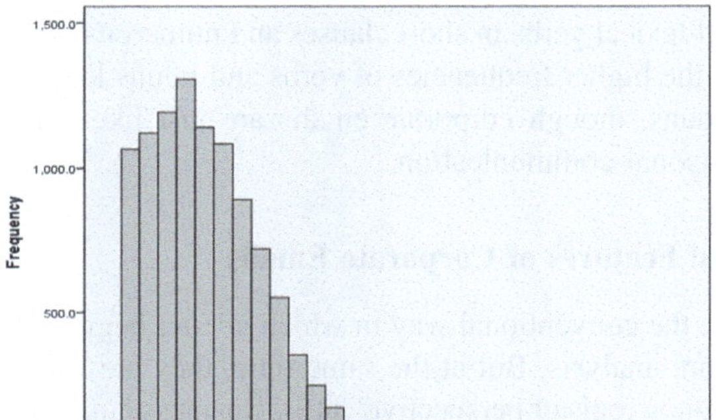

Fig. 1. The length distribution of all emails in corpus.

From Fig. 1, we can see that the distribution of email length is right skewed and most mails are around 100 words. This is much shorter than personal email messages, which is averaged a little more than 208 words [16]. Since "the linguistic features within a register vary depending on specific situational characteristics" [16], corporate emails are used between colleagues in the same working environment. Therefore, the senders and receivers share more background information than personal social communication, which leads to the brevity and conciseness of email writing.

4.2 Grammatical Features of Corporate Emails

Three basic grammatical features – lexical verbs, pronouns, and nouns – in email messages are examined in Biber and Conrad [16] to find out the register difference between email and other registers including conversation and academic prose. The same features are identified and extracted from corporate email corpus and compared with personal email messages.

Table 2. The use of major word classes in corporate email messages (Freq. per 1,000 words).

Word classes	Verbs	Pronouns	Nouns
Frequency	153.36	97.24	241.14

Table 2 shows that the distributions of all word classes are similar to those of personal email messages [16] except that verbs and nouns are a little more and pronouns a little less. This means the overall grammatical features of corporate email are quite similar to personal emails, but there are still relative variations in frequency distributions. On one hand, the use of lexical verbs and nouns is similar to personal emails, but due to the more intense task concentration in corporate email messages, the frequencies are a little higher than personal emails. A browsing of corporate emails can easily reveal a lot of lexical verbs in short clauses and numerous nouns in lists or tables. On the other hand, the higher frequencies of verbs and nouns lead to the relatively low frequency of pronouns, though corporate emails are just like personal emails in the function of interpersonal communication.

4.3 Conventional Features of Corporate Emails

Genre markers, like the conventional way in which a letter begins and ends, are largely associated with genre analysis. But at the same time, they are "also a text variety that can be described from a register perspective" [16]. Business email messages are usually more formal than the email messages sending to family and friends [17]. However, the business emails here refer mainly to inter-company communication. It may be different for intra-company communication. The formality can be shown in various ways including grammar and punctuation rules, etc. The genre markers of email opening and closing are also one important textual feature for email register analysis. The types of corporate email salutations are listed in Table 3.

Table 3. Distribution of corporate email salutations.

Salutation types	NONE	Hi	Hi NAME	NAME	Dear NAME	OTHER
Percent (%)	14.40	20.50	31.10	9.64	22.10	2.30

Compared with [16], it can be seen that the distribution of corporate email salutation types is right between the two sub-registers of emails, i.e., emails to friends and family and professional emails to colleagues. According to [16], email sending to friends and family mainly start with no addressing terms or Hi at the beginning, and followed by Hi NAME and NAME, respectively; emails to colleagues start with Hi NAME, Dear Name, and NAME mainly, with low frequency of NONE and Hi. Table 3 shows that corporate email messages adopt high frequency of Hi NAME and Dear NAME, which shows the features of professional communication. The frequencies of these two types of salutation are much higher than emails to friends and family, but lower than professional emails. On the other hand, Hi and NONE, which show high frequencies in emails to friends and family, also show high frequencies in corporate emails. Though they are lower than emails to friends and family, the frequencies are much higher than those in professional emails. From the frequency distribution, it can be seen clearly that corporate email messages are a sub-register of emails with special textual features.

5 Implications for Business English Teaching

Email, among many other new digital communication tools – such as instant messaging, blogs, microblogs, social networks, and shared workspaces – has its own compelling advantages that will keep it in steady use in many companies [17]. As a specific register or a sub-register of emails, corporate email messages used in international business communication have specific textual features that business English learners have to pay special attention to. As stated in Stroud and Heugh [8], in order to fully re-understand language education and best serve the contemporary globalized world, it is necessary to re-examine and reconfigure various language courses so as to provide students with opportunities to participate in local, regional and global human activities. This is especially true for business English teaching, including the teaching of business email writing.

In fact, even business email can be divided into different sub-registers, such as inter-company emails and intra-company emails. The corpus based corporate email investigation shows that intra-company email messages have linguistic features of both personal communication and professional information exchange, which should be taught to business English learners with careful designed syllabus and courses.

Certainly, this study covers only the body part of email writing and leaves many important sections aside, such as the subject line writing and email formatting. These are also crucial in business communication and business English learners cannot ignore their functions.

Acknowledgements. This work is financially supported by Graduate Education Innovation Plan of Guangdong Province (2018JGXM39).

References

1. Whitmell, C.: Business writing essentials: how to write letters, reports and emails. Kindle edn. Clare Whitmell (2013)
2. Hudson, R.A.: Sociolinguistics, 2nd edn. Cambridge University Press, Cambridge (2001)
3. Wang, L., Zhang, F.: Conceptualizing and theorizing business english as a sub-discipline. Foreign Lang. Res. **3**, 63–66 (2016)
4. Young, D.J.: Business English: Writing for the Global Workplace. McGraw-Hill Higher Education, New York (2008)
5. Ellis, M., Johnson, C.: Teaching Business English. Oxford University Press, Oxford (1994)
6. Durrell, M.: Sociolect. In: Herausgegeben von Ammon, U., et al. (eds.) Sociolinguistics: An International Handbook of the Science of Language and Society. 2nd edn., vol. 1, pp. 200–205. Walter de Gruyter, Berlin, New York (2004)
7. Duranti, A.: Linguistic anthropology: the study of language as a non-neutral medium. In: Mesthrie, R. (ed.) The Cambridge Handbook of Sociolinguistics, pp. 28–46. Cambridge University Press, Cambridge (2011)
8. Stroud, C., Heugh, K.: Language in education. In: Mesthrie, R. (ed.) The Cambridge Handbook of Sociolinguistics, pp. 413–429. Cambridge University Press, Cambridge (2011)
9. Altenberg, B.: Causal linking in spoken and written English. Stud. Linguist. **38**, 20–69 (1984)
10. Kjellmer, G.: "The lesser man": observations on the role of women in modern English writings. In: Aarts, J., Meijs, W. (eds.) Corpus Linguistics II, pp. 163–167. Rodopi, Amsterdam (1986)
11. McEnery, T., Wilson, A.: Corpus Linguistics. Edinburgh University Press, Edinburgh (1996)
12. Hunston, S.: Corpora in Applied Linguistics. Cambridge University Press, Cambridge (2002)
13. Baker, P.: Sociolinguistics and Corpus Linguistics. Edinburgh University Press Ltd., Edinburgh (2010)
14. Halliday, M.A.K., Hasan, R.: Language, Context and Text: Aspects of Language in a Social-Semiotic Perspective. Oxford University Press, Oxford (1989)
15. Spolsky, B.: Sociolinguistics. Oxford University Press, Oxford (1998)
16. Biber, D., Conrad, S.: Register, Genre, and Style, 2nd edn. Cambridge University Press, Cambridge (2019)
17. Bovee, C.L., Thill, J.V.: Business Communication Essentials, 7th edn. Pearson Education Limited, Essex (2016)

The Design and Implementation of a Virtual Reality Program for Improving Situated Consecutive Interpreting Skills

Isabelle Chou(✉), Cheng Zhang, and Yuemeng Yu

University of Electronic Science and Technology of China, Chengdu, China
isabellecchou@uestc.edu.cn

Abstract. In this paper, we present the Platform of Immersion Interpreting Learning in Virtual Reality and AI Assessment, embedded with three classical virtual scenarios in consecutive interpreting, aiming at improving situated consecutive interpreting skills for student interpreters at both undergraduate and postgraduate levels. This proposed project also integrates the AI assessment system into the virtual environment. It can evaluate students' performance based on their reaction time, level of completeness, accuracy, and fluency of the interpreting task. The project supports both visual and auditory channels of interaction, with learner's interpreted speech as the input and outputs include graphical scenes and speech from the speaker in the scenario. Moreover, the project allows interpreting trainers to design their training drills based on their teaching plans. Throughout a pre-and post-test of 10 graduate students, it is found that their average reaction time was shorter in post-test, and the level of interpreting fluency and accuracy were also improved after the training.

Keywords: Virtual reality · Immersive learning environments · Consecutive interpreting training · AI assessment system

1 Introduction

As a synonym of "cyberspace", virtual reality (VR) "is a completely spatialized visualization of all information in global information processing systems, along pathways provided by present and future communication networks, enabling full presence and interaction of multiple users, allowing input and output from and to the full human sensorium, permitting simulations of real and virtual realities, remote data collection and control through telepresence, and total integration and intercommunication with a full range of intelligent products and environments in real space" [1]. It allows users' virtual interactions with the VR system, so that under the social context, VR has been considered as "some form of immersive, synthetic environment which creates a feeling of presence or suspension of disbelief which is sufficient to make the user feel that the artificial world which they appear to inhabit is 'real'" [2].

VR has been integrated into language learning in a comprehensive way since late 20th century [3, 4]. And it has also been an integral part in computer-assisted language learning (CALL) which refers to the adoption of computer tools for learning foreign

C. Pang et al. (Eds.): SETE 2020/ICWL 2020, LNCS 12511, pp. 401–410, 2021.
https://doi.org/10.1007/978-3-030-66906-5_38

language or native language. VR tools fall into the third stage in the timeline of evolution of CALL, also known as integrative CALL, during which collaborative multimedia and social interactions prevail [5]. VR is a high-level, human-computer interface that produces immersive and interactive synthetic environments. It produces digital environments in which users' visual, auditory, and haptic perceptions are very close to those of the actual environments [6]. In second language acquisition, problems persist in the process of practice, like being lack of face-to-face communication in the target language, and no effective feedback in evaluating and monitoring learners' performance. VR programs then established attempting to build a comprehensive and interactive training scheme [7]. Similar problems can also be found in acquiring more complicated second language skills like interpreting. Traditionally, interpreter trainers prepare working sessions in the classroom guiding students into acquiring the appropriate strategies and developing the relevant skills. But stress in the marketplace differs from stress in the classroom, not to mention kinds of distractions may happen during the task [8]. Bearing all the above in mind, the Platform of Immersion Interpreting Learning in Virtual Reality and AI Assessment (PIIL) was developed, aiming at providing full interaction between interpreting students in a more intuitive and real environment, and monitoring their performance in the AI assessment system, integrated with the recognized text of students' interpreting performance and feedbacks.

The paper is organized as follows: Sect. 2 is an overview of VR tools in interpreting training. In Sect. 3, we describe the architecture of PIIL. Some pilot experimental results performed by PIIL are presented in Sect. 4. Section 5 concludes the paper with the devotion of the platform and its future development.

2 VR Tools in Interpreting Training

At present, the most influential VR project in interpreting is IVY (Interpreting in Virtual Environment). It is consorted by seven universities in Europe, where the rise of migration and multilingualism requires professional interpreters in many settings like business, legal, and medical. The project was designed mainly for students of interpreting, future users of interpreting services in higher education, as well as for vocational training and adult learning. These future interpreters have to face an ever-broadening range of interpreting scenarios which in turn ask for different interpreting skills. Against such a backdrop, IVY creates for interpreting training and simulation in an intuitive and user-friendly interface. It supports a range of virtual interpreting scenarios that can run in three different modes, i.e., interpreting mode, where students can practice using dialogues and monologues; exploration mode, where clients can learn about interpreting; live interaction mode, where both groups can engage in role plays[1].

In general, IVY is a simulated platform of social interaction and education, used by numerous institutions, universities, therefore, no specific training drill can be implemented into the platform fitting into the pedagogical purpose of the trainer or the language proficiency level of the student. IVY is more for self-training than for

[1] http://virtual-interpreting.net/.

systematic and professional training for undergraduate and postgraduate interpreting programs in universities. Comparatively, PIIL is designed for interpreting classes during which the trainer can upload the prepared drill to the platform for class works as well as pre-interpreting preparations. Details in design and implementation of PIIL are in the next section.

3 Platform of Immersion Interpreting Learning in Virtual Reality and AI Assessment

It has been widely acknowledged that the integration of virtual reality and AI on distributed platforms can provide a promising framework for the next generation of language learning technology [9]. We define PIIL as a scenario-based, the next generation virtual platform. Interpreting students can be immersed into VR scenarios simulating interpreting events in real market and interact with the system by high-end VR headsets. The architecture, features of PIIL and its implementation in interpreting training will be introduced in this section.

3.1 Features and the Architecture of PIIL

Features of PIIL are well-fitted into the purpose of training. They are categorized into technical features and pedagogical features, as seen in the following two tables (Table 1).

Table 1. Technical features of PIIL

Technical features	
Hardware	Computer+HTC VIVE focus headset
Software	CS+BS framework
Static immersive scene	3 DoF and 6 DoF
Type of task	Scenario-based simulation
Type of feedback	Multi-sensory

For the hardware, at present, the lab was equipped with VIVE Focus headset, which is the simple all-in-one solution for enterprise. It enjoys the instant wireless VR with high-resolution 3K AMOLED screens, Qualcomm® Snapdragon™ 835 processing, and world-scale tracking right out of the box[2]. As for the software, other than accessing the platform in the lab with the headset, authorized users can also log into the system on the website. All three scenarios are designed with pictures of 3 DoF and 6 DoF. In Table 2, some pedagogical features are defined with reference to previous proposals [10].

[2] https://enterprise.vive.com/ca/product/vivefocus/#: ~ :text=VIVE%20Focus%20is%20the%20simple,base%20stations%2C%20or%20sensors%20required.

Table 2. Pedagogical features of PIIL

Pedagogical features	
Principal objective	Making deep cognitive model of target skills
Curriculum	Syllabus is implemented in the relevant scenarios
Lesson	One scenario per lesson
Teacher's role	Collaborator or absent
Teacher's attitude	Helping accomplish the objectives of scenario

Table 2 describes some general pedagogical features of the platform. It aims at making deep cognitive model and creating VR interpreting scenarios for an immersive training and acquiring knowledge through interaction. To achieve this goal, we implemented three software layers and one hardware layer as to realize data collection and evaluation, and interaction with learners, as seen in Fig. 1.

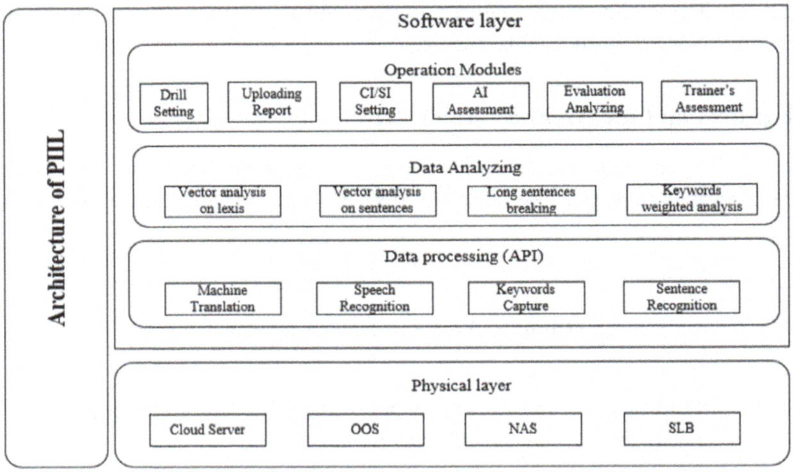

Fig. 1. Architecture of PIIL

Three software layers were then realized in four modules: Editing Module, Class Management, Training Platform, and AI Assessment, as illustrated in Fig. 2.

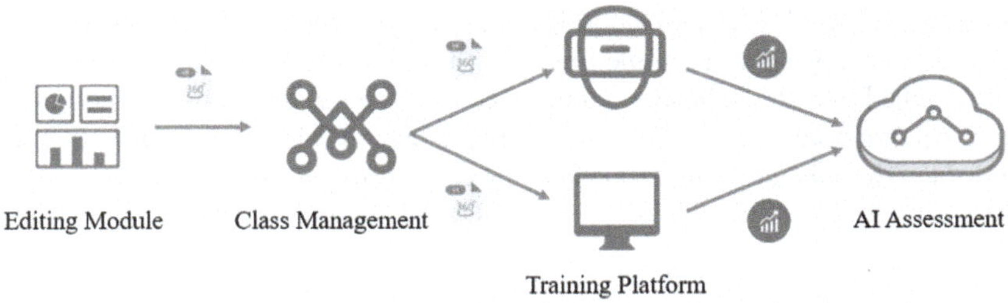

Fig. 2. Modules for PIIL

Among these modules, three of them are for trainers, including Editing Module, Class Management, and AI Assessment. For Editing Module, the trainer can upload any materials to any of the three scenarios that fit the syllabus of the course and his/her plan towards the training. Class Management allows the trainer to select and download any recorded data from the platform. And AI Assessment provides a score out of 100 for each training student performs, together with the transcription in verbatim. This information can serve as the reference to the trainer.

3.2 Integrating PIIL in Interpreting Training

Interpreting has seen to be complex cognitive activity and asks for various kinds of knowledge and technical skills [11–14]. Over the years, components of translation competence have also been described in details in many studies [15–17]. Even without any consensus, the scholarly agreement was achieved that interpreting is a highly situated activity. Interpreter's performance is heavily affected not only by linguistic factors but also extra-linguistic factors including speakers' performance (the speed of the speech, accent, information density, etc.) as well as the environment of the interpreting site (noises, settings of the conference hall, etc.). If the interpreter does not have sufficient knowledge of the themes and subject matters of the interpreting events, as well as good command of two working languages, these factors may even determine the quality of the interpreting performance.

The traditional interpreting training often prioritizes working language skills as it is unable to simulate or replicate any working environment. However, all three components of knowledge and skills (thematic knowledge, linguistic competency and environmental factors) must be fully integrated that can best training results be achieved. Therefore, Gile suggests that "[f]or the sake of optimizations, variability in these parameters would call for a variety of training programs" [12]. PIIL is such an innovative VR platform which provides simulated interpreting environment for students that the trainer is able to design more interactive programs for both consecutive and simultaneous interpreting training. By integrating it into existing interpreting training programs or creating a new one, PIIL's merits can be seen in four perspectives:

First, like other VR training programs, it immerses learners in the training. The PIIL creates the immersive illusion of being an interpreter in the conference site. It restores three typical scenarios that interpreters work, namely, public speech, international conference and news release. Each scenario simulates the real working environment of the interpreter to a large extent, including the position of interpreter, the layout of a common conference or press release, the number of audiences, to name a few. On a wearable VR headset, students have access to these scenarios working as a professional interpreter. It will provide a more real working environment close to the stress and tension like professional interpreters feel.

Second, PIIL is open and can be adapted to any interpreting course or program. As previously stated, PIIL has an "Editing Module" which allows the trainer to upload any audio files which fit his/her syllabus. Any available training materials can be created into a VR program which transports learners to a different world difficult to repeat in the classroom.

Third, AI Assessment of the PIIL can help the trainer to evaluate students' performances. The PIIL will transform the traditional way of evaluating students' interpreting performance. The System will keep a recording of each students' performance as well as a transcribed text of the recording. Based on Baidu and Tencent AI Translation engines, the System itself will provide a score of the performance with weighted analysis of pronunciation, accuracy, fluency and reaction time.

Fourth, PIIL facilitates the data analyzing for interpreting training studies. The System allows the authorized user to download selected recordings based on different criteria such as the time of training, specific class or specific students' ID. In that case, it can have a continuous assessment to a particular student or class' performance. And it also allows us to evaluate the validity and effectiveness of VR training program.

4 Experiments and Discussions

The scenario of public speaking under the drill of consecutive interpreting (CI) was implemented to evaluate the performance of the PIIL. This scenario contains one input channel and two output channels. The I/O channel is the auditory channel. The learner first listens to the speaker's speech in the scenario via the headphone of VR headset. He/she then interprets what was heard into the designated language over the microphone. Learners can also see the virtual world through the headset they wear. The sample scene of the created virtual environment of a public speech site is shown in Fig. 3.

Fig. 3. Virtual environment of a public speech

These experiments aim at testing the effectiveness of VR platform in interpreting training to improve learners' consecutive interpreting skills, in particular when environmental factors are involved. Thus, experiments reported here are not thorough and systematic evaluations of every aspect of PIIL, rather than a demonstration of the feasibility of the platform in interpreting training. It is expected that after several drills, we may witness a decrease in the average reaction time in CI, and a rise in learners' average accuracy and fluency level. The test subjects were 10 graduate students (7 females and 3 males) in Master of Translation and Interpreting program. All have a good command of interpreting. The average age of test subjects was 23.6 years old with the standard deviation (SD) of 1.4. And their interpreting experience was 2.3 years with the SD of 3.9. The subjects were asked to declare their interpreting skills based on China's Standards of English Language Ability. The result is shown in Table 3.

Table 3. Subjects' self-declared interpreting skills

Level	Interpreting skill (%)
−2 (very poor)	0
−1 (poor)	0
0 (average)	70
1 (good)	10
2 (very good)	20

All participants participated a pre-test and a post-test before and after the training. Two sets of tests (English-Chinese) are controlled by information density, the length of speech, and speakers' accent. The reaction time was computed and the level of accuracy and fluency were first evaluated by the AI Assessment, then by two independent raters. The mean of two scored are taken as the final results of the tests. Participants were then asked to finish ten CI drills in public speech with the VR headset within two weeks. The results are illustrated in Fig. 4, 5 and 6.

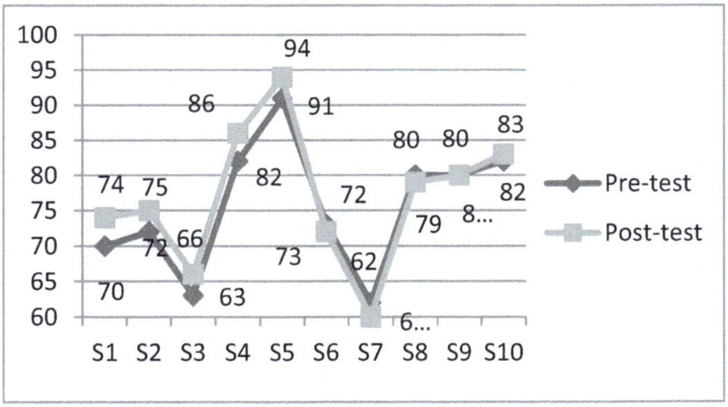

Fig. 4. Accuracy (pre- and post-test)

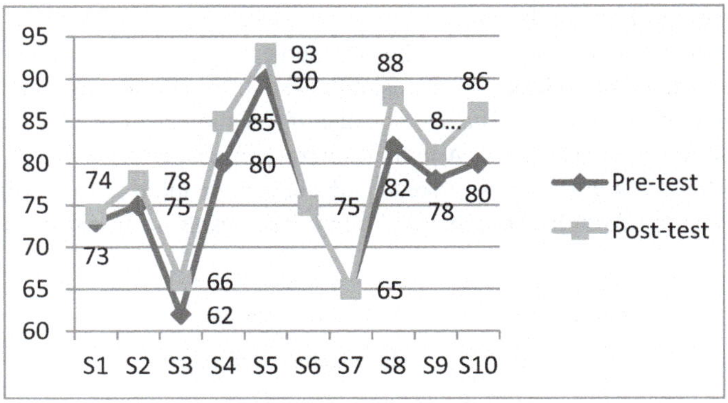

Fig. 5. Fluency (pre- and post-test)

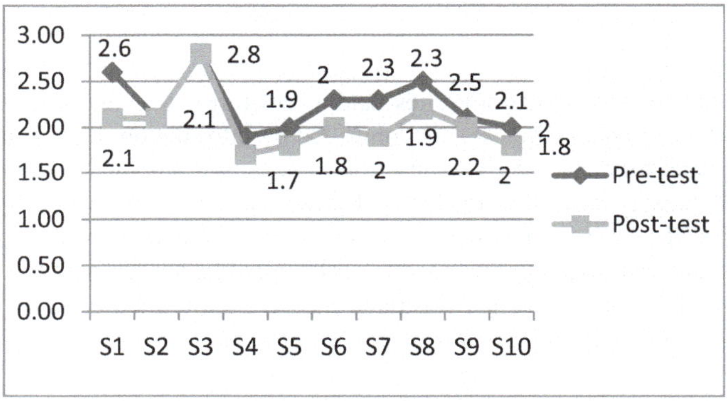

Fig. 6. Reaction time (pre- and post-test)

As depicted in Fig. 4, the accuracy level of some subjects slightly increased while some of them remained the same, with the mean of 75.5 and the SD of 8.674 in the pre-test and with the mean of 76.9 and the SD of 9.310 in the post-test. The average score of accuracy recorded a slight rise but the corresponding SD increased. In other words, the effectiveness of the training varied across subjects. Comparatively, majority of subjects received higher score in the post-test in terms of their fluency level, as illustrated in Fig. 5. In the pre-test, the mean is 76 and the SD is 7.720, and with the mean of 79.1 and the SD of 8.791 in the post-test. The gap of SD of fluency level between pre-and post-testing is larger than that of the accuracy level. It can be explained that students' cognitive management skills were also improved after the training. In the post-test without being distracted environmental factors from the VR scenario, they can allocate more of their attention to the fluency. Subjects also reacted faster after the training, as shown in Fig. 6, with the mean of 2.26 s and the SD of 0.28 in the pre-test and with the mean of 2.04 s and the SD of 0.29 in the post-test. Scholarly attention was given to the fact that interpreting training may influence working memory and executive function [18]. Our findings are in consistent with previous studies which in turn proves the effectiveness of VR training.

5 Concluding Remarks

In this paper, a VR interpreting training platform PIIL is presented. It creates an immersive environment for interpreting students, with three typical event sites, namely, international conference, public speech, and press release. It synchronizes visual and auditory channels if wearing a VR headset during the training. It has four modules, three of which help the trainer to prepare a VR drill and manage his/her class. It is also equipped with an AI assessment algorithm that can evaluate students' performance based on their pronunciation, accuracy and fluency level. Moreover, we proposed the general architecture of PIIL including three software layers and one hardware layer. Furthermore, we presented one implemented scenario as a prototype to evaluate the effectiveness of PIIL. Evaluation of the implemented system on 10 graduate students by pre-and post-test, showed that their average accuracy level and fluency level slightly raised, and the reaction time of subjects narrowed. More tests are needed to further explore the effectiveness of the platform in training interpreters. And more improvements can be made in terms of user's interface, deeper integration between scenarios and contents, and the algorithm of AI assessment.

References

1. Markley, R. (ed.): Virtual Realities and Their Discontents. Johns Hopkins University Press, Baltimore, London (1996)
2. Foster, D., Meech, J.F.: Social dimensions of virtual reality. In: Carr, K., England, R. (eds.) Simulated and Virtual Realities, pp. 209–224. Taylor and Francis, London (1995)
3. Rose, H., Billinghurst, M., Sayu, Z.: An immersive educational environment for learning Japanese. Human Interface Technology Laboratory, University of Washington, Seattle (1995)
4. Rose, H.: Design and construction of a virtual environment for Japanese language instruction. Unpublished Master's thesis, University of Washington, Seattle (1996)
5. Warschauer, M.: CALL for the 21st century. In: IATEFL and ESADE Conference, Barcelona, Spain (2000)
6. Zhao, Q.: A survey on virtual reality. Sci. China Ser. F: Inf. Sci. **52**(3), 348–400 (2009)
7. Schwienhorst, K.: The state of VR: a meta-analysis of virtual reality tools in second language acquisition. Comput. Assist. Lang. Learn. **15**(3), 221–239 (2002)
8. Gile, D.: Teaching conference interpreting. In: Tennent, M. (ed.) Training for the New Millennium: Pedagogies for Translation and Interpreting, pp. 127–151. John Benjamins Publishing, Amsterdam (2005)
9. Chang, B., Lee, S., Si, M., Hand, A.: Foreign language learning in immersive virtual environments. In: Proceedings of the SPIE the Engineering Reality of Virtual Reality 2012, Burlingame, CA, vol. 8289, pp. 2–9 (2012)
10. Bax, S.: CALL – past, present and future. System **31**(1), 13–28 (2003)

11. Christoffels, I., De Groot, A.M.B.: Components of simultaneous interpreting: comparing interpreting with shadowing and paraphrasing. Bilingualism: Lang. Cogn. **7**(3), 227–240 (2004)
12. Gile, D.: Basic Concepts and Models for Interpreter and translator training. John Benjamins Publishing, Amsterdam and Philadelphia (2009)
13. Seeber, K.G.: Cognitive load in simultaneous interpreting: existing theories - new models. Interpreting **13**(2), 176–204 (2011)
14. Setton, R.: Simultaneous Interpretation: A Cognitive-Pragmatic Analysis. John Benjamins, Amsterdam and Philadelphia (1999)
15. Hansen, G.: Success in translation. Perspect.: Stud. Translatol. **5**(2), 201–210 (1997)
16. Hansen, G.: Retrospection methods in translator training and translation research. J. Spec. Transl. **5**(1), 2–41 (2006)
17. Schäffner, C., Adab, B. (eds.): Developing Translation Competence. John Benjamins Publishing, Amsterdam and Philadelphia (2000)
18. Dong, Y., Liu, Y., Cai, R.: How does consecutive interpreting training influence working memory: a longitudinal study of potential links between the two. Front. Psychol. **9**, 875 (2018)

Knowledge-Network-Based Inter-institutional Collaborative Teaching of Translation

Yuanyuan Mu[1,2] and Wenting Yang[3](✉)

[1] School of Foreign Studies, Hefei University of Technology, Hefei, China
[2] Center for Translation Studies and Specialized Corpora, Hefei University of Technology, Hefei, China
[3] School of Foreign Languages, Anqing Normal University, Anqing, China
1339957716@qq.com

Abstract. Knowledge Network plays an essential role in translation teaching and learning, especially assisted with e-learning approaches and data-driven design. This paper focuses on inter-institutional co-construction of knowledge network for translation learning facilitated by Textwells, an online platform for translation teaching and learning. Translation teachers from five universities in China have participated in inter-institutional collaborative teaching of translation by adopting the knowledge network derived from the Textwells Platform and designing an online teaching template together in the 2020 spring semester during the COVID-19 pandemic. With this online collaborative teaching experience shared in a national-wide webinar, a short online survey has been conducted afterwards to investigate the perception of knowledge-network-based teaching among a group of teachers briefly trained for this new teaching approach. Therefore, it has been proved that knowledge-network-based collaborative teaching is deemed effective and will be applied to translation teaching and learning more widely in the future.

Keywords: Knowledge network · Collaborative teaching · Textwells · COVID-19 pandemic

1 Introduction

Since December 2019, the COVID-19 pandemic has gradually evolved from a local public health emergency to a national one and even to an international one, and hence has exerted great influence on all sectors of the whole society within several months. In particular, it poses a huge challenge to the higher education field. While facing challenges, colleges and universities are also meeting significant opportunities, that is, how to shift from traditional teaching to the innovative combination with online teaching, break through the limitation of time and space to guide "ubiquitous" learning, take full advantages of "Internet Plus", and encourage the integration of learning resources and collaborative innovation.

During the pandemic, the challenges of online teaching are as follows: first is "at the microlevel, mainly teachers' competence, including their guidance methods and strategies, and application of new media and technologies"; second is "at the meso

C. Pang et al. (Eds.): SETE 2020/ICWL 2020, LNCS 12511, pp. 411–419, 2021.
https://doi.org/10.1007/978-3-030-66906-5_39

level, mainly the construction of model ecology, including the mode and model of home learning and online teaching, the online teaching environment construction, and online learning mechanisms and guarantees"; third is "at the macro level, mainly education reform and development in the context of pandemic" [1]. For language courses, as most teachers are generally insufficient in technological knowledge, more may need to be invested in online guidance methods and learning mechanism construction. Thus, the core of teachers to transform from traditional teachers to online teachers is to "recognize their transition from knowledge authorities to knowledge reconfigurators with a sharing awareness and pushers of high-quality educational resources, from knowledge instillers to personalized learning consultants and interactive teaching facilitators" [2]. Similarly, learners "are not only knowledge acquirers, but also the providers and producers of knowledge, experience and methods" [3].

Based on the latest researches on online learning during the COVID-19 pandemic, we find that knowledge sharing and collaborative learning are the core for online learning, and therefore propose a translation knowledge network co-construction and collaborative teaching model based on the Textwells translation teaching platform.

2 Literature Review

Researches on database-based or corpus-based translation teaching and knowledge system construction have started relatively late, and has only developed in academia in the past decades [4–8]. In recent years, researchers in China have begun to pay attention to applying corpus into translation teaching, but those studies haven't tackled the interlinking problem between corpus and teaching details like the construction of translation knowledge system [9], which decoupled "corpus" from "translation teaching" instead of merging into teaching effectively. The lack of empirical investigation and reporting and evaluation on corpus usage [10] is a gap in the translation teaching field, and also a bottleneck restraining this research direction from further development. Therefore, there is an urgent need to improve the knowledge system in translation teaching corpus, which is, the knowledge network.

2.1 Researches on Knowledge System Construction of Online Translation Learning in China

At present, Chinese researches on the application of knowledge network to translation learning is very rare. When we search the researches published in core journals with "knowledge network" and "translation learning" (or any possibly related concepts like "translation teaching", "translation training") simultaneously as topics or keywords, the amount of papers being published is almost zero. However, the awareness of the above problem has sprouted up recently. Although there are still rare researches that clearly propose the concept of "knowledge network", researchers have begun to employ corpus in constructing knowledge system construction in translation teaching [11, 12]. And translation academia has made certain developments in the application of corpus technology and vigorously advocates the construction of corpus to facilitate the exploration of translation teaching, dictionary compilation and other related issues

[5, 8]. The development of translation teaching corpus and knowledge base has provided advantages for the construction of translation knowledge network, as high-quality teaching resources and annotations in the corpus and knowledge base are the foundation and source of knowledge resource library, knowledge correlation can be thus generated through data mining technology. This is of particular applicability for the teaching during the pandemic period when teachers face the challenges of effectively delivering knowledge to students by online resources. They need to track students' individualized knowledge in real time, and urge students to learn "ubiquitously" unconstrained to time and space. However, most of the current published researches on online learning during the pandemic are from the macro aspect or the technical application perspective without going further to knowledge system construction, especially in the language learning field.

2.2 Researches on Translation Knowledge Network and Collaborative Learning Worldwide

The Open University, UK has carried out a variety of knowledge-network-based teaching and researching activities and organized the results into book since 1990s [13], which, mostly concern about the relationship between knowledge network and collaborative learning. In terms of applying knowledge network to language researches, the overseas academia has made some achievements only in the recent decade, and still needs to be improved on. Such overseas publications focus more on interdisciplinary integration, such as the utilization of technology to cognitive psychological analysis in the language field, particularly knowledge acquisition in the lexical cognition [14], neural network knowledge model in natural language processing [15, 16] and cross-language information extraction from bilingual analogical corpora [17], etc., from which we can observe the innovative application of data mining technology and method innovation in the analysis of cognitive psychology. Nevertheless, researches on the application of translation knowledge network construction are yet in the development stage. When we search at the EBSCO host database with "knowledge network" and "translation learning" (or other related concepts like teaching or training) simultaneously as topics or keywords, few papers can be retrieved. In fact, the outbreak of COVID-19 pandemic has also exerted great pressure of online teaching on overseas colleges and universities, but no detailed research about online translation teaching has been searched out.

Therefore, practice and research on teaching reform are urgently needed in both China and overseas, and the scale and depth of knowledge base co-construction and collaborative teaching in online teaching should be paid particular attention to in the translation academia. In face of the sudden pandemic, it is still of necessity for the educational field to perfect emergency research directing at public crisis, implement the innovative research of online learning and form a long-term mechanism.

3 Research Design

3.1 Knowledge Network Construction

Facilitated by the corpus-based Textwells online translation teaching platform, we propose translation knowledge network co-construction and inter-institutional collaborative teaching to achieve distribution and integration of online resources. Facilitated by the big data algorithm, we aim to dig out the correlations among knowledge nodes from the annotated corpus, form the translation knowledge network, explore students' learning behaviors on a knowledge-network-based learning guidance.

As a corpus-based online translation teaching/self-learning platform, Textwells platform refines keywords from various translation texts, cultural phenomenon and translation strategies and methods as a series of knowledge nodes, which can be further employed in the construction of domain knowledge in language, culture and translation. There are altogether 56 knowledge nodes, covering six categories of Word Structure, Sentence Structure, Information Structure, Text Structure, Rhetorical Device and Intertextuality. Based on the 56 knowledge nodes, the platform provides detailed annotation of translation corpus, or, the descriptive explanation of bilingual textual phenomenon and translation process (Fig. 1).

Fig. 1. Some of the knowledge nodes

A large amount of annotation points to the co-occurrence relation and centralization of different knowledge nodes, and with the help of data mining algorithm, the platform can reveal the distribution of different nodes. The closely linked nodes would combine together and form dynamic clusters, and those clusters, when mapping out, generate what we call "knowledge network" of translation discipline, which, can then be widely used in the guidance of translation teaching and learning.

With the knowledge nodes as the basis of data retrieval and learning decision making, there is also a learning analysis system embedded in the platform. Through data cleaning and miming, platform can achieve real-time monitoring and feedback of students' learning condition. The platform can record their learning behaviors, including their study records and exercise records, analyze their learning trends and generate their individualized learning "map". For teachers, they can check the learning data analysis of any individual student or the whole class, access their teaching data and evaluate their teaching efficiency, which, provide the foundation of their teaching decision making. The platform would provide basic modules as references for teachers to create their own courses. All the teaching and learning records and analysis are closely linked with the translation corpus and knowledge nodes to improve the degree of corpus utilization and help the generation of translation knowledge network (Figs. 2 and 3).

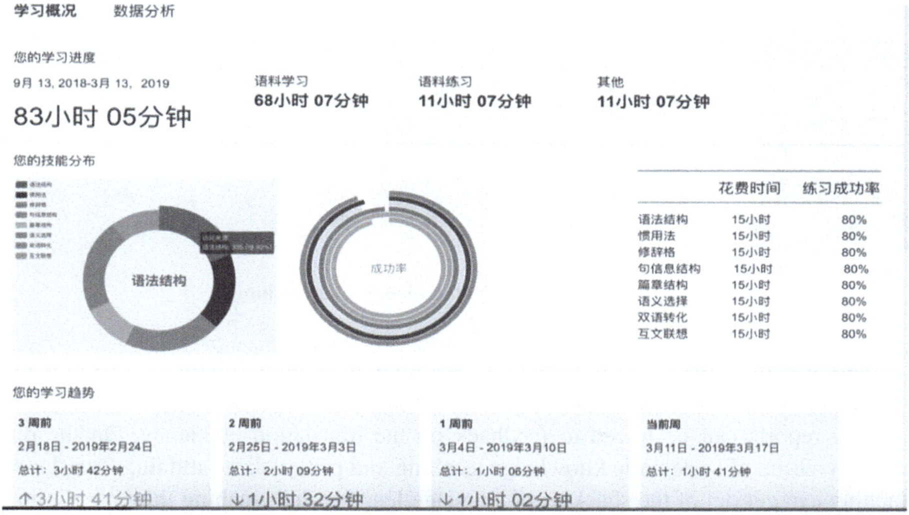

Fig. 2. Data analysis of learning knowledge nodes

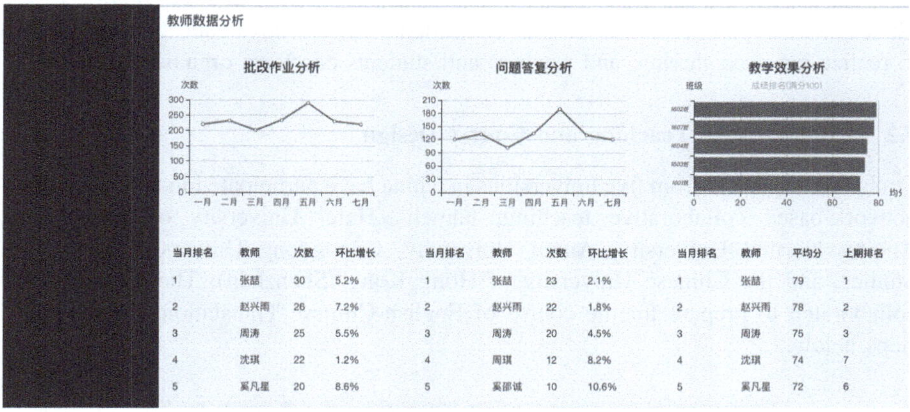

Fig. 3. Teaching records and data analysis

Underpinned by the concept of inter-institutional collaborative teaching, teachers from different colleges and universities interact with the platform developers to promote resource sharing and distribution. They can share teaching resources varying from lectures and corpus annotation to course design to teaching resource database enriching. Still, abundant learning and teaching data further enhance the accuracy of data mining, which, in turn, facilitate personalized learning (Fig. 4).

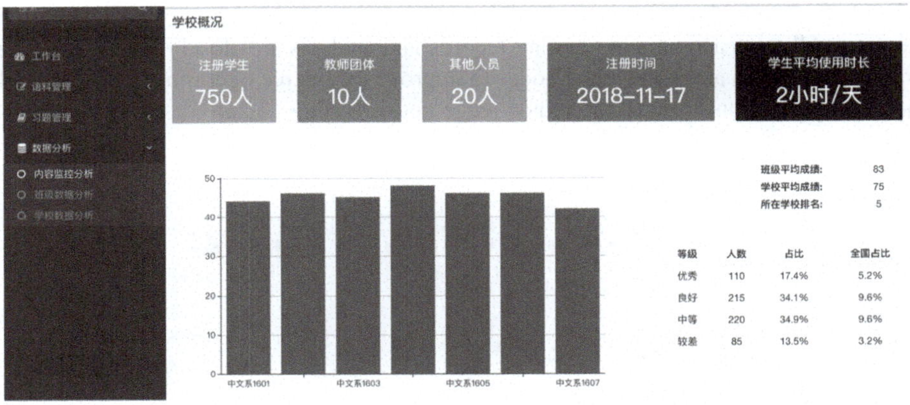

Fig. 4. Data analysis in collaborative teaching

On the knowledge-network based collaborative teaching, large-scaled and in-depth empirical translation studies can be conducted. And detailed corpus-based long-time learning reports can be traced to feedback on the translation efficiency. On the one hand, by virtue of systematic knowledge teaching and personalized training, translation teaching can get out of the shackles of language-learning-led teaching thinking, limited corpus, isolated course training modules, impressionistic criticism without theoretical support and singular classroom teaching model, and can improve students' critical thinking ability. On the other hand, through the exploitation and exploration of the platform, we hope to establish an inter-institutional collaborative education mechanism to realize resource sharing, and teachers and students can form organic interaction.

3.2 Collaborative Teaching and Course Design

Translation teachers from five universities in China have participated in the knowledge-network-based collaborative teaching, namely, Hefei University of Technology, Anqing Normal University, Anhui University, Guangdong University of Foreign Studies, and the Chinese University of Hong Kong (Shenzhen). The teachers have collaborated to prepare for the course of English-Chinese Translation, following the steps below.

Step 1: Mapping of the Knowledge Network

The English-Chinese Translation course is a basic course for undergraduate students. The teachers from the above-mentioned five universities all have the experience of teaching such a course as an introductory one. Although they had different syllabus and course design previously, in this experiment all the teachers work together to map the knowledge network into this course, in order to select the basic knowledge nodes, search for appropriate texts annotated with these knowledge nodes, and design relevant exercises derived from these knowledge nodes. In this way, the teachers conduct an experiment to form a collaborative teaching mode with reference to the basic knowledge nodes.

Step 2: Collaborative Course Design

The five teachers have worked together to design the course and add an additional function to the original Textwells platform, i.e., collaborative course template module. As shown in the following, the teachers adopted the main knowledge nodes for each session and introduced text annotations and exercises to facilitate teaching. Teachers from other universities can apply this teaching template to their teaching activities and modify their own teaching. Then, the teachers can click the button of applying this modified template and send it automatically to their students (Fig. 5).

Fig. 5. Collaborative teaching design

Step 3: Exercises in accordance with knowledge nodes

The teachers cooperated to design exercises in accordance with the knowledge nodes from the knowledge network. In order to help students form a more systematic understanding of the knowledge nodes, the teachers post the exercises to collect the results from the online translation learning platform (Fig. 6).

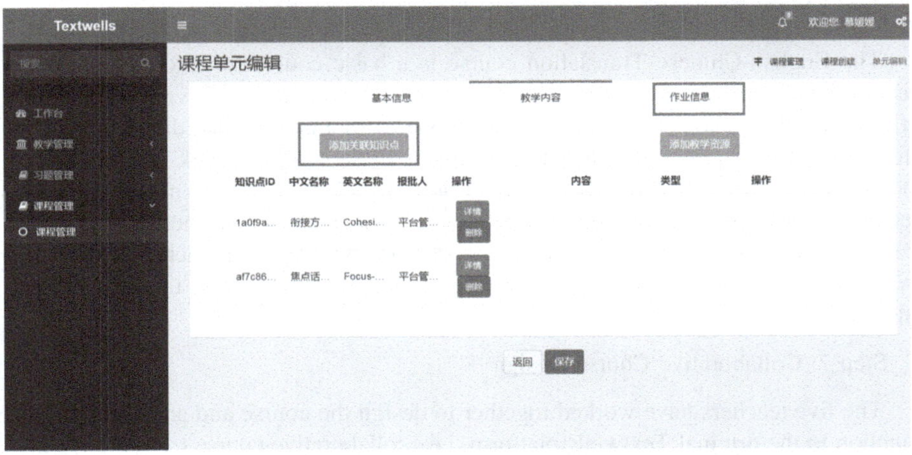

Fig. 6. Exercises in accordance with knowledge nodes

4 Perception of Knowledge-Network-Based Collaborative Teaching

With this online collaborative teaching experience shared in a national-wide webinar, a short online survey has been conducted afterwards to investigate the perception of knowledge-network-based teaching among a group of teachers briefly trained for this new teaching approach. In terms of the acceptability of such a new mode of teaching, 76% of the them chose 'helpful' and 'very helpful', with the percentage being 52% and 24% respectively. In terms of the effectiveness of this knowledge-network-based collaborative teaching, while 24% of them chose "neutral", none gave negative comments. 90.5% of them tended to have positive attitudes towards the Platform, and 9.5% of them gave neutral comments. However, none of them held negative attitudes.

5 Conclusion

The platform is designed to realize a better regulated, systematic methodology and an efficiency-motivated, theory-informed pattern to counteract the labor-intensive, time-consuming and space-constrained teaching of translation.

It will develop a ground-breaking and cost-effective educational paradigm for the teaching/(self-)learning of English-Chinese bilingual text-production in classroom/web-based settings to alleviate the pressure on language/translation courses. It also proposes an approach complementary to the mainstream lexico-grammatical paradigm in corpus-based research by focusing on textual management across languages, especially in terms of the relationship between information structuring and discourse function as seen in the context of translation. Co-construction of knowledge network and collaborative teaching will also make the online translation teaching and learning platform more powerful and effective.

Acknowledgements. The work was substantially supported by The National Social Science Fund of China (Project No. 19BYY125) and Anqing Normal University (Project No. 2019aqnujyzc089).

References

1. Yang, J.: Online teaching practice and experience during the COVID-19 pandemic. China Educ. Technol. **4**, 29–41 (2020)
2. Mu, S.: Deepening and Reconstructing: Mobile Learning and Online Education. Higher Education Press, Beijing (2019)
3. Fu, W., Zhou, H.: Challenges brought by 2019-nCoV epidemic to online education in China and coping strategies. J. Hebei Norm. Univ. (Educ. Sci.) **22**(2), 14–18 (2020)
4. Bowker, L.: A corpus-based approach to evaluating student translations. Translator **6**(2), 183–210 (2000)
5. Wang, K.: The use of parallel corpora in translator training. Media Foreign Lang. Instr. **6**, 27–32 (2004)
6. Johns, T.F., Lee, H., Wang, L.: Integrating corpus-based CALL programs in teaching English through children's literature. Comput. Assist. Lang. Learn. **21**(5), 483–506 (2008)
7. Zhu, C., Mu, Y.: Towards a textual accountability-driven mode of teaching and (self-) learning for translation and bilingual writing: with special reference to a CityU online teaching platform. Chin. Transl. J. **2**, 56–62 (2013)
8. Hu, K.: Corpus-based study of translation teaching. In: Hu, K. (ed.) Introducing Corpus-Based Translation Studies, pp. 177–191. Springer, Heidelberg (2016). https://doi.org/10.1007/978-3-662-48218-6_7
9. Qin, H., Wang, K.: Parallel corpus in translation teaching: theory and application. Chin. Transl. J. **5**, 49–52+95–96 (2007)
10. Kenny, M.A.: Discussion, cooperation, collaboration: the impact of task structure on student interaction in a web-based translation exercise module. Interpreter Transl. Trainer **2**, 139–164 (2008)
11. Miao, J., Sun, X.: On the construction of e-learning platform for translation instruction featured with knowledge systems. In: Han, Z., Li, D. (eds.) Translation Studies in China. NFTS, pp. 203–214. Springer, Singapore (2019). https://doi.org/10.1007/978-981-13-7592-7_11
12. Hu, K.: Corpus translation studies: connotations and implications. J. Foreign Lang. **5**, 59–70 (2012)
13. Eisenstadt, M., Vincent, T. (eds.): The Knowledge Web: Learning and Collaboration on the Net. Kogan Page, London (1998)
14. Szymański, J., Duch, W.: Information retrieval with semantic memory model. Cogn. Syst. Res. **14**(1), 84–100 (2012)
15. Li, Y., et al.: Incorporating knowledge into neural network for text representation. Expert Syst. Appl. **96**, 103–114 (2018)
16. Li, P., Mao, K.: Knowledge-oriented convolutional neural network for causal relation extraction from natural language texts. Expert Syst. Appl. **115**, 512–523 (2019)
17. Hashemi, H.B., Shakery, A.: Mining a Persian-English comparable corpus for cross-language information retrieval. Inf. Process. Manage. **50**(2), 384–398 (2014)

Corpus-Based Semantic Prosody Study of English-Chinese Translation: Taking Trump's Popular Saying "It Is What It Is" as an Example

Tiantian Wang[1] and Shili Ge[2](✉)

[1] Center for Linguistics and Applied Linguistics, Guangdong University of Foreign Studies, Guangzhou 510420, China

[2] Laboratory of Language and Artificial Intelligence, Guangdong University of Foreign Studies, Guangzhou 510420, China

geshili@gdufs.edu.cn

Abstract. US President Trump once again uttered a "popular sentence", "It is what it is", causing widespread complaints. Although its frequently reporting by major media abroad, Chinese English learners are so unfamiliar with the expression. Besides, the official translation given by the "China Daily" is "情况就是这样", which is unlikely to be a satisfied translation. To find empirical support for the argument, this paper focused on the frequency, diachronic trend and semantic prosody tendency of it and its official translation in two corpora (COCA and BCC). The research result shows that major differences exist between "It is what it is" and "情况就是这样". In contrast, "事已至此" and "It is what it is" are more comparable in all dimensions. Thus, "事已至此" is a better version of "It is what it is". This study regards semantic prosody as an indispensable part of translation, providing new perspectives for corpus-assisted translation studies.

Keywords: Translation studies · Corpus · Semantic prosody

1 Introduction

On August 3rd, 2020 local time, in an interview with Axios on HBO host Jonathan Swan, US President Trump uttered a "many quotable sentence", "It is what it is", causing widespread complaints. In contrast to the frequent appearance of this expression in news reports of major media abroad, the strangeness of Chinese English learners to it is much more eye-catching. Besides, the official translation given by the "China Daily" is "情况就是这样", which is unlikely to be a satisfied translation.

Thus, with the assistance of two established corpora, this paper not only explored a better translation of it, but regarded the semantic prosody tendency as an integrated part of translation.

C. Pang et al. (Eds.): SETE 2020/ICWL 2020, LNCS 12511, pp. 420–429, 2021.
https://doi.org/10.1007/978-3-030-66906-5_40

2 Literature Review

2.1 Corpus-Based Translation Studies

When research content is concerned, corpus-based studies fall into two categories: corpus-assisted teaching and corpus-guided practice.

Focusing on corpus-assisted teaching, while paying attention to the important role of corpus in overall English teaching [1], domestic scholars have put much emphasis on the merits of corpus in teaching specific courses, ranging from translation [2] to Chinese-English business translation [3]. Apparently, corpus-assisted translation teaching have played a dominant role among a series of corpus-assisted teaching studies, demonstrating academia's full awareness of the crucial role of corpus in translation teaching.

Compared with studies on corpus-assisted teaching, papers concerning corpus-guided practice were more microscopic, with exploring the usage of a specific word based on a corpus as main objective [4–10]. Moreover, there were studies centering on the role of corpus in the discrimination of synonyms [11], and data processing and discussion was primarily composed of frequency, collocation, distribution in a certain corpus and its sub-corpus. Among them, a quantitative research was carried out on "Super Bowl" [8], which is highly consistent with the starting point of the current research.

Besides, in terms of research tools, studies abroad [12, 13] were mostly based on multiple corpora, while domestic researchers mainly applied a single corpus. In other words, studies based on multiple corpora were fairly rare, let alone studies conducted on the basis of English corpus and Chinese corpus synchronously.

2.2 Application of Semantic Prosody in Translation

The term "semantic prosody" was created by Sinclair [14] and was a term specifically used in corpus linguistics; in short, semantic prosody is the semantic tendency shown by the its collocation. Currently, the most common method for the classification of semantic prosody is to divide it into three categories: positive, neutral and negative semantic prosody. As the name suggests, positive semantic prosody referred to words whose collocations are mostly positive; the same was true for negative semantic prosody and neutral one respectively [15].

Semantic prosody has been studied by a plethora of scholars. From 2000 to 2009, most relevant studies can be classified into three categories:

1) theoretical introduction;
2) empirical researches;
3) expansion of register [16].

The current study belongs to the second category.

Besides, a scholar have combed the development history of semantic prosody in the past 20 years, and held that studies of semantic prosody was transforming from pure linguistic research to more applied disciplines [15], especially in translation teaching. Not only that, more and more scholars were aware of the influence of semantic prosody on the selection of equivalent words in translation [17–19]. Partington [18] found that the seemingly similar words between two similar languages actually owned much different semantic prosody, posing risks for translators, let alone disparities in semantic prosody between quite diverse languages, English and Chinese. Likewise, Stewart [19] also emphasized that translators should try to ensure the semantic prosody are reflected in the target language.

Moreover, the semantic prosody features of translation activities and translated texts have emerged as a hot topic of comparative research [20], which serves as theoretical basis and methodological guidance for our research.

To sum up, there has not been any research that compares the semantic prosody of the original text and the translated text of a certain expression based on multiple corpora with the aim to assist translation teaching.

3 Research Design

3.1 Tools and Methods

Our study is based on two corpora; detailed information is shown in Table 1:

Table 1. Overview of corpora.

Name	Size	Time span	Fields covered	Developer
COCA	560 million words	1990–2017	Blogs, web pages, film and TV programme, spoken language, novels, magazines, newspapers and academic journals	Brigham Young University
BCC	15 billion Chinese characters	Until 2015	Newspapers, literature, Weibo, technology, comprehensive and ancient Chinese	Beijing Language and Culture University

Three common methods have been used for studying semantic prosody: data-based method, data-driven method, and a combination of the first two methods [21]. As the current study was based on large corpora, a data-driven method was adopted.

3.2 Research Questions

The study intended to answer the following research questions:

1) What are the features of “It is what it is” and “情况就是这样” in specific con-text?
2) Are there differences in the semantic prosody tendency of “It is what it is” and “情况就是这样”?
3) What is the better translation?

3.3 Procedures

1) Search for "It is what it is" in COCA and "情况就是这样" in BCC, and extract once every two or more concordances to ensure the representativeness of the sampling[27], and investigate theirs diachronic trend in respective corpora.
2) By checking the context of concordances, clarify the field of application and semantic prosody tendency of "It is what it is" and "情况就是这样".
3) Combining the concordances and the context where they are located, compare and explain the changes in the frequency of "It is what it is" and "情况就是这样" in use.
4) Taking account of the specific context of Trump's use of "It is what it is", pro-pose a better translation.

4 Results and Discussion

4.1 "It Is What It Is"

Frequency. Table 2 shows that a total of 912 concordances are retrieved in COCA, and "It is what it is" is frequently used in blogs, films, web pages, newspapers and spoken language.

Table 2. Frequency of "It is what it is" in COCA.

	Blog	Film	Web	Newspaper	Spoken	Total
It is what it is	280	198	180	139	115	912

Diachronic Trend. What can be observed from Table 3 is that the frequency of "It is what it is" in COCA is increasing year by year.

It suggests that the acceptance and popularity of the expression in the native English community is increasing, so Chinese English learners are recommended to use the expression in their daily communication.

Table 3. Diachronic distribution of frequency of "It is what it is" in COCA.

	1990–1994	1995–1999	2000–2004	2005–2009	2010–2014	2015–2019
It is what it is	11	30	31	106	138	188

Semantic Prosody. For answering research questions, extract broadly representative concordances for analysis according to different fields of application.

Table 4. Application of "It is what it is" on blog.

No.	Concordance
1	As far as Elaine, she probably should have waited, but she didn't. **It is what it is**! Lighten up folks!
2	A cruise vacation can be summed up best as: **It is what it is** and what you make of it
3	I don't like the idea of what I am pointing out, but **it is what it is**

It can be seen from Table 4 that 1) The location of "It is what it is" in a sentence is flexible; it can not only be placed at the beginning, middle and end of a sentence, but can be used as an independent sentence; 2) The semantic prosody tendency of "It is what it is" is not varied with context, and shows negative semantic prosody tendency.

Table 5. Application of "It is what it is" in film.

No.	Concordance
1	**It is what it is**. But it wasn't because I didn't trust you
2	**It is what it is**. Don't even try to lie about this

Observing Table 5, we can find that 1) "It is what it is" mostly appear as an independent sentence in films; 2) "It is what it is" in films shows negative semantic prosody tendency.

Table 6. Application of "It is what it is" on web pages.

No.	Concordance
1	**It is what it is**, unfortunately, right now
2	So knowing that **it is what it is**, you have a tough decision to make
3	Yeah, I know it sounds crazy but **it is what it is**

It is easy to find from Table 6 that 1) The location of "It is what it is" is also flexible on web pages; and 2) presented negative semantic prosody.

Table 7. Application of "It is what it is" in newspaper & spoken.

No.	Concordance	Source
1	"I hate that we have to go down this route, but **it is what it is**," Burns said	Newspaper
2	"I know he's a little frustrated, but **it is what it is**. It was major surgery," Cora said	Newspaper
3	It's just something that, okay, **it is what it is**	Spoken

Table 7 shows that 1) "It is what it is" appears as the form of dialogue in newspapers, which further proves the wide application of this expression in spoken language; 2) "It is what it is" in newspapers and spoken language presented negative semantic prosody.

To sum up, "It is what it is" has a negative semantic prosody tendency in most fields, and it always appears as an independent sentence.

4.2 "情况就是这样"

Frequency. Table 8 shows that "情况就是这样" was principally used in newspapers, technology texts and Weibo.

Table 8. Frequency of "情况就是这样" in BCC.

	Multi-field	Newspaper	Tech	Weibo	Total
情况就是这样	170	158	61	33	422

Put information together, it can be inferred from Table 8 that 1) "情况就是这样" is most frequently used in newspapers, while "It is what it is" on web pages and films, demonstrating that "情况就是这样" and "It is what it is" varies markedly in fields of application.

Diachronic Trend. Figure 1 shows that the frequency of "情况就是这样" has display an overall downward trend. Conversely, the frequency of "It is what it is" in COCA increase year by year. Thus, judging from the inconsistency of diachronic trend, "情况就是这样" is not the ideal translation of "It is what it is".

Fig. 1. Diachronic trend of "情况就是这样" in BCC.

Semantic Prosody. To explore the semantic prosody tendency of "情况就是这样", decide to select fields of high frequency (that is, newspapers and technology texts) of the expression, and extract the broadly representative concordances for further analysis. It can be concluded from Table 9 that:

1) "情况就是这样" can be used to describe a wide range of topics, such as social issues and national defense;
2) "情况就是这样" has an obvious negative semantic prosody tendency and is often appeared in spoken language;
3) "情况就是这样" is likely to appear as an integrated semantic unit of a complete sentence.

Table 9. Application of "情况就是这样" in newspaper.

No.	Concordance
1	最近海军在东京湾交锋的**情况就是这样**。到目前为止没有全面叙述这次交锋的材料。
2	不解决这个问题，人类进步就会受到阻碍。**情况就是这样**严峻。
3	说到这里，中山大学某附属医院外科副主任医师张晓东不免苦笑，"我知道这很荒唐，但现实**情况就是这样**——论文决定一切。"

Table 10 indicates that the usage of "情况就是这样" in technology texts take on the following features: 1) "情况就是这样" always exists as a part of a complete sentence; 2) "情况就是这样" demonstrates a neutral semantic prosody tendency.

Table 10. Application of "情况就是这样" in technology texts.

No.	Concordance
1	假如不是实际上存在这些先天纯粹综合知识的话，就会认为先天综合知识是完全不可能的。实际上休谟所遇到的**情况就是这样**，虽然他远远没有体会到问题在这里所提出的以及所必须提出的普遍性。
2	西部的工业也发生得比较迟，普遍落后于东部的工业。意大利的**情况就是这样**，高卢的情况更是如此。

From the above, "情况就是这样" is likely to vary with the application field (from neutral semantic prosody to negative semantic prosody), and it always exists as an indispensable part of a complete sentence.

4.3 "事已至此"

We proposed that "事已至此" is a more suitable translation for "It is what it is" than "情况就是这样". To support this argument, this paper first reported the search results of "事已至此" in BCC, and then compared it with those of "It is what it is" and "情况就是这样".

Frequency. Through the comparison and analysis of Table 2, Table 8 and Table 11, we can see that: 1) the total frequency of "事已至此" in BCC is more than twice that of "情况就是这样", which suggests that the former is more commonly used and more authentic in Chinese; 2) the most frequently used field of "It is what it is" is blog, while the official translation of "情况就是这样" is least frequently used on blog's counterpart, Weibo; and fortunately "事已至此" is more commonly applied on Weibo, so "事已至此" is a better translation when frequently used field is concerned.

Table 11. Frequency of "事已至此" in BCC.

	Multi-field	Weibo	Tech	Newspaper	Total
事已至此	738	133	88	35	994

Diachronic Trend. Observing Fig. 2, it is easy to conclude that, unlike "情况就是这样", the frequency of "事已至此" shows a normal up and down dynamic trend. Thus, "事已至此" is also a more suitable translation for "It is what it is" from the perspective of diachronic trend.

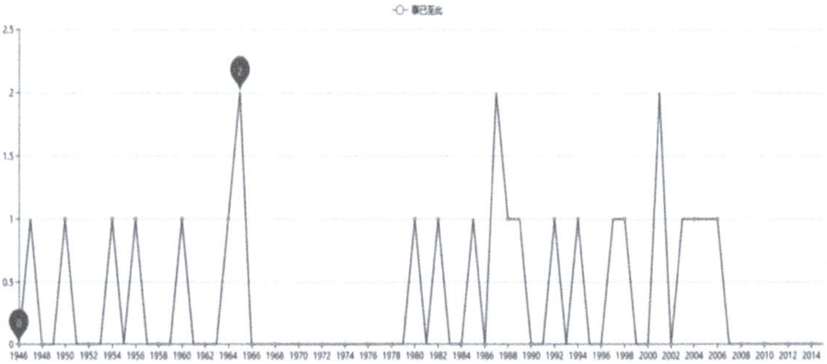

Fig. 2. Diachronic trend of "事已至此" in BCC.

Semantic Prosody. To check whether "事已至此" is a better translation in terms of semantic prosody tendency, we decided to select the most frequently used domains of "It is what it is" and "事已至此" (blog and Weibo) to conduct comparison.

Table 12. Application of "事已至此" on Weibo

No.	Concordance
1	无论如何，**事已至此**，我们也只能接受了。
2	突然一阵轻松，这是久违的感觉。既然**事已至此**，也没什么大不了的，不是吗？
3	这结果........我真的不太满意。但是，**事已至此**，我还有的选么？

Comparing Table 4 and Table 12, it can be easily found that the context of the Concordance 3 in Table 4 and that in Table 12 was quite similar: 1) both are collocated with "but" ("但是"); 2) both embody negative semantic prosody tendency.

In a nutshell, combining detailed analysis and comparison of frequency, diachronic trend and semantic prosody, "事已至此" is a better translation for "It is what it is".

5 Conclusion

Under the assistance of two corpora (COCA and BCC), whether in terms of frequency, diachronic trend, collocation, or in semantic prosody tendency, there are obvious disparities between "情况就是这样" and "It is what it is". In contrast, "事已至此" and "It is what it is" are not only more similar in all dimensions, but often used independently. It is more appropriate to translate "It is what it is" into "事已至此" hence.

Despite the many contributions contained, limitations still exist. For example, as the study was based on online corpora, it is not easy to get much more detailed information, if not impossible. This study is hoped to attract more experts to focus on the application of corpus in translation teaching, urging domestic translation teaching scholars to pay more attention to differences in language use between second language learners and native speakers, and in particular, promoting the core of domestic translation teaching to gradually transform from surface language structure to in-depth language use, thus enhancing the overall translation teaching level in China.

References

1. Zhang, J., Yu, Y.: A corpus-based study of be + being + adj.-on the application of corpora in foreign language teaching. J. Xi'an Int. Stud. Univ. **22**(1), 57–60+74 (2014)
2. Zhu, X.: On the employment of COCA and CCL in translation teaching. Foreign Lang. Learn. Theory Pract. **1**, 32–37 (2011)
3. Liu, X., Li, D.: Using the monolingual English corpus COCA in Chinese-English business translation teaching. Chin. Sci. Technol. Transl. J. **33**(1), 29–32+61 (2020)
4. Chen, S.: Corpus-based contrast and translation study of English and Chinese verbal metaphors-a case study of the "smile" construction and related translation examples of ulysses. Shanghai J. Transl. **4**, 75–79 (2017)
5. Hu, C.: Corpus-based research on the metaphor of "bubble". J. PLA Univ. Foreign Lang. **37** (1), 18–30 (2014)

6. Luo S.: Semantic prosody and pragmatic features of the causative GET: a study based on COCA. J. PLA Univ. Foreign Lang. **34**(1), 19–22+127 (2011)
7. Ma, Z.: Semantic prosodies of HAPPEN: a contrastive study. Foreign Lang. Learn. Theory Pract. **4**, 20–27+7 (2010)
8. Meng, G., Cao, Z., Yue, X.: A study of american super bowl from the perspective of mass media: based on COCA corpus. J. Beijing Sport Univ. **41**(7), 33–42 (2018)
9. Zhang, Y., Yu, C., Wu, X.: A COCA-based study of the cognition of "way"-constructions with mismatched motion meanings. J. PLA Univ. Foreign Lang. **42**(4), 85–93+160 (2019)
10. Zhao, R.: A corpus-based study on transdisciplinarization of ecological study and ecologicalization of disciplines. Foreign Lang. China **15**(4), 54–60 (2018)
11. Wu, X., Xiao, K.: A corpus-based cognitive linguistic study on the discrimination of English verbs' synonyms. Foreign Lang. Res. **33**(5), 46–51 (2016)
12. Adam, S.: Newly emerging subordinators in spoken/written English. Aust. J. Linguist. **34**(1), 118–138 (2014)
13. Peña, M.S.: Cognitive mechanisms underlying fake reflexive resultatives. Aust. J. Linguist. **36**(4), 502–541 (2016)
14. Sinclair, J.M.: Corpus, Concordance, Collocation. Oxford University Press, Oxford (1991)
15. Zhang, Y.: Overview of twenty years of research on corpus-based semantic prosody. Foreign Lang. Res. **6**, 23–28 (2012)
16. Zhai, H., Fang, H.: Review of domestic research on semantic prosody. Shandong Foreign Lang. Teach. **30**(2), 8–11 (2009)
17. Morley, J., Partington, A.: A few frequently asked questions about semantic or evaluative prosody. Int. J. Corpus Linguist. **14**(2), 139–158 (2009)
18. Partington, A.: Patterns and Meanings-Using Corpora for English Language Research and Teaching. John Benjamins, Amsterdam/Philadelphia (1998)
19. Stewart, D.: Safeguarding the Lexicogrammatical Environment: Translating Semantic Prosody. John Benjamins, Amsterdam/Philadelphia (2009)
20. Dai, J.: Review of domestic research on semantic prosody: trend, problems and suggestions. Technol. Enhanc. Foreign Lang. Educ. **3**, 20–25+52 (2018)
21. Wei, N.: The general methods of studying semantic prosody. Foreign Lang. Teach. Res. **4**, 300–307 (2002)

Language Teaching and Technology

Is Flexible Learning Flexible for Teachers? An Investigation of the Current Situation of China's Online Flexible EFL Teaching During the COVID-19 Outbreak

Min Yuan, Xiaojing Wu, Fengxin Xiong, Wenwen Li, Qiuzhu Xie, and Xiaobin Liu(✉)

South China Normal University, Guangzhou, China
liuxiaobin@m.scnu.edu.cn

Abstract. This study investigated the current situation of online flexible teaching in Chinese primary and middle schools, its impact on EFL teachers and the corresponding measures. An online survey regarding the basic condition of flexible teaching was conducted to 2,051 EFL teachers from different provinces in China. The results showed that online flexible teaching has been carried out in primary and middle schools nationwide in China. Varied online platforms and materials were applied by EFL teachers. Though the experiences of flexible teaching have had positive impact on their professional development, they have been confronted with lots of difficulties from different aspects. Therefore, to achieve effective flexible teaching, EFL teachers should be familiar with different teaching tools, prepare abundant teaching resources, and design effective teaching activities. Besides, other parties are supposed to provide sufficient support for them.

Keywords: COVID-19 · Flexible teaching · EFL teachers

1 Introduction

New semester of spring in 2020 has been postponed to contain the spread of COVID-19 at school. In order to maintain students' learning, Chinese Ministry of Education (MOE) has issued an initiative entitled "Disrupted classes, undisrupted learning". All kinds of schools at different levels actively responded to this initiative, providing flexible learning to over 270 million students from their homes. Online flexible teaching is regarded as teaching experiences in synchronous or asynchronous environments utilizing different devices with internet access [1], which can provide access to educational experiences that are more flexible in time and in space than campus-based education [2]. Therefore, this flexible teaching can effectively address the challenges that students can't go to campus to study in a regular way during the COVID-19.

This study investigates the overall situation of online flexible EFL teaching in Chinese primary and middle schools, its impact on EFL teachers and the corresponding measures to achieve flexible teaching by online questionnaire.

C. Pang et al. (Eds.): SETE 2020/ICWL 2020, LNCS 12511, pp. 433–443, 2021.
https://doi.org/10.1007/978-3-030-66906-5_41

2 Literature Review: Flexible Learning

Flexible learning and teaching have gained significant attention for decades. Flexible learning is defined as teaching and learning that is student-centered in respect to teaching and learning methods and resources and that is free from the limitations of the time, place and pace of study [3–5]. In this learning process, flexibility is representing through the choices given to learners with relation to these dimentions: time of course participation, content in the course, entry requirements, instructional approaches and learning materials, and course delivery and logistic [6]. As the information and communication technologies develops, newly enabled leaning modes have emerged that can bring compliment to flexible learning [2].

As an "anytime anyplace" learning, flexible education has been considered to possess obvious nature and strengths. First, Goode et al. [3] pointed out that learners are provided with various learning choices from multiple dimensions of study. Second, according to Goode [3] and Lewis & Spencer [7], it adopts learner-centered constructivism approach, indicating a shift of learning responsibility from the teachers only to both sides. Third, Collis [8] noted that as learners are granted rich choices and more responsibilities are transferred to students accordingly, learners are required to be more skilled at self-regulation in goal setting, self-monitoring, and teachers are asked to promote active learning. Fourth, flexible learning supported by technology offers people, especially those with geographic limitations, more language learning opportunities.

While flexible learning offers many opportunities, implementing flexible learning may bring with it many problems and challenges. Lobry de Bruyn [9], Rennie [10] and Robson [11] held that a major challenge in flexible learning is geography and access to and limits of developing technology. Beckman [12] added that the limits to the capacity of online technologies, such as high costs for high speed Internet access, also block flexibility. Another main challenge goes to self-regulation of and extra work required from students. Houlden and Veletsianos [13] pointed that some researchers concerned about the efficiency of the mobilization of flexibility and extra labour it requires from learners for succeeding. There are also challenges that lies in the part of instructors. Gregory and Lodge [14] and James and Beattie [15] suggested that flexible learning is time-consuming for instructors and inefficient if unsupported. Therefore, Nikolova and Collis [16] noted that flexibility calls for new roles, such as consultant, collaborator, facilitator, for teachers and imposes higher demands on learner's self-initiative and self-control. The last challenges lies down on instruction. Samarawickrema [17] argued that flexible learning would be more likely to be successful in the key of instructional design which determines things such as students' high teacher reliance and high goal orientation. Correspondingly, Huang, et al. [2] and Gordon [18] brought forward that to achieve effective teaching and learning, flexibility can be enabled by offering learners several ways of studying, such as individually and collaboratively, as well as several instructional approaches, such as discussion, seminar groups, and debates, etc.

3 Method

3.1 Research Questions

This study was driven by investigating three research questions as:

RQ1. What is the current situation of online flexible EFL teaching in primary and middle schools in China during the COVID-19 outbreak?
RQ2. What is the impact of online flexible teaching on EFL teachers during the COVID-19 outbreak?
RQ3. What further measures should be taken to achieve flexible learning during educational disruption?

3.2 Questionnaire Respondents

2,051 EFL education practioners filled out the questionnaire in February 2020 during a webinar open to more than 12,000 teachers in China. They come from different provinces across China, especially from Shandong, Guangdong, Inner Mongol, Beijing, Guizhou, Hubei provinces (Fig. 1). 96.49% of the respondents are primary and secondary school teachers with a total number of 1,979. The rest are 60 education and teaching management personnel holding a percentage of 2.93%, educational teaching researchers 3.36%, and employees of educational information-related enterprises 0.15% (Fig. 2).

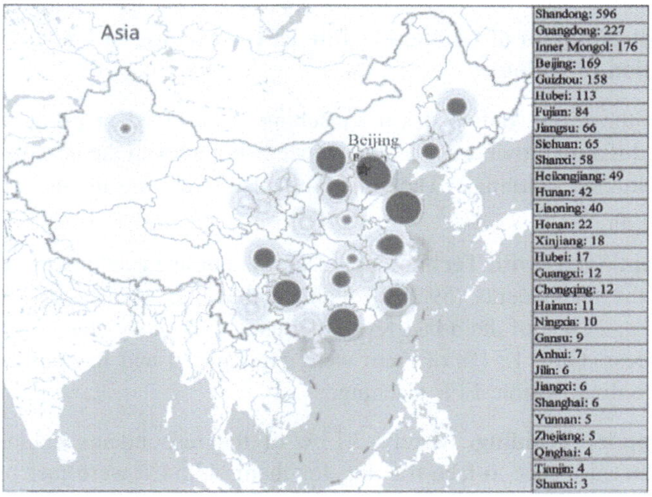

Fig. 1. Heat map of respondents' source provinces

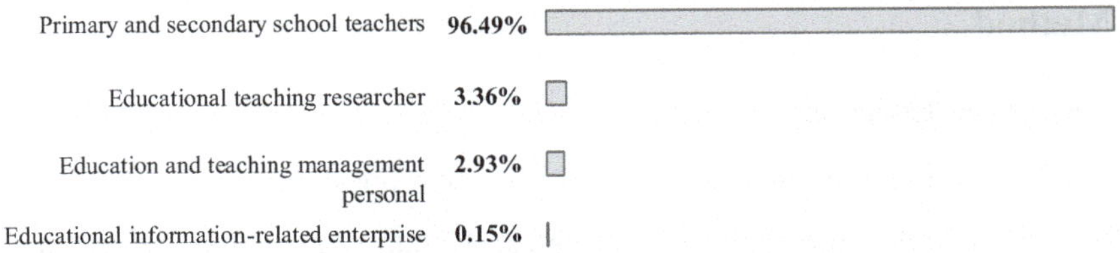

Fig. 2. Identity of questionnaire respondents

3.3 Questionnaire

As Table 1 is shown, the questionnaire measured the following main constructs:

Table 1. Brief description of the contents of the questionnaire

Contents	Questions
Respondents demographic data	Q1–Q5
Respondents' familiarity with technology	Q6–Q9
Ways of online flexible teaching and teaching content	Q10–Q15
Respondents' perceptions to online flexible learning	Q16–Q21

4 Results

4.1 Current Situation of Online Flexible EFL Teaching in Basic Education of China

According to the result, online flexible teaching is provided in primary and middle schools from different provinces in China, which is in line with the initiative "Disrupted classes, undisrupted learning". The following results show different perspective of current situation.

Teacher's Familiarity with Technology. As for the online teaching proficiency, more than half of the respondents (59.29%) claimed that they were complete novices in online teaching. 27.4% of them had few experience previously and 11.8% understood the basic process. Only 1.51% of them were experienced and have already got good command of online flexible EFL teaching.

School's Support Training. Merely 2.13% of the respondents claimed that their schools conducted a face-to-face training for them. 43.54% were just provided with documents and guidance. Whereas 16.12% stated that there were not any guidance or training for them. It is understandable because it is inconvenient and impossible for most of the schools to launch off-line conference during the epidemic. However, since a large number of teachers are not skillful in online flexible teaching, schools are supposed to provide training and guidance in high-quality teaching design, teaching method and teaching ability in online flexible EFL teaching.

Ways for Flexible Teaching. Though teachers may not be familiar with online teaching, almost 95% of them planned to carry out online tasks, among whom 55% choose to conduct live stream class, while 36.5% tend to employ recording class. Meanwhile, 78.6% of them would also give out the learning materials for students and assign autonomous learning tasks for them. Giving students flexible online learning patterns will allow them to explore more thus developing their independent learning ability.

Teaching Content. Intriguingly, the highest proportions of the teaching content chosen by respondents are the important knowledge points that students have learned before (66.52%) and that would be learned in the new term (63.98%). While knowledge that related to the outbreak (42.35%) and other extracurricular English knowledge (41.61%) also stood out prominently which means most respondents would prefer to combine various kinds of learning materials as their teaching contents for flexible teaching.

Tools for both Synchronous and Asynchronous Flexible Teaching. The teaching tools that teachers utilized are various according to their teaching ways. As shown in the results, the order of the five platforms was as follows: DingTalk (45.25%), Wechat Group (35.23%), QQ Group (21.51%), Tencent Ketang (12.14%), CCTalk (4.52%). During the period of COVID-19, DingTalk offers diverse and powerful functions including online classes, students' daily management and conference for schools and teachers. Considering that students, especially those who are primary students, are insufficient in information technology and some advanced products educators will first consider the relatively easy-handled tool. QQ group as a popular online learning platform, is characterized by simple operation, stable connection and mature system.

Source of Flexible Digital Learning Materials. During the delay of the new semester, 50% of the teachers choose to recommend the platforms and resources for EFL learning for students. The platforms and resources that teachers chosen are mainly recommended by the MOE (40.07%), the English Department in their schools (36.06%), the web resources and authoritative publishers that they found the materials themselves (21.15%) and other channels (2.72%). To support the flexible learning during the outbreak, the Chinese MOE provides various online educational resources from general education to specific subjects' learning materials and electronic books for primary and secondary students.

4.2 The Impact of Online Flexible Teaching on EFL Teachers

Teachers' Self-improvement. 60.02% of respondents claimed that they take full advantage of online resources to improve themselves during flexible teaching, though 39.98% failed to do so. Among those who achieve self-improvement, they believed that the top three abilities of informational technology are ability of designing online flexible EFL teaching (62.34%), ability of retrieving and accessing teaching information (60.98%) and ability of live stream (57.14%). These abilities closely relate to flexible teaching, which enable teachers to promote information technology literacy. Ability of producing tutorial videos (39.53%), ability of editing video/audio (37.66%), ability of making courseware (35.63%) and ability of processing images (32.19%) follow after them.

Obstacles Faced by Teachers. Obstacles of online flexible EFL teaching faced by teachers can be roughly divided into four categories. In terms of teacher, the heavy workload for teachers, teachers' unknown of what to do, family chores for teachers and low motivation of teachers accounts for 49.83%, 36.67%, 26.13%, 12.87% respectively. In terms of student, 58.90% thought that obstacle mainly come from the low motivation of students, while 48.76% believed that the challenge is unavailable hardware for students. In terms of technology, the proportion of unreliable network is 55.29% and of unreliable tools and platform is 24.67%. 22.33% of respondents claimed that the existence of misunderstanding will hinder flexible learning, while 11.70% stated that the impractical arrangement of educational bureau and school negatively influences on online flexible EFL teaching. Besides, 11.95% of them expressed that the English subject doesn't fit for online flexible teaching.

4.3 Measures to Achieve Online Flexible Learning

Ensuring Reliable Network Infrastructure. Reliable network infrastructure is a crucial premise in supporting flexible learning activities. Nevertheless, network is still a major challenge for flexible learning. Tens of EFL teachers reported:

"Due to the unreliable network, I can't conduct flexible teaching smoothly".

"Living in remote and rural areas without network, some students have difficulty in online flexible learning".

Undoubtedly, schools and MOE should take measures to ensure reliable network infrastructure. Schools should boost internet connectivity service for online education, especially for the under-served regions. Local educational bureau should organize to open broadcasting of classes on TV to provide learning experiences for those in remote areas without Internet [2].

Utilizing Suitable Learning Tools. The unreliable tools and platforms is one of the major problems of flexible teaching. To handle it, EFL teachers should get familiar with different kinds of platforms and prepare plan B in case of the instability of certain platform. Besides, as a respondent said:

"There are various learning tools and platforms to be chosen. But I tend to use those which are user-friendly and suitable for learning activities".

It's essential to effectively select learning tools since a variety of tools for different functions may cause EFL teachers getting lost if unclear of teaching aims. EFL teachers should bear teaching aims in mind and choose the most suitable tools and platforms that target the aims.

Adopting Suitable Learning Resources. Adopting different kinds of learning materials is also one of the center issues of flexible teaching. These materials are conducive to students' EFL learning with teachers' proper guidance. Some teachers also regard the resources as valuable teaching helper.

"Using resources from the Internet is really a great idea! They can help us to enrich our teaching content and innovate our teaching which sometimes seems boring for Chinese students".

While there are so many available digital resources for students, it is necessary for teachers to evaluate and choose the high-quality ones for students. There are five criteria for teachers to take consideration: suitability of content, suitability of difficulty, suitability of structure, suitability of the media and suitability of resource organization [2]. Therefore, EFL teachers should make full use of such free learning materials from Internet to foster their flexible teaching [19].

Designing Effective Teacher-Centered Activities. To motivate students to engage in flexible learning and to make sure their improvement, educators have provided some practices. Most of them choose to combine different ways with certain teaching content to best suit learners. While the synchronous instruction is adopted by nearly half of the teachers, they are also thinking of various ways to attract students' attention.

"I feel embarrassed if I just keep talking myself without children's timely response. And I just cannot know whether they get command of the knowledge".

If the teaching activities designed by teachers are fun enough while educated, students will be more willing to participate in flexible learning actively even without the supervision from parents and teachers.

Preparing Creative Student-Centered Activities. Learning sometimes can be conducted by students themselves especially in today's Internet age. It is a good opportunity for teachers to organize individual, partner and group-based learning.

Instructors are supposed to provide creative learning activities for learners like seminar groups, debates, student-led discovery approaches and educational gamification [18]. While teachers clarify the key and difficult points in the synchronous instruction, they can assign individualized and collaborative learning activities for students [2]. Considering that this is the first time for students to have online lessons, teachers carefully choose the teaching content for them to accept such kind of learning pattern. It is advisable for teachers to tailor different learning tasks or aims according to different style or preference of students [20]. Just as a teacher mentioned:

"I will give students freedom to identify the suitable learning materials for the courses rather than providing them the resources directly".

The fundamental way is to make good use of the available resources and learn from other good examples so as to come up with better teaching design to raise students' motivation.

Providing Sufficient Support and Training. Being complete novices or having few experience in flexible teaching, EFL teachers have difficulty after receiving guidance in the documents or even not any guidance or training. Therefore, it is necessary for the government and universities to provide sufficient supports and training for online teaching [21]. As a respondent mentioned:

"I want to know how to make micro lessons (tutorial videos) and how to design an effective online lesson. I just have few experiences on it".

Lectures related to flexible teaching from the experts in universities and interaction between EFL teachers from different schools is useful ways to improve flexible teaching.

> *"I want to learn from the effective lessons from other teachers, which is a significant way for me to improve the quality of my class".*

5 Discussion

Different from teaching and learning in brick-and-mortar schools, both educators and learners will find it hard to take advantage of the tools to work and learn smarter. In physical classrooms, students can have pair-work with classmates timely and finally achieve their learning goals. While in online classes they are unable to interact with group members for the constraints of space [22] Students also will have difficulty in asking teachers for help in real time when they come across troubles in the process of flexible learning. Most importantly, students' attitude towards online flexible learning also matters as willingness, self-direction and self-discipline are vital for students' online classes productivity [23].

With respect to these problems, as the result of our survey shows that EFL teachers tend to change their original teaching plans and conduct online flexible teaching to fit the current situation and minimize the inconvenience and disadvantages brought by flexible teaching. Due to the limited interactions of online flexible teaching, it is hard to check whether students have mastered new knowledge. If teachers just keep teaching new contents without timely consolidation and follow-up, then the learning effect will certainly be heavily discounted.

In the present study, teachers conduct flexible teaching by reviewing the previous knowledge, such as winter holiday homework, assigning tasks and answering students' questions online and launching group discussions. Teachers are also supposed to provide more flexible resources for learners. It's also a good opportunity for students to learn more than the textbook and campus-based EFL learning materials. The conventional teaching contents combine with the epidemic-related ones can enrich students' extra-curriculum knowledge and cultivate their awareness of social issues.

Both learners and teachers benefit greatly from a variety of resources and information as well as the connection with each other through new media. However, it will result in confusion sometimes. For learners, it will cause information overload; for teachers, barriers include: how to identify, select and adopt online flexible teaching practice to reach the teaching aims flexibly [18]. There still needs more research to figure out which combinations of technological affordances, subject domains, roles of adults and instructional and assessment approaches work best for particular types of students [24].

With regard to the impact of flexible teaching on teachers, a great number of EFL teachers improve themselves by developing abilities of information technology. It becomes necessary for teachers to be technologically competent, to effectively use and manage technology for teaching, and also to create and upload learning materials to students [25]. In flexible learning, EFL teachers need to design activities suitable for flexible learning, make courseware, utilize tools for live stream skillfully, and edit the videos of lessons. Through all the preparation and practice of flexible teaching, teachers become a skillfully technical service provider, thus promoting information technology literacy. And this is congruent with the idea that all kinds of technologies applied in the teaching process are important core qualities for online education teachers [26],"educators are largely required to be their own education technologists" [12].

However, the inclusion of technology into instruction thereby creating the online component has brought some level of unease to teachers, students and educational institutions [25]. For teachers, their awareness of flexible teaching and application skills are uneven [27]. The training and preparation of online flexible teaching, daily reports of students' health increase the burden of workload for teachers. Moreover, the survey reflects technological illiteracy and resistance due to the lack of support training and the failure to improve information technology abilities, which might contribute to teachers' low motivation of flexible teaching. And this is in accordance with the identified challenges from teacher's perspective as higher workload [28], lack of technological competency [29], resistance to technology [30]. For students, getting accustomed to traditional teaching, quite a lot of students feel unease and lack in self-regulation, leading to low motivation for them. The studies [30, 31] identified self-regulation as a key challenge faced by students during flexible learning. Besides, not all the students are easily access to tools for flexible learning simply because of uneven level of family support. Similarly, studies [32, 33] posed emphasis on the need for technological competency for flexible learning students. For technology, reliable network infrastructure is crucial to support different activities, such as synchronous cyber teaching using video conferencing, and asynchronous cyber learning by accessing or downloading digital learning resources, etc. [2]. Even though all major telecom service providers are required to boost internet connectivity service for flexible learning, millions of students simultaneously logging in cannot absolutely ensure the network to become reliable.

6 Conclusion

As shown in the present study, online flexible EFL teaching has been carried out in primary and middle schools nationwide in China. Different kinds of online platforms and materials were applied by EFL teachers. From teachers giving the positive feedback, we can tell that such kind of flexible teaching model has been already accepted by them. However, EFL teachers have faced lots of obstacles during flexible teaching, which not only from the teachers, students and Internet, but also from their unfamiliarity with technology. Therefore, to achieve flexible teaching, EFL teachers should be familiar with different tools, prepare abundant teaching materials and design effective teaching activities. Besides, it is important that society, schools, parents and students should give sufficient and positive support for EFL teachers. More significantly, EFL teachers ought to improve their information technology competence in ordinary teaching. As more and more creative teaching methods and activities pouring into flexible EFL teaching and learning, flexible teaching and learning that has inevitably shock the conventional way will absolutely not occur merely during the outbreak, but shed light on future teaching and learning.

However, this study was a preliminary investigation, so future studies were expected to have a further exploration on flexible teaching. Further study can be conducted to explored the reasons and methods on enhancing the motivation of EFL teachers and students so as to improve the better effect of flexible teaching and learning. Besides, the other topics about flexible teaching mentioned by the respondents, such as the effective way to arouse the learners' enthusiasm of flexible learning and to implement the teaching assessment are worth investigating.

Acknowledgements. This work is supported by the Center for Language Cognition and Assessment, South China Normal University. It's also the result of the following funds: Guangdong Academic Degrees & Graduate Education Reform Project (2019JGXM37), and University-Industry Collaborative Education Program (201801191006 & 201801191027).

References

1. Singh, V., Thurman, A.: How many ways can we define online learning? A systematic literature review of definitions of online learning (1988–2018). Am. J. Dist. Educ. **33**(4), 289–306 (2019). https://doi.org/10.1080/08923647.2019.1663082
2. Huang, R.H., Liu, D.J., Tlili, A., Yang, J.F., Wang, H.H., et al.: Handbook on facilitating flexible learning during educational disruption: the chinese experience in maintaining undisrupted learning in COVID-19 Outbreak. Smart Learning Institute of Beijing Normal University, Beijing (2020)
3. Goode, S., Willis, R., Wolf, J., Harris, A.: Enhancing IS education with flexible teaching and learning. J. Inf. Syst. Educ. **18**(3), 297–302 (2007)
4. Van den Brande, L.: Flexible and Distance Learning. Wiley, Chichester (1993)
5. Naidu, S.: How flexible is flexible learning, who is to decide and what are its implications? Dist. Educ. **38**(3), 269–271 (2017)
6. Collis, B.: Flexibility combinations TeleScopia project deliverable UT/DL1001/WP 1.6. Deutsche Telecom Generaldirektion, Bonn (1995)
7. Lewis, R., Spencer, D.: What is open learning?, Open Learning Guide 4, London Council for Education Technology, pp. 9–10 (1986)
8. Collis, B.: New didactics for university instruction: why and how? Comput. Educ. **31**, 373–393 (1998). https://doi.org/10.1016/S0360-1315(98)00040-2
9. de Bruyn, L.L.: Monitoring online communication: can the development of convergence and social presence indicate an interactive learning environment? Dist. Educ. **25**, 67–81 (2004). https://doi.org/10.1080/0158791042000212468
10. Rennie, F.: The use of flexible learning resources for geographically distributed rural students. Dist. Educ. **24**, 25–39 (2003). https://doi.org/10.1080/01587910303052
11. Robson, J.: The effectiveness of teleconferencing in fostering interaction in distance education. Dist. Educ. **17**, 304–334 (1996). https://doi.org/10.1080/0158791960170207
12. Beckmann, E.A.: Learners on the move: Mobile modalities in development studies. Dist. Educ. **31**, 159–173 (2010). https://doi.org/10.1080/01587919.2010.498081
13. Houlden, S., Veletsianos, G.: A posthumanist critique of flexible online learning and its"anytime anyplace" claims. Br. J. Educ. Technol. **50**, 1005–1018 (2019). https://doi.org/10.1111/bjet.12779

14. Gregory, M.S.-J., Lodge, J.M.: Academic workload: the silent barrier to the implementation of technology-enhanced learning strategies in higher education. Dist. Educ. **36**, 210–230 (2015). https://doi.org/10.1080/01587919.2015.1055056
15. James, R., Beattie, K.: Postgraduate coursework beyond the classroom: issues in implementing flexible delivery. Dist. Educ. **17**, 355–368 (1996). https://doi.org/10.1080/0158791960170209
16. Nikolova, I., Collis, B.: Flexible learning and design of instruction. Br. J. Edu. Technol. **29** (1), 59–72 (1998)
17. Samarawickrema, R.G.: Determinants of student readiness for flexible learning: some preliminary findings. Dist. Educ. **26**, 49–66 (2005). https://doi.org/10.1080/01587910500081277
18. Gordon, N.A.: Flexible Pedagogies: technology-enhanced learning. High. Educ. Acad. (2014).https://doi.org/10.13140/2.1.2052.5760
19. Collis, B., Moonen, J.: Flexible Learning in a Digital World, 2nd edn. Routledge and Falmer, London (2004)
20. Sadler-Smith, E., Smith, P.J.: Strategies for accommodating individuals' styles and preferences in flexible learning programmes. Br. J. Edu. Technol. **35**(4), 395–412 (2004)
21. Grace, L.J., Smith, P.J.: Flexible delivery in the Australian vocational education and training sector: barriers to success identified in case studies of four adult learners. Dist. Educ. **22**, 196–211 (2001). https://doi.org/10.1080/0158791010220202
22. Chang, R.I., Hung, Y.H., Lin, C.F.: Survey of learning experiences and influence of learning style preferences on user intentions regarding MOOCs. Br. J. Educ. Technol. **46**(3), 528–541 (2015)
23. Grow, G.O.: Teaching learners to be self-directed. Adult Educ. Quart. **41**(3), 125–149 (1991)
24. Bakia, M., Shear, L., Toyama, Y.: Understanding the implications of online learning for educational productivity. Office of educational technology us department of education (2012)
25. Rasheed, R.A., Kamsin, A., Abdullah, N.A.: Challenges in the online component of blended learning: a systematic review. Comput. Educ. **144**, 1–17 (2020)
26. Wang, W., Ning, L.: A research on teachers' core quality in online education. Softw. Guide: Educ. Technol. **17**(12), 10–12 (2018)
27. Chen, Y.: Challenges in the college English online teaching and its countermeasures. In: Proceedings of the 5th International Conference on Education, Language, Art and Intercultural Communication (ICELAIC 2018) (2018)
28. Sage, M., Sele, P.: Reflective journaling as a flipped classroom technique to increase reading and participation with social work students. J. Soc. Work Educ. **51**(4), 668–681 (2015)
29. Boelens, R., De Wever, B., Voet, M.: Four key challenges to the design of blended learning: a systematic literature review. Educ. Res. Rev. **22**, 1–18 (2017)
30. Lightner, C.A., Lightner-Laws, C.A.: A blended model: simultaneously teaching a quantitative course traditionally, online, and remotely. Interact. Learn. Environ. **24**(1), 224–238 (2016)
31. Chuang, H.H., Weng, C.Y., Chen, C.H.: Which students benefit most from a flipped classroom approach to language learning? Br. J. Edu. Technol. **49**(1), 56–68 (2018)
32. Jensen, J.L., Kummer, T.A., Godoy, P.D.D.M.: Improvements from a flipped classroom may simply be the fruits of active learning. CBE-Life Sci. Educ. **14**(1), ar5 (2015)
33. Akçayır, G., Akçayır, M.: The flipped classroom: a review of its advantages and challenges. Comput. Educ. **126**, 334–345 (2018)

Exploring Virtual Team Project in an English for Science and Technology Course

Jianqiu Tian(✉)

School of Foreign Languages, Peking University, Beijing 100871, China
tianjq@pku.edu.cn

Abstract. Project-based learning (PBL) is increasingly heralded as an effective means to develop the students' 21st century skills, including collaborative problem solving. In this research, with the content and language integrated learning (CLIL) course on science and technology, the students worked online on projects that produce video presentations on a topic in the frontiers of science and technology informed by original research papers as a result of the COVID-19 pandemic. Taking a teamwork perspective, the research describes the dynamics of the teamwork, explores the relationship between the team dynamics and the team products and analyzes the benefits and challenges of the virtual team projects. The results indicate that successful projects were associated with harmonious team process and hard work, students benefited from the collaborative research experience and faced challenges such as the time limit, the quality of the product and the cohesion between the team member's parts of the final product. Pedagogical suggestions are training for research and collaboration before project, setting clearer timeline of the team process and making more active interventions.

Keywords: Virtual teamwork · Project-based learning (PBL) · English for science and technology

1 Introduction

Project-based learning (PBL), a systematic teaching and learning method which engages students in complex, real-world tasks that result in a written product or presentation to an audience, has been heralded as an effective means to develop the students' 21st century skills for them to thrive in this increasingly complicated and changing world. PBL has been widely applied in various subjects including science, e.g. [1, 2], mathematics, e.g. [3, 4], technology, e.g. [5, 6] and social sciences, e.g. [7, 8].

In the field of language learning, PBL is referred to as project-based language learning (PBLL) and has primarily been deployed to develop the students' linguistic and cultural skills. PBLL has been researched from multiple perspectives, including Activity Theory. In addition to students' academic achievement, there has also been research on factors such as teacher's role, student motivation and learner autonomy. Recently, Cox and Montgomery [9] investigated how PBLL might help students develop 21st century skills and how student engagement is affected by the deployment of PBLL in the classroom.

C. Pang et al. (Eds.): SETE 2020/ICWL 2020, LNCS 12511, pp. 444–453, 2021.
https://doi.org/10.1007/978-3-030-66906-5_42

The current research probes into the PBLL from a teamwork perspective. With the content and language integrated learning (CLIL) course on science and technology, the students were required to work on projects that produce video presentations on a topic in the frontiers of science and technology informed by original research papers. As a result of the COVID-19 pandemic, the teamwork was exclusively conducted in the cyberspace. After a brief literature review, this paper will describe the dynamics of the teamwork, discuss the relationship between the team dynamics and the team product and analyze the benefits and challenges of the virtual project.

2 Literature Review

2.1 Project-Based Language Learning (PBLL)

Project-based learning (PBL) is a student-centered approach to learning that involves learners in solving meaningful real-world tasks. Buck Institute for Education [10] suggests a seven-element gold standard for PBL projects: (a) a challenging problem or question, (b) sustained inquiry, (c) authenticity, (d) student voice and choice, (e) reflection, (f) critique and revision, and (g) a public product.

Most PBLL focuses on the development of the linguistic and cultural skills. The most frequently targeted linguistic skill is writing skill, which has been reported in [11–13] and the works they cited. PBLL has also been deployed to improve the learners' speaking skill (e.g., [14–16]). In some cases, all four skill areas are developed [17, 18]. Moreover, cultural and other skills are also the focus of PBL; for example, [19] addresses all the five Cs standards set by the American Council on the Teaching of Foreign Languages (ACTFL), namely communications, cultures, connections, comparisons and communities.

In addition to academic achievements, there is research that focuses on the participants and their activities in PBLL. Madoyan [20] discusses the multiple roles of teachers in the project as the designer, facilitator, monitor and evaluator. Cox and Montgomery [9] measure engagement through the lens of Csikszentmihalyi's [21] flow theory and conclude that authentic, real-world tasks that require students to make use of 21st century skills facilitate their behavioral, cognitive, and emotional engagement.

2.2 Team Processes

The most cited model of team processes is provided by Marks, Mathieu and Zaccaro [22], whose taxonomy of team processes involves ten process dimensions nested within three superordinate categories: (1) transition phase processes, (2) action phase processes, and (3) interpersonal processes. Processes such as Mission analysis, Goal specification and Strategy formulation are more likely to occur during transition periods, while Monitoring progress toward goals, Systems monitoring, Team monitoring and backup behavior and Coordination are more likely to occur during action periods. Interpersonal processes including Conflict management, Motivation and confidence building and Affect management are expected to occur throughout transition and action phases. Moreover, teams use different processes simultaneously and over performance episodes in order to multitask effectively.

2.3 Collaborative Problem Solving

Collaborative Problem Solving (CPS), defined as solving problems in collaboration with others, is a psychological construct that combines the cognitive components of problem solving [23] on the one hand and social collaboration [24] on the other. More specifically, it is defined by Programme for International Student Assessment (PISA) 2015 as "the capacity of an individual to effectively engage in a process whereby two or more agents attempt to solve a problem by sharing the understanding and effort required to come to a solution and pooling their knowledge, skills and efforts to reach that solution" [25, 26].

Due to its increasing significance in diverse aspects of 21st century life, CPS is being assessed by PISA 2015 and other educational authorities. PISA 2015 assessment [26] distinguished 12 CPS skills, with each conceptualized on the basis of four individual problem solving processes, i.e., (A) exploring and understanding, (B) representing and formulating, (C) planning and executing, and (D) monitoring and reflecting, crossed with three social collaboration dimensions, i.e., (1) establishing and maintaining a shared understanding, (2) taking appropriate action to solve the problem, and (3) establishing and maintaining team organization, as is shown in Table 1.

Table 1. The 12-cell matrix illustrating the 12 CPS skills in the PISA 2015 assessment

	(1) Establishing and maintaining shared understanding	(2) Taking appropriate action to solve the problem	(3) Establishing and maintaining team organisation
(A) Exploring and Understanding	(A1) Discovering perspectives and abilities of team members	(A2) Discovering the type of collaborative interaction to solve the problem, along with goals	(A3) Understanding roles to solve problem
(B) Representing and Formulating	(B1) Building a shared representation and negotiating the meaning of the problem (common ground)	(B2) Identifying and describing tasks to be completed	(B3) Describe roles and team organisation (communication protocol/rules of engagement)
(C) Planning and Executing	(C1) Communicating with team members about the actions to be/being performed	(C2) Enacting plans	(C3) Following rules of engagement, (e.g., prompting other team members to perform their tasks.)
(D) Monitoring and Reflecting	(D1) Monitoring and repairing the shared understanding	(D2) Monitoring results of actions and evaluating success in solving the problem	(D3) Monitoring, providing feedback and adapting the team organisation and roles

Note. Drawn from the OECD CPS Draft Report in PISA 2015 (2013).

3 Methodology

This research takes a teamwork perspective on the virtual project. The team dynamics will be analyzed from the taxonomy of team process and the students' CPS skills will also be addressed.

The specific research questions are: 1. How did the students do their team project? In particular, what are the team dynamics? 2. How do the team dynamics relate to the performance of the team? 3. What are the benefits and challenges of the virtual project?

3.1 Course Context

The research site was an undergraduate English for science and technology course offered at a comprehensive university in China. The course focused on topics in frontiers of science and technology and was offered to students across the university but most students majored in science and technology. The students were either placed in a level C class upon university entrance with a test of vocabulary, listening and reading or were in an equivalent level (roughly B2 or C1 with CEFR) after semesters of study. Owing to the COVID-19 pandemic, the classes met online instead of face-to-face once a week for two hours. The course was taught in English, half of the texts were taken from original English language research journals, the medical articles for COVID-19 from *The Lancet* or *New England Journal of Medicine* and articles for other fields from *Science* or *Nature*, and the other half were from popular science journals such as *Scientific American* and comprehensive journals such as *The Economist*.

3.2 Participants

Thirty-six undergraduate students participated in the study, 25 males and 11 females, ranging in age from 17 to 22 years old, with Mandarin Chinese as their native language. The students were in two classes, with 15 in class 1 (9 males and 6 females), and 21 in class 2 (16 males and 5 females).

3.3 Canvas, the Learning Management System (LMS)

Canvas is a web-based learning management system that not only allows course instructors to post courseware, assignments and grades online, but offers convenient means of communication including discussion boards for asynchronous discussions and chat rooms for live discussions, and provides interfaces to external applications.

Groups, a fascinating feature of the Canvas LMS, can be used as a collaborative tool for students to work together on group projects and assignments. As a small version of a course, Groups allow students to store and share files, start a discussion, send a message and create group collaborations.

3.4 Data Collection and Procedures

The data for this study was collected during the Spring 2020 semester. The project required the students to work in teams of three to five to work on a topic of their choice on frontiers of science and technology. In particular, they were assigned the task of making a video on the topic based on their readings of original research papers, for which each student is responsible for at least one.

The students were provided with sample videos and they were to work collaboratively on the group pages of Canvas and/or WeChat groups. Some teams also held conferences using Zoom or Tencent meetings.

The videos were submitted to the Studio Plugin on the Canvas LMS. Following that, the students did self assessment, assessment of team members and assessment of video product of students from other teams according to the rubrics the instructor attached to the task. For the peer assessment, each student was assigned 3 other students' work randomly by the Canvas platform.

Some students also addressed their teamwork in the final review for the whole semester.

The composition of the teams, with gender, major and year is shown in Table 2.

Table 2. Information of students in teams

Team	TM	G	Mjr	Yr	Team	TM	G	Mjr	Yr
C1T1	M111	F	Phy	2nd	C2T1	M211	M	Egn	4th
	M112	M	IT	2nd		M212	M	Phy	3rd
	M113	M	IT	2nd		M213	M	Phy	3rd
	M114*	M	IT	2nd		M214*	M	IT	3rd
						M215	F	Econ	2nd
C1T2	M121	M	Phy	4th	C2T2	M221	M	IT	4th
	M122	F	Egn	1st		M222*	M	IT	4th
	M123	M	Chem	1st		M223	M	Math	2nd
	M124*	F	Med	1st		M224	M	Math	1st
C1T3	M131	F	Med	1st	C2T3	M231	M	Egn	3rd
	M132	F	Med	1st		M232	M	Geo	3rd
	M133*	F	Med	1st		M233	M	IT	2nd
						M234	M	Phy	2nd
						M235*	F	Bio	1st
C1T4	M141	M	Phy	4th	C2T4	M241	F	Geo	3rd
	M142*	M	Geo	3rd		M242	F	Lng	1st
	M143	M	IT	3rd		M243	F	Med	1st
	M144	M	Math	2nd		M244*	M	Med	1st
					C2T5	M251	M	Phy	2nd
						M252*	M	Chem	2nd
						M253	M	Sci	2nd

C = Class; T = Team; TM = Team Member; G = Gender; Mjr = Major; Yr = Year
Bio = Biology; Chem = Chemistry; Egn = Engineering; Geo = Geophysics/Geochemistry/Geography; IT = Information Technology; Lng = Language; Math = Mathematics; Med = Medicine; Phy = Physics; Sci = Sciences
* team leader

4 Result and Discussion

4.1 RQ1: The Project with a Focus on the Team Dynamics

Team Composition and Communication Means. All the teams used WeChat group to communicate for teamwork. Some teams have video calls on Zoom or Tencent meeting. Most of the teams had discussions on the Group page of the Canvas platform, which the instructor can participate in, and the ranking together with the number of posts is shown in Table 3.

Table 3. Team topics and use of communication means

Team	Topic	Group video call	Canvas discussion 1	Canvas discussion 2
C1T1	Big data in COVID-19 countermeasures	Not mentioned	1/22	1/17
C1T2	COVID-19 Medicine	No	7/4	5/4
C1T3	Epidemiology of COVID-19	Yes	N/A	7/1
C1T4	Machine Translation	Tencent meeting	5/6	N/A
C2T1	The use of AI in Speech Recognition: From ASR to NSR	Tencent meeting	8/3	5/4
C2T2	Introduction and some technologies of blockchain	Tencent meeting	6/5	N/A
C2T3	How consciousness generates and connects to areas of brain?	Tencent meeting	3/7	4/5
C2T4	Applications of mathematical models in epidemic simulation and control	Probably	2/10	2/10
C2T5	Nucleic acid detection of COVID-19	Probably	4/6	3/7

Team Process. More than half of the students reported that their teams worked effectively toward the goal in the self assessment. Inevitably, there are students who mentioned issues in teamwork. Nevertheless, there are also compliments on excellent team leaders and members.

In terms of the team process, students have mentioned all the processes in the transition and action processes, as is shown in Table 4.

Table 4. Process dimensions and examples

Process dimensions	Examples
Transition processes	
Mission analysis formulation and planning	*At first, we communicated with each other about how to search for the right articles. Then, we divided the task into several parts and each of us chose our own job.* (M122)
Goal specification	*We had set goals at the beginning of the project and revised it once.* (M124*)
Strategy formulation	*We had set goals at the beginning of the project and revised it once.* (M124*)
Action processes	
Monitoring progress toward goals	*In order to stay in tune, we shared our progress in the group.* (M124*)
Systems monitoring	*How to use Adobe Premiere* (M124*)
Team monitoring and backup behavior	*M222 gave us opinions and some of the materials.* (M224)
Coordination	*We solved the problems in the process of making videos and tried to polish the synthesized video together.* (M122)

The interpersonal processes were addressed in some of the comments by students in the final review of the whole semester or the worries part of the self assessment of teamwork.

One of the team members from the middle ranking team remarks on the positive affective factors that promote team dynamics:

In team work I learned how to cooperate and when the team was in bottleneck, we should step forward and make breakthrough, which requires bravery and spirit of devotion. (M244)

On the other hand, the team leader of C1T2, a low-ranking team, describes the difficulties in leading a team in terms of "asking members to do exactly what you want or realize what's in your mind" and coordinating the members who have "different abilities" and set different priorities.

Collaborative Problem Solving (CPS). Students demonstrated different degrees of CPS skills. The leader of C1T2 (M124) serves as a good example of almost all the aspects of the 12 elements as shown in Table 1, which can been seen from the words of M122 and the fact that M124 adjusted their plans and managed an intricate division of labor between the team members.

Our group leader M124 shared the introduction part on the WeChat group as a good model for us. After that, we all finished our parts and put them on the WeChat. We solved the problems in the process of making videos and tried to polish the synthesized video together. In short, I think that our group were able to communicate very well with each other and this was a good experience of teamwork. (M122)

4.2 RQ2: Relationship Between Team Dynamics and Team Performance

Figure 1 provides a ranking of the team performance of the nine teams, by the average score and adjusted score given by the peer assessments, with the adjusted score resulting from removing one of the lowest scores to offset the effect of the randomness in scoring. It indicates that C2T4 ranks in the middle in both rankings. C2T5, C1T1, C1T2 and C2T3 are teams that perform well in both rankings; on the other hand, C1T3, C1T4, C2T1 and C2T2 are teams that did not perform so well in either ranking.

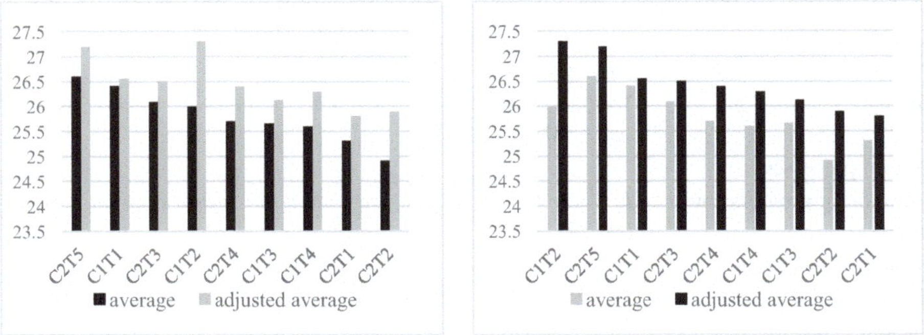

Fig. 1. Ranking of the team scores according to the average and to the adjusted average

First of all, excellent team performance is the result of hard work. One of the students from C2G5, a team of best performance stated that they "investigated (*sic*) more than 10 articles, including reviews, research articles and experimental protocols". A student from C2G4, relates that they "spent almost one month" to choose papers for their presentation. The overall team performance might be mediocre, but hard work compensated for the adverse factors.

Harmonious relationships among the team, at least most members of the team, is another crucial factor. For the well performing teams, although there may be minor issues, most of them enjoyed the collaboration and were satisfied with their product. Most of the teams that didn't perform very well had obvious communication problems. For C1T4, all the team members except the leader reported communication problems; for C2T2, the leader bemoaned the difficulties of teamwork.

The depth of topic is another contributor to the performance. C1T1 and C2T3 benefit from their attractive topic. On the other hand, although there is harmonious relationship among C1T3, their result is not very desirable, because all are freshmen and their discussion of the topic is a bit superficial. Aside from communication issues, C2T1 also suffers from a topic that has only one or two elements of the recent development.

Last but not least, participating in Canvas group discussions is also helpful, with the instructor's input being one of the facilitating factors. C1T1, the team with the most frequent discussions, is one of the best performing teams. A member from C2T3, another well performing team, relates that they "sincerely thank for our teacher's helpful suggestions". (M232)

4.3 RQ3: The Benefits and Challenges of the Virtual Project

More than half of the students mentioned the benefits of the virtual team project. In particular, most of the team leaders cherished the experience of leading the team.

The leader of C1T1 detailed the benefits of the project for scientific research, including reading and writing skills and the teamwork ability:

> *The ability to quickly get started with scientific research ... was mainly done through the Group Project, which allowed us to truly practice the process of identifying scientific research goals, researching literature, and reviewing industry progress. ... This training exercises our academic English reading and writing skills, and at the same time enhances our confidence in future scientific research (will not be frightened by a large number of English content full of professional terminology), broaden our academic horizons, improve teamwork ability, etc. In short, this is a good training, which will be of real help to the future scientific research process.* (C1T1)

A leader of C1T3, emphasized the importance of teamwork for future life and career:

> *In addition, this course is the most in-depth and demanding course for teamwork among all the courses I have studied in the year I came to university. ... I believe that whether it is in my future study life or work life, the adaptation to teamwork and the training of language skills are equally important.* (C1T3)

Nevertheless, the teams also encountered challenges, which were mostly addressed in the question of worries in the self-assessment. The greatest challenge the students reported is the time limit. There is mentioning of this factor in all the teams except C1T3. In C2T1 where one member didn't submit his part in time, three other members worried about the deadline. Next to it is the cohesion between the individual contributions. Members from four teams related their worries about the connection, combination or logical transition between their parts, including three well-performing teams and C1T3. There are also several students who worry about the quality, relevance or depth of their content.

In addition to the above top three worries, some students also mention barriers in communication, particularly those caused by the lack of face-to-face communication means in the virtual collaboration, including difficulty in understanding each other and delay in reply.

5 Conclusion and Implications

This study reveals how the students conducted teamwork in a virtual environment to produce a video introduction to a topic in frontiers of science and technology with original research paper as sources. Students used WeChat groups, video conferencing, and group page on the LMS system Canvas to work collaboratively. Successful projects were the result of hard work, harmonious team process, depth of topic, and input from the instructor. Students benefited from the project not only in terms of gaining experience in research but by improving their ability to be a team player, a crucial 21st century skill. Nevertheless, there are challenges such as the time limit, the quality of the product and the cohesion between the team member's parts of the final product.

To enhance the benefits of such projects, the instructors could provide explicit training for research and collaboration before project, set clearer timeline of the team process and intervene more actively in the team process. As teamwork and project-based learning is emphasized in the 21st century, more research is expected in this promising area. Future research could involve collecting more process data, e.g., discussions on WeChat, videoconferencing recordings, if any, and more inquiry into theoretical, pedagogical and technological issues.

References

1. Rogers, M.A.P., Cross, D.I., Gresalfi, M.S., Trauth-Nare, A.E., Buck, G.A.: First year implementation of a project-based learning approach: the need for addressing teachers' orientations in the era of reform. Int. J. Sci. Math. Educ. **9**, 893–917 (2011)
2. Schneider, R.M., Krajcik, J., Marx, R.W., Soloway, E.: Performance of students in project-based science classrooms on a national measure of science achievement. J. Res. Sci. Teach. **39**, 410–422 (2002)
3. Han, S., Capraro, R., Capraro, M.M.: How science, technology, engineering, and mathematics (STEM) project-based learning (PBL) affects high, middle, and low achievers differently: the impact of student factors on achievement. Int. J. Sci. Math. Educ. **13**, 1089–1113 (2015)

4. Holmes, V.-L., Hwang, Y.: Exploring the effects of project-based learning in secondary mathematics education. J. Educ. Res. **109**, 449–463 (2016)
5. Domínguez, C., Jaime, A.: Database design learning: a project-based approach organized through a course management system. Comput. Educ. **55**, 1312–1320 (2010)
6. Mioduser, D., Betzer, N.: The contribution of project-based-learning to high-achievers' acquisition of technological knowledge and skills. Int. J. Technol. Des. Educ. **18**, 59–77 (2008)
7. Chang, L.-C., Lee, G.C.: A team-teaching model for practicing project-based learning in high school: collaboration between computer and subject teachers. Comput. Educ. **55**, 961–969 (2010)
8. Johnson, D.R., Renzulli, L., Bunch, J., Paino, M.: Everyday observations: developing a sociological perspective through a portfolio term project. Teach. Sociol. **41**, 314–321 (2013)
9. Cox, C.B., Montgomery, C.: A study of 21st century skills and engagement in a university Spanish foreign language classroom. Foreign Lang. Ann. **52**, 822–849 (2019)
10. https://www.pblworks.org/what-is-pbl/gold-standard-project-design
11. Aghayani, B., Hajmohammadi, E.: Project-based learning: promoting EFL learners'writing skills. LLT J.: J. Lang. Lang. Teach. **22**, 78–85 (2019)
12. Alotaibi, M.G.: The effect of project-based learning model on persuasive writing skills of Saudi EFL secondary school students. English Lang. Teach. **13**, 19–26 (2020)
13. Grant, Sean: Implementing project-based language teaching in an Asian context: a university EAP writing course case study from Macau. Asian-Pac. J. Second Foreign Lang. Educ. **2**(1), 1–13 (2017). https://doi.org/10.1186/s40862-017-0027-x
14. Amarullah, A., Rachmawaty, N.: Exploring students' experience in project-based drama learning and their perception on their speaking anxiety. In: 2nd Educational Sciences International Conference (ESIC 2019), pp. 37–39. Atlantis Press (2019)
15. Rohmahwati, P.: Project-based learning to raise students' speaking ability: its' effect and implementation (a mix method research in speaking II subject at STAIN Ponorogo). Kodifikasia **9**, 199–222 (2016)
16. Sirisrimangkorn, L.: The use of project-based learning focusing on drama to promote speaking skills of EFL learners. Adv. Lang. Liter. Stud. **9**, 14–20 (2018)
17. Poonpon, K.: Enhancing English skills through project-based learning. English Teach. **40**, 1–10 (2011)
18. Simpson, J.: Integrating project-based learning in an English language tourism classroom in a Thai university. Australian Catholic University, Sydney (2011)
19. Mikulec, E., Miller, P.C.: Using project-based instruction to meet foreign language standards. Clear. House: J. Educ. Strategies Issues Ideas **84**, 81–86 (2011)
20. Madoyan, L.: Authenticity and teacher's role in project based learning. Armen. Folia Anglistika Ereván **2**, 109–114 (2016)
21. Csikszentmihalyi, M.: Creativity: Flow and the Psychology of Invention and Discovery. HarperCollins, New York (1996)
22. Marks, M.A., Mathieu, J.E., Zaccaro, S.J.: A temporally based framework and taxonomy of team processes. Acad. Manag. Rev. **26**, 356–376 (2001)
23. Mayer, R., Wittrock, M.: Problem solving. In: Anderman, E., Winne, P., Alexander, P., Corno, L. (eds.) Handbook of Educational Psychology, pp. 287–304. Erlbaum, Mahwah (2006)
24. Griffin, P., Care, E.: The ATC21S method. In: Griffin, P., Care, E. (eds.) Assessment and Teaching of 21st Century Skills, pp. 3–33. Springer, Dordrecht, The Netherlands (2015)
25. OECD: PISA 2015 results (volume V): Collaborative problem solving. OECD, Paris (2017)
26. OECD: PISA 2015: Draft collaborative problem solving framework. OECD, Paris (2013)

Designing Rain Classroom-Based Blended Learning Model for English Public Speaking Course

Chenxi Pan[1,2], Xuan Min[1,2](✉), and Yuanyuan Mu[1,2]

[1] School of Foreign Studies, Hefei University of Technology, Hefei, China
11352079@qq.com
[2] Center for Translation Studies and Specialized Corpora, Hefei University of Technology, Hefei, China

Abstract. Various emerging technologies have transformed education in the past decades. Blended-learning combining the strengths of both traditional face-to-face teaching approaches and online learning activities enhances the effectiveness of teaching and learning. Taking English Public Speaking Course as a case in point, the paper introduces the design of a blended-learning model which adopts Rain Classroom, a smart teaching tool to facilitate effective teaching and learning. A blended learning model as such demonstrates a new approach to training public speaking skills by improving students' learning autonomy, enhancing interactions and fostering a learning community.

Keywords: Blended learning · Rain classroom · Bloom's taxonomy · English public speaking course

1 Introduction

Mastering public speaking skills is acknowledged as a core competence for highly-educated professionals. It has been integrated into the educational standards of several countries such as Germany or the United States. As an act of strategic communication, English Public speaking involves English as a working language-the language employed everyday by people in business, law, government, and education [12, 13]. In a world shaped by globalization, the need for EFL students to have effective skills of English public speaking (EPS) is also growing over years. Previous studies have analyzed how various pedagogical approaches and tools like video-based blogs [15], Ted talks [11], Virtual Reality [10] can be adopted in teaching English public speaking. But few researches explore the possibilities of adopting blended-learning to facilitate English Public Speaking learning.

Blended learning (BL), which integrates online and face-to-face teaching and learning activities [5, 7], is likely to become the main teaching model in the future. Over the decades, academic research findings have proven that the BL approach which combines the advantages of both traditional face-to-face teaching approaches and online learning activities, not only helps better achieve desired teaching objectives and learning outcome, but also improves efficiency of teaching and learning in various disciplines [5, 6, 14].

C. Pang et al. (Eds.): SETE 2020/ICWL 2020, LNCS 12511, pp. 454–460, 2021.
https://doi.org/10.1007/978-3-030-66906-5_43

To successfully implement BL in a particular course, one of the most crucial prerequisites is to design the optimal configuration of the classes which achieve a harmonious balance between online access to knowledge and face-to-face human interaction. However, this aspect of research remains under-investigated. Much of the literature concerning blended course design focuses on blending, the technological aspects rather than the learning aspects [1]. However, a mixture of technology and content does not necessarily guarantee effective learning. It is how objectives, content and activities are designed, how students are engaged, and how learners' performance is evaluated that yield the best learning outcomes [8].

Therefore, this research aims to design a systematic blended learning model for EPS course and investigates how Rain Classroom, a smart teaching tool can be integrated into the traditional classroom to achieve the best learning outcome.

2 Literature Review

2.1 Blended Learning in EFL Context

Different studies suggest that BL approach facilitates language teaching and learning in the context of teaching English as a foreign language (EFL). Empirical and comparative studies have explored and proved the effects of blended learning in promoting EFL learners' specific language proficiency [3], i.e. vocabulary enhancement [17], reading and grammar competence [1], speaking and writing [16]. The advantages of BL approach over traditional language learning environment also lie in its potential to generate interest in the learning process, foster learner motivation and a collaborative and engaging environment amongst language learners. By extending language learning beyond the class setting, BL can help develop autonomous language learning [3, 4] and provide tailored solutions to individuals' language learning needs. So compared with traditional learning environments, EFL courses in the BL environment can achieve the desired language outcomes more successfully.

2.2 Rain Classroom

Rain Classroom is a smart mobile teaching tool newly developed by Tsinghua University in 2016. It was designed as a mini-app on WeChat, the most popular instant messenger in China. Because of the ubiquitous nature of mobile technology, Rain Classroom effectively connects teachers with students in and outside classroom. In the pre-class and post-class stage, teachers can prepare, upload and send the preview and review materials including videos, audio, texts, PPTs to Rain Classroom platform and monitor students' self-learning progress. In the class, students can interact with teachers and participate in activities through posting comments in Bulletin Boards which will be synchronously presented to the teachers and finishing quizzes assigned by teachers in limited time. It facilitates students to learn by making the whole learning process full of participation and engagement, and enhances students' learning enthusiasm by providing real-time feedback and evaluation. [9]. For teachers, Rain Classroom automatically collects and analyzes students' learning statistics, thereby evaluating teaching outcomes and enabling teachers to adjust the teaching content timely.

3 Blended Learning Model for EPS Course

3.1 English Public Speaking Course

The EPS course was first introduced to Chinese tertiary education curriculum in the early 21st century. Since then, the popularity of EPS teaching and learning continue to grow in China. The purpose of the course is to develop ESL learners' English public speaking knowledge and skills. Learners are required to complete three main speech tasks: an introductory speech, an informative speech and a persuasive speech, around which the course is centered and organized. Through the stages of developing a speech from preparation to presentation, all assigned oral and written activities and collateral readings for each speech assignment are designed with an intention to enable the students to not only learn about and practice how to choose an appropriate speech topic, how to write thesis statements, how to collect materials, how to organize a speech, and how to deliver the speech effectively, but also to develop students' competence to manage speech anxiety, to analyze audience and to incorporate feedback effectively, to become a better listener, and most importantly to think critically. Constrained by time and technology limit, a typical traditional teaching pattern is direct instructions given by teachers, followed by speech writing and revising and finally speech presentation [19]. However, in order to achieve the above-mentioned objectives, the best teaching approach for English public speaking course is "experiential rather than contemplative, active rather than passive, personal rather than impersonal" [18].

3.2 Blended Learning Model Designing

Bloom's Taxonomy and Blended-Learning Model Objectives. Clearly defining course objectives are core to a successful blended learning model, since objectives inform approaches to teaching practice, methods of content delivery, and requisite amount and locations for class meetings and interactions. So, first and foremost learning objectives geared toward a traditional classroom teaching should be redesigned. Bloom's revised Taxonomy [2] arranges learning into six cognitive levels in order of hierarchy and complexity. It has been often viewed as a pyramid with the most basic cognition at the bottom and the higher-level cognitive skills on the top. This equates to the amount of time spent on cognition process in the traditional EPS classroom which is restricted by limited teaching time (once a week, a single 16-week semester), and a large class scale, with a lot of on remembering and understanding and little on analyzing, evaluating and creating, which are assigned for homework. The redesigned learning objectives and their corresponding cognitive activities are presented in Table 1. The blended model, shifting the learning of factual and conceptual knowledge about English Public Speaking from F2F classroom to Rain Classroom platform, enables the F2F classroom to dedicate enough time to practicing, presenting, analyzing, and evaluating so as to help students' internalization of the procedural and metacognitive knowledge.

Table 1. English public speaking learning objectives in blended-learning environment.

Learning objectives	Cognitive activities	Potential delivery approaches
Understand and explain communication process	Remembering and understanding	A self-paced e-learning module and in class discussion
Understand basic principles and techniques for the research, composition, organization and delivery of speeches	Remembering, understanding and analyzing	A self-paced e-learning module and case study +demonstration
Plan and prepare speeches that introduce, inform and persuade	Creating and applying	Demonstration and presentation
Use presentation aids to enhance speeches	Understanding and applying	A self-paced e-learning and presentation
Conduct meaningful research on a variety of topics	Understanding, analyzing and applying	In class discussion and case study
Analyze your audience and design speeches to reflect your analysis	Understanding and analyzing	Case study and demonstration
Evaluate speeches according to verbal and non-verbal criteria	Evaluating	Case study

Teaching Process. In line with the re-designed learning objectives and their corresponding potential delivery approaches, the teaching process is designed in this way as presented in Fig. 1.

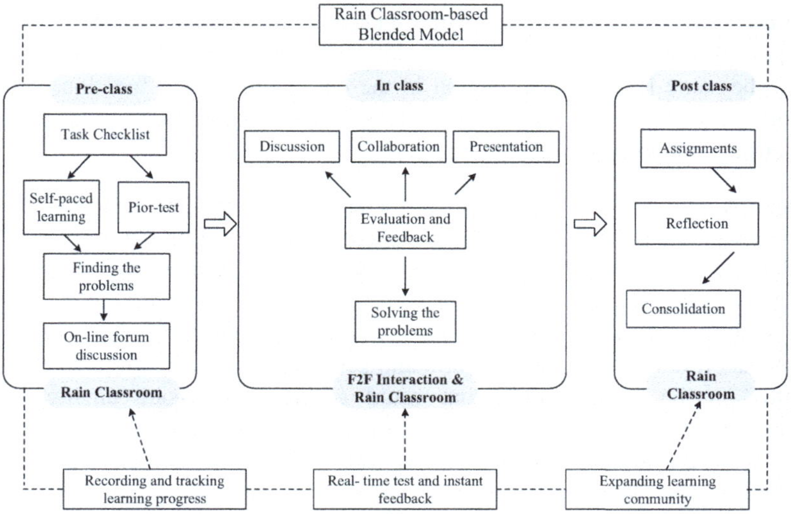

Fig. 1.

Pre-class: The teacher uploads high-quality previewing and teaching videos together with a task checklist on Rain Classroom. Assignments include watching videos, reading chapters in the assigned textbook *The Art of Public Speaking* and other specific tasks. Students are able to pause, rewind, and preview materials at their own pace. Learning the foundational knowledge by understanding and remembering about EPS skills at their own pace online allows for deeper learning to take place while in the classroom. Before class attendance, students are required to complete prior-tests on the Rain Classroom to assess their knowledge of the material. The prior-tests usually consist of multiple-choice, true/false, and short-answer questions, which are designed mainly for checking students' understanding of factual knowledge, the lower domains of Bloom's taxonomy. Through automated grading system, students can get an instant feedback about how they know about public speaking, types of speeches, strategies of delivery and etc. In this way, prior to the class, the students have mastered the factual or procedural knowledge and identified specific problems which could be discussed in the online forum and later in the F2F classroom. The dynamic tracking of academic performance within the Rain Classroom platform can also help teachers timely obtain learners' learning behaviors and demands.

In-class: In-class sessions are used for application and synthesis of knowledge. One of the major problems in English Public Speaking learning lie in the acquisition of conceptual knowledge which require critical thinking abilities, for example differentiating facts from opinions, the strategies to persuade, fallacies in persuasion and etc. By showing and discussing samples persuasive speeches with commentary, teachers would help students build the competence of analyzing and evaluating. Apart from discussion, much of the in-class time will be allocated to presentations of the three speech assignments followed by peer feedback and teachers' feedback. Traditional EPS course loaded with lectures provide less opportunity for feedback and instruction. Blended-learning model makes instant and robust commentaries possible, so that students have a chance to evaluate speeches and apply the evaluation criteria. The F2F interaction between teachers and students and student participation are strengthened by the support of Rain Classroom. Pre-class activities do not necessarily promote students to make the same level of contribution in the classroom, so the function of posting comments (anonymously) on the Rain Classroom platform that can be projected to PPT can level participation and interaction.

Post-class: After the class, students can digest and further strengthen learned content by applying English public speaking skills learned online and offline in accomplishing a particular speech task. By doing research on speech topics, writing speech manuscript, recording speech rehearsal, recognizing speech anxiety and figuring out strategies for coping with the anxiety, students create, apply and reflect. Students can submit assignments on Rain Classroom platform and get instant feedbacks from teachers, post comments on the online forum that could generate discussion among learners. Extending the interaction from the classroom to the online platform, Rain Classroom helps to build a community-supported learning environment which is conducive to enhancing students' learning engagement and interaction.

Evaluation. Each English public speech is generally evaluated in terms of matter, method and manner. However, the evaluation method for English Public Speaking course is mainly formative which means the whole process of preparing and presenting three speeches is assessed. Students also need to complete a peer evaluation that contributes to their final mark. Rain Classroom facilitates formative evaluation by recording students' performance in various prior tests and post tests and providing statistics that reflect students' interaction frequency and learning engagement etc.

4 Conclusion

The advent and proliferation of digital technologies like PC, the Internet, and mobile devices have been transforming approaches to teaching and learning in an unprecedented way. In this paper, we elaborate on the combination of Rain Classroom with the traditional F2F interaction to facilitate blended learning and the application of the model in the English Public Speaking Course. Thoughtfully designed, the integration between online activities conducted on Rain Classroom platform and F2F courses can promote students' participation and engagement in in-class activities; promote their understanding of challenging course concepts; and foster a sense of community within the course.

Acknowledgements. The work was substantially supported by Anhui Provincial Education Commission (Project No. SK2017A0049) and The National Social Science Fund of China (Project No. 19BYY125).

References

1. Alammary, A., Sheard, J., Carbone, A.: Blended learning in higher education: three different design approaches. Australas. J. Educ. Technol. **30**(4), 440–454 (2014)
2. Anderson, L.W., et al. (eds.): A Taxonomy for Learning, Teaching, and Assessing: A Revision of Bloom's Taxonomy of Educational Objectives. Pearson Education Group, Boston (2001)
3. Banditvilai, C.: Enhancing students' language skills through blended learning. Electron. J. e-Learn. **14**(3), 220–229 (2016)
4. Benson, P., Voller, P.: Autonomy and Independence in Language Learning. Routledge, London and New York (2014)
5. Garrison, D.R., Kanuka, H.: Blended learning: uncovering its transformative potential in higher education. Internet High. Educ. **7**, 95–105 (2004)
6. Garrison, D.R., Vaughan, N.D.: Blended Learning in Higher Education: Framework, Principles, and Guidelines. Josey-Bass, San Francisco (2007)
7. Karkour, I.: A blended learning model for teaching reading in English as a Foreign Language. Teach. Engl. Technol. **14**(4), 17–31 (2014)
8. Kozma, R.B.: Counterpoint theory of 'learning with media'. In: Clark, R.E. (eds.) Learning from Media: Arguments, Analysis, and Evidence, pp. 137–138. Information Age Publishing Inc., Greenwich (2001)
9. Li, X.M., Song, S.Q.: Mobile technology affordance and its social implications: a case of "Rain Classroom". Br. J. Edu. Technol. **49**(2), 276–291 (2017)

10. Li, Y.: An analysis of the principle of applying Virtual Reality to the professional training in English public speeches. Foreign Lang. China **15**(6), 73–78 (2018)
11. Li, Y., Gao, Y., Zhang, D.: To speak like a TED speaker: a case study of TED motivated English public speaking study in EFL teaching. High. Educ. Stud. **6**(1), 53–59 (2016)
12. Lucas, S.E.: The role of public speaking in China's English language curriculum. In: Wang, L. (eds.) English Public Speaking in Global Context: Challenges and Innovations, pp. 3–24. Foreign Language Teaching and Research Press, Beijing (2009)
13. Lucas, S.E.: English public speaking and the cultivation of talents for Chinese college students. Chin. J. Appl. Linguist. **36**(2), 163–182 (2013)
14. Picciano, A.G.: Blending with purpose: the multimodal model. J. Asynchronous Learn. Netw. **13**(1), 7–18 (2009)
15. Shih, R.: Blended learning using video-based blogs: public speaking for English as a second language students. Australas. J. Educ. Technol. **26**(6), 883–897 (2010)
16. Tomlinson, B., Whittaker, C.: Blended Learning in English Language Teaching: Course Design and Implementation. British Council, London (2013)
17. Tosun, S.: The effects of blended learning on EFL students' vocabulary enhancement. Procedia – Soc. Behav. Sci. **199**, 641–647 (2016)
18. Wan, J.B.: Localization of the English public speaking course in China's EFL curriculum. Asian Soc. Sci. **9**, 94–104 (2013)
19. Zhang, X., Ardasheva, Y.: Sources of college EFL learners' self-efficacy in the English public speaking domain. Engl. Specif. Purp. **53**, 47–59 (2019)

Computer Assisted Student Interpreters' Self-assessment: Ways and Inspiration

Menglian Liu(✉)

Laboratory of Language and Artificial Intelligence,
Guangdong University of Foreign Studies, Guangzhou 510420, China
lmenglian@gdufs.edu.cn

Abstract. This article explores self-assessment of student interpreters by using a CAIT online platform. The aim of this study is to investigate the effective supporting ways of online platform in the development of self-assessment skills for student interpreters. The results indicate that the conceptual, procedural and tactical ways take an active role of improving students' self-assessment skills. While practicing interpreting, these methods can help student interpreters better understand interpreting standards and be aware of them so as to improve their self-assessment competency. The study explores that the student self-assessment development process with the CAIT platform is from a way of disoriented to relieved, and then to natural. Results also show that teacher's feedback on students' self-assessment takes an important role in guiding students' training and help them keep training. The findings show some inspirations for both teachers and learning designers. Firstly, conceptual, procedural and strategic supporting ways can be put into E-learning platform to improve students' self-assessment ability of interpreting. But they should be well designed into different places. Secondly, some positive ways should be placed at the beginning of CAIT E-learning self-assessment. Thirdly, teacher should guide students' self-assessment in this online interpreting practice and give feedback to students' self-assessment so as to motivate their self-assessment performance.

Keywords: Computer assisted interpreter training (CAIT) · Self-assessment · E-learning platform · Supporting methods

1 Introduction

Computer assisted interpreter training (CAIT) is a relatively new field in interpreting studies. It explores the implementation of information and communication technologies (ICT) in the training of interpreters. The international research on the application of CAIT in the development of self-assessment skills has focused mainly on the effectiveness of its implementation as a tool in the training of student interpreters. It is rare to find studies on the methods of how to design and develop CAIT E-learning platform effectively with considering self-assessment competency development for the training of student interpreters. This paper is to explore the effective ways of designing and developing CAIT E-learning platform for student's self-assessment in interpreting training and the inspirations for the implementation of them.

C. Pang et al. (Eds.): SETE 2020/ICWL 2020, LNCS 12511, pp. 461–471, 2021.
https://doi.org/10.1007/978-3-030-66906-5_44

2 Theoretical Background

2.1 CAIT and Self-assessment

Sandrelli & de Manuel Jerez [1] indicate that since the 1990s several independent information and communication technology-based projects were undertaken that shaped the gradual development of what has come to be known as CAIT(Computer assisted interpreter training). The interest of using CAIT in interpreter training is from the belief that part of the solution may lie in the more widespread use of appropriate technology, and also an attempt to make interpreter training more in line with market requirements [2].

Regehr et al. [3] define self-assessment as the ability of each individual to identify his or her own relative strengths and weaknesses. Within interpreting training, assessment and self-assessment procedures play an important role for interpreters by encouraging self-regulation processes in interpreting trainees [4]. Lee [5] mentioned that "self-assessment is not only important during the training phase of interpretation, but it is critical to professional interpreters as well." Sandrelli & de Manuel Jerez [1] has also stated that "self-assessment skills…are essential for trainees, both to ensure progress and to maintain quality standards in their future careers as professional interpreters".

The development of information and communication technology (ICT) has promoted positive developments in interpreting training and self-assessment projects. Witter-Merithew [6] studied the "Education Interpretation Certificate Course (EICP)" distance learning program and found that guiding students to conduct effective self-assessment is the key to successful completion of EICP learning of remote students. Bartłomiejczyk [7] indicates that self-evaluation by means of critically listening to one's own recorded interpreting has often been suggested as a useful method of quality control. Berber [8] concluded that ICTs in general support the efforts presented in the Gile's Effort Model and that information technology in the form of interpreter training tools are specifically aimed at the second effort the Effort Model, where the student can "listen to him/herself repeatedly for self-evaluation and improvement of production skills". Deysel [9] studied that CAIT may prove a viable tool also for in-house training and development of self-assessment skills of professional interpreters.

The aforementioned studies led to insights that the implementation of CAIT in the training of interpreters may be desirable and an appropriate addition to traditional training methods.

2.2 Psychological Methods

Psychologists think that self-assessment is both an important component process and a manifestation of self-learning ability [10]. Self-assessment can provide a self-regulated system and motivation for behavior. It's a unique ability of humans to regulate one's own behavior through self-assessment [11].

Educational psychologists concluded some ways which can promote students' self-assessment as the followings [12]. Firstly, let students set specific behavioral goals before learning and have them check their accomplishments against these goals. Secondly, Show the specific criteria to students so that they can evaluate their learning performance. Thirdly, in some cases, teacher can delay giving feedback to students so as to give them opportunities to evaluate their own learning. Fourthly, encourage students to make realistic assessments of their own learning. When students' self-assessment match teachers' assessment, it can reinforce the student's learning.

2.3 Cognitive Science Model

From cognitive science, three different types of tools are identified to support a cognitive process which are conceptual, procedural and strategic tools [13]. Each type of tool supports a different type of knowledge involved in a process: concepts, procedures and rules or strategies. Brisebois et al. [14] constructed a "competency self-management process model" which based on cognitive science. Three main steps have been identified in this model. Firstly, the learner must describe his/her actual competencies. Secondly, from this self-assessment evaluation, results bring out the learner's strengths and weaknesses compared to entry and target competencies as prescribed by the course designer. Thirdly, the learner identifies within a list of learning activities, those who best suits him/her.

3 Methodology

3.1 Research Questions

This article is to study the effectiveness of the supporting ways of conceptual, strategic, and procedural tools of a designed CAIT E-learning platform. It also wants to get some inspirations of how to help student interpreters improve their self-assessment competency and also what should teachers do to guide their students' self-assessment while practicing both consecutive and simultaneous interpreting. The following research questions were explored:

- Do the supporting conceptual, strategic and procedural ways of the CAIT E-learning platform effective in improving students' self-assessment competency? And to what extend?
- What is the development process of students' E-learning platform-based self-assessment? What is the inspiration?
- What is the teacher's role for this kind of self-training way? And what's the inspiration?

Based on this, this study carried out experiment and survey. The study takes it as a goal of improving student interpreters' self-assessment competency. It also takes the use of ICT as a method of implementation. The aim is to explore the effective ways to support student interpreters' self-assessment by using the E-learning platform. The focus is how to embed the conceptual, procedural and strategic ways into the platform to make sure that it can help student interpreters practice self-assessment efficiently after interpreting self-training.

3.2 Participants

The research participants are the students of two classes whose major are French. The number is 54. They are junior university students who have French interpreting course for a-year study.

3.3 Research Design

Interpreting training E-learning platform develop with the conceptual, strategic and procedural self-assessment supporting ways. Design and develop the interpreting training E-learning platform. The E-learning self-assessment supporting designing framework is as the showing of Fig. 1.

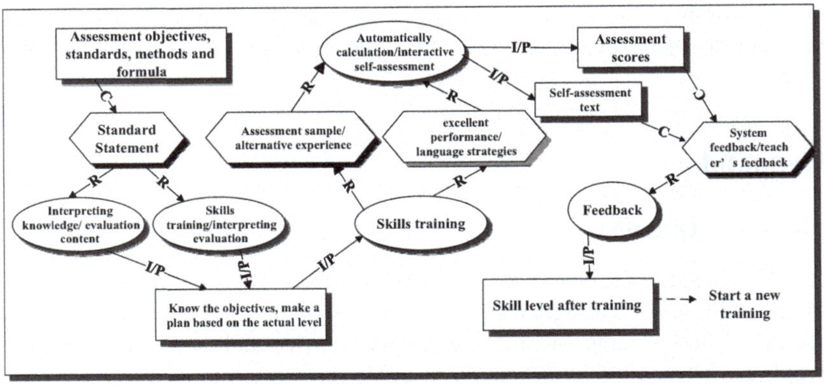

Fig. 1. E-learning platform Self-assessment design framework. (Procedures are identified with ovals; concepts with rectangles and strategies with hexagons. Arrows show links between knowledge objects and give the direction of the flow of information. Letters associated to arrows specify the nature of the link: C for composition, R for regulation, and I/P for input/output.)

Assessment supporting ways have been set up in many places of the interpreting E-learning platform.

Conceptual Supporting Ways Setting. In the "Basic knowledge of interpreting" section, put interpreting quality evaluation information including objectives, criteria, methods, evaluation formulas, etc. When students open the pages of this section, they can clearly see the assessment information and try to remember and use them during interpreting training.

Conceptual and Strategic Supporting Ways Setting. Design scaffolding training activities. During each interpreting skill scaffolding training activity, show assessment criteria of each interpreting skill, such as the formula, along with assessment sample texts of relevant skills. By using this method, it tries to help student interpreters deep understanding the assessment criteria of each interpreting skill and make clear predictions about their learning goals. Also for strategic supporting ways, set "Icon Area" section and upload many wonderful videos of live interpreting by many expert interpreters.

Procedural Supporting Ways Setting. Put the assessment formula into the interpreting self-assessment procedures so that self-assessment scores can be automatically calculated. At the same time, design and develop an interactive text self-assessment module which can support students check and reflect on their interpreting performance so as to make a realistic self-assessment. The ways of self-assessment in the study include both scoring and text assessment.

Assessment Criteria. The research objects of this study are university students. The main form of assessment for this kind of students are mainly performance and diagnostic assessment. The purpose of this assessment is to check the application of the educational program and assess the students' academic performance. Assessments are used to find out what mistakes students make and how to remedy them in order to improve their self-assessment competency and interpreting skills.

The assessment criteria are divided into five main categories of Fidelity of information, Accuracy of language delivery, Fluency of interpreting (time variable), Flexibility and Effectiveness of communication strategy use, and Satisfaction. The parameters of these categories will change according to different skill items. By assigning certain weights to the parameters, a formula for assessing can be derived. The formula is:

$$\text{Total score} = \text{Information} * V\% + \text{Language} * W\% + \text{Fluency} * X\% + \text{Strategy} * Y\% \\ + \text{Satisfaction} * Z\% \ (V, W, X, Y, \text{and } Z \text{ represent different percentages})$$

Questionnaire Design. Design a questionnaire. The questionnaire was divided into different sections, all of which aimed to collect data on:

- the participants' perceived knowledge about the effectiveness of the self-assessment support ways
- the perceived knowledge of the participants as pertaining to their self-assessment activities and their awareness of his/her strengths and weaknesses in interpreting performance.

3.4 Procedures

Self-assessment Practice. Practical study is carried out. Students conduct self-training and self-assessment by using this CAIT E-learning platform. The interpreting training records and self-assessment scores and text are automatically stored in the database. Teachers can log in and see the self-assessment and learning results. Teachers can also give feedback to every self-assessment of each student.

Data Collection. Check all the learning and assessment record data in the database by the classification of different requirements, such as "interpretation skills", "student number", "exercise questions", etc. Import the various types of data respectively into excel sheets for further data analysis.

Survey. Distributed 54 questionnaires and received 53 valid ones. The questionnaire return rate was 98.1%.

4 Results and Discussion

4.1 The Effectiveness of Conceptual, Strategic and Procedural Supporting Ways of the CAIT E-learning Platform is Great in Helping Student Interpreters Improving Self-assessment Competency

For the question "The value of interpreters' self-assessment supporting ways in the interpreting E-learning platform in improving students' interpreters' self-assessment skills", 58.4% of the students thinks the effect was great, 13.2% thinks it was very large. The result indicates that the self-assessment supporting methods of computer-assisted interpreter self-assessment used in the study are effective in improving students' self-assessment ability of interpretation.

The final exam results comparison also shows this point. The following Table 1 reflects the comparison of the final exam scores of the first-semester interpreting courses between the experiment class and the two comparison classes. The results show that the scores of the students in the experiment class are generally higher than those of the two comparison classes.

Table 1. Table of final exam comparison.

Proportion of achievements (%) (1st Semester)						
Class	Rate	Excellent	Good	General	Pass	Failed
Practice class	100	5.6	53	34	7.4	0
Comparison 1	95.7	3	30.4	48	17.3	4.3
Comparison 2	94	1	26	40	24	6

The Conceptual Supporting Ways Can Help Students Better Understand the Assessment Criteria and Realize the Distance Between Themselves and the Standards. The conceptual ways are most useful in preparing students for self-assessment and in understanding the interpreting assessment operation.

Conceptual ways are set in two places. The first place is in the "Interpreting Knowledge" column. Here place the interpreting training goals, standards, methods and assessment formula. The second place is in the interpreting skills training process. For the role of these two conceptual ways, the questionnaire shows that 56.8% of the students believe the role of these ways is great. For the details, the results show that the No. 1 was most useful in helping students prepare for self-assessment (71.3%),

Table 2. Table of the role of conceptual ways for self-assessment

Items	Effect of improving students' self-assessment competency
Conceptual knowledge of interpreting assessment in Interpreting Knowledge Column	Establish assessment awareness <Understand assessment knowledge <Prepare for self-assessment
Conceptual knowledge such as assessment criteria is presented in advance in skill training	Understand the assessment of specific skills <Enhance assessment consciousness <Deeply understand the specific operation steps of interpreting assessment

No. 2 has the greatest effect in deeply understand the specific operation steps of interpreting assessment (69.7%). Table 2 shows the results.

The Strategic Ways Can Help Students Obtain Alternative Experiences and Improve Their Interpreting Self-assessment Abilities. The strategic ways provided by the E-learning platform is mainly in the following two areas. No. 1 area is in the scaffolding training process, sample texts of assessment cases are provided. No. 2 area is to place famous interpreters' live interpreting videos and opinions on interpreting in the "Icon Area" section. The survey results show that 70.2% believe the sample text of interpreting in No. 1 can help students to further understand the evaluation criteria for interpreting. By reading the assessment sample texts, students can gain alternative experiences and further understand the specific operation steps of interpreting assessment, thus improving their interpreting ability. 52.5% think that expert interpreting videos and related opinions in No. 2 place can help students obtain alternative experience, and also help students understand the distance between themselves and the interpreting Icons, so as to make a correct assessment of themselves.

The Procedural Ways Can Help Students Understand the Rules of Interpreting Assessment Standards By Participating Every Step of the Assessment Procedure. The procedural ways are mainly embodied in: No. 1. Students assign scores at the corresponding assessment parameter indicators according to the interpreting quality assessment formula. No. 2. students enter their own assessment text in the self-assessment interactive fill-in box with the base of their own interpreting performance. The survey shows that 88.6% of the students think that the procedural ways of No. 1 1 helps them directly evaluate their interpreting performance according to the standard, which not only reinforce their understanding of the evaluation criteria, but also experience the concrete operation of self-assessment and improve their self-assessment skills quickly. 91.2% think that Procedural ways on No. 2 can help students reflect on their interpreting performance by reference to the assessment standard as also as to improve their self-assessment abilities.

4.2 The Student Self-assessment Performance Process Through this CAIT E-learning Platform is from a Way of Stunned to Relieved, and then to Natural

Export the learning record form named "Student ID" from the system's database to understand the students' learning and self-assessment. The self-rating scores reflect that when the students start a certain skill training, the self-scores are not optimistic due to the tension, etc., but as the training progresses, the self-scores of the students gradually increase. This aspect indicates that students are more and more satisfied with their interpretation training performance, and they are more and more confident in their ability to interpret. On the other hand, they also show that students' self-assessment ability of interpretation is gradually increasing. The student's self-assessment text is a good testimony to this view. If you evaluate yourself at the beginning, you are so nervous that the information you hear is forgotten when you interpret it; the information can be translated, but the fluency of the expression is not good, and some proper nouns or terms are unfamiliar. They should be well prepared for translation so as to grasp the translation information. This shows that students' self-assessment ability is constantly improving.

4.3 Teachers' Feedback on Students' Self-assessment Can Actively Promote Student to Develop Self-assessment Ability

The study compares the number of questions, self-assessment scores, self-assessment content, and final exam scores of students whose teachers gave evaluative feedback with those who did not give feedback. Research found that students whose teachers gave self-assessment feedback were higher (better) than those who did not get feedback. As shown in Table 3, the number of students who received feedback was almost double the number of students who did not receive feedback; the average score of students who received feedback was 12 points higher than the average of those who did not respond.

In terms of self-evaluation, students who received feedback from teachers were more adept at evaluating their interpreting performance holistically in terms of assessment parameters with the base of teachers' feedback. This confirms that teacher's feedback on students' self-assessment can positively contribute to students' self-assessment competency, which in turn motivates students to learn and improves the effectiveness of training.

Table 3. Comparison of learning performance between have-feedback and non-feedback

Items	Students with teacher's feedback	Students without teacher's feedback
Training Number (average)	42	23
Self-assessment score (average)	83	70
Self-assessment content	Good at comprehensive assessment of interpreting from the perspectives of standard evaluation criteria	Evaluation is single, one-sided, lack of professional perspectives
Final score (average)	90	78

5 Inspiration

Conceptual, procedural and strategic supporting ways can be put into E-learning platform to improve students' self-assessment ability of interpreting. But they should be well designed into different places. Only two levels of headings should be numbered. Lower level headings remain unnumbered; they are formatted as run-in headings. Each type of ways supports a different type of knowledge involved in a process such as concepts, procedures and rules or strategies. Specific supporting ways are set in different places of the E-learning platform according to different learning purpose. The most important in designing CAIT E-learning experience is to understand the links between knowledge objects and the direction of the flow of information.

More positive ways should be placed at the beginning of CAIT E-learning self-training and self-assessment. When students conduct self-assessment through the E-learning platform, the self-assessment competency process experience a process of stunned → relieved → natural. These means many positive self-learning and self-assessment ways should be carried out to help students fell comfort and familiar so that they will not be stunned at the beginning of self-learning. Special ways also should be put into the E-learning platform to guide and motivate students' CAIT E-learning training.

Teacher's role in CAIT E-learning is important. Teachers should take this role and guide students' self-training and interpreting self-assessment. Teacher should also give feedback to students' self-assessment so as to motivate students' self-training and self-assessment. The teacher's tracking and feedback on students' self-assessment is the direction to guide students' training and the motivation to promote students' continuous training and self-assessment.

This study discovers the ways and inspirations of CAIT's self-assessment in supporting student interpreters' training. The use of CAIT tools to support student interpreting training and promote students' interpreting skills as well as improving their self-assessment ability is an interdisciplinary and complicated issue. Text analysis study of students' self-assessment will be made in the future with the base of this study.

Acknowledgements. This work is financially supported by the National Key Project of Social Sciences (18AYY004) the Research Project of the Translation Studies Center of Guangdong University of Foreign Studies (CTS201717B).

References

1. Sandrelli, A., Jerez, J.D.M.: The impact of information and communication technology on interpreter training: state-of-the-art and future prospects. Interpret. Transl. Trainer **1**(2), 269–303 (2007)
2. Sandrelli, A.: New technologies in interpreter training: CAIT. Textologie und translation (translation and textology), pp. 261–294. gnv, Tubingen (2003)
3. Regehr, G., Hodges, B., Tiberius, R., Lefchy, J.: Measuring self-assessment skills: an innovative relative ranking model. Acad. Med. **71**(10), S52–S74 (1996)
4. Arumí Ribas, M., Esteve Ruescas, O.: Using instruments aimed at self-regulation in the consecutive interpreting classroom: two case studies. Electron. J. Foreign Lang. Teach. **3**(2), 158–89 (2006)
5. Lee, Y.H.: Self-assessment as an autonomous learning tool in an interpretation classroom. Meta: journal des traducteurs/Meta: Translators' J. **50**(4) (2005)
6. Witter-Merithew, A., Taylor, M., Johnson, L.: Guided self assessment and professional development planning: a model applied to interpreters in educational settings. In: Nettles, C. (ed.) Tapestry of Our Worlds: Proceedings of the 17th National Conference of the Registry of Interpreters for the Deaf, pp. 153–226. RID Publishing, Silver Spring (2001)
7. Bartłomiejczyk, M.: Interpreting quality as perceived by trainee interpreters. Interpret. Transl. Trainer **1**(2), 247–267 (2007)
8. Berber, D.-C.: The use of pedagogical and non-pedagogical ICT in conference interpreter training. In: Pellat, V., Griffiths, K., Shao-Chuan, W. (eds.) Teaching and Testing Interpreting and Translating, pp. 229–249. Peter Lang, Bern (2010)
9. Deysel, E., Lesch, H.: Experimenting with computer-assisted interpreter training tools for the development of self-assessment skills: national Parliament of RSA. Interpret. Technol. **11**, 61 (2018)
10. Pang, W.: Self-regulated Learning-Principles and Strategies of Teaching and Learning. East China Normal University Press, Shanghai (2003)
11. Bandura, A.: Social cognitive theory: an agentic perspective. Annu. Rev. Psychol. **52**(1), 1–26 (2001)
12. Paquette, G.: Designing virtual learning centers. In: Adelsberger, H., Collis, B., Pawlowski, J. (eds.) Handbook on Information Technologies for Education & Training, within the Series International Handbook on Information Systems, pp. 249–271. Springer, Heidelberg (2001). https://doi.org/10.1007/978-3-662-07682-8_16

13. Brisebois, A., Ruelland, D., Paquette, G.: Supporting self-assessment in a competency approach to learning. In: E-learn: World Conference on E-learning in Corporate, Government, Healthcare, and Higher Education. Association for the Advancement of Computing in Education (AACE), pp. 2828–2835 (2005)
14. Cai, X.: Research on the performance of interpreting interactive teaching model. Chin. Transl. **4**, 4–7 (2008)

Teacher and Technology in Foreign Language Education During the COVID-19 Pandemic: A Perspective from the Philosophy of Technology

Lin Cheng[1] and Yuan Yang[2](✉)

[1] Guangdong University of Foreign Studies, Guangzhou 510420, Guangdong, China
[2] Huazhong University of Science and Technology, Wuhan 430074, Hubei, China
yuan_yang@live.de

Abstract. During the COVID-19 Pandemic, online learning became a contingency plan for the abrupt school closure. A whole range of problems arose due to the lack of preparedness for "online learning for all". This paper focuses on the problems of online language teaching and learning and attempts to address the issues of the relationship between teacher and technology, and the potential of technology, by adopting new ideas from the philosophy of technology in recent years, such as "being rejected by technology" and "the mediating role of technology". Only through collaboration with technology can teachers become complete actants and agents in online teaching. In order to fully explore the potential of technology in online learning, the positive mediating role of technology needs to be kept in mind. Rather than the conventional theory of "technology as tool", "human-machine coupling" is a better attitude for us to adopt.

Keywords: Online teaching and learning · Philosophy of technology · Technology-mediated · Rejected by technology

1 Introduction

During the COVID-19 pandemic, online teaching and learning became the mainstream and the norm in many countries and regions. And many teachers have been caught up in it. In this period, its strengths were highlighted, and its problems also magnified. Compared with other subjects, foreign language teaching and learning strongly demand the "presence" of both teachers and students, i.e., verbal and eye contacts, exchanges of facial expressions between teachers and students. Educational technology should help us meet these demands. The Dutch philosopher of technology Peter-Paul Verbeek states that, "in our times of rapid technological development, doing ethics of technology has become a much needed yet also challenging activity. This is especially true in China, where the pace of technological innovation has been very fast over the past decades and will remain so in the future. How to deal with technology innovations and

C. Pang et al. (Eds.): SETE 2020/ICWL 2020, LNCS 12511, pp. 472–477, 2021.
https://doi.org/10.1007/978-3-030-66906-5_45

their impacts on society in a responsible way" [1]? During the pandemic, China succeeded in launching online teaching within an extremely short time period, which fitted the situation at the time, but it also urged us to think over the issues that Verbeek raised. To answer the question of how teacher and technology influence each other and how the two can better work together is particularly important for the study of foreign language teaching.

In the first section of this paper, we present the impact of online teaching technology on the teacher from the perspective of philosophy of technology; in the second section, we raise the awareness of the phenomenon of being "rejected" by technology through showing examples and data during the pandemic; in the third section, we attempt to show the positive potential of online learning technology, and last but not least, we put forward our suggestions on the research and practice for online language teaching and learning.

2 Teacher "in Relation to Technology"

In 1990, the American philosopher Don Ihde described that technology is everywhere in our daily life at the beginning of his book *Technology and the Lifeworld: From Garden to Earth.* Thirty years later, technology has become more pervasive in our lives and has entered the field of education. However, its products are often merely recognized as tools. It is important to note that what is being referred to here is not conscious AI technology, but rather educational technology that is commonly used by everyone. Online courses during the pandemic have transformed the conventional "teacher-student" bilateral relationship into a "teacher-technology-student" trilateral relationship. This means that technology is not just a simple tool anymore; it has become an indispensable participant in the teaching and learning process, even if such technology is not autonomous.

As educators often say, the roles and functions of teachers are changing in modern concepts of education. But, given the impact of technology in shaping our teaching and learning activities, we would like to offer another perspective: with the involvement of technology, the teacher's role and function are not the only things that are changing in the developing process from "teacher-student" to "teacher-technology-student", but also the teacher himself or herself. Technologies are not "mere things that, like inert matter, do nothing in themselves", in fact, they "affect the very ways we act, perceive, and understand" [2]. Or as Verbeek states, "technologies cannot be seen as fully fledged moral agents [...] neither can humans, if we do not take into account how technologies help to shape human interpretations, practices and decisions" [1]. Indeed, technology shapes humans constantly. Take the relationship between daily users and smartphones for example, the smartphone is shaping the user constantly. Together the user and the smartphone form a "unity" which is still changing. However, will the user whose smartphone suddenly is taken away still be a complete "him" or "her"? Clearly, technology participates in the construction of our subjectivity and has become a part of our existence. But this does not mean that we are "hapless victims of technology", nor does it mean that we should try to "escape from its influence". On the contrary, we must "shape our existence in relation to technology" and "learn to live with it" [3].

The massive teaching experience gained during the pandemic reminds us that it is high time that we should reflect on technology from a different perspective. Just like a driver can only drive a car if he or she operates it the way it supposes to be, teachers must collaborate with online technology to accomplish online teaching activities successfully. This collaboration does not indicate the simple transference of the teaching plan originally designed for offline courses to the online world. What it should mean is the exploration of the advantages and disadvantages of online teaching from a new perspective, and the design of new teaching models and strategies based on technology. For example, in foreign language teaching, we should consider how to combine task-based, collaborative, and project-based learning models with technology tools to suit specific teaching contexts and different types of students.

The teaching practice during the pandemic turned the teacher into the one who constantly kept underlying the use of online teaching technology. Yet "technology alone does not create language learning any more than dropping a learner into the middle of a large library does" [4]. While technological support helps to improve the effectiveness of digital teaching, its role can only be achieved through the lesson plan and strategies meticulously designed and successfully implemented by teachers. In that sense, teacher and technology are so tightly bound ("human-machine entanglement") that the interaction between the two creates a "united" agent. The reflection on this "unity" will be the grounds for online teaching.

3 "Rejected" by Technology

As we know, nowadays, a specific intelligent technology often separates the world into two parallel parts: those who are within the technological system and are able to use technology and those who are outside the system and are unable to use technology. This has resulted in vastly different situations for individuals and created "separation" among the individuals involved. Although online teaching technology has helped reduce the spatial distance between educators and students, the "barriers" to effective interaction and student engagement have not been lifted. As the Chinese ethicist Daixing Tang states, "technology is always intended to serve people, but once it is created, it gains its own existence und unfolds itself" [5]. For example, a large number of teachers have the experience of "talking" to themselves in front of the screen rather than "communicating" with their students, which will have a significantly adverse effect on foreign language classes, where communication is considered as a major learning goal. Furthermore, we may be "separated" by technology tools or Apps for online courses, and by the time we become proficient in one App, we find that our students have adapted to another one in other classes given by other teachers.

Even more worrisome than being "separated" by technology is being "rejected" or even "abandoned" by it. During this COVID-19 pandemic, online learning became a contingency plan for the abrupt school closure. A whole range of problems arose due to the lack of preparedness for "online learning for all", which is understandable and inevitable. No one has expected that the digital and online education could have expanded so quickly. In such "unprepared" situations, we will encounter the frustration of being "rejected by technology" more easily and frequently. This is a notion that some Chinese philosophers of technology like Hongxiu Yan, Qingfeng Yang, Yongmou Liu, Weiwen Duan [6] have been discussing since the outbreak of the pandemic. There are two main reasons for being rejected during the online-teaching: objectively speaking, the lack of technical conditions, or the difficulty in using the tools; subjectively speaking, the inadequacy of teachers' ability, preparedness, and experience in using technology for online teaching. Even in countries with highly advanced technologies like Germany, lots of schools and teachers are not well prepared. For instance, some do not have unified technological tools: according to the analysis of Institut der Deutschen Wirtschaft (IW) [7] which was published on the German Website "News4teachers", only about 28% of 12-year-old children in Germany possess a computer, and 42% of 14-year-old children have one. The interviews and questionnaire surveys conducted by Xianfeng Luo [8] on the college English course during the pandemic yielded the finding that 80% of teachers who were interviewed admitted that, due to their unfamiliarity with modern information technology and lack of proficiency in the operation of various educational software, they were not prepared to teach 'fully online'. In addition, 55.47% of students pointed out that the poor network status and software emergencies often resulted in their failure to log onto the teaching platform or to complete the learning tasks, which severely affected the effect of online learning.

As for being "abandoned" by technology, the fact the Zoom withdrew from the Chinese market is an extreme example. During the pandemic, many Chinese teachers chose Zoom as the major tool for their online courses. And they felt abandoned when this tool which they had become accustomed to suddenly withdrew from the Chinese market. Technology has a strong "presence"; therefore, we can feel frustrated and even powerless when we are separated, rejected or even abandoned by it. To answer the question of how we can prepare the educators and students to avoid this situation will be essential for online education model building.

4 Potential or "Mediating Role" of Technology

As aforementioned, it is incredibly important to explore the collaboration between teachers and technology and to determine how technology shapes or profoundly affects teachers in the teaching process. The findings can contribute to the software development, where the ideas of philosophers of technology such as Verbeek [3] inspire us and require us to refine and materialize more educational ideas into technological design. We frequently hear educators say that we need to motivate students, to use blended teaching, to integrate offline and online teaching etc. These are educational concepts that we have already been familiar with or educational visions that we have fully endorsed. But is there a small design of online teaching tool that can materialize our educational philosophy and make online teaching and learning more consistent with our educational values? This technology should probably be designed like speed bumps on the road to meet our safety need, to enable us to slow down immediately and to minimize the negative consequences of driving. It means that technology should not only function as an inactive tool, but also play the role of an active mediating, as Verbeek said, "the often-found instrumentalist approach to technology is replaced with an approach that focuses on the actively mediating role of technologies in human actions and perceptions" [3].

Lamy & Hampel [9] believe that computer-mediated online task development should be guided by the theoretical framework and concepts (e.g., conversational tasks guided by cognitive theory and simulations task guided by community building concepts); it should consider the class type, the role of both teachers and students, and the medium to ensure the efficiency of teaching and learning activities. Moreover, it also needs to think about how to organize the task in order to facilitate learner interaction and improve learners' communicative skills, their engagement and the ability to work together to solve problems. This theory could also be the guidance for online language learning technological development.

Some Apps or tools have been created. The discussion feature in the Chinese App Weizhujiao (MicroTutor) offers the teacher the possibility to get timely feedback from students in the virtual classroom. The teacher can initiate class discussions at any time to match the teaching schedule. The anonymity of online discussion has proved to be more motivating than the physical classroom. Ideas are eventually projected and displayed, with word cloud dynamics and key information zoomed in, so that students can inspire each other, which is hard for our traditional classroom to achieve. During the pandemic, video conferencing software such as DingTalk, Tencent Meetings, Teams, Zoom and others made it possible for offline courses go online in a short period of time. But while these software programs which were originally developed for business meetings did allow teachers and students to overcome spatial distance, some of the teacher-student and student-student interaction that a physical classroom can provide were lost, which is detrimental to teaching and learning activities, especially in foreign language education - a flaw that needs to be remedied by other designs. The breakout room feature of Zoom, designed for pair work or group work, offers the possibility of role-play technique, sandwich method, self-directed learning, cooperative learning, etc., which compensates to some extent the lack of interaction in the virtual classroom.

Meanwhile, the use of technology in teaching practice, in turn, influences the development of technological design. The practitioner (teacher) needs to answer the question whether a "moralized" technology tool achieves its goal in the teaching and learning activities or has negative effects in the teaching process.

5 Closing Words

During the COVID-19 pandemic, online teaching was rapidly mainstreamed and normalized. In the post-pandemic era, it may not be the only option for mass education anymore, but it will never be out of the picture either. In terms of foreign language teaching and learning in the digital age, the pandemic has given us a good opportunity to rethink the relationship between technology and teacher, to practice online language teaching model, and to facilitate the development of theory and technological design. Considering the indispensability of technology and the cooperative and interactive character of the relationship between humans and online learning technology, being "rejected by technology" should not simply be regarded as humans (teachers) being "rejected" by the tool, but by the "cooperator", in other words, by one part of "agency".

Only through collaboration with technology can teachers be complete actors in online teaching. Instead of the conventional theory of "technology as tool" or an inactive co-existence of humans and technology, "human-machine coupling" is a better attitude for us to adopt in order to take advantage of the malleability of technology, to create better designs of educational technology, and to work better with technology. Currently, the time is not ripe for talking about "liberating" nor "slaving" teachers by technology. We reject an entirely technological optimism or pessimism. We believe that the key to online language teaching and learning success is to fully explore the potential of technology and to observe and reflect upon technology.

Acknowledgement. We would like to express our thanks to Ms. Dixi Ke, Ms. Yang Liu, Professor Xiaoqing Qiu and Ms. Dr. Rirong Li for their suggestions.

References

1. Verbeek, P.-P.: Preface of Chinese translation. In: Moralizing Technology. Understanding and Designing the Morality of Things (Yan, H.X., Yang, Q.F., Trans.), pp. 1–2. Shanghai Jiao Tong University Press, Shanghai (2016)
2. Ihde, D.: Technology and the Lifeworld: From Garden to Earth. Indiana University Press, Bloomington (1990)
3. Verbeek, P.-P.: Moralizing Technology. Understanding and Designing the Morality of Things. The University of Chicago Press, London (2011)
4. Healy, D.: Classroom practice: communicative skill-building task in CALL environment. In: Egbert, J., Hanson-Smith, E. (eds.) CALL Environments: Research, Practice and Critical Issues, pp. 116–136. TESOL, Alexandria (1999)
5. Tang, D.X.: The crisis of social justice driven by the development of artificial intelligence. J. Hum. **8**, 19–28 (2020). (in Chinese)
6. Duan, W.W., Liu, Y.M., Yan, H.X., Yang, Q.F.: The era of technology and the future of humanity: being rejected by technology, should we do nothing but smile? (in Chinese). https://www.thepaper.cn/newsDetail_forward_7848124. Accessed 10 Sept 2020
7. The Inequality of Opportunity is Aggravated by the Corona Crisis – A Guest Commentary. (in German). https://www.news4teachers.de/2020/05/die-chancenungleichheit-wird-durch-die-corona-krise-verschaerft-ein-gastkommentar/. Accessed 13 Sept 2020
8. Luo, X.F., Ma, W.N., Yao, Y.H.: A practical research of online teaching of college english after the outbreak of COVID-19 pandemic: problems, measures and effects. Technol. Enhanc. Foreign Lang. Educ. **3**, 30–35 (2020)
9. Lamy, M.N., Hampel, R.: Online Communication in Language Learning and Teaching. Palgrave Macmillan, Basingstoke (2007)

Author Index

The manufacturer's authorised representative in the EU is Springer
Nature Customer Service Centre GmbH, Europaplatz 3, 69115 Heidelberg,
Germany. If you have any concerns regarding our products, please
contact ProductSafety@springernature.com

Printed and bound by CPI Group (UK) Ltd, Croydon, CR0 4YY
30/09/2026
02225439-0019